Joint IMEKO TC1-TC7-TC13 Symposium: Metrology Across the Sciences: Wishful Thinking?

Journal of Physics: Conference Series Volume 772

Berkeley, California, USA
3 – 5 August 2016

Editors:

Mark Wilson
William Fisher

ISBN: 978-1-5108-3529-0
ISSN: 1742-6588

Printed from e-media with permission by:

Curran Associates, Inc.
57 Morehouse Lane
Red Hook, NY 12571

proceedings.com

Some format issues inherent in the e-media version may also appear in this print version.

Printed by Curran Associates, Inc. (2016)

For permission requests, please contact the Institute of Physics
at the address below.

Institute of Physics
Dirac House, Temple Back
Bristol BS1 6BE UK

Phone: 44 1 17 929 7481
Fax: 44 1 17 920 0979

techtracking@iop.org

Additional copies of this publication are available from:

Curran Associates, Inc.
57 Morehouse Lane
Red Hook, NY 12571 USA
Phone: 845-758-0400
Fax: 845-758-2633
Email: curran@proceedings.com
Web: www.proceedings.com

TABLE OF CONTENTS

Journal of Physics: Conference Series **772** (2016) 011001 doi:10.1088/1742-6596/772/1/011001

2016 IMEKO TC1–TC7–TC13 Joint Symposium: Metrology Across the Sciences: Wishful Thinking?

PREFACE

The 2016 IMEKO (International Measurement Confederation) TC1-TC7-TC13 Joint Symposium was organized and hosted by the BEAR Center (the Berkeley Evaluation and Assessment Research Center) in the Graduate School of Education at the University of California, Berkeley on 3-5 August 2016. The work of this symposium is reported in this volume. The scope of the symposium includes the main topics covered by the three participating Technical Committees:

TC1 Education and Training in Measurement and Instrumentation
TC7 Measurement Science
TC13 Measurements in Biology and Medicine

The excellence of previous events in this well established series sets a high standard. Maintaining that standard was vitally important for the 2016 event, as this was the first IMEKO event in the United States since the U.S. rejoined IMEKO, and because this was the first joint symposium hosted by psychometricians.

The focus was on metrology across the sciences, a recurring theme of past Joint Symposia reinvigorated recently by new collaborations of engineers and psychometricians. This larger theme was developed by keynote lectures on uncertainty in measurement, and on philosophic concerns with objectivity. Participants repeatedly remarked on useful interactions with scientists from different disciplines. The Symposium was attended by experts working across a wide range of measurement-intensive fields from 18 countries, including Brazil, Japan, and China, and provided a useful forum for them to share and exchange their work and ideas.

In total over sixty papers are included in the volume, organised according to the presentation sessions. Each paper was independently peer-reviewed by at least two reviewers from a distinguished international panel.

The Symposium was held at the University of California - Berkeley's Clark Kerr Campus, a beautiful Spanish-style conference complex, highlighted by terra cotta accents, tiled fountains and landscaped courtyards. The Clark Kerr Campus was originally the site of the California Schools for the Deaf and Blind. It is listed on the National Register of Historic Places, and is in close proximity to both the Lawrence Berkeley National Laboratories, and San Francisco Bay, where Joint Symposium participants enjoyed tours and a dinner cruise. The site proved appropriate for an international symposium concentrated on new possibilities for supporting scientific innovation.

We would like to take this opportunity to thank the members of the Symposium Steering and International Programme Committees, many of whom provided reviews of the papers included here. We would also like to extend special thanks to our sponsors, Fluke, Meaningful Measurement, MetaMetrics, and Modus Outcomes for their financial support of the Joint Symposium.

The editors hope that this volume will contribute to advancing the state of the art in measurement science and technology across engineering, education, and health care applications.

Mark Wilson and William Fisher
Editors

BEAR Center
Graduate School of Education
University of California, Berkeley

Peer review statement

All papers published in this volume of *Journal of Physics: Conference Series* have been peer reviewed through processes administered by the proceedings Editors. Reviews were conducted by expert referees to the professional and scientific standards expected of a proceedings journal published by IOP Publishing.

Journal of Physics: Conference Series **772** (2016) 012001 doi:10.1088/1742-6596/772/1/012001

How Does Measuring Generate Evidence? The Problem of Observational Grounding

Eran Tal[1]
[1]Assistant Professor, Department of Philosophy, McGill University, Montreal, Canada

E-mail: eran.tal@mcgill.ca

Abstract. The epistemology of measurement is an area of philosophy that studies the relationships between measurement and knowledge. One of its central aims is to explain how measurement can function as a reliable source of scientific evidence. Key to such explanation is a clear characterization of the dependence of measurement on observation, but such characterization has remained elusive. This article traces the recent historical trajectory of views on the observational grounding of measurement, clarifies the current state of the problem, and proposes new directions for progress. Specifically, I argue in favour of viewing measurement outcomes as the best predictors of observed instrument indications under a given theoretical-statistical model of the measurement process. The evidential efficacy of measurement outcomes is explained by their relatively high epistemic security, rather than by their inferential or structural closeness to observation.

1. Introduction

Measurement outcomes are commonly reported as unconditional factual claims. Examples are statements such as "the electric current in this wire at time t_1 is 1.0±0.3 ampere" and "the CO_2 concentration in this air sample at time t_2 is 400±10 ppm". Claims of this form are habitually used as evidence for testing scientific hypotheses, for designing and testing instruments, and for making decisions in areas such as healthcare, public policy and everyday life.

And yet, as is well-known to anyone who designs measuring systems, no measurement outcome is unconditional. Assumptions are always involved in designing instruments, in executing measurement procedures, in interpreting and analyzing data, in evaluating bias and uncertainty, and in calibrating measuring devices against standards. Despite this, scientists do not usually report measurement outcomes in the form: "if assumptions $\{A_1, A_2 \ldots A_n\}$ about the measurement process hold, then the electric current in this wire at time t_1 is 1.0±0.3 ampere". What warrants the neglect of background assumptions in the final report, and is the epistemic status of measurement outcomes as unconditional evidence justified?

Explaining the evidential efficacy of measurement is a central task of the epistemology of measurement, an area of philosophy that studies the relationships between measurement and knowledge [1][2][3]. This task becomes particularly interesting when one compares measurement to other kinds of estimation, such as the derivation of quantity values through theoretical prediction and computer simulation. Much like measurement, these other forms of estimation involve a combination

[1] To whom any correspondence should be addressed.

of observed data, inference and background assumptions. Unlike measurement, scientists rarely treat the results of these other kinds of estimation methods as unconditional evidence. Instead, the results of theoretical predictions and computer simulations are usually viewed as conditional on the assumptions that informed the estimation – at least until such results are confirmed through measurement.

What warrants this seemingly privileged evidential status of measurement? Traditionally, answers have appealed to a special sort of connection between measurement and qualitative observation. In what follows I will discuss and reject the traditional view, formulate a problem that arises from the rejection of the traditional view, and outline a solution to that problem. Ultimately, I will argue that the warrant for treating measurement outcomes as unconditional evidence is their high epistemic security, rather than their inferential or structural closeness to observation.

2. Is measurement a kind of observation?

2.1. Foundationalism about measurement

Measurement has been traditionally viewed as quantitative observation, namely as an activity in which numbers are assigned to objects or events based on observation [4]. The claim that measurement is a kind of observation is trivially true if the term 'observation' is used so loosely as to encompass all measurement by default. In some contexts such loose usage may be warranted [5]. However, the classification of measurement as observation only becomes philosophically interesting when 'observation' is understood more narrowly, for example, as knowledge one obtains through immediate and unaided sense perception. Although the precise explication of such a concept of observation is riddled with philosophical difficulties, for current purposes we can suppose that scientists observe the indications of their measuring instruments, such as the positions of pointers and numerals appearing on displays. There is no doubt that measurement *involves* observation in this sense. Measurement also typically involves inference, theory, statistics, abstraction, idealization, prediction, calculation and instrument-building, but one is not usually drawn to claim that measurement is a species of any of these. Why, then, is it tempting to believe that measurement is a *kind* of observation, and is this belief justified?

A central advantage of classifying measurement as observation is that it immediately becomes clear why scientists are justified in using measurement outcomes as evidence for testing hypotheses and designing new instruments. If measurement outcomes are no more than observational reports cast in mathematical language, the empirical content of any measurement outcome is reducible in principle to some set of observations. The most comprehensive attempt to analyze measurement along these foundationalist lines is due to Norman Campbell [6]. Campbell's 'fundamental measurement' is a type of procedure that exhibits the additive structure of a quantity, such as length, weight or time duration, based on qualitative observations alone and without requiring antecedent measurements. Other, 'derived' procedures can then extend the discovery of structure to quantities such as density and temperature, for which no fundamental measurement procedure is known. Viewed in this way, the entire edifice of physical measurement is ultimately grounded in nothing but qualitative observation.

Despite their appeal, foundationalist views of measurement suffer from several interconnected problems, of which two are especially relevant to the current discussion. The first problem concerns idealization. Campbell and his followers assumed that qualitative structures were given in observation, but in practice such structures are rarely observed in a reproducible manner. For example, consider any two clocks that are meant to instantiate the same frequency, that is, to 'tick' at the same rate. Even the most stable atomic clocks exhibit a systematic frequency drift when compared accurately enough. It is easy to show that systematically drifting clocks provide inconsistent estimates of duration, and in some cases even assign the same time intervals a completely reverse order of duration. Despite this, practicing scientists do not infer the existence of multiple, fundamentally measureable kinds of quantity, 'time-A', 'time-B' etc. Rather, they suppose an ideal ordering of time intervals and assume that concrete clocks merely approximate this ideal. This is by no means special to clocks. Indeed, nearly every comparison between the indications of measuring instruments reveals nonlinear systematic discrepancies if carried out accurately and for a sufficiently large sample. In practice, then, Campbell's qualitative structures are not given by observation, but are idealizations of the measurement process in the hypothetical absence of error.

An additional difficulty is that the choice of idealized structure for the underlying quantity is not uniquely determined by observation. The underdetermination of measurement outcomes by observation is the second major problem facing foundationalist views of measurement[2]. Going back to the clock example, no amount of data obtained through clock comparisons can by itself reveal which clock has the more stable frequency. Multiple ways of distributing errors among different clocks exist that are all compatible with observation, but yield inconsistent estimates of time duration. The inferential gap between instrument indications and measurement outcomes must be closed by other means. Theoretical considerations, along with considerations of convenience, fill this gap [7].

Taken together, the problems of idealization and underdetermination undermine the foundationalist attempt to reduce the content of measurement outcomes to raw, qualitative observations. This point must have been partially recognized by the authors of the Representational Theory of Measurement [8]. RTM provides an axiomatic treatment of qualitative structures, but calls such qualitative structures 'abstractions' and admits that not all of the axioms are empirically testable [9]. This is a step in the right direction, because it implicitly recognizes that extra-observational (such as theoretical, statistical, and pragmatic) assumptions inform choices of measurement scale and systematic error distribution. On the other hand, RTM says little about how its axioms may be justified, or about what sorts of observations would provide evidential support to those axioms that are empirically testable.

2.2. Observational grounding

The ambiguity surrounding the relationship between measurement and observation in post-foundationalist accounts gives rise to the problem of observational grounding [10][11]. The problem may be construed as a logical inconsistency among three compelling claims:

(1) Measurement is a more reliable source of evidence than other forms of quantitative estimation, such as those based on theoretical prediction and computer simulation
(2) If measurement is a more reliable source of evidence than other forms of quantitative estimation, it is because observation plays a special role in justifying measurement outcomes
(3) Observation does not play a special role in justifying measurement outcomes compared to other forms of quantitative estimation, such as those based on theoretical prediction and computer simulation.

At least one of these claims must be rejected to avoid contradiction. Foundationalists like Campbell would reject (3), and argue that the content of measurement outcomes mirrors the structure of qualitative observation, whereas the content of quantitative estimates derived through theoretical prediction and simulation does not. The challenges of idealization and underdetermination block this move: at least some of the content of measurement outcomes is imposed from the 'top down', by adjusting inconsistent observations in light of idealized background assumptions. Observations partially constrain measurement outcomes but do not fully determine them. The same may be said of other forms of quantitative estimation. After all, observation plays important roles in theoretical and statistical modelling, prediction and simulation, such as providing raw data and determining initial and boundary conditions. In all of these activities, much like in measurement, observation is combined with inference and background assumptions to generate knowledge claims. What, then, is the special role observation plays in justifying measurement outcomes, if such unique role exists? And if such unique role does not exist, how does one explain the elevated epistemic status measurement is commonly thought to enjoy as a reliable source of evidence?

[2] This problem should not be confused with the more familiar underdetermination of *theory* by observation.

A good first step is to consider the theoretical context in which measurements are performed. Historical examples show that the measurement of a certain kind of quantity and the theory behind that kind of quantity are co-dependent and progress through a process of mutual refinement. Hasok Chang has provided an illuminating account of how this process unfolded in the case of temperature [12]. Bas van Fraassen similarly describes the 'empirical grounding' of scientific theories as a "process of simultaneously, harmoniously extending both the theory and the range of relevant evidence" [13]. According to van Fraassen, a theory is empirically grounded when its significant parameters can be uniquely determined through measurement. Such determination must itself depend on the theory for modelling the measurement apparatus and analyzing instrument indications. Moreover, such determination must be refutable, in the sense that the theory does not simply fix measurement outcomes in advance but allows the production of outcomes that are inconsistent with the theory's own predictions.

Van Fraassen's criterion of empirical grounding is meant to apply to scientific theories rather than to measurement procedures. Nonetheless, the co-dependence of measurement and theory suggests that the observational grounding of measurement and the empirical grounding of theories are but two sides of the same epistemic coin. Measurement and theory have the right sort of relationship with observation when the production and analysis of observations supports a fruitful, progressive coherence between measurement and theory. A plausible criterion of observational grounding for measurement may therefore be obtained by inverting Van Fraassen's criterion of empirical grounding for theories:

> **Preliminary criterion of observational grounding.** A measurement outcome is grounded in a set of observations if and only if those observations were obtained and analyzed in a manner that determines the values of relevant theoretical parameters in a unique, coherent and refutable way.

2.3. A model-based perspective

The preliminary criterion of observational grounding is still very general. Before one can discern whether this criterion can resolve the problem of observational grounding, one needs to answer the question: what sort of relation between measurement outcomes and observations satisfies the criteria of uniqueness, coherence and refutability?

An answer may be gleaned from studying the methods metrologists use to calibrate measuring instruments and to trace uncertainties to measurement standards. Calibration is a kind of modelling activity, that is, the activity of constructing and testing a theoretical and / or statistical model of a measurement process [14][15]. The kinds of models constructed during calibration vary widely, from simple input-output ('black-box') representations of a measuring instrument to detailed and modular ('white-box') representations. Despite their diversity, all calibration procedures have a common aim, which is to reliably extrapolate backwards from the observed indications (or 'end-states') of the measuring procedure to values of the quantity intended to be measured.

Due to uncertainties, there is rarely a single value that would exclusively predict the observed indications. Calibration therefore aims at selecting the best range of predictors of each possible set of observed indications. The 'best' set of predictors is coherent with background theoretical and statistical assumptions, maximally accurate, and invariant ('unique') under changes to the measurement procedure and environment [15]. In addition, the best predictors need to be selective (or 'refutable') in the sense that they only predict a proper, and preferably small, subset of the possible indications of the measurement apparatus. In other words, the criteria for the selection of best predictors during calibration coincide with the criteria for the observational grounding of measurement, which in turn mirror the criteria for the empirical grounding of theories.

We are finally in a position to offer an answer to the opening question: is measurement a kind of observation? The answer is that measurement is not a kind of observation if by that one means that the empirical content of measurement outcomes is epistemically or semantically reducible to relations among qualitative observations. At the same time, measurement is observationally grounded, in the

Journal of Physics: Conference Series **772** (2016) 012001 doi:10.1088/1742-6596/772/1/012001

sense that measurement outcomes have a specific sort of relationship with qualitative observations. Measurement outcomes constitute the *best predictors* of the observed end-states of a measurement process relative to a particular theoretical and statistical model of that process. The best predictors are those that maximize the predictive power of the model as well as that of underlying theories, and thereby enable the measurement and theory of the relevant quantity to cohere in the most accurate and universal manner practically possible.

3. How does measuring generate evidence?

The analysis of measurement outcomes as model-based predictors strengthens claim (3). There is no reason to think that observations constrain measurement results in a fundamentally different way than they do other sorts of model-based, quantitative scientific estimates. This does not mean that theoretical prediction and computer simulation are in general equally reliable sources of evidence compared to measurement. Although observation does not play a special role in justifying measurement outcomes, there are good reasons for thinking that measurement outcomes are usually more *secure*. The notion of secure evidence has been proposed by Kent Staley [16][17]. Roughly speaking, the security of a knowledge claim is higher for an epistemic agent when there are fewer possible scenarios in which that agent would deem the claim incorrect. Security is closely (and inversely) tied to the likelihood that, for all that a given scientific community knows, a knowledge claim would require revision in the foreseeable future. Note that security is not simply the inverse of uncertainty: a knowledge claim can be reported as having a low uncertainty but rest on assumptions that are likely to be revised in the foreseeable future. Such a claim would have a low security.

Measurement outcomes, especially those reported with low uncertainties, are never fully secure. At least some of the sources of information that contribute to the production of high-quality measurement outcomes are fallible. Theories are occasionally revised, statistical assumptions could be proven wrong, simplifying assumptions may be refined, and so on. To counter these sources of insecurity, scientists and engineers who design measuring systems try to rely only on uncontroversial and well-tested portions of theory and statistics, and to err on the side of caution when evaluating uncertainties. Moreover, scientists habitually compare measurement outcomes formed under different circumstances and assumptions to each other and test them for consistency within the limits of their respective uncertainties. Indeed, such comparison is implicit in the process of calibrating a measuring instrument against a standard. Consistency with a standard strengthens the likelihood of consistency with other measuring instruments calibrated against that standard, and recursively with an entire web of instruments linked to the standard through mutual comparisons [15]. These various precautions and robustness tests do not render measurement outcomes completely immune to revision, but they make it less likely that measurement outcomes will require revision in the foreseeable future given the current state of knowledge in the scientific community.

These considerations suggest that the key to understanding the source of evidential efficacy of measurement outcomes may lie in their relative security. The epistemic credentials of measurement are not different in kind from those of other modes of quantitative estimation, such as theoretical prediction and computer simulation. However, the outcomes of measurement procedures tend to be more secure, because they are obtained by following the precautions and passing the robustness tests detailed above. This partially explains why scientists are warranted in neglecting background assumptions when reporting measurement outcomes. The assumptions underlying measurement outcomes are usually selected in a conservative manner, and calibration further guarantees that the outcomes are invariant under changes to as many background assumptions as is practically possible to test. This is not generally the case for theoretical predictions and simulation-based estimates, which are often meant to explore speculative theoretical scenarios and are often not calibrated with the same level of rigour.

At the same time, nothing in principle prevents theoretical predictions and computer simulations from matching or even surpassing the degree of security of the best measurements available for a given quantity. This already appears to be the case for some *ab initio* calculations of the properties of simple atoms and molecules, such as the static electric dipole polarizability of helium [18]. Such results are not only accurate, but also considered sufficiently secure to underwrite the official determination of the values of fundamental physical constants, such as the Boltzmann constant [19]. In such cases it may be useful to think of the computation as a kind of measurement process, inasmuch as it generates evidence that is as secure as the best equivalent 'traditional' measurement outcomes.

4. Conclusions

Measuring generates evidence, not by preserving the structure of 'raw' qualitative observations, but by constraining the region on parameter space where the best predictors of observed indications are likely to lie. Such constraining involves observation along with other sources of information, including theories, tools of data analysis, simplifying assumptions, and computational techniques. Measurement outcomes are model-based predictors that have attained a high degree of security through various strategies, such as robustness tests. It is their security, rather than inferential or structural closeness to observation, that makes measurement outcomes suitable for serving as unconditional evidence claims.

This way of thinking about measurement leads to a novel solution to the problem of observational grounding. Claim (3) is retained, and no unique role is attributed to observation in justifying the outcomes of measurement. Instead, claim (2) is rejected, and the grounds for a difference in evidential reliability shift to a difference in epistemic security. Additionally, claim (1) is qualified, so that the elevated evidential status of measurement outcomes depends on their security. This proposed solution is but a sketch for an eventual, comprehensive solution to the problem of observational grounding. A full solution would require careful explication of the relevant notions of measurement, observation, evidence and information, as well as a clear account of the ways in which the epistemic security of measurement outcomes may be assessed and enhanced.

Acknowledgments
Work on this article was supported by a Marie Curie Intra-European Fellowship within the 7th European Community Framework Programme, held at the University of Cambridge.

References

[1] Mari L 2003 *Measurement* **34** 17-30
[2] Leplège A 2003 *Social Science Information* **42** 451-462
[3] Tal E 2016 *British Journal for the Philosophy of Science* **67** 297-335
[4] Bunge M 1998 *Philosophy of Science Volume 2* (New Brunswick and London: Transaction)
[5] Shapere D 1982 *Philosophy of Science* **49** 485-525
[6] Campbell N R 1920 *Physics: The Elements* (London: Cambridge University Press)
[7] Poincaré H 1898 The measure of time *The Value of Science* (NY: Dover, 1958) pp 26–36
[8] Krantz D H, Luce R D, Suppes P, and Tversky A 1971 *Foundations of Measurement volume 1* (San Diego and London: Academic Press)
[9] Luce R D, Krantz D H, Suppes P, and Tversky A 1990, *Foundations of Measurement volume 3* (San Diego and London: Academic Press), section 21.8
[10] Tal E 2013 *Philosophy Compass* **8** 1159-1173
[11] Tal E 2015 Measurement in science *Stanford Encyclopedia of Philosophy* ed E N Zalta
[12] Chang H 2004 *Inventing Temperature* (Oxford: Oxford University Press)
[13] van Fraassen B C 2012 *Philosophy of Science* **79** 773-784
[14] Mari L 2005 Models of the measurement Process *Handbook of Measuring Systems Design* vol 2 ed P Sydenham and R Thorn (Wiley) chapter 104
[15] Tal E Measurement, calibration and inference to the best prediction (under review)
[16] Staley K W 2004 *Philosophy of Science* **71** 467-488.
[17] Staley K W 2012 *European Journal for Philosophy of Science* **2** 21-43
[18] Lach G, Jeziorski B and Szalewicz K 2004 *Phys. Rev.* 92 233001
[19] Fischer J 2015 *Metrologia* **52** S364–S375

Journal of Physics: Conference Series **772** (2016) 012002 doi:10.1088/1742-6596/772/1/012002

A novel algorithm for bad pixel detection and correction to improve quality and stability of geometric measurements

R Celestre, M Rosenberger and G Notni

Ilmenau University of Technology
Department of Quality Assurance and Industrial Image Processing
Gustav-Kirchhoff-Platz 2, 98693 Ilmenau, Germany

E-mail: `rafael.celestre@tu-ilmenau.de`

Abstract. An algorithm for detection and individually substitution of bad pixels for further restoration of an image in the presence of such outliers without altering overall image texture is presented. This work presents three phases concerning image processing: bad pixel identification and mapping by means of linear regression and the coefficient of determination of the pixel output as a function of exposure time, local correction of the linear and angular coefficients of the outlier pixels based on their neighbourhood and, finally, image restoration. Simulation and experimental data were used as means of code benchmarking, showing satisfactory results.

1. Introduction

Image based measuring techniques rely on the most precise estimation of dark to bright transitions, the so called edge detection, to extract geometric features from the examinee. Going towards the subpixel resolution, values delivered by pixels, as means of their grey values, play a much important role for precise geometric measurements [1]. For accurate, improved and stable measurements it is vital that bad pixels are not only identified, but also corrected locally - not to alter overall image texture, before the image is used for metrology or quality control applications.

2. Bad pixel definition

A bad pixel can be defined as a pixel that does not behave as expected, producing anomalous values and therefore, no valuable information can be extracted from them. The data provided by them is not only less relevant but also less reliable than the produced by its neighbourhood. Bad pixels occur either isolated or in clusters, the latter being more difficult to handle.

2.1. Classification of outliers

There are plenty of ways in which a pixel may not deliver reliable information: from defects on electronics to reading and accessing the data generated by the sensors. It is out of the scope of this work to discuss the origins of bad pixels, since for this application it is enough to group outlier pixels by their behaviour on the resulting image, i.e. resulting grey value:

- **Linear pixels with false bias**: a category where the pixel output behaves linearly with illumination intensity or exposure time, but due to either a low or high bias compared to its neighbours, values displayed by this pixel differs considerably from the ones around it, given an homogeneous illumination;
- **Nonlinear pixels**: a subset of defect pixels, where the output signal displayed on an image does not relate to the illumination intensity or exposure time in a linear fashion;
- **Dead pixels**: are those who have very low sensitivity to illumination intensity variations, consistently presenting low grey values;
- **Hot pixels**: this category is highly sensible to illumination intensity variations, presenting persistent high grey values.

3. Bad pixel mapping

For applications where subpixel precision is aimed, it is important to be able to perform pixel correction without disturbing healthy parts of the image, i.e. altering the texture or smoothing edge transitions. This approach can be done by mapping unhealthy pixels and performing localised correction operations. The criteria used on this work for tagging a pixel as an outlier is its linearity as a function of sensor integration time, measured by the coefficient of determination [2] calculated from the linear regression done pixel-wise by the method of the least squares [3] to each pixel of this image stack. The procedure for generation of a so called bad pixel map is as follows:

Firstly, multiple images are taken at a fixed integration time (t_{int}). The exposure time is incrementally increased until saturation of the detector is reached, so that at the end of this calibration procedure a set of multiple images is obtained for each integration time - see [4] for further guidance. This set is then reduced by averaging the images obtained at each integration time. Doing so is important in order to suppress temporal noise, which can occasionally disturb bad pixel mapping. This happens because the variance of the temporal noise ($\sigma_{temp.} \approx 1\%$) is normally larger than the variance of spacial inhomogeneities of the image acquisition device ($\sigma_{spac.} \approx 0.3 - 0.5\%$) [4]. From this reduced set, *averaged image stack*, a set of three matrices is generated based on the assumption that for zero intensity until 90% of the saturation intensity, the grey values are a linear function of the integration time, as shown in fig-1(right). Linear coefficient (LC), angular coefficient (AC) and coefficient of determination (R^2) are 2D pixel maps of the quantities they are named after.

A binary pixel map for unhealthy pixels, *bad pixel map*, is generated based on a threshold for R^2. The closer this parameter is to one, the better the linear model is for the data set in question [2]. For each matched pixel, i.e. pixel whose R^2 is lower than the threshold, its coordinates are saved on a coordinate array, called *bad pixel location vector.*

4. Image restoration in the presence of bad pixels

The recovery of the lost information due to unhealthy pixels is done in two steps: generation of the corrected matrices and the image restoration in the presence of bad pixels. Firstly a map of irrecoverable pixels is generated from the *bad pixel map* by visiting each location from the *bad pixel location* array and substituting the outlier with the median of the values of its Moore neighbourhood:

$$\text{corrected pixel value} = \text{median}_{nxn}\{\text{bad pixel neighbourhood}\}, \tag{1}$$

where n is related to the Moore neighbourhood range r by: $r = (n-1)/2$ [1].

After this stage, all remaining low values on the *irrecoverable pixel map* cannot be corrected by the presented method. Generally, this matrix highlights pixel clusters that are too big for correction. The procedure of local substitution of bad pixels with the aid of the *bad pixel location array* is also applied for the linear and angular coefficient matrices, as well as for the acquired image. Those resulting matrices are called *corrected* matrices and will be used for image restoration.

The second step is done under the following assumptions: grey values assumed by a pixel are a linear function of the integration time of the imaging sensor:

$$\text{grey value} = \text{linear coefficient} + \text{angular coefficient} \times t_{int}, \tag{2}$$

and that the average number of photons per squared unity of area at the detector (A), e.g. pixel surface, is a multivariable function of exposure time, wavelength (λ) and irradiance (E) on the sensor surface, the latter, being given in power per area unity [4]:

$$\frac{\text{number of photons}}{A} = \frac{t_{int}\lambda E}{hc}, \tag{3}$$

where h is the Planck's constant and c is the speed of light. By letting both wavelength (ideally monochromatic) and irradiance constant, the number of photons per squared unity of area is a linear function of the exposure time. One can, therefore, relate the pixel grey value directly and linearly to the number of incident photons. This is the core idea of the correction algorithm: pixel-wise recover the number of photons as a function of the grey value. From equations 2 and 3:

$$\text{number of photons} = \frac{\text{grey value} - \text{linear coefficient}}{\text{angular coefficient}} \cdot \frac{\lambda E}{hc}, \tag{4}$$

without losing generality, the constant term can be dropped, hence:

$$\text{number of photons} \propto \frac{\text{grey value} - \text{linear coefficient}}{\text{angular coefficient}}, \tag{5}$$

The image recovery, performed on the *corrected image* from the previous step, is done as follows: the *corrected linear coefficient* matrix is extracted from the *corrected image* and the resulting matrix is pointwise divided by the *corrected angular coefficient* matrix. Those operations lead to the restored image, that is, a mapping of the number of photons per pixel - a more realistic description of the scene measured:

$$\text{recovered image} = \frac{\text{corrected image} - \text{corrected linear coefficient matrix}}{\text{corrected angular coefficient matrix}}. \tag{6}$$

This image can be pointwise multiplied by the *irrecoverable pixel map* to make visually clear where bad pixels that are unable to be corrected are located.

5. Results

The benchmarking of the algorithm was performed in three phases: bad pixel mapping and recovery of lost information with simulated data, in order to explore the code sensitivity in a controlled environment; bad pixel mapping and recovery of lost information with measured data in accordance with the EMVA standard 1288 guidelines; and a probe measurement to evaluate image restoration and its effectiveness.

5.1. Bad pixel mapping and recovery of lost information with simulated data

Were done as a proof of concept and to investigate the algorithm sensitivity. A set of images created to mimic the experimental process as described in sec. 3. Four types or pixel misbehaves, namely the ones presented at subsec. 2.1, were introduced at random positions on the images, as shown in fig. 1(left). The methodology described in sec. 3 was applied to the artificial images.

Journal of Physics: Conference Series **772** (2016) 012002 doi:10.1088/1742-6596/772/1/012002

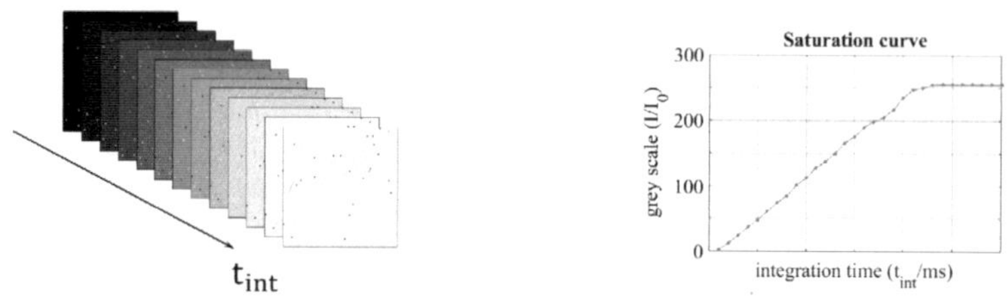

Figure 1. (left) Shows stack of averaged images for increasing exposure times (from dark to saturated images). (right) Integration time is increased until saturation is reached, so that a suitable integration time can be chosen as a sound operation point.

The aforementioned routine for bad pixel mapping is sensitive to most types of pixel misbehaves (non-linear, hot and dead pixels). Nonetheless, false bias and false angular coefficients - i.e. angular coefficients that differ significantly from those of its neighbourhood, but nonetheless are still linear - are often not ruled out as bad pixels. The only conditions where the latter misbehaves are tagged as bad pixels, is when the grey values reach saturation prematurely, that is, below the operation point and have a behave similarly to a hot pixel.

5.2. Bad pixel mapping and recovery of lost information with measured data

In order to test the algorithm on real imaging sensors, a calibration setup was assembled following guidelines displayed at [4]. A schematic can be seen on fig. 2, where the main components are displayed. The sensor under test is a EV76C661 with CMOS technology from the company *e2v*. The procedures followed here are detailed at sec. 3. The light intensity was chosen so that the saturation of the sensor was reached at 1s of integration time. This working point is particular interesting for low light applications and high dynamic range imaging. The light required for the experiment is monochromatic, which was obtained by means of optical filters. Although a multispectral characterisation of the sensor is out of scope, measurements were performed from 400nm to 950nm in 50nm steps. The results described next were taken at 650nm - fig. 3. The Moore neighbourhood range used was of a unity.

The angular coefficient (fig. 3(a)) gives an idea of how *fast* a pixel increases its grey value, in other words, how fast it reaches saturation. The linear coefficient (fig. 3(c)) can be understood as the pixel value without an illumination source or as an offset. To get a glimpse of how bad pixels behave, one must take a look at the angular and linear coefficient distribution after the proposed treatment (figs. 3(b) and (d), respectively). The former has a *lump* right before 0.35 (grey value/s), which is smoothed after treatment (figs. 3(a) and (b)). Low angular coefficients seem to be unaffected by the algorithm. The latter shows alarmingly high offsets: from 20 to over 100 grey values. After correction, all linear coefficients are under 20 - figs. 3(c) and (d). It is clear that the *lump* found on the angular coefficient histogram and the spikes over 20 grey values at the linear coefficient histogram are as bad pixels - they correspond to less than 0.35% of the sensor surface. As for the generation of the *irrecoverable pixel map*, no bad pixels were identified. Since there are no clusters of foul pixels on this sensor, one is lead to believe in a complete elimination of bad pixels.

5.3. Evaluation of image restoration in the presence of bad pixels

After the successful benchmarking of the bad pixel mapping and recovery of lost information, it was necessary to study the performance of the image restoration algorithm - described at sec. 4. The experimental setup shown in fig. 2 was kept unaltered, except by the introduction of an objective lens between (6) and (7). The results can be seen at fig. 4. The reader might point out

Journal of Physics: Conference Series **772** (2016) 012002 doi:10.1088/1742-6596/772/1/012002

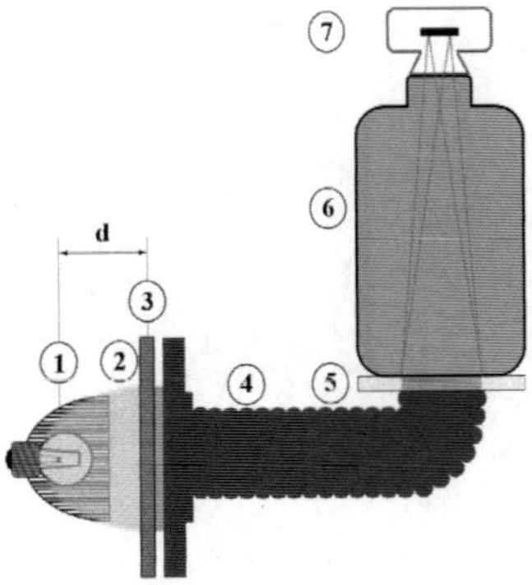

Figure 2. Shows a schematic of the measurement setup for sensor characterisation, where: 1) is a halogen lamp, 2) broadband light, 3) metallic filter for wavelength selection (bandwidth of ±10nm), 4) optical fibre, 5) diffuser plate to ensure homogeneity of the illumination field, 6) tube to prevent contamination from ambient line and 7) is the sensor being probed. The d stands for the variable distance between light source and filter, used to adequate the light intensity to the detector, which was chosen to make the sensor reach saturation at around 1s of acquisition time.

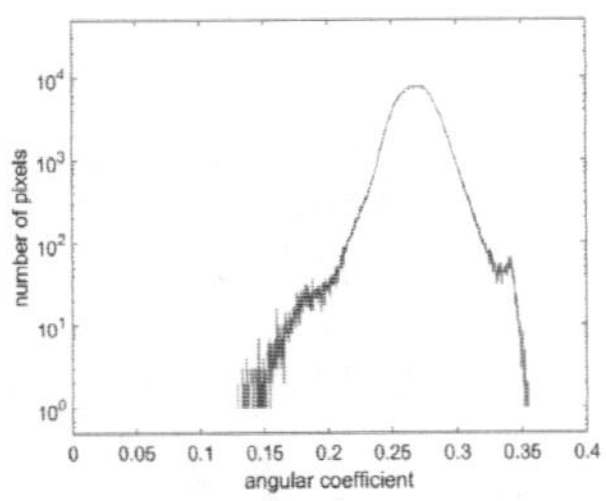

(a) untreated angular coefficient distribution.

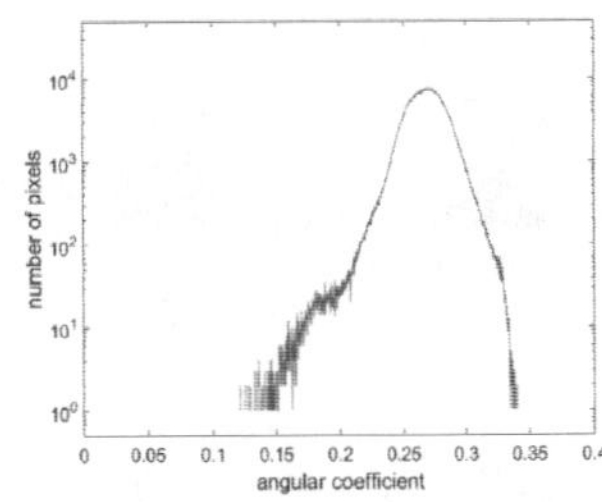

(b) angular coefficient distribution after treatment.

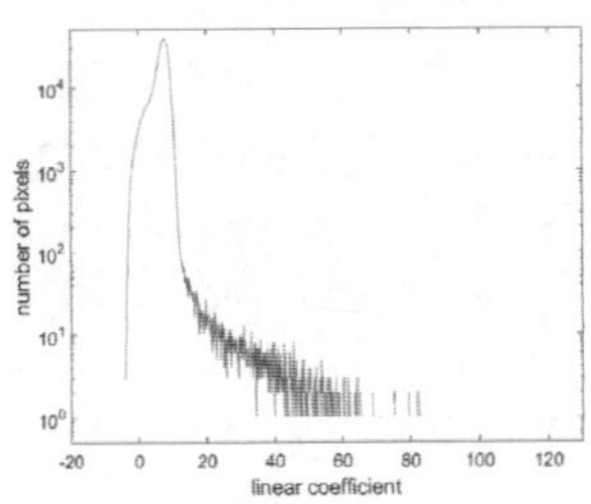

(c) untreated linear coefficient distribution.

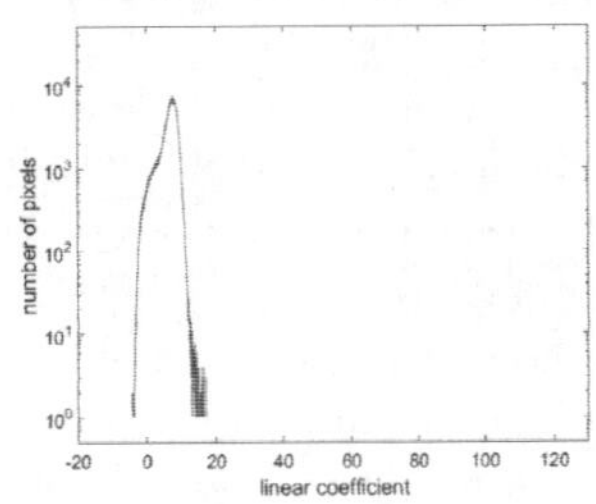

(d) linear coefficient distribution after treatment.

Figure 3. Shows real data of a sensor characterisation with the proposed algorithm. Histograms from angular and linear coefficients calculated with the least squares method are displayed.

that the treated image, fig. 4(c), appears to be noisier than its untreated counterpart, fig. 4(a), although seemingly true, it can be accounted as different pixel sensitivities - figs. 3(a) and (b) show how broadly distributed they are. On the other hand, the contrast is increased by a factor of 3 - from a contrast of 7 times bright to dark values (untreated image) to 22 times bright to dark values (treated image), which is very positive for geometric measurements. Nonetheless, new studies are meant to tackle better algorithms for image restoration, in order to suppress the aforementioned inhomogeneities.

6. Conclusion

The presented work can be seen as two modular and independent methods of image processing: bad pixel identification, value correction and image restoration. The first block is able to identify outliers and locally substitute their values (*in situ* action), without altering overall image texture:

Journal of Physics: Conference Series **772** (2016) 012002 doi:10.1088/1742-6596/772/1/012002

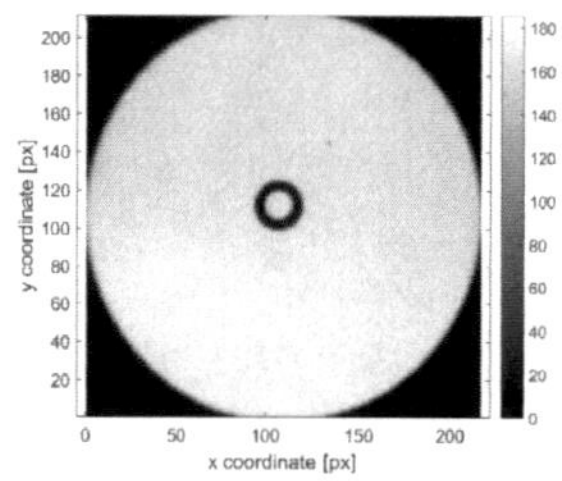

(a) untreated image.

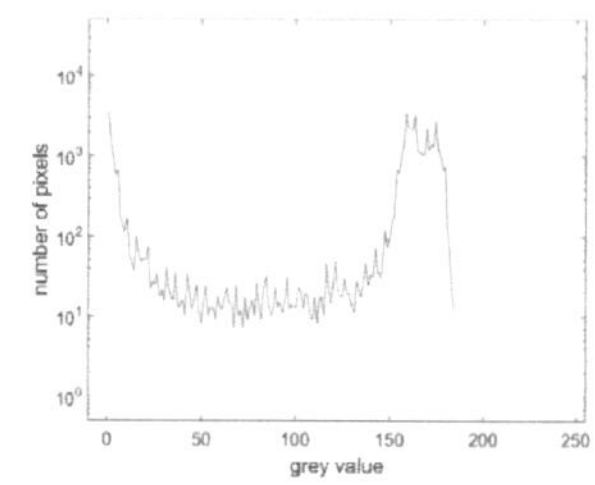

(b) grey value histogram.

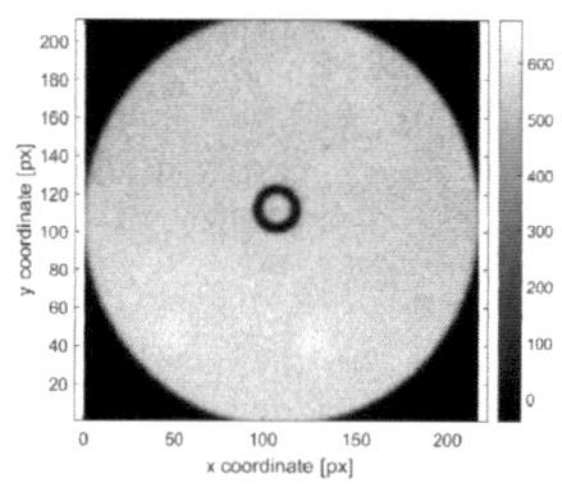

(c) treated image.

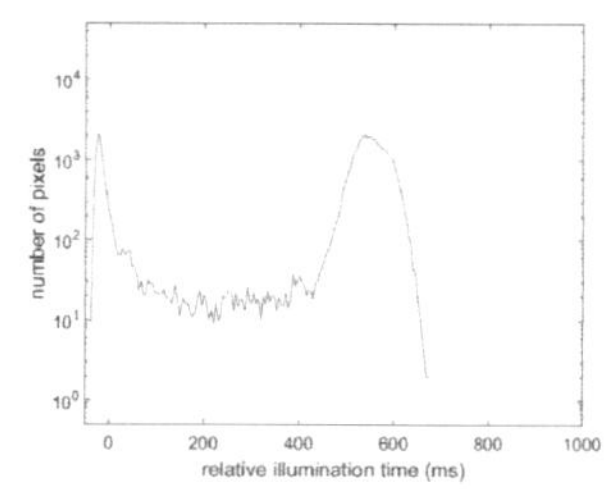

(d) grey value equiv. histo.

Figure 4. Shows (a) untreated image and (a) the distribution of grey values in a histogram. (c) image treated with the proposed algorithm and the (d) distribution of relative illumination time histogram.

its counterparts not only do not generate a list with outliers position (eg. EMVA1288 only gives the number of outliers), but also often disturb healthy pixels (see filtering operations [1]). There seems to be a correlation between high offset-pixels and *fast* ones - fig. 3, which leads to an interesting conclusion for pixel misbehaves: false bias and unusually high angular coefficients are often linked. The second method described is somehow similar to the shading correction (compensation for illumination), widely used for image processing [1], but done in a more reliable way for being able to account for bad pixels, whereas the latter is done mainly to account for illumination inhomogeneity. The recovered image seems to be more noisy, but contrast between bright and dark spots is increased. The resulting image is theoretically a better description of the measured scene. The current method is appealing to low light applications and HDR imaging, where integration times are often high and bad pixels have a more prominent role on image quality. Although multispectral analysis is out of scope, it could be seen - in preliminary analysis - that for the different colour channels, the behaviour of this algorithm is quite similar.

7. Acknowledgments

The authors are thankful for the financial support from the InnoProfile project "ID2M QualiMess Next Generation" from the German Federal Ministry of Education and Research. Special mention to Mr. Pavel Votyakov for helping with measurements.

8. References

[1] W. Abmayr. *Einfuehrung in die digitale Bildverarbeitung.* Teubner, 1994.
[2] Richard Anderson-Sprecher. Model comparisons and r^2. *The American Statistician*, 48(2):113–117, 1994.
[3] I. Kreyszig. *Advanced Engineering Mathematics.* John Wiley & Sons, Inc., 2006.
[4] European machine vision association. EMVA Standard 1288 A3.0 - Standard for Measurement and Presentation of Specifications for Machine Vision Sensors and Cameras, 2010.

States measurement in a context of intelligent connected furnitures

C. Deffo Sikounmo, E. Benoit, S. Perrin
LISTIC – Universite Savoie Mont Blanc
B.P. 806, 74960 Annecy-le-Vieux, France

E-mail: cedric.deffo-sikounmo@univ-smb.fr, eric.benoit@univ-smb.fr, stephane.perrin@univ-smb.fr

Abstract. The concept of intelligent connected furnitures opens new possibilities for the interaction between non animated objects. Actually intelligent connected furnitures are able to perceive their environment and can globally act as a measuring system. This paper presents how a community of objects acquire its own state. In this system, the measurement value is a weighted spatial ontology.

1. Introduction

The Internet of Thing is perceived as a pervasive presence around us of a variety of things or objects interacting between each other [1]. The given *things* are physical individuals, biological ones or virtual ones. The physical ones are mainly connected objects, the biological ones are connected humans or animals and virtual ones have no existence in the concrete world but are also able to propose services. A set of interacting things is designed as a community and a set of physical or biological interacting things is designed as a concrete community. The purpose of this paper is based on the principle that the elements of a community must know the potential and the concrete interactions between each other. This knowledge is mandatory inside a community in order to define the of services it can propose. Indeed, a community of objects may propose a set of services larger than the union of the sets of services proposed by its individual objects. More precisely, a subset of objects can propose services depending on the state of this subset. Such state is defined by the state of the objects and by the relations between objects. It is materialized by the relations inside this subset of objects and is represented by a physical ontology where the node entities are the physical objects or their parts, and the vertices are the relations [2].

The application field of our studies is defined by our partnership with a furniture manufacturer specifically concerned by a family of things that are the furnitures. In this context, the IoT community is made of smart furnitures like tables, chairs or sofa and of smart object already present in the furnitures environment: home automation devices that equip the house or the building, wearable devices and smart phones hold by the furniture users. The two main options to identify the state of an IoT community are : the acquisition of this information from an expert who is supposed to own it and who is able to transfer it, or the acquisition through a measuring system. In the specific context of

Journal of Physics: Conference Series **772** (2016) 012003 doi:10.1088/1742-6596/772/1/012003

furniture, it is not possible to ask users to update the knowledge of smart furnitures each time he moves a chair. Then the second option is to give perception capabilities to the furnitures. It has been defined that the main need for perception for smart furnitures is the proprioception of the orientation. Then all furnitures are equipped with at least a communication device a processor and a 3DoF accelerometer. Due to cost limitations, the accelerometer is imposed.

To sum up, the goals of this study, is to embed a form of intelligence commonly named *smartness* onto furnitures in order to promote them as smart connected objects in order to build a community of object in a context of the IoT. This paper presents a solution to measure the situations, defined as the states of a community of objects.

2. A model for the situations

As given in the introduction, the general model used for the representation of the community states is a weighted spatial ontology, more precisely a mereotopology [3][4][5]. Ontologies are made of entities and relations. In our concern, the entities are physical objects or physical parts of objects. The relations are limited to the relations *part_of* and *close_to*. The *part_of* relation defines the link between an object and his parts. This relation is a transitive and reflexive relation. The *close_to* relation is a fuzzy version of the topological relation *connected*.

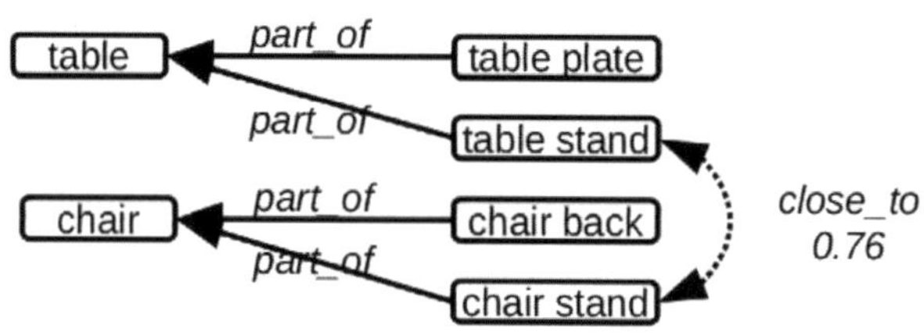

Figure 1. Ontology representation including 2 instances of furnitures with their parts.

The *part_of* relation is defined by the structure of objects that are supposed to be immutable during their use. Then, as each element of the *part_of* relation links an object or a part with a part inside an immutable structure, the *part_of* relation is also immutable.

Conversely, the *close_to* relation is able to change during the life of the object community. Indeed, the objects can be added, removed, or moved. As the *close_to* relation leads to a closeness concept, we chose to weigh the elements of the relation with a fuzzy closeness degree that represents the truth of the assertion "instance *a* is close to instance *b*".

Formally, a given ontology graph as shown in figure 1. is already a representative of the state of the object community. In order to facilitate the use of the results, the global ontology graph is compared with reference situations. Each reference situation is defined by a set of relations between objects and represented by an ontology. When the ontology graph of a reference situation is a sub-graph of the global ontology graph then the situation is realized. The state of the object community is then a set of realised situations. As the general ontology graph is weighted with fuzzy degrees, the realisation itself is also weighted by a fuzzy degree and the measurement result is a fuzzy subset of situations. Practically, the degree of realisation of the situation "a chair close to a table" is the truthfulness of the assertion "a table is close to a chair" that we represent by the weight of the *close_to* relation between a table and a chair.

The goal of this paper is to propose a method to perform the measurement of the closeness between instances, i.e. objects or parts, on the base of incomplete and uncertain direct measurements of footsteps. Then to produce a result for measurement of the global state.

3. The seismic measurements

As the physical distance needs expensive measuring systems in the context of furnitures, we propose to define the closeness as an indirect measurement derived from near environmental measurements. One of these measurements is the seismic one. The idea is to perform the correlation between several measurements of a same vibration event produced by the environment of furnitures. Actually the smart furnitures already hold a 3DoF accelerometer ADLX345 associated to a small Arm7 plate-form with communication capabilities in order to get the 3 main respective functionalities: sensing, processing, and communicating. The goal of the study is then: is the embedded existing system of the smart furnitures enough to perceive the proximity of other furnitures?

Figure 2. presents an example of one of the seismic measurements performed by a table instrumented with an accelerometer. During this experiment, a chair is moved close to the smart table. In this example, 4 large events appear on the 3 axis. They are due to the direct contact of the chair with the table. We can remark also that smaller events appear only on the vertical axis (z). These last events are characteristic from seismic s-waves and fits with the impact of the chair on the floor.

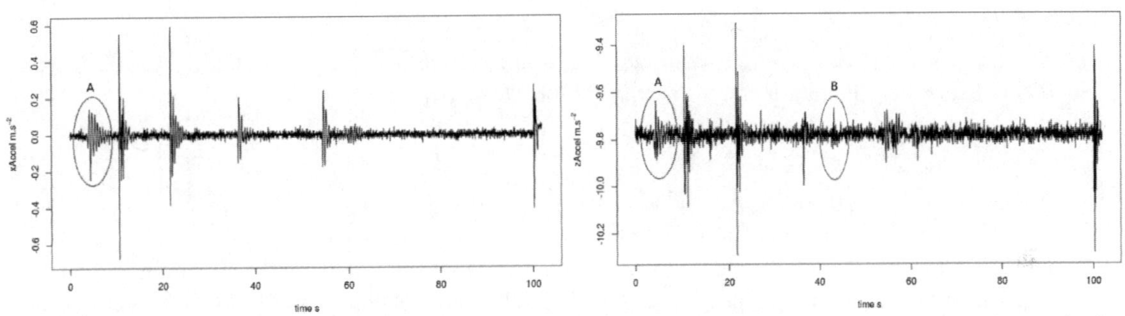

Figure 2. The accelerometer is sensitive on all axis to direct impacts like event A, and on the vertical axis only to indirect impacts like event B.

If we suppose that the objects doesn't move during a long time, and that peoples are moving around the furnitures. The goal is to perceive the impact of shoes on the floor produced by human footseps in order to identify the correlation between impact events, then to estimate the closeness between smart furnitures. An other experiment where an impact of shoe is performed close to a smart table had been performed.

The difficulty of such measurement is given by the amplitude of the signal that is close to the quantization noise. Nevertheless, as it has been seen that the seismic s-waves signal produced by shoes impact has a low frequency under 100Hz. Then to extract the signal from quantization noise, the acquisition frequency had been fixed at 3200Hz that is the maximum available with the given accelerometer.

Considering footstep as seismic events they produce an elastic surface wave that can be detected an analysed respectively with sensors and algorithms devoted to the detection of seismic events [6][7]. The wavelet transformation is a privileged for that analysis of surface waves signals produced by seismic events . We choose use the Morley wavelet transformation [8] to analyse the signal given by footstep events. indicates that seismic events can be extracted from the signal.

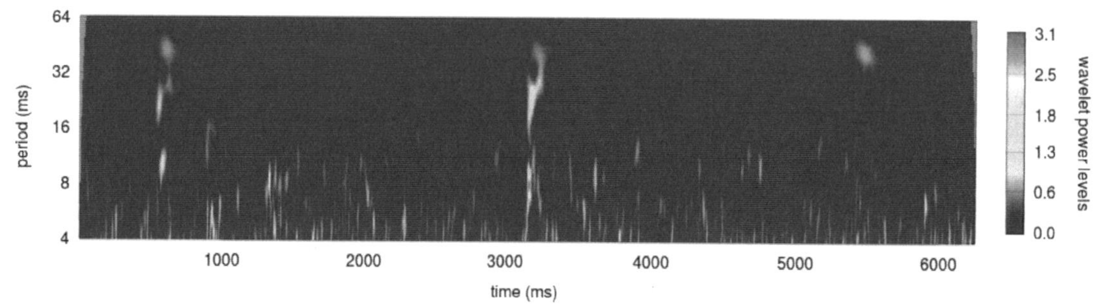

Figure 3. Wavelet coefficients power level computed with a Morley wavelet. 3 shoes impacts are measured. The vertical axis is the Fourier period and the horizontal axis is the translation coefficient.

The figure 3 shows the result of a Morley wavelet transformation of the signal produced by the vertical axis of the accelerometer during an experiment where a feet knock the floor close to the smart table. The Morley wavelet is defined by the complex function:

$$\psi(t)=\pi^{-1/4} e^{i\omega t} e^{-t^2/2}$$

And the wavelet coefficients W are computed by a convolution of the signal with a set of wavelet daughters obtained by dilatation and translation of the Morley wavelet.

$$W(\tau,s)=\sum_t x_t \frac{1}{\sqrt{s}} \varphi^*\left(\frac{t-\tau}{s}\right)$$

During this experiment, 3 events are produced every 2.5 seconds.

The wavelet transformation of a seismic wave packet indicates that it is made of 2 mono-frequency wave packets with frequencies respectively close to 24Hz and 40Hz and a third mono-frequency wave packet with frequencies between 100Hz and 130Hz (see figure 4). The second frequency of 40Hz fits with the results of the studies of Ekimov [9] on the vibration signature of the human footstep in buildings. The first one (24Hz) doesn't appear in the same studies. It is possibly related to the material of the shoe. The third mono-frequency wave packet (between 100Hz and 130Hz) doesn't appear in all events, as shown in figure 5, and is possibly related to the way the footstep is performed. These 2 hypothesis are not yet tested and are under study.

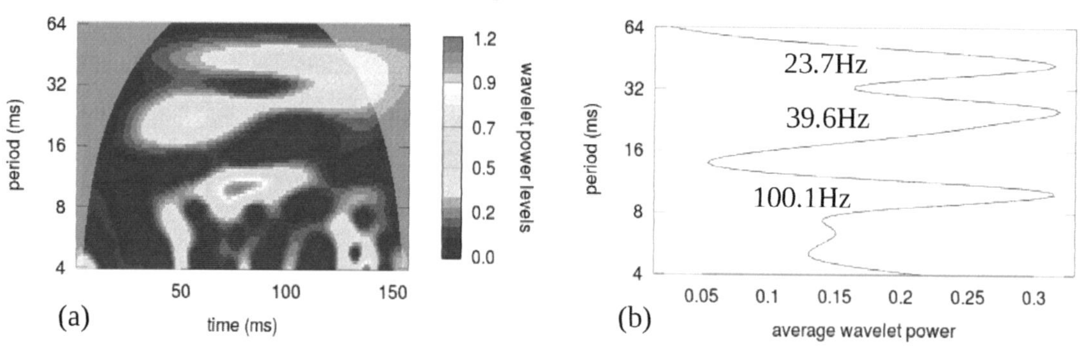

Figure 4: Details of the 1st event in the plane time/frequency (a). The wave packet is made of 3 mono-frequency wave packets as shown on the average of wavelet power on a 150ms window (b).

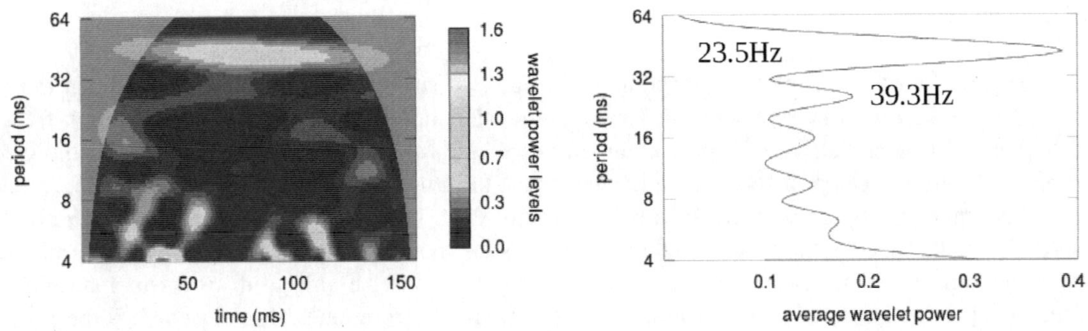

Figure 5. Details of the 3rd event in the plane time/frequency (a). The wave packet is made of 2 mono-frequency wave packets as shown on the average of wavelet power on a 150ms window (b).

At this step, the seismic events appear as wave packets with frequencies 24Hz and 40Hz. A threshold and a normalisation of the wavelet power for these 2 frequencies gives a fuzzy value that represent the truth that the event "footstep" occurred. The measurement result is a fuzzy value.

4. The closeness measurement

The definition of closeness needs another paper but is resumed below.
The closeness fuzzy value related to a *close_to* relation between 2 physical entities is supposed to value the truth of the assertion "entity 1 is close to entity 2" at a given time. We suppose that a period of time where the entities doesn't move can be identified. We want then to reinforce our knowledge about the truth of this assertion during this period of time.
The closeness between objects is related the conflict between the measurement sources. Indeed, when the smart furnitures are close, they perceive similarly their environment, and the conflict between the information they produce is low. When they are far from each other, the conflict is also low for strong far events like a door slam but become high for small events like footsteps that are detected by one object but not by another one.
Within the field of uncertain information fusion approaches [10], the possibility theory and the evidence theory are potential candidates for the management of conflict .
The possibilistic approach qualifies an assertion with a measure of possibility and a measure of necessity. It is able to represent the impossibility, the lack of knowledge and the necessity. In the context of this study, the *close_to* relation is weighted by a couple possibility/necessity (Π, N) . At the beginning of the stable period, the couple is initialised with the couple (1,0) that represent the lack of knowledge. Each footstep event detected simultaneously by the 2 sources maintains a high possibility measure but a footstep event detected by one source but not the other improves the impossibility then reduces the possibility. This method is optimistic and doesn't distinguish between the lack of knowledge and a totally possible *close_to* relation.
The evidence theory applied on the context of this study distributes a unitary amount of belief masses on the set $\{\emptyset, \{a\}, \{\bar{a}\}, \Omega=\{a, \bar{a}\}\}$ where a is the assertion "entity 1 is close to entity 2". This approach is more precise and distinguishes the full conflict (with belief masses on $\emptyset$) the certainty on the assertion or on the negation of the assertion, and the lack of knowledge (with belief masses on Ω). At the beginning of the stable period all belief masses are on Ω . A footstep event detected simultaneously by the 2 sources transfers masses on $\{a\}$. A footstep event detected by one source but not the other transfers masses on $\emptyset$.

5. Conclusion

The measurement of the states of a community of connected smart furnitures is a challenge that must be faced to be able to propose a solution for the home-care support. This paper proposes a link from the higher level of modelling of object communities to the raw measurements. Despite the limited performance of the accelerometers embedded into smart furnitures, this study shown that the detection of small seismic events can be performed by the way of continuous wavelet transformation. The first results shown that the footstep can be recognized. This last point is important for the fusion process that will estimate the closeness. Indeed the smart furnitures must be able to distinguish small close seismic events that improve the belief into the closeness, from strong ones. The closeness is the pivot concept between the numerical measurement and the abstract representation of object communities. Its determination is the basic point to identify situations defined as states of object communities. The fusion at the higher level of imprecise information like the closeness is still under study.

This research is supported by the Universite Savoie Mont Blanc, the Association des Pays de Savoie and Miliboo.

6. References

[1] Atzori L and Iera A and Morabito G 2010 The Internet of Things: A survey *Computer Networks* **15(54)** doi:10.1016/j.comnet.2010.05.010 2787—2805

[2] Varzi A C 1998 Basic Problems of Mereotopology *Formal Ontology in Information Systems:Ios Press* 29–38

[3] Ye J, Stevenson G and Dobson S 2011 A top-level ontology for smart environments *Pervasive and mobile computing* **7(3)** 359—378

[4] Benoit E, Dapoigny R 2006 Mereotopological approach for measurement software *Advanced Mathematical and Computational Tools in Metrology VII* 13--22

[5] Jones C B, Alani H and Tudhope D 2001 Geographical information retrieval with ontologies of place *Spatial information theory: Springer* 322—335

[6] Rim B and Michel O and Frassati F and Carmona M and Lacoume JL 2013 New algorithm for footstep localization using seismic sensors in an indoor environment *J. of Sound and Vibration*, ***333*** 1046--1066.

[7] Antoine Roueff A and Mars J and Chanussot J and Pedersen H 2005 Dispersion Estimation From Linear Array Data in the Time-Frequency Plane *IEEE Trans. on Signal Processing* **53(10)** 3738—3748.

[8] Aguiar-Conraria L and Soares MJ 2011 The Continuous Wavelet Transform: A Primer. *NIPE Working Paper Series* **16/2011**.

[9] Ekimov A and Sabatier JM 2006 Vibration and sound signatures of human footsteps in buildings *The Journal of the Acoustical Society of America*, ***120*(2)**, 762—768.

[10] Dubois D and Liu W and Ma J and Prade H 2016 The basic principles of uncertain information fusion. An organised review of merging rules in different representation frameworks *Information Fusion* **32(2016)** 12–39

IMEKO2016 TC1-TC7-TC13 IOP Publishing
Journal of Physics: Conference Series **772** (2016) 012004 doi:10.1088/1742-6596/772/1/012004

An Optical Phased Array for LIDAR

Y Wang, M C Wu

School of Engineering, University of California, Berkeley

E-mail: youmin.wang@berkeley.edu

Abstract. We have previously demonstrated the development of an Optical Phased Array (OPA) micromechanical system (MEMS) used for beam steering, which shows great advantages over previous mechanisms such as opto-mechanical, acousto-optical (AO) or electro-optical (EO). We aim to integrate the OPA MEMS system into the application of automobile navigation, which is currently primarily dominated by opto-mechanical scanning based systems. Opto-mechanical scanning devices are usually bulky and relatively slow, while competing technologies (AO, EO) utilize devices that while small in size, cannot provide the steering speeds and versatility necessary for many applications. In drawing from phased array concepts that revolutionized RADAR technology by providing a compact, agile alternative to mechanically steered technology, the OPA based LIDAR program seeks to integrate thousands of closely packed optical emitting facets, precise relative electronic phase control of these facets, and all within a very small form factor. Comparing with other competing LIDAR system, the OPA based LIDAR system will have multiple degrees of freedom for phase control which enables not only agile beam steering but also beam forming and multiple beam generation, greatly expanding the diversity of applications.

Traceability of high focal length cameras with diffractive optical elements

L Lages Martins[1], A Silva Ribeiro[1] and J Alves e Sousa[2]

[1]LNEC - National Laboratory for Civil Engineering, Lisbon, Portugal
[2]LREC - Regional Laboratory for Civil Engineering, Funchal, Madeira, Portugal

Corresponding author's e-mail address: lfmartins@lnec.pt

Abstract. This paper describes the use of diffractive optical elements (DOEs) for metrological traceable geometrical testing of high focal length cameras applied in the observation of large-scale structures. DOEs and related mathematical models are briefly explained. Laboratorial activities and results are described for the case of a high focal length camera used for long-distance displacement measurement of a long-span (2278 m) suspension bridge.

1. Introduction

The use of digital images for dimensional and geometrical measurements is increasing both in Science and Industry due to the technological developments occurred over the past decades in digital cameras and computers, which are now more accessible, reliable and with higher performance [1]. A camera permits a high quantity of information in a single image, including the dimension and geometrical shape of the observed object within a reduced timeframe interval, without contact, using a robust, real time automatized procedure.

In order to achieve high accuracy measurements, cameras are subjected to geometrical testing in order to estimate their intrinsic parameters – focal length, principal point image coordinates and distortion coefficients – which are required to be used as input quantities in the measurement process, with a direct impact on the metrological quality of the measurand.

In areas such as Photogrammetry and Computer Vision, which usually deals with reduced or normal focal length cameras (lower than 100 mm), this task is accomplished by conventional methods [2-5] supported in reference patterns or points. However, in the case of high focal length cameras, these methods often proved to be numerically unstable, providing unrealistic estimates for the intrinsic parameters. To overcome this limitation, several alternative approaches [6-7] have recently been proposed, e.g., the use of diffractive optical elements or DOEs.

This paper presents an innovative approach based on the use of DOEs to establish an accurate metrological traceability for geometrical testing of high focal length cameras, considering that this method has simplified setup requirements, if compared with laboratorial testing using high accuracy goniometers, and is more accurate than the complex approach of field testing with reference points [6]. Following a brief description about the geometrical testing of cameras using DOEs, the paper presents its application in the characterization of a high focal length (600 mm) camera used for displacement measurement in a long-span (2278 m) suspension bridge: the 25th of April Bridge (P25A), in Lisbon (Portugal). Estimates and measurement uncertainties, obtained by a Monte Carlo Method [8] in a non-linear, multivariable and complex optimization process, are discussed and a sensitivity analysis is performed to identify the main contributions for the measurement uncertainty evaluation.

2. Camera geometrical testing by the DOE method

The DOE method is supported in a collimated laser beam which passes through a DOE, creating a known spatial pattern of diffraction dots in the camera's focal plane. The DOE, usually a diffraction grating, has a complex microstructure with two or several depth levels produced by photolithography, which leads to a high accuracy spatial period.

The knowledge of the laser wavelength, the DOE spatial period and the diffraction dots image coordinates is used to determine the camera's intrinsic parameters – focal length, principal point image coordinates and distortion coefficients – through a non-linear optimization process which aims at the minimization of the following function [7]

$$\min_m \left\| \begin{bmatrix} \hat{u} - u_0 \\ \hat{v} - v_0 \end{bmatrix} - f \begin{bmatrix} x \\ y \end{bmatrix} [1 + k_1 \cdot (x^2 + y^2) + \cdots] \right\|^2 \quad (1)$$

being the unknown variables described m given by the following vector,

$$m = [f, u_0, v_0, k_1, \omega, \psi, \kappa, \alpha, \beta] \quad (2)$$

where $(\hat{u}, \hat{v})$ and (x, y) are, respectively, the diffraction dots distorted and ideal image coordinates, f is the camera's focal length, (u_0, v_0) are the principal point image coordinates, k_1 is the first-order radial distortion coefficient, (ω, ψ, κ) are the camera's orientation Euler angles and (α, β) are the DOE orientation angles with respect to the collimated laser beam coordinate system.

This method has proven to be suitable for the geometrical characterization of reduced focal length cameras [7]. However, no information is available about its application to high focal length cameras, thus justifying this study. This type of cameras is quite relevant for long-distance observation of large-scale structures since they contribute to a high displacement measurement sensitivity.

3. Experimental setup

Based on the main features of the tested camera (see Table 1), an experimental setup was developed in a laboratorial optical bench for the implementation of the DOE method, as shown in Figure 1.

Table 1. Camera components

Component	Description
Image sensor	Visible and near-infrared CCD; 7.4 μm squared pixel; (1920 × 1080) pixels
Frame grabber	PCMCIA interface; Imperx Framelink Application
Telephoto lens	Sigma, 300 mm/2.8 APO EX DG
Optical converter	Sigma, APO 2× EX DG

In a first stage, the experimental testing was made only with the 300 mm lens and in a second stage with the converter assembled in order to evaluate the influence of this component. Due to a manual infinity focus adjustment ring on the lens, additional studies were also performed – image acquisition with different focus positions – to determine its impact on the intrinsic parameters repeatability.

The DOE selected for this study was a diffraction grating (Holoeye, model DE-R-241, 8 mm diameter and a thickness of 1.2 mm) capable of producing a 21 × 21 dots matrix pattern, with a total diffraction angle of 4.8° and a spatial period estimate equal to 152.4 μm. This grating is optimized for a laser wavelength of 635 nm, which results in the lowest intensity increase (about 0.3%) of the central (zero order) diffraction spot. Since the total diffraction angle produced by the diffraction grating was higher than the camera's field of view, not all of the diffraction orders were observed, as shown in Figures 2 and 3.

The main operational concerns were related to laser beam attenuation by the use of a neutral filter (to avoid saturation and sensor damage) and the optical alignment which has a direct impact on the profile quality of the diffraction orders irradiance (absence of unwanted effects such as defocusing, reflections, diffraction fringes, non-collimated and non-Gaussian beam).

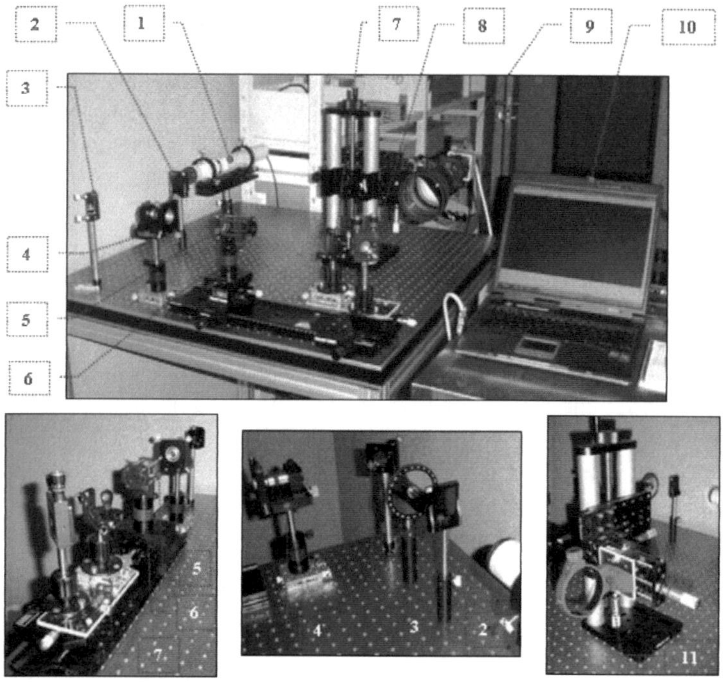

1 – Laser (λ = 633 nm)
2 – Neutral filter
3 – Mirror
4 – Spatial filter
5 – Reference lens
6 – Aperture
7 – Diffraction grating
8 – Mirror
9 – Tested camera
10 – Computer
11 – Camera support

Figure 1. Experimental setup for the laboratorial implementation of the DOE method.

Figure 2. Observed diffraction pattern for the telephoto lens (without converter assembly).

Figure 3. Observed diffraction pattern for the telephoto lens and converter assembly.

The obtained images were subjected to digital processing – thresholding for binary conversion, followed by opening morphological operation with a disk as the structuring element and Gaussian adjustment – aiming at the determination of the dots image coordinates. The obtained estimates were used as input quantities, as well as the laser wavelength and the diffraction grating spatial period estimates, in a non-linear optimization process.

This calculation task was supported in the Nelder-Mead simplex algorithm [9] dedicated to multivariable functions, such as equations (1) and (2), using a direct search method in which the calculation of numerical or analytical gradients are not required. Initial values for the unknown variables were defined based on nominal or approximated values, and were kept constant for both the two studied cases (300 mm or 600 mm focal length assemblies). A 10^{-6} pixel^2 stopping criterion was defined for the objective function given by equations (1) and (2).

4. Results and discussion

Tables 2 and 3 summarize the obtained results – average values and experimental standard deviations – for the camera's intrinsic parameters, noticing that a total of five images of the diffraction pattern were acquired for each infinity focus position. With respect to the lens distortion, only the first-order radial distortion coefficient was accounted in equations (1) and (2), since this type of optical aberration is known to be reduced for the case of high focal length lenses.

Table 2. Focal length and first-order radial distortion coefficient.

Intrinsic parameter	f / mm		k_1 / 10^{-6} m^{-2}	
∞ focus position	Lens	Lens & converter	Lens	Lens & converter
A	300.1±1.4	590.9±2.2	110±32	-102±51
B	300.9±0.5	604.8±4.9	69±9	76±32
C	301.7±0.2	597.0±2.1	80±11	9±17

Table 3. Principal point image coordinates.

Intrinsic parameter	u_0 / pixel		v_0 / pixel	
∞ focus position	Lens	Lens & converter	Lens	Lens & converter
A	532.6±2.4	566.2±2.1	996.5±2.9	954.2±3.7
B	536.3±0.8	533.8±1.7	1012.7±1.1	965.4±2.5
C	534.5±1.0	546.9±2.2	1012.5±1.3	963.3±4.4

The results obtained for the focal length parameter are closer to the expected nominal values, however, in the case where the converter is used, the deviation to the 600 mm nominal value and the corresponding experimental standard deviations are higher. This is justified by the reduced number of observed diffraction dots and by the mechanical assembly of the two optical components. The influence of the infinity focus position is also more noticed when the converter is mounted on the lens, namely, for position B.

Regarding the first-order radial distortion coefficient, estimates varied between -102×10^{-6} m^{-2} and 110×10^{-6} m^{-2} with experimental standard deviations ranging from 9×10^{-6} m^{-2} up to 51×10^{-6} m^{-2}. Although having a wide variation interval, the estimates magnitude is insufficient to affect the accuracy of the diffraction dot's image coordinates (lower than 0.01 pixel). This result confirms the reduced effect of radial distortion in high focal length lenses and allows the removal of this strongly non-linear component from the optimization processes (intrinsic parameterization and displacement measurement), thus increasing its numerical stability. Radial distortion differences between the two tested assemblies are mainly noticed in the extreme focus positions (A and C), with a reduced impact on the image coordinates accuracy.

Estimates for the principal point image coordinate in the x-direction are close to the expected 540 pixel nominal value, for both tested assemblies, and the magnitude of the corresponding experimental standard deviations is also similar for the studied infinity focus positions. A minimum dispersion value is always obtained for the intermediate focus position (B) for both principal point image coordinates (u_0, v_0).

Significant differences were found between the two tested assemblies in the estimates of the principal point image coordinate in the y-direction (v_0), noticing that the lens and converter assembly presents results close to the 960 pixel nominal value. This can be justified by incorrect centering of the observed diffraction pattern in the y-direction of the camera without the converter, as seen in Figure 2, where the zero-order diffraction dot is not correctly centered in the middle of the image acquired.

Due to the non-linear nature of the optimization process, a Monte Carlo method [8] was applied for the determination of the measurement uncertainty of the estimated intrinsic parameters. This computational algorithm was developed in Matlab using validated functions and the Mersenne Twister pseudorandom number generator [10]. For each simulation, a total of 10^4 trials were performed in order to obtain convergent solutions. For this case, an iterative procedure was implemented and convergent solutions were obtained considering that the result of a Monte Carlo simulation updated the initial values of the following non-linear optimization.

The standard uncertainties related to the diffraction dot's image coordinates (ranging from ¼ up to ¾ of a pixel) and to the diffraction grating spatial period (0.15 μm) were propagated to the output quantities (the laser wavelength quantity was considered constant). The 95% expanded uncertainty of the focal length parameter was found between 0.75 mm and 1.9 mm. Regarding the principal point coordinates, the 95% expanded uncertainty ranged between 0.06 pixel and 0.21 pixel. Correlation effects were noticed between intrinsic parameters (correlation coefficients between -0.35 and -0.25).

The performed sensitivity analysis revealed the diffraction dot's image coordinates as the major contribution (close to 75%) for the output measurement uncertainty, reflecting the impact of input uncertainties components related to digital image processing and laser beam collimation quality.

The studied camera (with the 300 mm lens and converter assembly) was later applied in the measurement of the displacement of the main span central section of the P25A Bridge (Fig. 4 and 5).

Figure 4. P25A suspension bridge in Lisbon (Portugal).

Figure 5. Camera installed in the P25A suspension bridge main span central section.

In this approach, the camera was rigidly connected to the lower region of the stiffness beam and orientated towards the tower foundation, establishing an estimated observation distance of 510 m. A geometric referential composed by four active targets, with known world coordinates, was placed in a bridge tower foundation (Figure 4), visible for the camera range in the main span central section.

In this condition, changes of the camera's projection center world coordinates are considered representative of the bridge displacement and can be determined by non-linear optimization using as input quantities the targets image and world coordinates (obtained from digital image processing and previous laboratory dimensional testing, respectively) and the camera's intrinsic parameters (by previous laboratory testing, as described in this paper).

Maximum vertical and transverse displacements of 1.62 m and 0.29 m were measured with a 95% expanded uncertainty comprised between 4 mm and 7 mm, for the observed environmental conditions (Summer and Winter measurement campaigns). For this case, the performed sensitivity analysis revealed that the contribution of the intrinsic parameters (focal length and principal point image coordinates) ranged from 15% up to 24% of displacement measurement uncertainty.

5. Conclusions

This study enabled the conclusion that the DOE method is suitable for metrological traceable geometrical testing of high focal length cameras, at least, up to 600 mm, without any numerical instability in the optimization process, as it often occurs when conventional methods are applied. Radial distortion revealed to be negligible not affecting significantly the image coordinates accuracy. This in turn permitted the removal of this strongly non-linear component from the optimization, therefore, contributing for the achieved numerical stability of the parameterization and measurement processes.

The camera with the 600 mm focal length assembly was used in the structural observation of the P25A Bridge. Although the obtained estimates and measurement uncertainties of the camera's intrinsic parameters contributed for about 15% to 24% of the final displacement accuracy, other input quantities of the mathematical model such as the targets image coordinates (55% - 58%) and target world coordinates (18% - 30%) coordinates showed higher contributions to the output uncertainty.

The DOE method can now be used to identify the occurrence of any damage or drift effect in the camera, namely, in permanent observation scenarios, due to the long-term exposure of the camera to an aggressive dynamic environment.

Acknowledgments

The authors gratefully acknowledge José Rebordão, Manuel Abreu and Fernando Monteiro from the Laboratory of Optics, Lasers and Systems of the *Faculdade de Ciências da Universidade de Lisboa* which contributed for the experimental setup and discussion of intrinsic parameterization results.

This work was funded by the Portuguese Ministry of Education and Science National Funds through FCT-Foundation for Science and Technology, under PhD research grant ref. SFRH/BD/76367/2011 assigned to L Lages Martins.

The project I&D UID/MAT/04674/2013 of FCT-Foundation for Science and Technology, supported J Alves e Sousa's contribution, and is grateful acknowledge.

References

[1] Schwenke H, Neuschaefre-Rube U, Pfeifer T, Kunzmann H, 2002 *CIRP Annals – Manufacturing Technology*, **51**, 2, 685-699
[2] Abdel-Aziz Y, Karara H, 1971 *Proceedings of the Symposium on Close-Range Photogrammetry*, 1-18
[3] Heikkilä J, Silvén O, 1997 *Conference on Computer Vision and Pattern Recognition – Proceedings CVPR 1997*
[4] Tsai R, 1987 *IEEE Journal of Robotics and Automation*, **RA-3**, 4
[5] Zhang Z, 2000 *IEEE Transactions on Pattern Analysis and Machine Intelligence*, **22**, 11
[6] Hieronymus J, 2012 *International Archives of Photogrammetry, Remote Sensing and Spatial Information Sciences*, **XXXIX-B5**
[7] Bauer M, Grießbach D, Hermerchmidt A, Krüger S, Scheele M, Schischmanow A, 2008 *Optics Express*, **16**, 25
[8] *S1-GUM Evaluation of measurement data. Supplement 1 to the "Guide to the expression of Uncertainty in Measurement". Propagation of distributions using a Monte Carlo method"*, 1st edition, JCGM, 2008
[9] Lagarias J, Reeds J, Wright P, 1998 *SIAM Journal of Optimization*, **9**, 1, 112-147
[10] Matsumoto M, Nishimura T 1998 *AMC Transactions on Modeling and Computer Simulation*, **8**, 1, 3-30

Assessment of the Metrological Performance of Seismic Tables for a QMS Recognition

A Silva Ribeiro[1], A Campos Costa[1], P Candeias[1], J Alves e Sousa[2], L Lages Martins[1], A C Freitas Martins[1], A C Ferreira[1]

[1]National Laboratory for Civil Engineering (LNEC), Lisbon, Portugal
[2]Regional Laboratory for Civil Engineering R. A. Madeira, Funchal, Portugal

E-mail: asribeiro@lnec.pt

Abstract. Seismic testing and analysis using large infrastructures, such as shaking tables and reaction walls, is performed worldwide requiring the use of complex instrumentation systems. To assure the accuracy of these systems, conformity assessment is needed to verify the compliance with standards and applications, and the Quality Management Systems (QMS) is being increasingly applied to domains where risk analysis is critical as a way to provide a formal recognition. This paper describes an approach to the assessment of the metrological performance of seismic shake tables as part of a QMS recognition, with the analysis of a case study of LNEC Seismic shake table.

1. Introduction

Seismic testing using large infrastructures, such as shaking tables and reaction walls, is performed worldwide using different technical approaches. Measurement and control of these systems is based on complex design of instrumentation, often unique, able to measure and dynamically process several quantities: length and time, electrical (transducers, data acquisition, signal processing and device control), and to provide information of dynamic phenomena under study (e.g., velocity and acceleration). To assure the accuracy of these systems, metrological management should be used, including traceability to SI through calibration procedures and assessment of metrological performance to verify compliance with standards or applications requirements as part of a Quality Management Systems (QMS). The growing importance of conformity assessment in domains of critical risk, leads to the need of promoting the recognition of Research Facilities [1] according to international standards [2-3]. The development of methods to perform the assessment of metrological performance of these infrastructures is, therefore, a relevant contribution to set up the conditions for its implementation. This paper will also describe the procedures and experimental results developed for the assessment of LNEC's shaking table.

2. Seismic Infrastructure of LNEC Earthquake Engineering Research Centre

LNEC Earthquake Engineering Research Centre operates since 1996 as part of the European Seismic Engineering Research Infrastructures, developing research for seismic experimental and analytical modelling of structural systems to seismic hazard and risk analysis applied in the Civil Engineering testing of structures and components. The facilities include a large 3D shake table with a complex set of measuring devices able to measure several quantities (e.g., pressure, force, acceleration, displacement and strain) [4].

The main infrastructure characteristics (metrological and physical) include its large payload capacity (able to support weights up to 400 kN) at extreme testing conditions (near collapse).

The testing platform is made of steel with size of (4,6x5,6) m^2 and weight of 392 kN, 3 axis hydraulic actuators, acquisition system, 8 A/D channels and 96 configurable digital input channels. The operation and control uses in-house based software.

Figure 1 – Top view of LNEC Earthquake Engineering testing room

3. Quality Management System

Quality Systems models, created in the 1950's by Deming, Juran and Crosby, have today huge impact on the Organizations, were integrated Total Quality Management (TQM) has a relevant role establishing management rules of custom-focused organizations. The principles of Management System can be found today in ISO 9000 and ISO 17000 Series standards, highlighting the role of ISO 9001 [2] with the requirements of Quality Management and ISO 17025 [3] with the general requirements for the competence of testing and calibration laboratories. In the first, the aim is to establish recognition to the development of international trade and in the second, to promote competence recognition.

In many of today's technical activities where measurement results are obtained and needed for the decision process, the growing interest in mutual recognition is due to higher expectations of consumers and the development of international trade. For this purpose, the level of engagement can be established at different levels:

- *Qualification* is a degree that applies examinations to meet specific experience requirements and to comply with Quality Management Systems (QMS).
- *Certification* (organization, person or product) is defined as the compliance with specific requirements, with the assessment made by an external entity;
- *Accreditation* is a formal, third party recognition of technical competence.

The introduction of QMS into activities related to large experimental infrastructures with R&DI is becoming more relevant, considering that technical competence is part of the qualification for testing and measurement. This implies, in many cases, the redefinition of the formal processes of management and the independent assessment of the QMS implementation.

For the development of the QMS based on ISO/IEC 17025 several technical requirements need compliance. In the context of this paper, the interest is focused on the metrological requirements specifically applied to the core equipment that holds the control and measurement of the infrastructure testing facility, namely: traceability and calibration procedures; metrological conformity assessment; measurement correction and uncertainty evaluation; data record management; and data analysis (acquisition & processing of data).

4. Metrological Testing Specifications for Conformity Assessment

The conformity assessment is the basis for the validation of testing specifications, namely, performed by seismic shaking tables (tests and requirements being presented on Table 1). The procedure adopted to conformity assessment compares the target measurement uncertainty (concept defined by [5] *as measurement uncertainty as specified as an upper limit and decided on the basis of the intended use of measurement results*) with the instrumental measurement uncertainty (defined as [5] *component of measurement uncertainty arising from a measuring instrument or measuring system in use*). The procedure was developed according to five stages:

- first stage establishes the measurand quantities to apply the conformity assessment to and the requirements considering the intended use (for R&D&I, testing, or others);
- second stage identifies the instrumentation that support the measurement of quantities needed for the system control and operation;
- third stage establishes the set of tests to be performed:
    - Cross-axis motion
        - Static position – scale accuracy, reversibility, discrimination, repeatability return to zero tests
        - Dynamic mono-axial excitation (response to input sine wave)
    - Rotational motion
        - Mono-axial and combined rotational motion
        - Rotational spurious motion
    - Channels acquisition and transducing (A/D)
    - Amplitude and frequency dynamic response
- fourth stage establishes testing procedures including reference standards and traceability to SI, sampling, statistics to be applied and measurement uncertainty evaluation [6].
- fifth stage performs conformity assessment by a comparison between experimental results and defined requirements based on the intended purposes of testing.

Table 1 – Description of testing set, main characteristics and assessment requirements

Test description	Instrumentation & measurement interval	Function	Reference method & standards	Traceability	Target meas. uncertainty
1 - Cross-axis motion: static position & Discrimination	Inductive LVDT (hydraulic actuators) $l = \pm$ 120 mm	Measurement & control	Direct comparison using laser interferometry	IPQ - Instituto Português da Qualidade (NMI)	< 1 mm
2 - Cross-axis motion: dynamic mono-axial excitation	Inductive LVDT (hydraulic actuators) Sine waves	Measurement & control			< 5 %
3 - Rotational motion: Rotational spurious motion	Inductive LVDT (hydraulic actuators) $\theta = \pm$ 10 °	System response to stimulus			< 5 °
4 - Channels acquisition and transducing	NI PXI Boards & Controllers 1 Hz – 100 Hz	Measurement	comparison using input signal generator, digital voltmeter and universal time counter	TAP Metrology Laboratory	According with A/D resolution
5 - Amplitude & frequency response	NI PXI Boards & Controllers 0 – 1 V/ 1 Hz –100 Hz	Measurement			< 5 % < $5{\cdot}10^{-3}$ Hz

In Table 1, the quantities and conformity requirements, related to the 1st stage, are found in column 6, obtained for the common protocol of measurement and control of LNEC shake table. The 2nd stage results are presented in column 2. The 3rd stage defines the set of testing to be applied, described in column 1. The 4th and 5th stages are detailed on next Section. The approach considers that accelerometers (ENDEVCO 7290A to control functions of the shaking table and PCB 337A62 for measurement purposes) used for measurement and control of as well as for measurement of the performance of tested objects, are calibrated under traceable conditions to SI.

5. Experimental procedures and results

The procedure to evaluate dimensional cross-axis motion and rotational motion across axis performances of the LNEC shaking table was based on interferometry, using a laser interferometer (λ = 633 nm) as reference, being able to perform linear and angular measurements

Journal of Physics: Conference Series **772** (2016) 012006 doi:10.1088/1742-6596/772/1/012006

with specific setups of the optical components. The procedure developed had two main parts (setup alignments and data acquisition). The critical contributions to measurement uncertainty found were alignment, synchronization and influence quantities (temperature and pressure). The experimental requirements included error minimization due to misalignment of optical elements with the axis of the actuators both vertical and horizontal (preliminary tests at full range were performed), signal synchronization and compensation of temperature and pressure influence.

Two tests were designed to evaluate scale accuracy and reversibility (test A, see Figures 2 and 3) and discrimination (test B, see Fig. 4). To perform data acquisition, input dynamic series were provided to establish calibration steps of 30 mm with low variance of displacement.

For discrimination, step transitions of 0,5 mm, 0,1 mm and 5 mm were given at 20 mm, 50 mm and 80 mm positions. Calibration was performed by sampling from interferometer and LVDT's data, with at least 500 sampling pairs to have a Gaussian representation of the probability distribution. Two of the geometrical setups are presented in Figures 5 and 6.

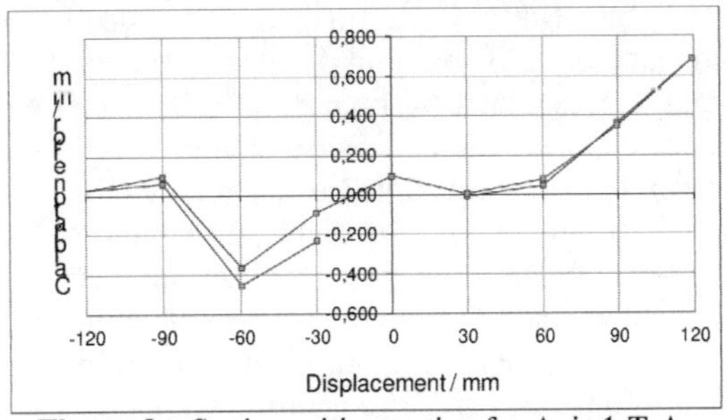

Figure 2 – Static position testing for Axis 1-T-A, scale calibration error with reversibility

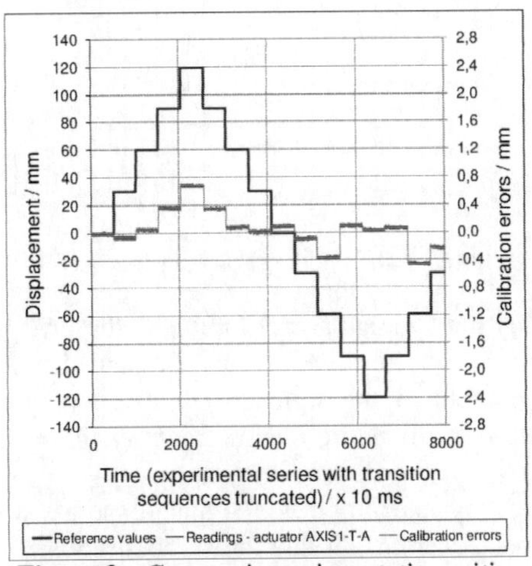

Figure 3 – Cross-axis motion: static position testing axis 1-T-A calibration errors

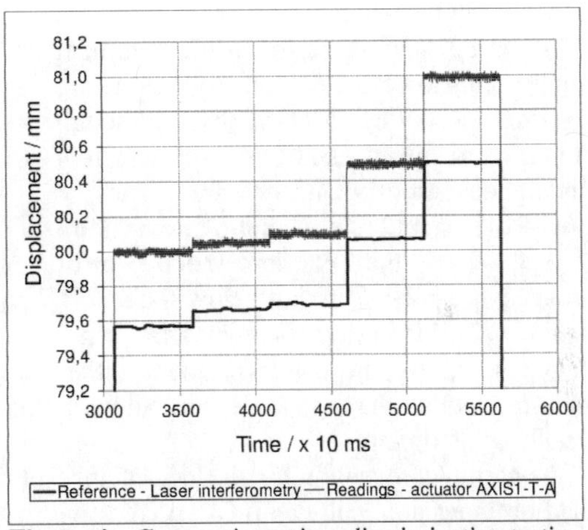

Figure 4 – Cross-axis motion: discrimination testing for axis 1-T-A at reference displacement of 80 mm

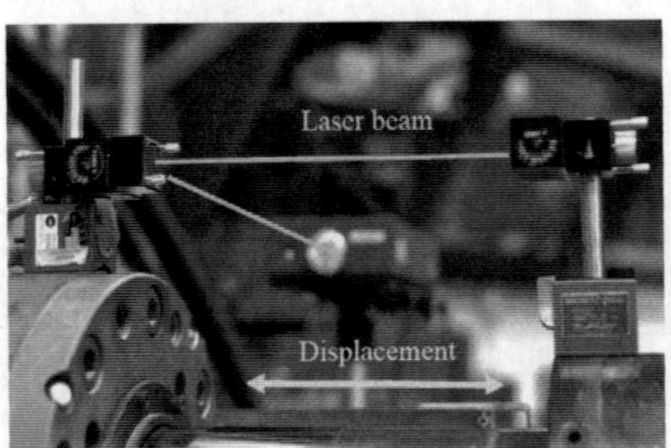

Figure 5 – Geometrical Setup for the cross--axis motion testing

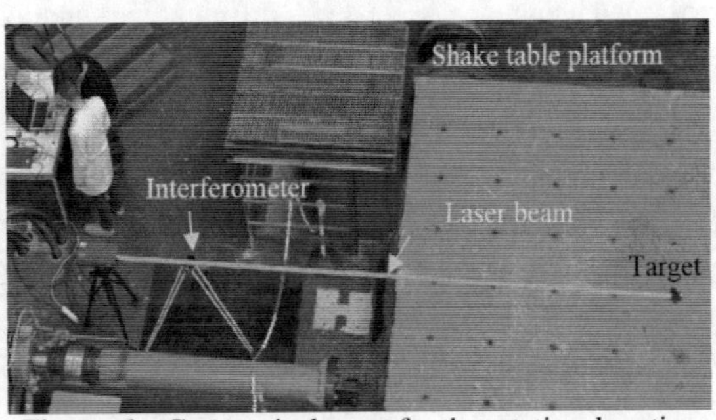

Figure 6 – Geometrical setup for the rotational motion testing

The assessment of the displacement error in dynamic motions were also carried out by imposing a time history of periodic random noise with amplitude of 27,30 mm (RMS=7,94 mm) and a frequency range with a low pass Butterworth filter with a cutoff frequency of 3,333 Hz and 4 poles. An error signal was obtained by taking the difference between the laser interferometer and measuring instrument time history records. In Figure 7 is shown the probability distribution function of the random error. A normal distribution was LSQ fitted to the distribution of the error random variable yielding a mean value of 0,02 mm and a standard deviation of 0,16 mm for the error.

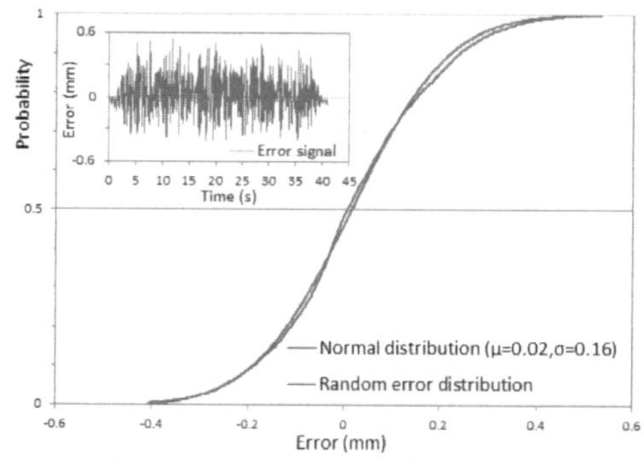

Figure 7 – Random error probability distribution

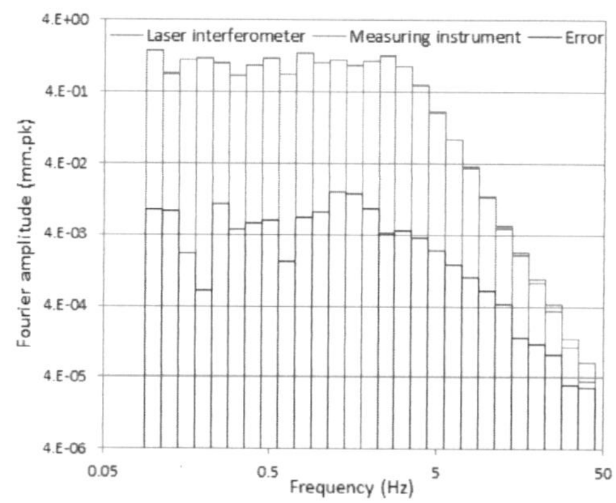

Figure 8 – 1/3 octave Fourier spectra

Figure 8 shows the 1/3 octave Fourier spectra ordinates computed from the laser interferometer and measuring instrument as well as the difference between them which is the error seen in spectral ordinates.

As can be seen, the error in measurement signals seems to be normally distributed along the time, showing stationarity, although without zero mean, and an expected low standard deviation when compared with the signal RMS (signal to noise ratio around 2%). In fact, from the Fourier spectra error ordinates, it can be seen that below the cutoff frequency the average signal to noise ratio error is around 0.7%, increasing for higher frequencies, above 40 Hz, to 45%. It must be said that for frequencies above the cutoff frequency, the energy content of the signal is greatly decreased which means that the amplitudes are below the resolution of the entire measurement chain of the measuring instrument.

The testing procedure for the compliance of frequency and amplitude signal response was based on input nominal electrical voltages with a signal generator, at laboratory conditions, and parallel measurement setup using a digital voltmeter reference standard. This allowed to evaluate the compliance of the low-pass filter of the acquisition system for the 8 channels. Tests were performed using sine waves with different parameters and a set of random noise signals.

6. Measurement uncertainties & Metrological conformity assessment

The conformity assessment procedure was based on the comparison of target measurement uncertainties with instrumental uncertainty. This approach, applied to voltage and frequency and the compliance is presented on Table 2 considering a confidence interval of 95 %.

Table 2 – Parameters required for the conformity assessment of voltage and frequency measurements

Quantity	Target uncertainty	Instrumental uncertainty
Voltage	According with A/D resolution	3,0x10-4 V
Frequency	< 5·10-3 Hz	1,0x10-6 Hz

Instrumental measurement uncertainty related to the calibration of the actuators LVDTs was calculated using contributions provided by the several tests presented combined using GUM [6]

and assuming moderate nonlinearity of the mathematical model. Table 3 includes these contributions using the uncertainty budget table as defined in the GUM.

Table 3 – Description of testing set, main characteristics and assessment requirements

Quantity	*PDF**	$u(x_i)$ /mm	c_i	$u_i(x)$ /mm	v
Resolution	R	0,05/√3	1	2,89 x 10^{-2}	50
Calibration	N	0,10/2	1	5,0 x 10^{-2}	500
Cal. Error correction	R	0,15/√3	1	8,67 x 10^{-2}	15
Repeatability	N	0,015/1	1	1,5 x 10^{-2}	500
Zero error	R	0,097/√3	1	5,6 x 10^{-2}	5
Discrimination	U	0,1/√2	1	7,1 x 10^{-2}	14
Reversibility	R	0,08/√3	1	4,6 x 10^{-2}	50
Acquisition & proc.	T	0,05/√6	1	2,0 x 10^{-2}	500
			$u_c(l)$ /mm	1,5 x 10^{-1}	
			v_{ef} / k	62 / 2,05	
			$U_{95}(l)$ /mm	3,1 x 10^{-1}	

* Probability distribution function rectangular (R), Triangular (T), ArcSin (U), Gaussian (N).

In this case the target uncertainty considered was to be lower than 1 mm, which is assured by the instrumental measurement uncertainty of 0,31 mm for a confidence interval of 95%.

7. Final Remarks, conclusion and further development

The metrological assessment of R&DI infrastructures is essential for a better knowledge of the system performance and capability, providing evidence of the compliance with technical and operational requirements. The implementation of QMS in this context leads to an instrumentation management improvement being an asset in promoting confidence and competence recognition. Furthermore, robustness of results is also achieved by its traceability of measurement results to SI and through the evaluation of measurement uncertainty, both needed to establish the conformity assessment in practice.

Further developments include establishing a framework that could be applied to the broader methods used on seismic shake tables testing, e.g., the metrological assessment of these infrastructures with criteria based on the comparison of target uncertainty with instrumental uncertainty. Finally, the same approach can also be applied to similar large testing infrastructures that perform R&DI in other domains of Civil Engineering.

The studies develop for the conformity assessment showed that the seismic simulator complies with the metrological requirements. Further studies will attempt to evaluate the effect of these uncertainties in the measurement process related to the study of physical models under seismic experimental conditions, which is a new approach in this field of knowledge.

8. References

[1] SERIES – Seismic Engineering Research Infrastructures for European Synergies - WP 3 NA 2.4 Deliv. 3.3 – *Draft Common Protocol for the Qualification of Research*, 2012.

[2] EN ISO 9001:2015 – *Quality management systems* – Requirements. ISO, Geneve.

[3] ISO/IEC 17025:2005 – *General requirements for the competence of testing and calibration laboratories.*

[4] Duarte, R. T., Rito-Corrêa, M., Vaz, C. T., Campos-costa, A. – *Shaking table testing of structures*, 1994 Earthquake Engineering, tenth World Conference, Rotterdam.

[5] International Vocabulary of Metrology – Basic and general concepts and associated terms (VIM 3rd edition) JCGM 200:2012 (JCGM 200:2008 with minor corrections).

[6] *Evaluation of measurement data - Guide to the expression of Uncertainty in Measurement*, JCGM 100:2008 (GUM 1995 with minor corrections).

The evaluation of activity of Technical Committees of Standardization for Metrology and Measurement on national level

O Velychko[1], T Gordiyenko[2]

[1]State Enterprise "Ukrmetrteststandard", 4, Metrologichna Str., 03680, Kyiv, Ukraine
[2]Odessa State Academia of Technical Regulation and Quality, 15, Kowalska Str., 65020, Odessa, Ukraine

E-mail: velychko@hotmail.com

Abstract. General issues and features are considered in relation to the evaluation of Technical Committees for Standardization (TC) activity is considered. The use of generalized criteria for evaluation of the national TC, which takes into account not only the number of standards developed by TC, but other indicators are proposals. Results comparative analysis evaluation of activities national TCs of Metrology and Measurement Industry is presented.

Introduction

A mutual exchange with experience and information in field of metrology is very important. Many International Organizations is developing of voluntary technical standards in field of metrology (measurement, testing etc.), in particular it International Organization for Standardization (ISO), International Electrotechnical Commission (IEC) and International Organization of Legal Metrology (OIML) [1–4].

1. National TCs activity in the field of Metrology and Measurement Industry

The effectiveness activity of the national TCs at national, regional and international levels remains one of the urgent issues for the needs of governmental and non-governmental organizations. Updating and actualization of standards in different sectors of the economy depend primarily on the performance of national TCs. So important is the revitalization of the TC, analysis and identification of the main factors that influence the effectiveness of their work.

Research to assess the effectiveness of national of TCs and considering only the number of developed National Standards (DSTU), incl. harmonized with International standards (IS) and European standards (ES) (further – harmonized DSTU), and some other parameters, did not gives objective results [1–3].

Priorities for national standardization to develop harmonized DSTU determined the laws of Ukraine and other regulations. Given Ukraine's preparations for the compiling with the European Union (EU) of the Agreement on Conformity Assessment and Acceptance of Industrial Products (ACAA), the introduction of European standards in Ukraine is top priority, because the use of them is seen as evidence of compliance with the directives of the "new approach" [4–6].

Journal of Physics: Conference Series **772** (2016) 012007 doi:10.1088/1742-6596/772/1/012007

The principles of technical harmonization in Europe are specified by the decisions of the EU, but its directives set out only some main conditions, required for the provision of measurements reliability, but technical requirements are based on standards (IS and ES) and recommendations. Dominating aim harmonization in EU is to ensure free trade relations.

International guidelines and standards should be a priority for national TC in the field of Metrology and Measurement Industry, along with the harmonization of standards of evidence as to European directives. Such standards should be implemented as harmonized DSTU as they are the basis for metrology and laboratory accreditation activities in Ukraine. Without their harmonization cannot even talk about international recognition of product testing at the national level.

It should be borne in mind that not all national TCs is "mirror" with international counterparts in ISO, IEC, OIML or European Organization of Standardization (CEN/CENELEC), has the opportunity to work with them and develop harmonized DSTU. Because of the specific scope of the some TCs develop only national DSTU, industry standards and technical specifications [1].

The results of the comparative analysis of participation of national TCs in field of Metrology and Measurement Industry in the appropriate TCs of International Organization and European Organization are shown in Table. 1.

Table 1. Participation of national TCs in field of Metrology and Measurement Industry in the TCs of International and European Organization

TCs of Ukraine	TCs of International Organizations			TCs of European Organizations
	ISO	IEC	OIML	CEN/CENELEC
63 General rules and regulations state system of traceability	12 Quantities, units, symbols, conversion factors 69 Applications of statistical methods	25 Quantities and units	2 Units of measurement 4 Measurement standards and calibration and verification devices	–
65 Devices Industrial Control	30 Measurement of fluid flow in closed conduits	65 Industrial-process measurement, control and automation	8 Measurement of quantities of fluids	–
66 Optical and optomechanical products	172 Optics and photonics	–	14 Measuring instruments for used for optics	170 Ophthalmic optics
77 Medical equipment	76, 84, 106, 121, 150, 170, 172, 173, 194, 198	62 Electrical equipment in medical practice 87 Ultrasonics	18 Medical measuring instruments	85 Eye protected equipment 140 in vitro diag-nostic madical device. 163 Sanitary appliances 205 Non-active medical devices
90 Means of measuring electric and magnetic values	–	13 Electrical energy measurement, tariff- and load control 38 Instrument transformers 66 Safety of measuring, control and laboratory equipment 85 Measuring equip-ment for electrical and electromagnetic quantities	12 Instruments for measuring electrical quantities	CLC/TK 38X Instrument transformers

TCs of Ukraine	TCs of International Organizations			TCs of European Organizations
99 Materials and devices for scinti-llation techniques associated with measurement of ionizing radiation	85 Nuclear energy	45 Nuclear instrumentation	15 Measuring instruments for ionizing radiations	430 Nuclear energy, nuclear technology and radiological protection CLC/TK 45AX Instrumentation for nuclear equipment
122 Analysis of the gases, liquid and solids	158 Analysis of gases 193 Natural gas, SC 1 Analysis of natural gas; REMCO Committee on reference materials	–	–	238 Test gases. Test pressures. Appliance categories and gas appliance
156 Devices for measuring mass, force, strain and determining mechanical properties of materials	–	9 Instruments for measuring mass and density 10 Instruments for measuring pressure, force and associated quantities	–	–

2. Generalized criterion for assessing the activities of TCs

Evaluation activity of national TCs should be carried out not only by the number of the developed DSTU, harmonized DSTU, but also taking into account the factors that most influence the effectiveness of the TCs.

Evaluation of the effectiveness activity of national TCs considering the number of developed standards and other indicators did not give objective results. Therefore, it is proposed to use the generalized criteria δ_Σ for evaluation of TCs what containing weight coefficients for different types of standards:

$$\delta_\Sigma = a \cdot \frac{K_{DSTU\,h}}{K_\Sigma} + b \cdot \frac{K_{DSTU}}{K_\Sigma} + c \cdot \frac{K_{SOU}}{K_\Sigma}, \tag{1}$$

where $K_{DSTU\,h}$ – the total number of harmonized DSTU developed of national TC; K_{DSTU} – the total number of DSTU developed of national TC; K_{SOU} – the total number of industry standards and technical specifications (further – no DSTU) developed of national TC; K_Σ – the total number of standards developed of national TC; *a* – Given the priority of developing harmonized DSTU, take a weight coefficient a = 1, and other weight coefficients evaluated by an expert at: b = 0,3; c = 0,05. of harmonized DSTU for national standardization; *b* – weight coefficients of DSTU for national standardization; *c* – weight coefficients of no DSTU for national standardization [1].

Given the priorities of national standardization are established three conventional levels of significance criteria δ_Σ (0 to 1) to develop harmonized DSTU, DSTU and other standards that define the three levels of performance of TCs for any industry or sector of the economy: "high", "medial", "low". Peculiarities of each of the levels and values of generalized criteria shown in Table 2. Given the priority of developing harmonized DSTU, take a weight coefficient a = 1, and other weight coefficients evaluated by an expert at: b = 0,3; c = 0,05.

Using the formula (1) and reported data of TC for a certain period, can calculate the value of the generalized criterion δ_Σ for any TC. Thus, for objective evaluation of TC should take into account all factors that can directly affect the efficiency of its activity.

Table 2. Levels, peculiarities of evaluation and values of the generalized evaluation criteria of TCs

Level evaluation of TCs	Peculiarity of evaluation of TCs	The value of the generalized criterion of evaluation of TCs, δ_Σ
"high"	Many number of harmonized DSTU in the field of TCs	$\delta_\Sigma \geq 0{,}7$
"medial"	The average number of harmonized DSTU in the field of TCs	$0{,}7 > \delta_\Sigma > 0{,}3$
"low"	Small number of harmonized DSTU in the field of TCs	$\delta_\Sigma \leq 0{,}3$

The evaluation values of generalized criteria δ_Σ and the level of efficiency of TCs in field of Metrology and Measurement Industry shown in Table 3.

Table 3. The evaluation values of generalized criteria and the level of efficiency of TCs in field of Metrology and Measurement Industry

TCs designation	The total number of standards developed by TC (in t. h. DSTU / harmonized DSTU / no DSTU)	The value of the generalized criterion, δ_Σ	Level of efficiency TC
90	9 (0/9/0)	1,0	"high"
99	33 (2/31/0)	0,96	
156	13 (3/10/0)	0,84	
122	41 (10/31/0)	0,83	
77	23 (8/15/0)	0,76	
63	280 (110/170/0)	0,73	
65	49 (32/17/0)	0,54	"medial"

Virtually all TC showed "high" level of efficiency (except TC 65, who showed "medial" level) because his work gave priority to the development of harmonized DSTU and did not develop the no DSTU Comparing data TC 90, which reopened in 2006, and TC 65, established in 1993, stipulates that TC 90, developing only 9 harmonized DSTU (100%), received the highest value of generalized criterion δ_Σ – 1. At the same time TC 65, almost 50 developing standards got "medial" level of efficiency, since concentrated its work on developing no DSTU.

Condition of introduction of IS and ES by TCs in Metrology and Measurement Industry shown in Figure 1. As shown in Figure 1, the rate of implementation of IS and ES in the field of Metrology and Measurement Industry at the national level remains quite low despite the commitments of Ukraine under the WTO.

Therefore, the question of harmonization of national standards with IS and ES in the field of Metrology and Measurement Industry is still relevant. Significant number of standards not yet implemented in Ukraine (only 53%). Should bear in mind periodic updating, revision and replacement of standards for the directives of the "new approach" from European authorities in the Official Journal of EC. In such circumstances, national TCs should work vigorously and almost simultaneously with the European TCs [7].

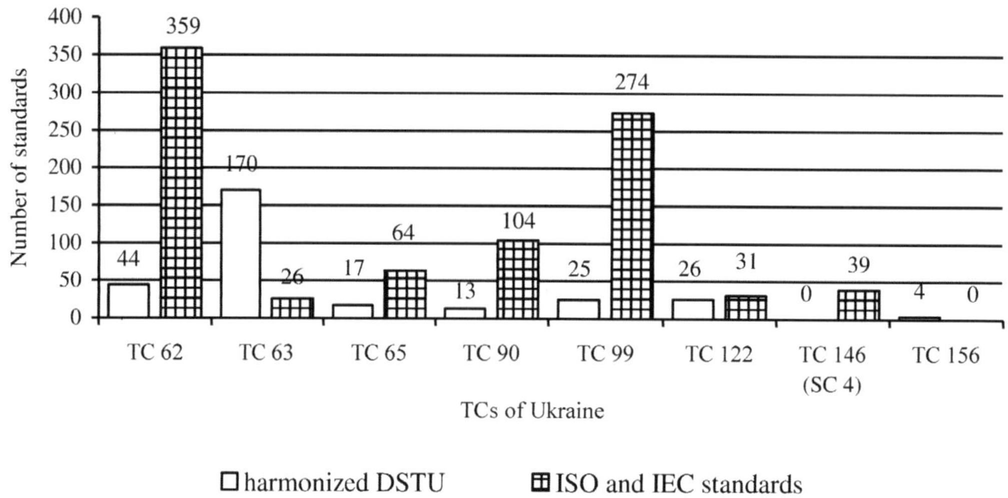

Figure 1. Number of standards developed by National and International TCs in field of Metrology and Measurement Industry

3. Conclusion

The results of TC identify of objective and subjective indicators, traditional of which are based on data on the number of developed standards (t. h. harmonized). But not always take into account other important factors such as participation in the International and European Organizations, the presence of overlapping fields with other TC, financial and technical capabilities and more. Therefore, evaluation of TC is necessary to take into account all influencing factors.

Found that national TC, which have "mirror" international counterparts in ISO, IEC, OIML or European Organization of Standardization have the better opportunity to work with them, to develop harmonized national DSTU, have more domestic demand sectors, have higher levels of performance evaluation.

The proposed general criterion for assessing TC includes weight coefficients for different types of standards and levels of approval, taking into account national priorities. The proposed approach allows us to objectively evaluate the activity of TC, despite the battery life and the absolute value of the number of developed standards.

4. References

[1] Gordiyenko T. B. Evaluation of the Technical Committees of Standardization using generalized criterion. Proceedings of the Odessa State Academy of Technical Regulation and Quality. – Odessa: OSATRQ, 2015, № 1. – P. 140–144 (in Ukrainian).

[2] T. Gordiyenko, O. Velychko. Joint scopes activity the IMEKO and International Organizations of Standardization Technical Committees in field of Metrology // XIX IMEKO World Congress "Fundamental and Applied Metrology". – Lisbon, Portugal. – 2009. – P. 1094–1098.

[3] T. Gordiyenko, O. Velychko. Joint scopes activity of IMEKO and International Metrological Organizations Technical Committees // XIX IMEKO World Congress "Fundamental and Applied Metrology". – Lisbon, Portugal. – 2009. – P. 1142–1147.

[4] Gordiyenko T. B., Kolomiets L. V., Velychko O. M. Activity the technical committees standardization of international organizations in field of metrology // Proceedings of the Odessa State Academy of Technical Regulation and Quality. – Odessa: OSATRQ, – 2012. – Is. 1 (1). – P. 11–16.

[5] Velychko O. M., Gordiyenko T. B., Kolomiets L. V. The basis for implementation of international and regional guides and standards in national standards in field of metrology // Proceedings of the Odessa State Academy of Technical Regulation and Quality. – Odessa: OSATRQ, – 2014. – Is. 2 (5). – P. 6–10.

[6] O. Velychko., T. Gordiyenko. Toward the harmonization of national documents in field of metrology // Proceedings of Intern. Metrology Conf. CAFMET 2010. – Cairo, Egypt. – 2010. – 7 p.

[7] Velychko O. M., Kolomiets L. V., Gordiyenko T. B. Amendments to the Technical Regulations for Measuring Instruments // "Competence". – 2013. – № 6 (106). – P. 45–47 (in Russian).

Examples of measurement uncertainty evaluations in accordance with the revised GUM

B Runje[1], **A Horvatic**[1,4], **V Alar**[1], **S Medic**[2] and **A Bosnjakovic**[3]

[1]University of Zagreb, Faculty of Mechanical Engineering and Naval Architecture, Ivana Lucica 1, 10000 Zagreb, Croatia
[2]Karlovac University of Applied Science, Trg J.J.Strossmayera 9, 47000 Karlovac, Croatia
[3]Institute of Metrology of Bosnia and Herzegovina, Augusta Brauna 2, 71000 Sarajevo,Bosnia and Herzegovina

E- mail: amalija.horvatic@fsb.hr

Abstract. The paper presents examples of the evaluation of uncertainty components in accordance with the current and revised Guide to the expression of uncertainty in measurement (GUM). In accordance with the proposed revision of the GUM a Bayesian approach was conducted for both type A and type B evaluations.The law of propagation of uncertainty (LPU) and the law of propagation of distribution applied through the Monte Carlo method, (MCM) were used to evaluate associated standard uncertainties, expanded uncertainties and coverage intervals. Furthermore, the influence of the non-Gaussian dominant input quantity and asymmetric distribution of the output quantity *y* on the evaluation of measurement uncertainty was analyzed. In the case when the probabilistically coverage interval is not symmetric, the coverage interval for the probability *P* is estimated from the experimental probability density function using the Monte Carlo method. Key highlights of the proposed revision of the GUM were analyzed through a set of examples.

1. Introduction

The Guide to the expression of uncertainty in measurement (GUM) was last amended in 2008. The revision included minor changes and corrections. The new revision JCGM 100 201X introduces an entirely new approach to estimating measurement uncertainty in order to address the problem of measuring uncertainty evaluation in situations not covered in the current JCGM 100 : 2008 version of the Guide. Key conceptual changes in the evaluation of measurement uncertainty with respect to the current version of the GUM include the application of the Bayesian approach and conditional probabilities for the estimation of type A standard uncertainty as well as the problem solving assessment of measurement for the cases when the number of measurements is less than four [3,4]. Changes in the GUM in accordance with the norm JCGM 100 201X are listed below.

[4]To whom any correspondence should be addressed.

2. Changes of the Guide to the expression of uncertainty in measurement in accordance with the document JCGM 100 201x cd

In accordance with the proposed revision, the simplest way to eliminate the disadvantages of the GUM method is through adopting the Bayesian approach and applying the Monte Carlo simulation in the process of calculating measurement uncertainty. Key changes can be stated as follows:

- The current approach to estimating type A uncertainty components relies on the frequentist view while the new version of the GUM uses the Bayesian approach [5].
  Calculation of type A standard uncertainty in accordance with the Guide from the year 2008 and in accordance with revised GUM from year 2015 is shown with expressions (1) and (2) where s_i represents the standard deviation.

$$u(x_i) = \frac{s_i}{\sqrt{n}} \tag{1}$$

$$u(x_i) = \left(\frac{n-1}{n-3}\right)^{1/2} \cdot \frac{s_i}{\sqrt{n}} \tag{2}$$

  From expression (2) it is clear that the minimum required number of measurements for the evaluation of the standard deviation is four. This requirement is considered to be an example of good measurement practice [5].
- In the cases where the number of measurements is between one and four, the evaluation of standard uncertainty is based on the results and experiences of some earlier measurements. Knowledge based on previous results is expressed through a pooled standard deviation estimated from a sufficiently large number of measurements $n_p$were sufficient means greater than or equal to 20 measurements. In the cases were $1 \leq n < 4$, standard uncertainty is estimated by the expression (3).
-

$$u(x_i) = \left(\frac{n_p-1}{n_p-3}\right)^{1/2} \cdot \frac{s_p}{\sqrt{n_p}} \tag{3}$$

- In the process of evaluating type B uncertainty components, where the low availability of input data is assumed, in the current GUM, the quantity is described using the symmetrical rectangular probability density function (PDF). Standard uncertainty is calculated as the standard deviation of the rectangular PDF with assigned degrees of freedom. The rectangular distribution is displayed through the use of fixed endpoints a and b. In the revised GUM, the degrees of freedom were not assigned to the standard uncertainty due to the fact that the distribution was based on available knowledge.

- Coverage interval is calculated from the probability density function of the output quantity y.
  - ✓ In the cases where the functional relationship between output and input quantities is linear and the central limit theorem is met, probability distribution described by y and $u(y)$ is approximately normal, one can assume that when $k = 2$ this results in an interval with a probability of about $P = 95\%$.
  - ✓ In other cases, Gaussian inequality ($k = 3$) is applied for the symmetrical distribution while the Chebisheva inequality ($k = 4.5$) is applied for the asymmetric distribution [5].
  - ✓ In the case of asymmetric distribution, the probabilistically symmetric coverage interval is not sufficient. It is necessary to apply the method of the Monte Carlo simulation and estimate coverage interval from experimental probability density function for a given probability P.

3. Examples of measurement uncertainty evaluations

The values of the uncertainty component obtained through type A evaluation in accordance with the GUM from the year 2008 and a revised Guide from the year 2015, depending on the number of measurements, are given in Table 1.

Table 1. The values of the uncertainty component obtained through type A evaluation, depending on the number of measurements.

Number of measurements, n	Standard uncertainty JCGM 100 : 2008	Standard uncertainty JCGM 100 : 201X	
5	$u^*(x_i) = \frac{s_i}{\sqrt{n}}$	$u(x_i) = \left(\frac{n-1}{n-3}\right)^{1/2} \cdot \frac{s_i}{\sqrt{n}}$	$u(x_i) = 1.41 \cdot u^*(x_i)$
10			$u(x_i) = 1.13 \cdot u^*(x_i)$
20			$u(x_i) = 1.06 \cdot u^*(x_i)$
30			$u(x_i) = 1.04 \cdot u^*(x_i)$

In accordance with the new revision of the GUM the standard uncertainty, calculated on the basis of $n = 20$, increased by 5.7% compared to the version of the Guide from 2008.

The following example illustrates the estimation of the standard uncertainty of the thermal expansion coefficient of the steel gauge blocks [6]. According to information available in the literature it is known that the thermal expansion coefficient of steel gauge blocks α equals $(11 \times 10^{-6} \pm 1 \times 10^{-6})\ K^{-1}$. According to the current GUM the input quantity will be described with a priori rectangular probability distribution, with the lower bound being $a = 10 \times 10^{-6}\ K^{-1}$ and the upper bound being $b = 2 \times 10^{-6}\ K^{-1}$. The question of the number of the degrees of freedom which should be attributed to standard measurement uncertainty begins to be raised. The probability density functions (pdf) of input quantity α, in accordance with current and revised GUM, obtained by using the Monte Carlo simulation (MCS) with a number of simulation $M = 100000$, are given in Figures 1 and 2.

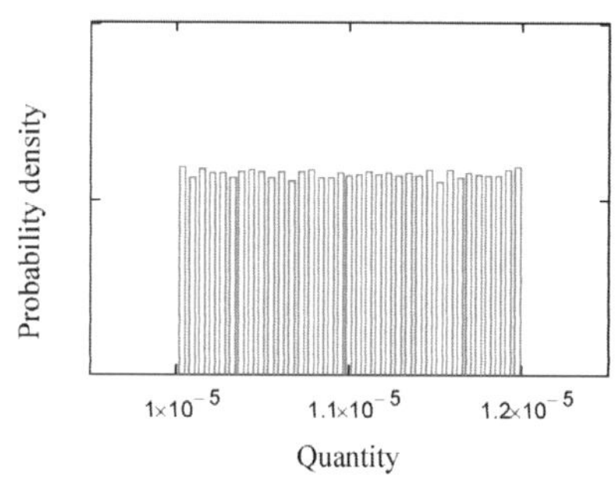

Figure 1. Probability density function of input quantity α, in accordance with the current GUM.

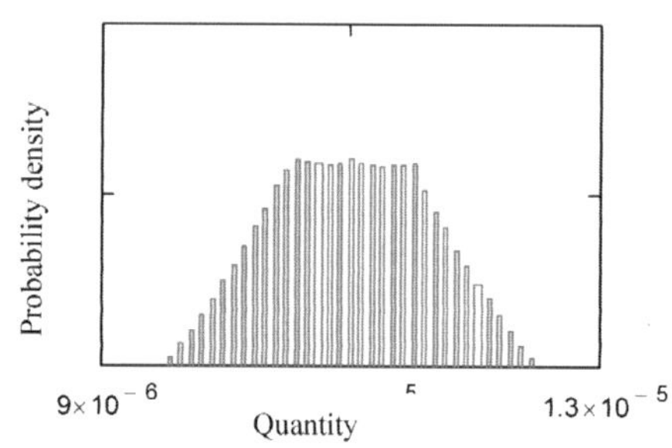

Figure 2. Probability density function of input quantity α, in accordance with the revised GUM.

In this case the standard deviation of input quantity α is characterized by a trapezoidal distribution and equals $6.241 \times 10^{-7}\ K^{-1}$. This value is larger than the standard deviation of α which is characterized by a rectangular distribution ($5.788 \times 10^{-7}\ K^{-1}$).

In the following section, the use of the current and revised GUM methods for the output quantity y, which is described using the equation $y = x^2$, are shown [5].
The input quantity is defined by the Gaussian probability density function (as shown in Figure 3) with the mean being 3 and the standard deviation being 5. Figure 4 shows the probability density function of the output quantity y. The input and output quantities obtained using the Monte Carlo simulation (MCS) with a number of simulation $M = 100000$, are given in Figures 3 and 4. Figure 4 also displays the shortest 95 % confidence interval evaluated according to the revised GUM. It is clearly visible that the 95 % confidence interval evaluated according to the current GUM is clearly not reliable because it includes infeasible (negative) values of the output y.

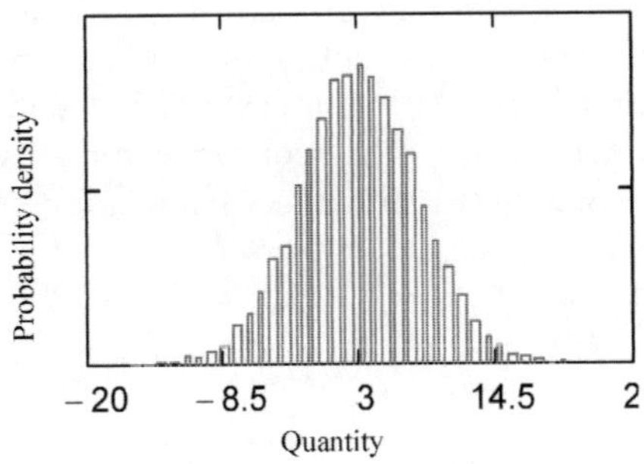

Figure 3. Probability density function for input quantity x.

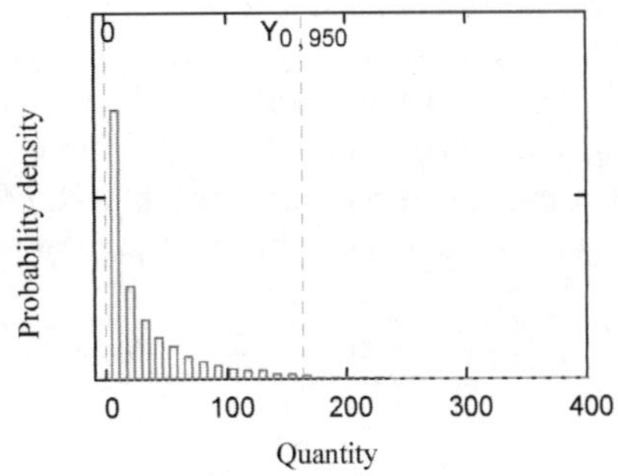

Figure 4. Probability density function and coverage interval for output quantity $y=x^2$.

The following example is an example of the evaluation of the measurement uncertainty that was performed for the calibration procedure of the micrometer setting rod. The mathematical model is provided in expression 4 while the input values and probability density functions in simulation of value L_X is provided in Table 2.

Mathematical measurement model:

$$L_X = L_{ix} + \delta L_{ix} + \delta L_T + \delta L_E + \delta L_A + \delta L_P \quad (4)$$

L_X	-	actual (corrected) length of the setting rod
L_{ix}	-	measured length of the setting rod
L	-	nominal length of setting rod
δL_{ix}	-	influence of the maximum permissible error
δL_T	-	influence of temperature
δL_E	-	influence of elastic deformation
δL_A	-	influence of Abbe error
δL_P	-	influence of misalignment of measuring probes

Table 2: Input values and probability density functions in simulation of value L_X

Input value x_i		Probability density function $g(x_i)$
measured length of the setting rod	L_{ix}	Normal distribution (M, 0; 0.17 μm)
influence of the maximum permissible error	δL_{ix}	Rectangular distribution (M; - 0.29 + 5.8 L; 0.29 + 5.8 L) μm
influence of temperature	δL_T	Normal distribution (M; 0; 0.20 + 0.7 L) μm
influence of elastic deformation	δL_E	Normal distribution (M; 0 μm; 0.21 μm)
influence of Abbe error	δL_A	Rectangular distribution (M; - 0.03 μm; 0.03 μm)
influence of misalignment of measuring probes	δL_P	Rectangular distribution (M;-0.11 μm; 0.11 μm)

It may be noted from Table 2 that the uncertainty element $u(\delta L_{ix})$, especially in case of longer setting rods, will have substantially greater contribution to uncertainty compared to others. In other words, for a certain length of the setting rod the conditions of the Central Limit Theorem cease to be valid. In applying the GUM method, it is difficult to predict at which moment the required conditions of Central Limit Theorem are not met any more. For the mentioned example, the measurement uncertainty was calculated by applying the MCS method.

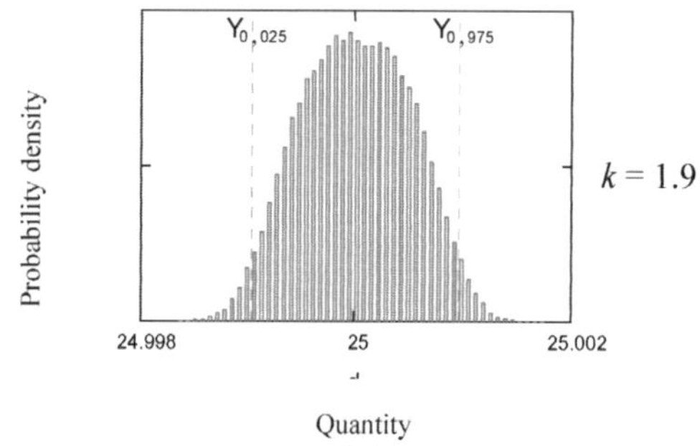

(a)

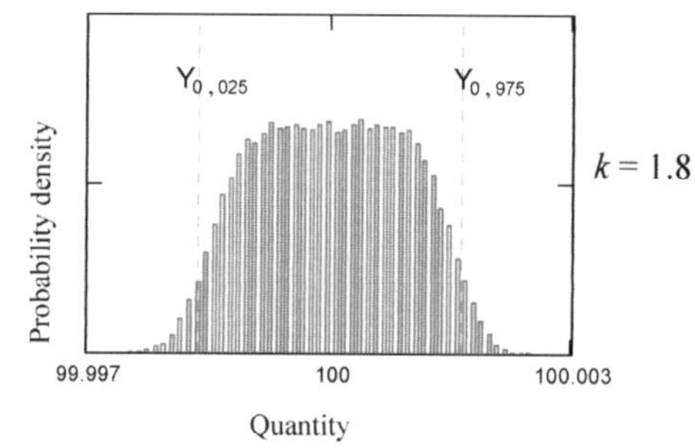

(b)

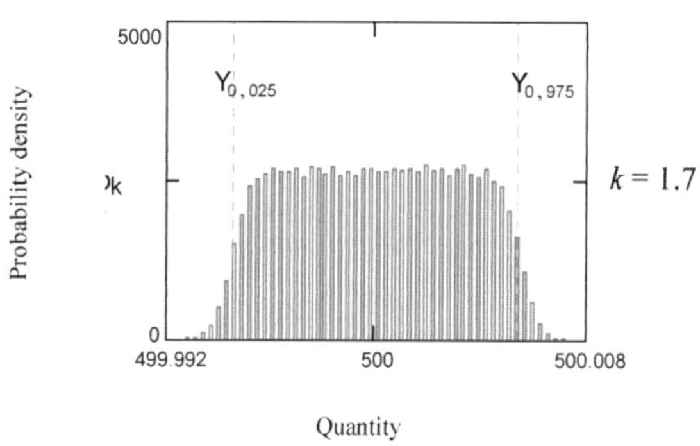

(c)

Figure 5. Probability density functions and coverage interval: a) for 25 mm setting rod, b) 100 mm setting rod and c) 500 mm setting rod

By applying the MCS method it has been determined that, depending on the length of the setting rod, the output distributions change their appearance from the normal-like to trapezoidal-like distributions as seen in figures 5a, 5b and 5c. For setting rods of nominal lengths 25 mm, 100 mm and 500 mm the coverage interval and coverage factor k are directly determined from the experimental probability density function. By MCS method it has been determined that the value of the coverage factor k changes regarding the length of setting rod. The coverage interval and coverage factor are directly determined from the experimental probability density function obtained by combining different probability density functions of input values. In measuring the length, the expanded measurement uncertainty is very often expressed in dependence on the length. This example indicates the problems that may result in expressing the expanded measurement uncertainty.

The following example, the evaluation of measurement uncertainty for the calibration of the vernier caliper given in document EA 4/02, also shows that the conditions necessary for the application of GUM method are not always fulfilled. The method used for calculating the coverage factor is clearly related to the fact that uncertainty of measurement associated with the result is dominated by two influences: the mechanical effects and the finite resolution of the vernier scale. Thus the assumption of a normal distribution for the output quantity is not justified [7]. The probability density function (pdf) for the 150 mm length vernier caliper is presented in Figure 6.

Journal of Physics: Conference Series **772** (2016) 012008 doi:10.1088/1742-6596/772/1/012008

In the final example (Figure 7) a comparison is made between the 95 % coverage intervals estimated using the GUM and MCS methods with an asymmetric PDF as the output quantity.

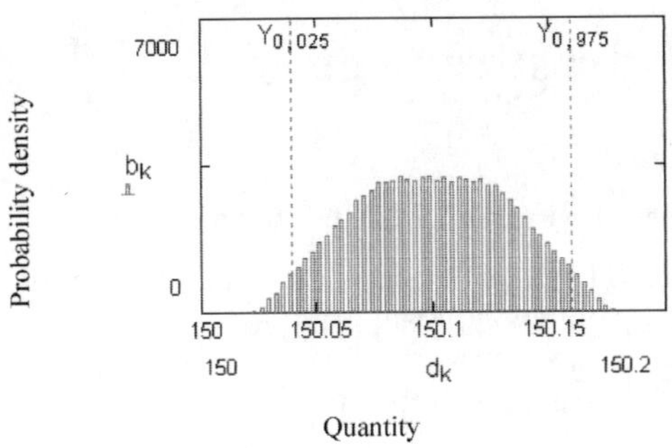

	$\overline{x}$	$\tilde{x}$	s	95% coverage interval
GUM	150.10	150.10	0.03	[150.04; 150.16]
MCS	150.10	150.10	0.03	[150.04; 150.16]

Figure 6. Probability density function and coverage interval for symmetric non-normal distribution.

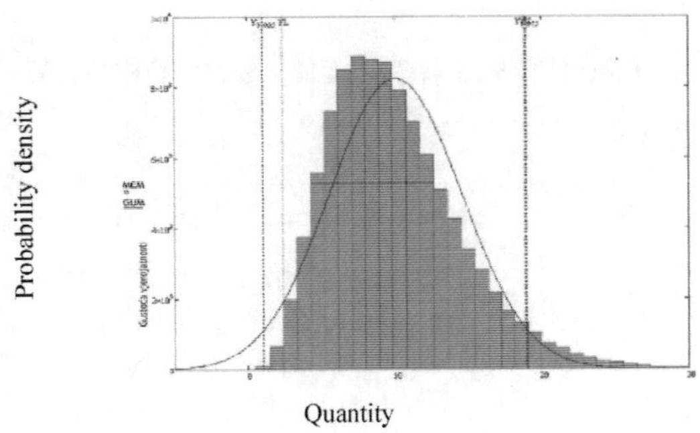

	$\overline{x}$	$\tilde{x}$	s	95% coverage interval
GUM	9.995	9.342	4.467	[1.061; 18.929]
MCS	9.995	9.342	4.467	[2.425; 18.830]

Figure 7. Probability density function and coverage interval for asymmetric distribution.

4. Conclusions

Key highlights of the proposed revision of the GUM can be stated as follows:

1. Adoption of the Bayesian approach in the calculation of the A and B types of uncertainty components as well as the application of the Monte Carlo simulation.
2. In accordance with the revision of the GUM, the value of the type A standard uncertainty components is higher than in the existing version of the Guide from the year 2008.
3. The coverage interval is calculated from the probability density function of the output quantity y.
4. The coverage interval no longer depends on the degrees of freedom calculated using the Welch-Satterthwaite formula.

5. References

[1] JCGM 100:2008, Evaluation of measurement data — Guide to the expression of uncertainty in measurement
[2] JCGM 101:2008, Evaluation of measurement data — Supplement 1 to the "Guide to the expression of uncertainty in measurement" — Propagation of distributions using a Monte Carlo method
[3] JCGM 100 201X CD, Evaluation of measurement data | Guide to uncertainty in measurement, Committee draft
[4] JCGM 110 201X CD, Evaluation of measurement data | Examples of uncertainty evaluation, Committee draft
[5] Cox M and Harris P, Revision of the Guide to the expression of uncertainty in measurement - impact on national metrology institutes and industry *UK CCRI BIPM*, (*Teddington, UK, 16-17 May 2013)*
[6] Lira I 2016, *Eur. J. Phys.* **37** 025803
[7] EA - 4/02 (2013) Evaluation of the Uncertainty of Measurement in Calibration

A meta-structural understanding of measurement

Luca Mari[1], Andrew Maul[2 1#], David Torres Irribarra[3], Mark Wilson[4]

[1] School of Engineering, Università Cattaneo – LIUC, Castellanza (VA), Italy
[2] Graduate School of Education, University of California, Santa Barbara, CA, USA
[3] MIDE, Pontificia Universidad Católica de Chile
[4] Graduate School of Education, University of California, Berkeley, CA, USA

E-mail: amaul@education.ucsb.edu

Abstract. It is not always clear to what extent the logic and vocabulary of measurement as used in different scientific disciplines are mutually coherent, nor how measurement can be demarcated from, say, opinion. In recent decades there have been a number of attempts to provide necessary and/or sufficient sets of conditions for when measurement is achieved, usually in terms either of inputs (e.g., whether an evaluated property is a quantity), or outputs (e.g., whether a procedure assigns numbers according to a rule). We argue instead that the public trust attributed to measurement is best justified in terms of the structural features of the process rather than of its inputs or outputs.

1. Introduction

In both scientific and lay discourse, the concept of measurement is commonly associated with precision, accuracy, and trustworthiness, yet it remains remarkably difficult to provide a fully satisfactory definition of the concept, as the vast array of proposed definitions witnesses (e.g., [1]). At least two basic reasons concur to explain this difficulty. First, as the scope of the concept of measurement has broadened, it is not always obvious what – if indeed anything – is common among all the processes claimed to be measurements. Second, the concept of measurement has come to be characterized according to purely formal criteria, thus abstracting from the concrete realization of the process, due in part to the fact the evaluation of non-physical properties cannot conform to the traditional structure of measuring systems operating on the basis of the transduction implemented by a physical sensor. As a consequence, theoretical interpretations of measurement have become so abstract that they may be unable to provide a convincing and useful demarcation of measurement from formally similar processes that are generally thought to lack epistemic authority, such as most instances of the expression of subjective judgments and opinions.

One may question whether working on the definition of 'measurement' is a worthwhile endeavor. Our position on this matter is practical: there is a social interest in sharing scientific and technical vocabulary across disciplines, particularly in the case of an infrastructural activity like measurement [2], and there is a social acknowledgment of the epistemic authority of measurement, which has

1 # To whom any correspondence should be addressed.
Authors' names are ordered alphabetically. One of the authors is a member of the Joint Committee on Guides in Metrology (JCGM) Working Group 2 (VIM). The opinion expressed in this paper does not necessarily represent the view of this Working Group.

critical consequences in particular in terms of public trust attributed to the outcomes of putative measurement processes and the resources devoted to such processes.

In synthesis, the position we propose here is that measurement has to be characterized as a process by its operative structure, not by the specification of the functional relationship connecting its inputs to its outputs. As a corollary, our claim is that any purely black-box (meta-)model cannot adequately account for fundamental features of measurement.

2. Defining measurement

Under the general hypothesis that measurement is a process that operates on inputs (at least the measurand, in the case of direct measurement methods) to produce outputs (at least the measurement result), the idea is that measurement is an instance of the meta-model:

input → [transformation] → output

specified as:

input → [evaluation] → output

given that measurement produces values of properties (instead of, e.g., physical transitions or objects). The acknowledgment that not all evaluations are measurement calls for a further specialization, and thus the characterization of what measurement is, obtained by differently focusing on one or more of these three entities [3, 4], and in particular claiming that the necessary and/or sufficient conditions for measurement can be located in one or more of these entities.

A first category of views, such as those given by Bridgman [5] and Dingle [6], could be characterized as focusing on the outputs of the process. Dingle, for example, defines measurement as "any precisely specified operation that yields a number," thus referring to an aspect of the output of the process (i.e., that it is numerical) as a necessary and sufficient condition for measurement. In the absence of an account of what forms of precision are necessary, the requirement that the procedure be "precisely specified" seems to demand sufficient clarity regarding what the operations of the procedure are, but places no demands concerning the inputs of the procedure nor the manner in which they are transformed into outputs.

A second category of views focuses instead, or sometimes additionally, on the inputs of the process. For example, Michell (e.g., [7]), in line with the Euclidean tradition, defines 'measurement' as the assessment of a quantity, the measurand, in comparison to a second quantity, the unit. According to this view, whether or not a given property is a quantity is an ontological issue, and is a pre-condition for measurement. In this view, a property that is not a structured as a quantity is a priori not measurable: thus a feature of the input, namely that it is a quantity, is a necessary condition for measurement. Arguably, Michell's view also requires that the output of the process be numerical – specifically, that it be expressed as a ratio of a quantity value to a standard unit.

A third category of views characterizes measurement in terms of the formal relationship between inputs and outputs, but is silent regarding the operative structure of the transformation by which the relationship is implemented as a causal mechanism that induces variations in the outputs from variations in the inputs. Representational theories of measurement (e.g., [8]), for example, define 'measurement' in terms of a morphism between an empirical relational system and a numerical relational system; thus, in this view, a particular feature of the formal dependence of the output on the input is a (necessary and) sufficient condition for measurement.

As we have argued elsewhere (e.g., [1], [9], [10]), while each of the views described above succeeds in capturing valuable intuitions about measurement, none of them provides a fully satisfactory set of conditions that could universally be used (a) to distinguish measurement from non-measurement processes, nor (b) to distinguish better (or more trustworthy, useful, etc.) from worse instances of measurement. Views in the first category tend to fail to disallow instances of rule-based numerical assignment that have no epistemic value, such as procedures based on subjective judgment such as (formalized) guesses or statements of opinion, or precisely-specified but arbitrary rules. Views in the second category tend to be too restrictive, insofar as they would disallow many widely accepted cases of measurement in the physical sciences in addition to, potentially, all cases in the social sciences, regardless of epistemic or pragmatic value. Finally, views in the third category tend to be incapable of distinguishing processes that generate dependable knowledge from processes that do not but are otherwise functionally similar evaluations.

3. Rethinking the definitional problem

As it has been hinted at in the previous section, debates concerning the definition of measurement may conflate two different questions: (a) how should measurement be defined, as distinct from other types of evaluations? and (b) how can we judge the dependability of the information provided by an evaluation? In principle the two questions could be considered redundant if measurement were defined as any evaluation that yielded dependable knowledge, but this would broaden the scope of the concept of measurement to potentially include essentially all forms of inquiry, and is *prima facie* inconsistent with every formal and lay conception of measurement of which we are aware – for example, it is commonly accepted that not all measurements guarantee the same high dependability, and not all opinions are flawed by the same low dependability. Hence, dependability of empirical information is definitely a worthwhile target, but the problem is how such dependability can be obtained and assessed as a stable and publicly justifiable feature of a given process.

Particularly when the required quality of the results is not so high (relatively low resolution, sensitivity, precision, etc.), an experienced, skilled individual may express her/his opinions in such a way that they would be accepted as dependable (and discovered to be morphically related to some set of empirical relations). But for an evaluation to be able to produce results that can be publicly trusted, and thus considered a measurement, the process must be at least in principle replicable, and even a successful result is not considered the outcome of a scientific or technological process if the way in which it has been obtained is not at least in principle publicly explainable.

As we have argued elsewhere [1, 3, 4, 10, 11], such a public trust is primarily grounded in the demonstrated capability of providing sufficient object-relatedness ("objectivity") and subject-independence ("intersubjectivity") of the information produced by those processes that are traditionally structured as measurements (and are traditionally called "measurements"). The extent to which a given process yields results with these epistemic virtues, in turn, depends on its operative structure, which includes the design, setup, and implementation of the replicable transformation of inputs into outputs, and therefore unavoidably on the structure of the process, which must be then construed according to an open-box (meta-)model. *Ceteris paribus*, the better researchers understand the causal chain leading from inputs to outputs, and thereby understand the conditions under which one could expect this causal chain to remain invariant, and the better such understanding is implemented in the actual process, the more trust can be placed in the objectivity of the results. And, *ceteris paribus* again, the better researchers understand the standards and the processes by which they can be compared with each other in a traceability chain, and thereby understand the conditions under which one could expect these standards to remain stable, the more trust can be placed in the results as for their intersubjectivity.

In context, the level of specificity in this understanding (and thereby, the degree of warranted trust) will inevitably depend on the context and purpose for which the measurement results are to be used, and, of course, on historical evidence that it has been found to be dependable in the past.

4. Two examples

Measurement instruments are designed for a wide variety of purposes, and even different instruments designed to measure the same property may take on different forms depending on the purpose[2] for which the results are to be used and the context in which the instrument will operate. For example, the

[2] For the case of physical properties and their measurement, this dependence is often related to the expected quality (precision, resolution, etc.) of the results, so that different instruments take on different form depending on the quality of the results demanded by a specific purpose.

property programming proficiency could be assessed for a number of distinct purposes, among them (a) the selection of candidates for a position in a professional setting, and (b) the diagnosis of specific patterns of misunderstandings for educational purposes.

In the first instance, the emphasis may be on deciding whether the candidate is able to successfully undertake the duties entailed by the position; as such the instrument could consist of a simulation of these duties, and an evaluation of the performance of the candidate on each of the relevant tasks. In this instance it may be that the relevant information is whether (or the degree to which) candidates are able to successfully execute the duties of the position, independently of the reasons they are able to do so – and it may be that there are different combinations of knowledge, skills, abilities, and other individual characteristics that would enable a candidate to be successful. The final evaluation could take any number of forms depending on the context: it could be a dichotomous classification (qualified / not qualified), with measurement uncertainty expressed in terms of the probability of an accurate classification; alternatively, it could take the form of rank-ordering all candidates from most qualified to least qualified, especially if some predetermined number of candidates were to be selected and the goal was to identify the most qualified individuals; as yet another possibility, each candidate might be given a score reflecting their overall level of proficiency, where interpretation of the score may be aided by reference to population norms and/or specific sets of criteria associated with different scores [12].

In the second instance, given the desire for utility of the test for pedagogical purposes, a much greater emphasis might be placed on understanding the specific elements of knowledge and skill that go into successful or unsuccessful completion of specific tasks; this in turn could lead to different sorts of tasks being designed and included in the assessment. In contrast to the first instance, which left the specific reasons for success or lack thereof relatively unexamined, this sort of test would likely foreground the need for a theory-based explanation of the processes leading to successful task completion at the level of the individual – that is, domain-specific cognitive theory. Again, the results of the assessment could take several forms depending on the context and the preferences of the stakeholders involved, ranging from categorical classification of individuals on a number of specific forms of knowledge and skill, to a continuous estimate of the individual's overall level of proficiency, aided by a theory-grounded interpretative device such as a 'construct map' [12].

In terms of the meta-model mentioned in section 2, these two scenarios plausibly have the same input—that is, programming proficiency of the individual.[3] The manner in which programming proficiency is transformed into the outputs of the test are different, as shaped by the selection of specific tasks, task features, and contextual considerations, but plausibly involve many of the same cognitive, motivational, and affective processes at the level of the individual. The outputs are distinct, and need not be quantitative or even numerical in either case—though they could be, if this was judged by the relevant stakeholders to be the most useful manner in which to present and communicate information about programming proficiency.

As discussed in the previous sections, at least two questions could be asked of each of these instances: (a) does the evaluation qualify as measurement? and (b) does the evaluation yield results that are dependable, for their intended purposes and in their intended contexts? The second question is clearly of greater interest to the relevant stakeholders than the first, and in turn depends, as previously noted, on the degree to which the design, setup, and implementation of the replicable transformation of inputs into outputs is well-understood and implemented.

As a second example, the assessment of wind speed in the context of sailing could either be by using a cup anemometer, a sonic anemometer or observation using the Beaufort Scale. The cup anemometer is "the most basic type of anemometer [which] consists of a series of cups mounted at the end of arms that rotate in the wind,[4]" while the sonic anemometer, a more recently-developed tool in

[3] Alternatively, it could be argued that the inputs are distinct, given that the pragmatic concerns motivating the creation and use of the instrument also serve to shape the definition of the target attribute itself [9].

[4] Cup anemometer. (n.d.) The American Heritage® Dictionary of Student Science, Second Edition. (2014).

the measurement of wind speed, has no moving parts and "measures the projection of the wind velocity vector on the acoustic path" [13]. Finally, the Beaufort Scale relies on a combination of observable environmental indicators plus the judgement of a human observer to produce an estimate of wind speed. We can use any of these three to answer a question about wind speed in order to set the sails accordingly.

The selection of one of these methods involves changes in the first three of the four components of the model introduced here: that is, changes in the process structure, the instrument, or the procedure. To begin with, these methods rely on different underlying theories—a sonic anemometer arguably relies on acoustics theory while the cup anemometer and the Beaufort Scale do not—tracing different causal narratives justifying each procedure. Of course, the instrument in each case is changed, varying from moving cups in one case, to a static device in the sonic anemometer, up to the use of environmental cues integrated by a human observer. It is also worth noting that the use of anemometers versus observations based on the Beaufort scale also changes the structure of the outcome, in one case modelling it as an ordinal attribute and in the other modelling it as quantitative. Similarly, each of these methods of assessing wind speed will require different procedures in order to obtain measurement results. Despite their differences, all these approaches can potentially be replicable, reliable and explainable measurements that allow us to achieve the goals defined by the context. Interestingly, varying the context could make some of these procedures more or less adequate, as a different context could make some approaches incompatible with the goals of that measurement need; for instance, if a context of application is that one that requires the use of wind speed as a quantity to be incorporated into an equation, then that would be a poor match for the results produced through the use of the Beaufort Scale. This variety of possible measurement approaches, which involve alternatives in the process structure, instrument, and procedure can be found both in the physical and the social sciences.

5. Concluding thoughts

Measurement can be characterized as a structured process, designed to be performed by given devices according to given specifications to produce information in a given context [11]. This results in a meta-structural understanding of measurement processes composed of the (1) process structure, (2) instrument, (3) procedure, and (4) context. This paper has not aimed to provide a conclusive demonstration of the thesis that this meta-structural understanding of measurement is superior to what we have referred to as the "black box" (meta-)models of measurement implied by some other perspectives, though we hope that we have laid the groundwork for an alternative way of thinking about measurement and measurement quality. Further scholarship will expand on this argument, in the hopes that by focusing on the extent to which structural features of the process yield results with epistemic value in given contexts, rather than on prescriptive stances regarding what should or should not count as measurement, coherent dialogues can take place across disparate areas of inquiry, while still respecting the fact that measurement results play a privileged role as a source of publicly accepted knowledge.

References

[1] L. Mari, A quest for the definition of measurement, Measurement, 46, 2889-2895, 2013.
[2] JGCM, International Vocabulary of Metrology (VIM) – Basic and General Concepts and Associated Terms (2008 edition with minor corrections), 2012.
[3] A. Frigerio, A. Giordani, L. Mari, Outline of a general model of measurement, Synthese, 175, 123–149, 2010.
[4] L. Mari, P. Carbone, D. Petri, Measurement fundamentals: A pragmatic view, IEEE Trans. Instr. Meas, 61, 8, 2107–2115, 2012.
[5] P. Bridgman, The logic of modern physics. Macmillan: New York, 1920.
[6] H. Dingle, A theory of measurement, The British Journal for the Philosophy of Science, 1, 5-26, 1950.

[7] J. Michell, The logic of measurement: A realist overview, Measurement, 38, 285–294, 2005.
[8] D. Krantz, R. Luce, P. Suppes, A. Tversky, Foundations of measurement, vol. 1: Additive and polynomial representations. New York: Academic Press, 1971.
[9] A. Maul, D. Torres Irribarra, M. Wilson, The philosophical foundations of psychological measurement, Measurement, 79, 311-320, 2016.
[10] L. Mari, A. Maul, D. Torres Irribarra, M. Wilson, Quantification is neither necessary nor sufficient for measurement, Journal of Physics: Conference Series, 459, 2013.
[11] L. Mari, M. Wilson, A structural framework across strongly and weakly defined measurements, proc. IEEE I2MTC 2015, 1522–1526.
[12] M. Wilson, L. Mari, A. Maul, D. Torres Irribarra, A comparison of measurement concepts across physical science and social science domains: instrument design, calibration, and measurement, Journal of Physics: Conference Series, 588, 2015.
[13] J.C. Wyngaard, Cup, Propeller, Vane, and Sonic Anemometers in Turbulence Research, Annual Review of Fluid Mechanics, 13, 399-423, 1981

Toward a formal theory of the measuring system

Giovanni Battista Rossi and Francesco Crenna

Università degli Studi di Genova – DIME/MEC, Genova, Italy

E-mail: gb.rossi@dimec.unige.it

Abstract. Measurement aims at obtaining a numerical description of objects/events/persons in real world by means of a measuring system. Measurement is widely used as a key way for obtaining high quality information from the real world, across disciplines. In the present day, there is growing consensus in holding that measurement is characterized by the use of something that qualifies as a "measuring system". Therefore, we discuss sufficient conditions for an empirical system to qualify as a measuring system.

1. Introduction

The notion of measurement (or measuring) system has been recently raised as a key issue in measurement science [3-4]. Curiously, at the dawn of measurement theory, Campbell noted that the way measurement is actually performed is of no foundational interest [1], which probably contributed to often identify the theory of measurement with the representational approach [6]. On the other hand others recognised the centrality of this issue in measurement. One of the earliest contributions in this direction was an important communication by the late Gonella, at the IMEKO World Congress in Huston [7], which unfortunately had much less resonance than it deserved. The focus on the measurement system as a way for going beyond the representational viewpoint, was well highlighted by Mari [9], although somewhat overlooking, in our opinion, the still permanent validity of the representational approach in addressing measurability issues [10]. Another curious fact is the almost total absence of such a notion in The Guide to the evaluation of uncertainty in measurement (GUM) [15], somehow reflecting, consciously or not, the initial Campbell's prejudice. On the other hand, anyone involved in measurement in real life, and/or in education in measurement, cannot fail to recognise the centrality of this concept, and, consequently, of the related modelling issues, as attested, in our experience, by all accredited textbooks on measurement, one for all being the excellent book by Bentley [8]. In consequence of this awareness, other papers have been published meanwhile on the notion of measurement (or measuring) system, including contributions by Sommer [13-14] and by Ruhm [12]. In fact, such a model is essential for measurement science to qualify as an *autonomous discipline,* and the development of a generally agreed view of measurement among disciplines [5], which constitutes the key focus of this IMEKO Joint TC1-TC7-TC13 event, strongly depends, in our humble opinion, upon that. Noteworthy, for example, the important and influential Rash model, defines *a kind of measurement system*, although of an essentially probabilistic nature [16-17]. Furthermore, from the application side, we think that one major difficulty for an extensive application of GUM's principles, very likely stems from the lack of reference, in the GUM and its related documents, to a proper modelling of the measuring system.

Therefore, as a part of such a debate, we try and contribute here by proposing a general model of the measurement process, which is based on the primitive notion of "empirical relations" which allows defining measurement in terms of a more primitive notion. As far as we know, what here proposed is totally new in the measurement-science panorama.

2. The proposed framework

2.1. The notion of cross order

To attain to a general description of the measuring system (MS), we consider an input-output description, where the input is an "object"[1] $a \in A$, in a (not directly observable) state[2] $\alpha \in A^*$, in respect to some quantity[3] x, and the output is a (directly observable) indication of the MS, $b \in B$, with respect to some quantity y.

One idea we wish to convey here, is that the measuring system makes the *originally unobservable* state of the measurand *actually observable*, by transducing it into another – generally, but not necessary, different – quantity, belonging to the MS. For example, a force to be measured may be transduced into the strain of an elastic bar and then into a voltage; or a temperature into a voltage, through the thermoelectric effect, and so on. Another possibility is that, e.g., a small displacement is transformed into an easy-to-detect rotation of a needle on a gage: here a mechanical quantity is made actually observable by an amplification transformation.

Another idea to be conveyed, is that a mapping from *a scale holding for the measurand* into *a scale holding for the output quantity* of the MS, is realised.

With these ideas in mind, let us then look for appropriate properties of the MS, that fully characterise it. For the principle of parsimony, generally accepted in science, such properties should be as weak as possible.

To achieve this goal, we firstly introduce the notion of *cross-order*, that we *define* as a weak order, denoted by "$\succsim$", that holds true over the *union* of two sets, $C = A \cup B$ in our case, and that also satisfies the following additional properties:

- its restriction to B is a *strict order,* denoted by "$\succ$";
- for each $a \in A$ there exists $b \in B$ such that $a \sim b$, where "$\sim$" denotes equivalence.

Then, it is possible to show that there exist order-preserving measure functions on both A and B,

- $m_A : A \to X$, and
- $m_B : B \to Y$,

where X and Y are numerical sets, that fulfil order-representation properties. In fact, since $A \subset C$, the structure $(A, \succsim)$ is a weak order, and thus, for each $a_1, a_2 \in A$, there exists a function m_A such that [11]

---

[1] Here the term "object" generically denotes the carrier of the property to be measured (this property is also called the "measurand" in the current terminology). Therefore it may be either a *material object*, such as a workpiece, whose length is measured for quality-control purposes, or an *environment*, such as an office, whose temperature, noisiness, ... are measured for checking the conformance to ergonomic standards, or an *event*, such as the taking off of an airplane, where the peak loudness of the emitted noise is controlled for safety reasons, or even a *person*, whose attitude to mathematics is tested, for allowing access to a specialized high-level college.

[2] The notion of "state" here considered corresponds to the notion of "quantity value", of the current terminology, at least in the interpretation recently provided by Mari and Giordani [4]. In a few words, the state of a quantity is a non-numeric entity that expresses the way in which an object manifests some (measurable, in our case) property.

[3] Here quantity means "measurable property".

$$a_1 \succsim a_2 \leftrightarrow m_A(a_1) \geq m_A(a_2). \quad (1)$$

Furthermore, since $(B, \succ)$ is a strict order, for each $b_1, b_2 \in B$, there exists a function m_B such that

$$b_1 \succ b_2 \leftrightarrow m_B(b_1) > m_B(b_2). \quad (2)$$

2.2. Transduction or observation

As already noted, one of the main ideas we want to express is that, in the input/output model we are considering, the output, y, constitutes an *observable entity*, whilst the input, x, *does not*. This idea can be expressed by regarding m_B as a *known* function, which somewhat implies that we assume to be able to express instrument indication by numbers, whilst m_A is *known to exist* but it *is otherwise unknown*: this is indeed the reason why *we need a measuring system for actually performing measurement*.

Regarding m_B as a *known* function, is equivalent to assume that we are always able to express instrument indications by numbers. Let us then briefly discuss this point. One could argue that y is also a quantity, and therefore it also needs measuring, which requires an additional transduction step associated to y. But this would lead to a *regressio ad infinitum*, since for the transduction of y we would also have an output stage needing measurement, and so on. Therefore, for the measurement procedure to be really applicable – as it actually is – we should require *that it ends with a transduction where the indication may be directly expressed by numbers*. As an example, in the case of temperature measurement by liquid-in-glass thermometers, we can assume that a expresses the thermal state of the measurand, and b the height of the liquid column in the thermometer. Such states (heights) of the column may be revealed by traits on the external surface of the glass tube containing the thermometric liquid, and a number may be associated to each trait. Now, although the number we *directly* read on the glass is linked to the height of the column through intermediate transformations – for example if we associate the level of the column to its representative trait by visual inspection, we may have an error due to misalignment, and so on – but the very final step, that is the association of the number to the traits, may be reasonably assumed as a direct, uncertainty-free step: the hypothesis that we may associate a greater number to a trait that precedes, rather than following, a trait with a lower number, would be an unreasonable speculation! Furthermore, for the reasoning above, this possibility of finally associating a number to an indication, *must* always be the case. Lastly, note now that it is immaterial how many transduction steps we have before this last one: we may thus always consider, without loss of generality, an overall transduction that corresponds to the chaining of all such transduction steps, followed by the final conversion-to-numbers step. This is exactly what we propose in this model.

After this premise, let us now go back to the characterisation of what we have called the "transduction" or, more generally, the *observation* process. This process may be described, *at different discourse levels*, by the following functions, *all of which may be called observation or transduction functions*, to avoid an unnecessary multiplication of terms that would be confusing.

- A function, φ', form "things" to "things", that maps objects into states[4] of the output quantity of the MS; this function may be formally defined and characterised as follows: $\varphi' : A \to B$ such that

$$b = \varphi'(a) \leftrightarrow a \sim b. \quad (3)$$

Note that such a function it is well defined, since the order on B is strict.

[4] The attentive reader may have noted that we describe the output by "indications" or by "states of the output". In fact, since a strict order is assumed for the output quantity, states and indications are mutually isomorphic.

- A function, φ, form "things" to "numbers", that maps objects into indications of the MS; that is, $\varphi: A \rightarrow Y$ such that

$$y = \varphi(a) \leftrightarrow y = m_B(\varphi'(a)). \tag{4}$$

- Lastly, a function, f, from "numbers" to "numbers", $f: X \rightarrow Y$ such that

$$y = f(x) \leftrightarrow y = m_B(\varphi'(a)) \wedge x = m_A(x). \tag{5}$$

In the application of this model to the description of real systems, under certain circumstances, f may be called "calibration function" (whilst a general acceptance of this definition may create problems, as has been discussed elsewhere [18]).

An overview of these transformations is provided in Fig. 1.

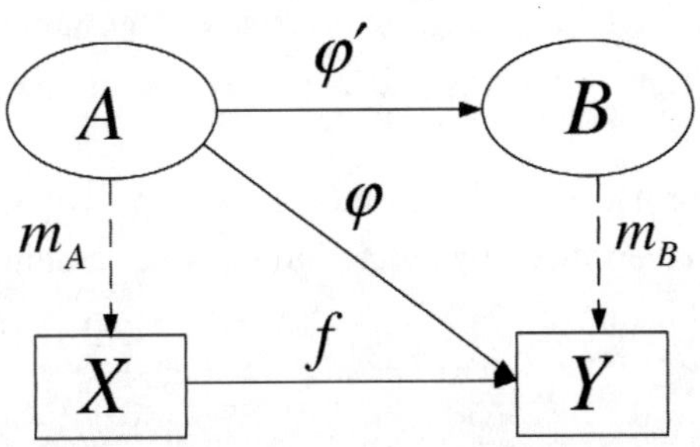

Fig. 1 Topology of the measuring system. Ellipses denote sets of "things", rectangles represent sets of numbers, solid-line arrows indicate real transformations, described at different discourse levels, and dashed-lines arrows express purely numerical correspondences, *which do not correspond to real transformations but nonetheless do contribute to understand the properties of some real transformations.*

Now, the key property of MS is that *the observation transformation is order preserving*, and *this holds true at all the description levels just considered.*

In formal terms, for each $a_1, a_2 \in A$,

$$a_1 \succsim a_2 \leftrightarrow \varphi'(a_1) \succsim \varphi'(a_2) \leftrightarrow \varphi(a_1) \geq \varphi(a_2) \leftrightarrow f(a_1) \geq f(a_2). \tag{6}$$

2.3. The overall measurement process

Let us now show why property (6) is so important: in fact *it ensures that the MS is capable of performing proper, consistent, measurements.*

To show this, let us introduce the *measurement function*

$$\gamma: A \rightarrow \hat{X}, \tag{7}$$

where $\hat{X}$ is the set of the possible measurement values[5]. Obviously, such a set coincides with the set of measure values, X, we already encountered. This simply means that if $x \in X$ is one of the possible measure values, it will also be one of the possible measurement values, that is $x \in \hat{X}$, and vice versa.

Then, interestingly enough, if the MS satisfies property (6), it also satisfies the following representational property, that holds true between objects and measurement values:

$$a \succcurlyeq b \leftrightarrow \gamma(a) \geq \gamma(b). \tag{8}$$

In fact such a property may be satisfied by taking, for each $a \in A$,

$$\hat{x} = \gamma(a) = f^{-1}(\varphi(a)) = f^{-1}(y), \tag{9}$$

which makes sense, since under the above assumptions, f turns out to be invertible.

Therefore, γ may be implemented, in the current practice of measurement, by

- Inputting the measurand to the MS and acquiring the (numerical) indication y: this is the *observation* phase;
- Applying the transformation f^{-1} to the indication, where f is known, typically by calibration, and is invertible, thanks to the above assumptions, which constitutes the *restitution* phase.

3. Conclusion

The notion of measuring system is gaining increasing interest in measurement science. Here we have presented a formal model of the MS, in terms of empirical relations among objects to be measured and the measuring device. We hope that this may help fill a gap in these studies.

By now, we have limited our investigation to order structures and related ordinal quantities, but we plan to extend this treatment to all the structures of metrological interest, as a further step of this study. Another possible development may consist in searching for a suited probabilistic counterpart of this model, which does not look like an easy task, at the present stage.

References

[1] N.R. Campbell, Physics – the Elements (1920). Reprinted as Foundations of Science, Dover, New York, 1957.
[2] G. B. Rossi, F. Crenna, A first-order probabilistic logic with application to measurement representations, Measurement 79 (2016) 251-259
[3] A. Frigerio, A. Giordani, L. Mari, Outline of a general model of measurement, Synthese (2010) 123-149
[4] L.Mari, A.Giordani, Quantity and quantity value, Metrologia 49 (2012) 756–764
[5] A. Maul, D. Torres Irribarra, M. Wilson, On the philosophical foundations of psychological measurement, Measurement 79 (2016) 311–320
[6] J. Michell, The logic of measurement: A realist overview, Measurement 38 (2005) 285–294
[7] Gonella, L. (1988). Measuring instruments and theory of measurement. Paper presented at the XI IMEKO World Congress, Houston.
[8] J. P. Bentley, Principles of measurement systems, Pearson-Prentice Hall, Harlow, 2005.

[5] This set is often assumed to be the set of the real numbers. In fact this very strong assumption, in our opinion, is not really needed in measurement science. This is why we leave it undefined in this presentation. See Ref. [11], or even Ref. [9], for additional ideas on this very important point, that cannot anyway be treated here.

[9] L. Mari, Beyond the representational viewpoint: a new formalization of measurement, Measurement 27 (2000) 71-84
[10] G. B. Rossi, Measurability, Measurement 40 (2007) 545-562
[11] G. B. Rossi, Measurement and probability, Springer, Dordrecht, 2014.
[12] K.H. Ruhm, Dynamics and stability – A proposal for related terms in Metrology from a mathematical point of view, Measurement 79 (2016) 276-284
[13] K.D. Sommer, Modelling of measurements, system theory and uncertainty evaluation. In: F. Pavese, A. Forbes (Eds), Data Modeling for Metrology and Testing in Measurement Science. Birkhäuser, Boston, 2009 (ISBN 978-0-8176-4592-2).
[14] K.D. Sommer, U. Hanebeck, M. Krystek, F. Sawo, Modelling of dynamic measurement systems for uncertainty analysis employing discretised state-space forms. In: F. Pavese, M. Bar, A.B. Forbes, J.M. Linares, C. Perruchet, N. F. Zhang (Eds), Advanced Mathematical & Computational Tools in Metrology & Testing VIII, World Scientific, Singapore, 2009, pp. 309-318.
[15] BIPM, IEC, IFCC, ISO, IUPAC, IUPAP; OIML, Guide to the expression of uncertainty in measurement, ISO, Geneva, (1993).
[16] L. Mari, M. Wilson, An introduction to the Rasch measurement approach for metrologists, Measurement 51 (2014) 315-327
[17] T. Bond, C.M. Fox, Applying the Rasch model: Fundamental measurement in the human sciences, Taylor and Francis, New York, 2015.
[18] G. B. Rossi, Toward a unified framework for static and dynamic measurements, 2015 IEEE International Instrumentation and Measurement Technology Conference (I2MTC) Proceedings, Pisa, 2015, pp. 1516-1521.

A conceptual roadmap for setting up a system of units in the "New SI" context

Luca Mari[#]

Università Cattaneo – LIUC, Castellanza (VA), Italy

lmari@liuc.it

Franco Pavese

formerly Consiglio Nazionale delle Ricerche, Istituto di Metrologia, Torino, Italy

Abstract. The definition of a system of units is analysed here as a structural problem, for which the physical theory provides the necessary conditions of realization but remains as backgrounder. The structurally more sophisticated option is provided by the so-called Global Constant Definitions, discussed in this paper through a conceptual roadmap intended to separate the components of the definitions based on theoretical foundation from those related to conventional features and, at the same time, to ensure the structural stability of the system and its partial modularity and therefore flexibility. The critical case of the so-called "New SI" is used as an illustration example.

1. Introduction

The quantity units used in measurement of ratio quantities provide an empirical reference to numerical values in mathematical equations that represent physical laws. Units have then the fundamental role of relating theory and experiment, and a system of units is a critical component of what metrology has to offer to empirical sciences and to their relation to societal needs. This role is crucial and cannot be under-evaluated, or taken for granted.

The so-called "New SI" [1], still in draft stage at the moment, is based on a structure, which we call here Global Constant Definitions (GCD),[1] that is significantly different from the one (either Explicit Unit Definitions (EUD) or Explicit Constant Definitions (ECD) [2]) that has been adopted in defining systems of units so far. This proposes an interesting problem: how to fulfil the societal and metrological need of stability in time of the system without hindering the fundamental requisite of science to maintain the principled possibility of evolving in time [3]. A requisite of metrology is indeed the continuity of the units with time, to avoid error-prone conversion factors or massive recalibration processes, and thus user disconcert. A manifest breach of this principle would be, for example, a length unit that, though maintaining its name, would change its definition so that the new metre were double of the previous one. Of course, nobody can guarantee that the presently available knowledge on the system of quantities and constant quantities on which the New SI is based will

[#] One of the authors is a member of the Joint Committee on Guides in Metrology (JCGM) Working Group 2 (VIM). The opinion expressed in this paper does not necessarily represent the view of this Working Group.

[1] In the BIPM documents the GCD is instead called "Explicit Constant Definition".

remain unchanged in future. On the contrary, the history of science and technology has taught that new knowledge will plausibly be produced that will modify our understanding.
The strategies of unit definition have evolved from (1) practical engineering requirements, related to the adoption of units defined as quantities of material standards, thus objects to be kept stable in time (e.g., the prototype of the kilogram), to (2) experimental physics or chemistry, where given kinds of objects have experimentally proved to have properties stable enough in time (e.g., the frequency of a given radiation of the caesium 133), to (3) stability in time to be proved by basic principles or laws of fundamental physics, which is the aim of the present evolution of the system of units. [2]
How to conciliate continuity and possibility of change is a structural challenge for all such strategies, but the third one, i.e., GCD, is specifically in a critical position on this matter, given that adapting unit definitions to new knowledge either might be formally impossible or might require changing the background theory. To this aim we propose here a modified version of GCD that makes it flexible to the embedding of new knowledge while maintaining all its structural merits.

2. A conceptual roadmap for unit stability under knowledge evolution

A critical issue of GCD is how possible future changes in physics could be taken into account in the system of units of which the system itself is based. It might happen that new basic physical equations between quantities will be established that produce a new structure of the system of quantities (as it happened with the introduction of relativity into the Newtonian mechanics) or that some quantities are found that are not to be time and space invariant as instead previously supposed. Of course such changes would produce a scientific revolution in Kuhn's sense [3], requiring a brand new approach to accommodate the changes in the SI system of units. However, less dramatic changes might more likely occur, which a sound structure of the system of units should be able to accommodate without any revolution provided that its conceptual structure — while (i) based on the principles and tools of and contents of fundamental physics, and thus in particular on the currently accepted system of quantities and the set of fundamental constants and (ii) linked to the current SI so that the principle of continuity is fulfilled — has some flexibility.
The paper intends to describe a conceptual roadmap that satisfies both these requirements, by construing a system of units according to a two-stage structure that explicitly includes a fundamental system and a conventional one:

- both a system of quantities and a set of constants corresponding to the dimensions of the base of the system are assumed; this represents the *fundamental* system where the numerical value of each constant is 1;
- a *conventional* system is then considered linked to the above fundamental system, where it is admitted that the numerical values of the constants can have values different from 1, assigned according to the best available present knowledge so that in the change to the new system the units maintain their values as expected according to the principle of continuity.

We start from considering, as the consistent and systematic the strategy used in the "New SI", called here "Global Explicit Constant Definition" (GCD). In such a system, the units are defined by taking as referent a set {Ci} of quantities modelled as constants, and taking a given numerical value k_i for each C_i. Each unit $[Q]$ is then defined as:

$$[Q] = \prod_i (C_i/k_i)^{n_i}$$

In other words, a system GCD is characterised as a system of units where the constants C_i have numerical values k_i. If C_i are constants, fundamental or not, but considered universally valid and invariant in time and space according to the current physical or chemical current theory, the system of units has its most general definition relative to the current state of knowledge.
However, this definition depends critically from the dependence on the set $\{k_i\}$.

[2] Actually the constants used in the "New SI" are, as indicated in [1], of four kinds: (i) fundamental constants of nature, (ii) specified atomic parameters, (iii) conversion factors, and (iv) technical constants.

Why it is critical? In a definition constrained by the choice of the set $\{k_i\}$, the quantities C_i modelled as constants have, in fact, a double role: (i) they connect quantities, (ii) they are parameters in physical/chemical laws, being themselves quantities of kind of objects. For example, the quantity linking energy and mass in the equation $E = mc^2$ is the speed of light in vacuum. The role (ii) makes in principle the quantities C_i measurable, via an empirical interaction with their kind of objects, the light in vacuum in the case of c. Actually, the numerical value k_i of C_i has been obtained by means of measurements, making use of a system of units established on standards based on the magnitude of objects (strategy EUD) or of kind of objects (strategy ECD). The numerical value of these standards that will be based on GCD is mandated to remain consistent with the former according to the continuity principle indicated hereinbefore.
Being the quantities C_i possible objects of empirical interaction, thus allowing measurements referred to standards of units pre-defined, it is, in principle and in practice, possible that the measurement results are affected by systematic errors. Therefore, one must admit, for example, as possible the fact of new experimental evidence that the numerical value of speed of light is different from the currently stipulated one. If the units are defined independently from the relevant quantity (c for length), the situation is not problematic: one only has to correct the consensus value, expressed in the relevant unit, associated to the quantity, from k_i to k_i^*.
On the contrary, what to do when the required change occurs after a system of units of the type GCD has been established also based on C_i is a delicate issue. One could

- keep the system unchanged, admitting that in fact it is not based on constants assumed as invariant in time and space;

or

- change the whole definition, where in the new system the numerical value associated to C_i is k_i^*.

However, adjusting the strategy GCD could produce the required flexibility to a minor cost.
The problem is clearly structural, not involving physics or chemistry: thus, in the following we will refer to an hypothetical simplified system, only including units of time (duration), length and mass. In such a system, the “New SI” scheme would bring to the following type of definition (using data from the CCU 2016 Draft of SI Brochure):

The SI is the system of units in which the unperturbed ground state hyperfine splitting frequency of the caesium 133 atom $\Delta\nu_{Cs}$ is 9 192 631 770 Hz, the speed of light in vacuum *c* is 299 792 458 m s^{-1} and the Planck constant h is 6.6260693 × 10^{-34} J s.

We can abstain from making use of any structural feature by writing instead:

The SI is the system of units in which the frequency C_1 is 9 192 631 770 Hz, the speed C_2 is 299 792 458 m s^{-1} and the action C_3 is 6.6260693 × 10^{-34} J s.

There is still the problem that this definition makes reference to units (hertz, metre, second, joule) which are the object of the definition itself, known only by admitting that the user already knows the previous definition. The simple solution is to rewrite as follows:

The SI is the system of units in which the frequency C_1 is 9 192 631 770 units of frequency, the speed C_2 is 299 792 458 units of speed and the action C_3 is 6.6260693 × 10^{-34} units of action.

One interesting issue of this type of definition is that the quantities referred to, as the “fundamental bases” of the system, are frequency, speed and action, not time (duration), length and mass. The latter are anyway retrievable from the former by means of the quantity equations of physics, which do not depend on any specific unit.

The roadmap of deconstruction of the definition can now be completed, by understanding that the numerical values k_i do not have a structural role, thus by rewriting the definition in the most 'fundamental' form:

The SI-fundamental is the coherent system of units in which C_1 is the fundamental-unit of frequency, C_2 is the fundamental-unit of speed and C_3 is the fundamental-unit of action.

In this way, according to this definition, for example the Planck constant has a value 1 fundamental-unit of action.
Now it is possible to rebuild the system, linking the SI-fundamental with a SI-conventional (the present one), the latter with the aim to implement the continuity principle:

The SI is the coherent system of units in which
(A) the unit of time (duration) is the second, symbol s, the unit of length is the metre, symbol m, and the unit of mass is the kilogram, symbol kg;
(B) the frequency $\Delta\nu_{Cs}$ is k_1 s^{-1}, the speed c is k_2 m s^{-1}, the action h is k_3 kg m^2 s^{-2};
(C) the numerical values are $k_1 = 9{,}192631770\times10^{-10}$, $k_2 = 2{,}99792458\times10^{-9}$, $k_3 = 6{,}6260693\times10^{-34}$
(or whatever values CODATA will propose when the system is decided; clause (C) specifies stipulated values, i.e., exact, derived from experimental uncertain values).

3. Conclusions

In summary, a system of units of this modified GCD type is defined according to the following logical scheme:

- a system of equations between quantities and a set of constant quantities $\{C_i\}$ are assumed, their number corresponding to the dimension of the base of the system;
- from these components a fundamental system is derived first, where each C_i has a numerical value 1;
- a conventional system is linked to it, where one assume that each C_i can assume a value different from 1, in order to make possible the continuity with the previous system;
- in agreement with the best knowledge available at the time of the change, numerical values are assigned to the k_i such that the continuity is ensured. They are in fact those accepted in the previous system, under the hypothesis that the best knowledge available is the one immediately preceding the formulation of the new system.

The benefits of this conceptual structure are twofold:

- a *structural stability* of the system, which needs to be modified only in the event of a radical novelty, e.g., the re-formulation of fundamental equations between quantities, or the gained evidence that a quantity assumed to be invariant is not;
- a *partial modularity* of the system, thanks to the possibility of re-parametrisation of the structure, so that the correction of the value of a constant would only require to adjourn the stipulated value.

This conceptual structure of the definition clearly distinguishes between the components of the definition that arise from theory and the components that instead are conventional. Such a distinction also makes the whole system easier to understand in its fundamental structure, an issue extremely important due to the importance of the SI in our societies.

Acknowledgements
The authors thank Prof. Ingvar Johansson, for some clarifications he provided to a draft of this paper, and appreciate his manuscript [4] and Prof. Alessandro Giordani for the fruitful discussions on this subject.

Journal of Physics: Conference Series **772** (2016) 012011 doi:10.1088/1742-6596/772/1/012011

References

[1] BIPM, CCU: 9th Brochure of the SI, draft of 11 December 2015, published on January 4, 2016 (2016 Draft), http://www.bipm.org/en/measurement-units/new-si (consulted on February 12, 2016).
[2] I.M. Mills, P.J. Mohr, T.J. Quinn, B.N. Taylor, E.R. Williams, Redefinition of the kilogram, ampere, kelvin and mole: a proposed approach to implementing CIPM recommendation (CI-2005), Metrologia 43, 227–246, 2006.
[3] T. Kuhn, The structure of scientific revolutions, University of Chicago Press, 1962
[4] I. Johansson, Constancy and Circularity in the SI, (2014) pp. 25, in http://metrologybytes.net/PapersUnpub/OpEds/Johansson_2014.pdf.

Journal of Physics: Conference Series **772** (2016) 012012 doi:10.1088/1742-6596/772/1/012012

Ability evaluation by binary tests: Problems, challenges & recent advances

E Bashkansky[1] **and V Turetsky**[2]
ORT Braude College of Engineering, P.O. Box 78, Karmiel, 2161002, Israel
[1]Industrial Engineering and Management Department
[2]Department of Applied Mathematics

E-mail: ebashkan@braude.ac.il

Abstract. Binary tests designed to measure abilities of objects under test (OUTs) are widely used in different fields of measurement theory and practice. The number of test items in such tests is usually very limited. The response to each test item provides only one bit of information per OUT. The problem of correct ability assessment is even more complicated, when the levels of difficulty of the test items are unknown beforehand. This fact makes the search for effective ways of planning and processing the results of such tests highly relevant. In recent years, there has been some progress in this direction, generated by both the development of computational tools and the emergence of new ideas. The latter are associated with the use of so-called "scale invariant item response models". Together with maximum likelihood estimation (MLE) approach, they helped to solve some problems of engineering and proficiency testing. However, several issues related to the assessment of uncertainties, replications scheduling, the use of placebo, as well as evaluation of multidimensional abilities still present a challenge for researchers. The authors attempt to outline the ways to solve the above problems.

Introduction

Qualitative testing aimed to measure the latent ability and resulting in a binary output of type "good/bad", "pass/fail", "true/false", "yes/no", "positive/negative", "detected /non detected", "1/0" etc. is often used in various fields of science, educational measurement, quality control, industry and healthcare, including analytical chemistry and microbiology.

In this paper, we deal only with the basic and simplest issue of so-called one-dimensional ability, when the test item performance of the object-under-test (henceforth OUT) can be explained by a single latent ability. We consider the case when a qualitatively homogeneous population of OUTs is tested using a set of non-destructive test items having different, (known or unknown beforehand) levels of difficulty, and we need to evaluate/compare the intrinsic abilities of these OUTs and the levels of difficulty of the test items if the latter are unknown beforehand. This type of set hereinafter will be called the test and it can include any —but must be the same for all OUTs—number of test items. For instance, in the psychometrics test the OUT is an examinee, a separate question on the exam is a test item and the examination as a whole is a test. We assume [1] that the test item response is estimated on the binary scale base ("good/bad", "pass/fail", "true/false", "yes/no", "positive/negative", "detected /non detected", "1/0" etc.) and the results of different test items, applied to the same OUT,

[1] Corresponding author.

are conditionally independent (i.e., the response to one test item does not affect the response to another). It is also assumed that the inherent ability of the OUT is independent of the test item difficulty. Homogeneity here means that the same item response model is applied to all members of the population, but in any case does not imply equality of the tested abilities.

Such qualitative testing is often used in various fields of science, educational measurement, quality control, industry and healthcare, including analytical chemistry and microbiology [2-10]. As the response to each test item provides only one bit of information per OUT, for a full assessment of an OUT's ability (hereafter denoted by *a*), the test usually comprises a series of test items presenting various levels of difficulty (hereafter denoted by *d*). These levels can be known or unknown beforehand. The latter situation presents a case that is more complicated to analyze. The search for effective ways of planning and processing the results of such tests is highly relevant. In recent years, there has been some progress, generated by both the development of computational tools and the emergence of new ideas, which will be presented in this article.

2. Main definitions and assumptions

2.1 Definitions

Consider m OUTs taking part in a testing procedure. Every OUT has to pass the test consisting of a set of n *test items*, having different levels of difficulty. Let a_i denote the *ability* of OUT i (i=1,2,…I) and d_j the difficulty of the test item j (j = 1,2…J) in relation to the studied ability. In some experiments every OUT can be tested by the same test item repeatedly. For each intersection $i \times j$ we will use r_{ij} to denote the number of such repetitions, $r_{ij}^{(+)}$ for the amount of positive results, $r_{ij}^{(-)}$ for the amount of negative results and $f_{ij} = r_{ij}^{(+)} / r_{ij}$ for the relative frequencies of positive results obtained in these repetitions. Also, we use 1 to indicate a positive result and 0 otherwise. Accordingly, the results of the testing can be presented by an $I \times J$ matrix $\boldsymbol{R}^{+}$, whose elements $r_{ij}^{(+)}$ may change from zero to r_{ij}.

Thus, results of testing every OUT by three test items can be presented as in Table 1.

Table 1. Results of testing OUT$_i$

	Test item 1	Test item 1	Test item 2	Test item 2	Test item 3	Test item 3
OUT$_i$	$r_{i1}^{(+)}$	$r_{i1}^{(-)}$	$r_{i2}^{(+)}$	$r_{i2}^{(-)}$	$r_{i3}^{(+)}$	$r_{i3}^{(-)}$

Further, let $P(d|a)$, customarily called the item response function (IRF), express the probability that the OUT with ability a will successfully pass the test item having difficulty d. The following reasonable and quite plausible main assumptions concerning the testing can be made [1].

2.2 Assumptions concerning the test results

2.2.1 The responses to different test items, applied to the same OUT, are conditionally independent, i.e., the response to one test item does not affect any other. This is admissible, if no direct link between the difficulty of the test item and the inherent ability of the OUT exists. In other words, the difficulty of the test item does not affect the inherent ability of the OUT.

2.1.2 Testing is organized in a such way that, thanks to randomization, OUT does not know whether it has to pass the same or a different test item, so the responses to repeated test items, applied to the same OUT, are also conditionally independent as in the Bernoulli scheme with probability of "success" $P(d_j|a_i)$

2.2.3 Two OUTs with the same results are considered indistinguishable under the given test and judged to have the same—most likely—ability.

2.3 Scale invariant item response model

We assume that ability and difficulty (directly, or through some transformation) may be brought to the same measuring scale. Then, applying considerations of dimensional analysis, we conclude that the dimensionless response function should be scale invariant. Mathematically, this means that for any reasonable $\lambda > 0$:

$$P(\lambda d|\lambda a) = P(d|a), \tag{1}$$

which is possible only if $P(d|a)$ is a function of the ratio *d/a* between difficulty and ability or vice versa. Accordingly, all limitations and solutions are defined and formulated only for relations *d/a* or *a/d*.

2.4 Novel process control based IRF

Let us interpret a standard statistical process control procedure [11] as the ability of the $\bar{x}$ control chart to detect a shift in process quality (a shift from the in-control value), when the standard deviation is assumed known and constant. Denote the magnitude of the shift, measured in units $\sigma / \sqrt{n}$ (where n means the sample size), as the k and a distance between control limit and centre line, measured in the same units, as L . Usually, the $\bar{x}$ control chart has two limits, thus essentially, it combines two detectors. To clarify, we will take into account only one of the limits—the upper one—for example. Then, according to [11], the probability of detecting the shift up of magnitude k is given by $1-\Phi(L-k)$, where Φ denotes the standard normal cumulative distribution function. Now, define:

- *Ability* a of the one-sided control chart to detect the change: $a = e^{-\frac{L}{\sqrt{2}}}$ (the farther from the center the control limit is, the lower the sensitivity of the control chart is),
- *Difficulty* d of detecting the shift: $d = e^{-\frac{k}{\sqrt{2}}}$ (the larger the shift is, the easier is to detect it and vice versa).

Both a and d are defined in the range of zero to one; nevertheless, their ratio can take any value. The maximal level of difficulty $d = 1$ is achieved for the case when the shift essentially does not exist ($k = 0$) so a positive response signifies a *false alarm*. Hereafter, such test items will be called *placebos*.

Thus:

$$P(d|a) = 1 - \Phi(-\sqrt{2} \cdot \ln a + \sqrt{2} \cdot \ln d) = 1 - \Phi(\sqrt{2} \cdot \ln \frac{d}{a}). \tag{2}$$

In our previous publications [1], we used a Rasch similar IRT:

$$P(d|a) = [1 + \tanh(1 - \frac{d}{a})]/[1 + \tanh 1] . \tag{3}$$

Figures 1.a and 1.b demonstrate that the IRF presented in Formulas (2) and (3) are very close one to another. Hereafter, we will mainly use the IRF in the form provided by Formula (2).

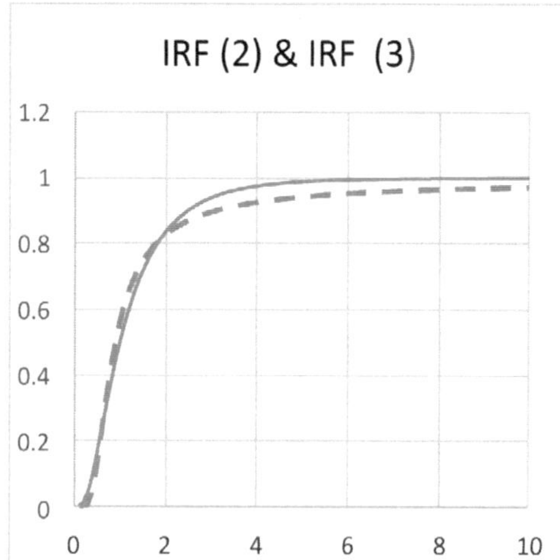

Figure 1.a IRF vs. *a/d* according to (2) and (3); the latter is presented by the dashed line.

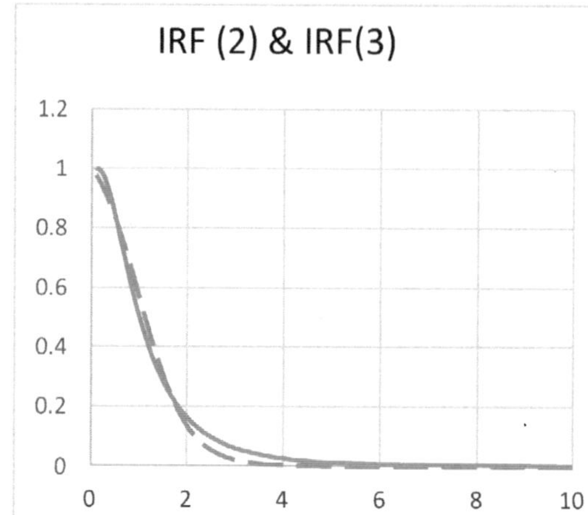

Figure 1.b IRF vs. *d/a* according to (2) and (3); the latter is presented by the dashed line.

3. Review of the types of problems and ways to solve them

We now discuss what types of problems researchers may face and the possible approaches for solving them.

3.1 Levels of difficulty of the test items are known beforehand

Suppose the levels of difficulty of the test items (d_1, d_2, … d_J) are given and results of testing every i-th OUT are represented in a form similar to Table 1. Then the problem of estimating i-th's ability can be resolved by applying the maximum likelihood estimation (MLE), i.e., by maximizing the likelihood function:

$$L(a_i;d_1,d_2,\dots d_J)=\prod_{j=1}^{J} P(d_j|a_i)^{r_{ij}^{(+)}}\cdot[1-P(d_j|a_i)]^{r_{ij}^{(-)}}. \tag{4}$$

It seems simple, but the problem is that if the test results themselves are not reliable, improbable and self-contradictory—the probability of passing the test items predicted by $P(d_j|a_i)$ may be very far from the observed relative frequencies $f_{ij}=r_{ij}^{(+)}/r_{ij}$. In this case, a chi-square or similar test can serve as indicator of confidence in the estimated ability.

In example presented below in Table 2, $d_1=0.2$; $d_2=0.6$; $d_3=1.0$ (placebo) and the most likely value of the ability equals 0.4. Predicted and observed frequencies are very close and $\chi^2=0.008$ is very small.

If for the same test items the obtained results appear as presented in Table 3 (only results of placebo test item are interchanged), the most likely value of the ability equals 0.53. Predicted and observed frequencies are very far and $\chi^2=39.3$ is very large. This result is not because of a bad assessment. It is the consequence of the inconsistency/dissonance of the testing results themselves: indeed, 9 from 10 positive responses to placebo looks beyond common sense!

Table 2. Results of testing OUT i given $d_1 = 0.2$; $d_2 = 0.6$; $d_3 = 1.0$ (placebo)

OUT i	Test item 1 $r_{i1}^{(+)} = 21$	Test item 1 $r_{i1}^{(-)} = 4$	Test item 2 $r_{i2}^{(+)} = 10$	Test item 2 $r_{i2}^{(-)} = 26$	Test item 3 $r_{i3}^{(+)} = 1$	Test item 3 $r_{i3}^{(-)} = 9$
Pred.	20.91	4.09	10.19	25.81	0.98	9.02

Table 3. Observed vs. expected/predicted frequencies

OUT i	Test item 1 $r_{i1}^{(+)} = 21$	Test item 1 $r_{i1}^{(-)} = 4$	Test item 2 $r_{i2}^{(+)} = 10$	Test item 2 $r_{i2}^{(-)} = 26$	Test item 3 $r_{i3}^{(+)} = 9$	Test item 3 $r_{i3}^{(-)} = 1$
Pred.	22.90	2.10	15.49	20.51	1.85	8.15

3.2 Levels of difficulty of the test items are unknown beforehand and OUT population is very large

Suppose we consider a very large population of OUTs tested without repetitions (i.e., $r_{ij} = 1$) by J test items with unknown beforehand levels of difficulty. "Large" means that it is big enough to consider a proportion of the OUTs that have successfully passed some test item as a probability of its successful passing by being randomly selected from the OUT population. So, if the studied ability a is distributed among the tested population of OUTs according to some probability density function $f(a)$, then the proportion $f(d)$ of OUTs that successfully overcome the test item having difficulty d is expected to be some convolution of $f(a)$ with $P(d|a)$:

$$p(d) = \int P(d|a) f(a) \mathrm{d}a\,. \tag{5}$$

Here, the OUTs with the same test results are considered indistinguishable under a given test and judged to have the same most likely abilities. Proposed in [1], the iterative procedure allows us, in this case, to find both spectra: J difficulties and 2^J abilities, as well as the distribution of these abilities among the studied population.

3.3 Levels of difficulty of the test items are unknown beforehand and the OUT population is limited

For this case, the OUTs with the same test results are also considered indistinguishable under a given test and judged to have the same most likely abilities. The likelihood function takes the form:

$$L(a_1, a_2, \ldots a_I; d_1, d_2, \ldots d_J) = \prod_{i=1}^{I} \prod_{j=1}^{J} P(d_j | a_i)_{r_{ij}^{(+)}} \cdot [1 - P(d_j | a_i)]_{r_{ij}^{(-)}} \,, \tag{6}$$

and depends on $I+J$ variables. Although it is usually much less than in $I + 2^J$, as in *3.2* , the problem of finding the maximum in $I+J$ space is rather serious, a bit easier because apparently this maximum is unique. For example, let us suppose that the results of the test are those that are shown in Table 2, but except for $d_3 = 1$(placebo) the rest of the difficult levels (d_2, d_3) are unknown. By maximizing (6), we get the following: $d_1 = 0.2, d_2 = 0.613$; $a_i = 0.4041$. For more than one OUT, the problem becomes more complicated and encounters computational difficulties [12].

Journal of Physics: Conference Series **772** (2016) 012012 doi:10.1088/1742-6596/772/1/012012

4. Challenges

The problem of binary test evaluation in the case when the levels of difficulty of the test items are unknown beforehand and the amount of tested OUTs is limited and large remains a challenge for researchers dealing with binary testing in different fields of science and engineering. A problem of optimal allocation of the experimental resources is also yet unsolved.

References

[1] Turetsky V and Bashkansky E 2016 Testing and evaluating one-dimensional latent ability *Measurement* **78** 348–57.

[2] AOAC (Association of Official Agricultural Chemists) Official Methods of Analysis 2013 Guidelines for Validation of Qualitative Binary Chemistry Methods, International Stakeholder Panel on Alternative Methods (ISPAM).

[3] Bashkansky E and Gadrich T 2013 Some statistical aspects of binary measuring systems. *Measurement* **46** 1922–27.

[4] Wehling P, LaBudde RA, Brunelle SL, Nelson MT 2011 Probability of detection (POD) as a statistical model for the validation of qualitative methods. *J AOAC Int.* **94**(1) 335–47.

[5] Wilrich C and Wilrich P 2009 Estimation of the POD function and the LOD of a qualitative microbiological measurement method *J AOAC Int.* **92**(6) 1763–72.

[6] Wilrich P 2010 The determination of precision of qualitative measurement methods by inter-laboratory experiments *Accreditation and Quality Assurance* **15** 439–44.

[7] Altman G, Bland M 1994 Diagnostic tests. 1: Sensitivity and specificity. *BMJ* **308** (6943) 1552.

[8] Uhlig S, Krugener S, Gowik P 2013 A new profile likelihood confidence interval for the mean probability of detection in collaborative studies of binary test methods *Accreditation and Quality Assurance* **18** 367–72.

[9] Pendrill L and Fisher W 2015 Counting and quantification: Comparing psychometric and metrological perspectives on visual perceptions of numbers *Measurement* **71** 46–5.

[10] Uhlig S et al. 2015 Qualitative PT data analysis with easy-to-interpret scores *Accreditation and Quality Assurance*, **20** 347–353

[11] Montgomery D 2009 *Introduction to Statistical Quality Control* 6th ed. (New York: John Wiley & Sons Inc.) 247 eq. (6.19).

[12] Bashkansky E and Turetsky V 2016 Proficiency testing: binary data analysis *Accreditation and Quality Assurance*, published online 21April 2016, DOI: 10.1007/s00769-016-1208-x

Evaluating the impact of a Quasi-ipsative scoring approach on the scoring of a VARK style questionnaire

E Fitkov-Norris and A Yeghiazarian

Kingston Business School, Kingston University, Kingston-upon-Thames, UK.

Email: e.fitkov-norris@kingston.ac.uk

Abstract. This paper demonstrated the application of a quasi-ipsative scoring approach to assess the relative strengths of individual preferences in a VARK style questionnaire. The approach identified a chi-squared test as more suitable method for analysing the type of data gathered by the VARK questionnaire. The results suggest that the quasi-ipsative chi-squared based approach does not appear to be as sensitive as the original t-test approach in identifying significant modalities. In order to increase the sensitivity of the test, the requirement for overall test significance has to be relaxed and individual cells considered in the analysis. The findings also put some doubt on the statistical validity of the original t-test approach as it also uses the deviation from the means (in the form of standard deviation) rather than statistical significance, as a tool for assessing the strength of individual preferences. The causes of the discrepancies between the two scorings techniques need to be examined further, before recommendations can be made on which approach is better suited to identifying the strength of individual preference for information input modality.

1. Introduction

This article assesses the impact of a new, quasi -ipsative scoring mechanism on the recommendations made by a multidimensional instrument measuring the strength of individual's learning style preference utilising dichotomous questions. The instrument, designed by Fleming [1] evaluates an individual's preference for Visual, Auditory, Read/Write and Kinaesthetic (VARK) mode of information input with the objective of identifying dominant traits. The instrument consists of 16 single stimuli questions, each presenting a different scenario and asks the user to tick all options that apply to them in a particular scenario. For example, a user with a single preference is expected to tick one option within each scenario, whereas a user with multimodal preferences is expected to tick several options within that scenario. The output from this scoring scheme is a 4 by 16 binary matrix indicating a user's preference for each of the 4 dimensions and the strength of each preference is identified by the frequency with which a particular dimension is chosen. Allowing uses to choose the number of ticks that apply in a particular scenario fits relatively closely with an intuitive understanding of the way preferences are defined, namely, "Relatively stable evaluative judgments in the sense of liking or disliking a stimulus, or preferring it or not over other objects or stimuli …" [2, p. 9]. The larger the number of ticks on each dimension, the stronger the user preference. Asking the users to tick all that apply also allows the instrument the flexibility to assess users with varying number of dominant preferences (unimodal, bi-modal, tri-modal and multimodal).

However, this scoring technique also has some disadvantages. Firstly, different users can tick a different number of ticks in total. As a result, it is not possible to compare individual responses to the population of all users, as the expected number of ticks in the population cannot be determined. One solution is to compare the number of ticks for a user in each dimension to the expected number of ticks in the sub-population of users who ticked the same number of ticks, see for example [3]. So if a user has ticked a total of 16 ticks, one can compare the number of actual ticks in each dimension to the expected number of ticks in each dimension for all users who ticked 16 options in total. Similarly, a user with 17 thick in total will be compared to the expected values sub-population of users who chose 17 options in total. By definition the relative number of ticks in each dimension is expected to monotonically increase as the total number of ticks given by the user increase. Violation of this property can lead to discontinuities in the assessment of preferences, as each additional tick changes the reference point for the individual and the respective expected values. In some cases an extra tick could also lead to a significantly different recommendation on the person's preferences as the mean expected number of ticks for a preference can go down despite the fact that the overall number of ticks has gone up. For example, in Table 1 individual A had a total of 8 ticks in category R and was categorised as strong R, while individual B, with the same number of ticks in that category was categorised as mild R, although their profile is almost identical. Similarly, individual C is classified as having two preferences (AR), while individual D has a very similar profile, but is classified as having all 4 (VARK). In some cases, adding an extra tick can lead to loss of preferences (for example, individual E is tri-modal while individual F has a mild single preference).

Table 1 Examples of original VARK scoring

	V	A	R	K	Total ticks	Original scoring
Individual A	0	3	8	4	15	strong R
Individual B	0	3	8	5	16	mild R
Individual C	2	4	5	2	13	VARK
Individual D	2	4	5	3	14	VARK
Individual E	2	5	4	6	17	ARK
Individual F	2	5	4	7	18	mild K

Secondly, the technique compares the average number of tick in the respective sub population to the number of ticks the user selected using a t-test and uses the number of standard deviations away from the mean the users score to determine the strength of their preference. The t-test is designed to be used with quantitative data, which has interval or ratio measurement scale. However, the data gathered by the questionnaire is binary, confirming the presence of a preference but not providing any information on the relative strength of that preference. The analysis used makes the implicit assumption that each tick (across the 16 scenarios) denotes equal difference in preference from any options that have not been ticked. So an implicit assumption is made that a user ticking Auditory and Kinaesthetic in a particulate scenario has equal preferences for the two and in addition, the difference in preference between the Auditory and Kinaesthetic preferences selected and the Visual and Read/Write preferences that were not selected are also the same. This assumption has not been tested explicitly and therefore, using the t-test in the way may be inappropriate.

One solution to this problem is to change the setup of the questionnaire to a fixed-sum ranking question, where we require users to provide a ranking or relative order of their four preferences within each scenario (1st, 2nd, 3rd and 4th), for example. This approach will provide more detailed and accurate information on the actual preferences the user has and can be analysed using non-parametric tests such as Friedman's Test to compare ranks given across the 4 dimensions. However, the approach will increase the burden for the tests users and may lead to underside response behavior [4]. In addition, fixed-sum questions can lead to an artificial reduction on the response range and distorted scale relationships [5]. Although some solutions to the problem of fixed-sum ipsative scales have been proposed [6], the suggested changes may lead to significant changes in the scales of the questionnaire and their introduction is not recommended without carrying to extensive testing and validation.

To ensure that the integrity of the existing questionnaire is not violated [3], this article proposes an alternative approach for the analysis of the existing scores using Chi-squared non-parametric test for association. The chi-square test is designed to compare frequencies of nominal data and does not make any implicit assumptions about the rating scale of the data. However, the total number of tick given by each participant is different as the design of the questionnaire allows each question to carry up to 4 marks (ticks). As one of the assumptions of chi-squared test for association is that each question can carry only one count towards the total, this test cannot be used for normative comparison with the general responses within the population. To ensure that this assumption is not violated, the test has to be applied at the level of the individual. The test will compare the proportion of ticks (out of 16) the user has given to each dimension and can be used to identify the relative importance of each dimension within the context of the individual. This is a quasi-ipsative approach to personality measurement as the VARK instrument does not meet the fixed-sum prerequisite of strictly ipsative approaches [7]. Despite their relatively recent introduction, meta-analysis into personality trait measurement instruments suggests that quasi-ipsative approaches provide a more robust and reliable estimation of the trait compared to instruments adopting normative and purely ipsative approaches [7]. Therefore, this article will suggest a methodology for using Chi-squared test to assess the information processing preferences of users for visual, auditory, read/write and kinaesthetic information and will compare the efficacy of the quasi - ipsative approach to the classification results using the existing scoring mechanism.

2. Description of the method

The analysis was carried out on sample of 51 observations of VARK data provided by Neil Fleming and collected through the VARK online questionnaire version 7.1. (http://vark-learn.com/the-vark-questionnaire/). 73% of the sample was female, 23% male and 4% did not specify. 78% of respondents were students and 14% teachers. 45% of students were under 25 years old. Over half of respondents (57%) were from USA and next largest group was from UK (21%).

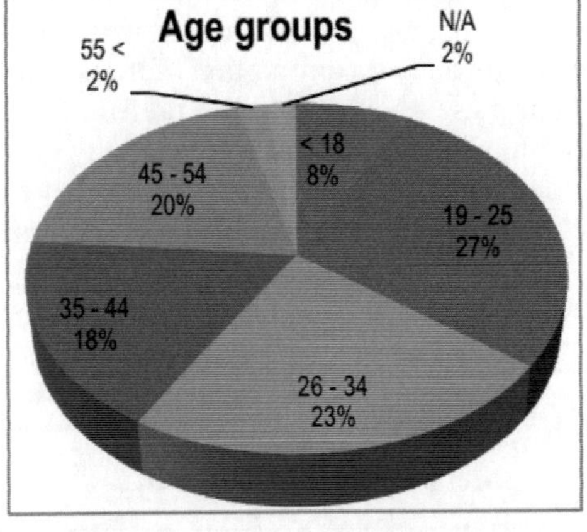

Figure 1 Age of participants

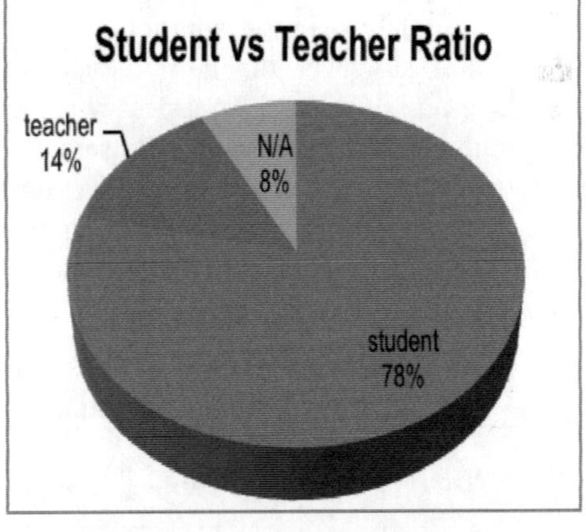

Figure 2 Occupation of participants

The data was analysed in three different ways. Method 1 was based on the Chi-squared test statistical significance. Observations with statistically significant number of ticks across all four dimensions (sig. < 0.05) were identified as having at least one preference. The strength of the preference was evaluated using the following criteria: Adjusted residual for a particular cell is in the range of (1.64 - 1.95) indicated a mild preference for that modality, adjusted residual for a particular cell is in the range of (1.96 – 2.56) indicated a strong preference and adjusted residual for a particular cell of 2.57 and above indicated a very strong preference for that modality. This approach for identifying the strength of a modal preference is identical to the one adopted by Fleming [8] although the standard residuals in the original research were derived using a t-test. Observations that did not have statistically significant dimensions across the 4 modes (i.e. sig >=0.05) were classified as using all 4 modes (VARK).

Method 2 is similar to method one but rather than look at the relative strength across all 4 dimensions as measured by significance, it examined the adjusted residuals in each cell for significance across the positive proportion of the preferences. In this case all observations that had an overall significance level less than 0.05 as well as observations that had at least one statistically significant residual cell were classified as having strong preference. However, if a particular cell contained significantly fewer than expected number of ticks, this cell was ignored (i.e. it was not taken to indicate a lack of preference in that modality).

Method 3 is adopted the same approach as Method 2 but considered significant adjusted residuals both in the positive and in the negative sense so that having significantly fewer ticks on a dimension was interpreted as lack of preference in that modality and therefore the preference was removed.

The result from the three methods of quasi-ipsative scoring were compared with standard VARK method of scoring and it can be seen that the degree of agreement between the new and old scoring approaches increased as the information taken into account on making the recommendation increased. (Table 2).

Table 1 Overall agreement of proposed scoring methods with original scoring approach

Row Labels	Method 1		Method 2		Method 3	
match	30	59%	35	69%	39	76%
not match	21	41%	16	31%	12	24%
Grand Total	51	100%	51	100%	51	100%

The discriminatory power across modalities and sensitivity of the different scoring techniques also increased as the factors taken into considerations increased. Method 1 was very poor at predicting bimodal and tri-modal preferences although it is relatively accurate at predicting users with unimodal preferences (77%). However, the method overestimated significantly the proportion of users with multimodal preferences. Method 2 does less well in predicting users with unimodal preferences but is more accurate in identifying multimodal users. Neither Method 1, nor Method 2 identify any tri-modal preferences. Method 3, which considers both positive and negative preferences, performs better than Method 1 and Method 2 in identifying users with both unimodal and multimodal preferences. Furthermore, Method 3 identified a number of users with tri-modal preferences, although the categories identified as significant are different from the categories identified using the original t-test method.

Table 2 Agreement of proposed scoring methods with original scoring approach by number of learning preferences

Row Labels	Original Method	Method 1	Method 2	Method 3
1-pereference	22	13 (77% match)	30 (67% match)	23 (83% match)
2-pereferences	3	1 (100% match)	1 (100% match)	1 (100% match)
3-pereferences	7	0 (0% match)	0 (0% match)	13 (54% match)
4-pereferences	19	37(52% match)	20 (70% match)	14 (93% match)
Grand Total	51	51	51	51

3. Discussion

The application of the quasi-ipsative approach using a Chi-squared test to identify the relative strength of input mode preferences for Visual, Auditory, Read/Write and Kinaesthetic information lead to different classification of preferences from the original t-test approach. The quasi-ipsative approach was less sensitive in identifying bi-modal and tri-modal preferences and its recommendations varied significantly from the recommendation made using a t-test. This was due to the fact that very few user profiles were identified as having statistically significant differences between their preferences overall. The Chi-squared test was less likely to flag statistically significant differences compared to the t-test. This is not surprising since the chi-squared test is designed for non-parametric data and relaxing the assumption of that the number of ticks are normality distributed throughout the population leads to an increase in the expected standard error term. Furthermore, as mentioned above the t – test assumes that the date gathered is measured on at least approximately interval scale and counted data does not necessarily satisfy this assumption.

The instrument's sensitivity and discriminatory power could be improved significantly by considering individual cell significance and identifying individual cells with significantly more and/or fewer than the expected number of ticks (Methods 2 and 3). However, adopting Methods 2 and 3 violates the statistical rigor of the Chi-Squared approach as a cell in a table could have more/fewer than expected counts simply by chance. On the other hand, this approach replicates very closely the t-test approach taken in the original scoring of the questionnaire which also does not require a significant results across the categories but instead uses 1, 2 and 3 standard deviations from the mean as the indicator for the preference strength. Given that the data is nominal in nature, the Chi-squared test is a better alternative for data analysis as it makes fewer assumptions about the properties of the data.

The adoption of a quasi – ipsative approach also deals with potential classification discontinuities brought on by the use of averages that are not monotonically increasing. The original and new classification the individuals highlighted in Table 1 are displayed in Table 4 below. The new classifications handle some of the inconsistent classifications but again highlights the issue with the sensitivity of the alternative scoring approach as for example individual E is identified as having 4 modalities of equal strength, while the original method identified that user with a tri-modal preference.

Table 3 Original vs. New scoring approach comparison

	V	A	R	K	Total ticks	Original scoring	New Scoring (Method 3)
Individual A	0	3	8	4	15	strong R	strong R
Individual B	0	3	8	5	16	mild R	strong R
Individual C	2	4	5	2	13	AR	VARK
Individual D	2	4	5	3	14	VARK	VARK
Individual E	2	5	4	6	17	ARK	VARK
Individual F	2	5	4	7	18	mild K	VARK

The introduction of quasi – ipsative scoring may have led to exaggeration of the negative preferences of individuals. By definition, the new classification is fixed sum, i.e. adding a tick to one category means that a tick was not given to another. The benefit from this approach is that it allows interpretation to both positive and negative preferences. So if a person had not ticked a V we can conclude that they are not visual in that scenario. However, the original test was not designed to identify negative preferences and it does not contain any control questions for negative preferences. Therefore, despite its ability to provide additional information, adopting the quasi-ipsative approach and using it to identify the lack of preferences without undergoing validation is not recommended.

4. Conclusions

This paper demonstrated the application of a quasi-ipsative scoring approach to assess the relative strengths of individual preferences in a VARK style questionnaire. The results suggest that the quasi-ipsative chi-squared based approach does not appear to be as sensitive as the original t-test approach in identifying significant modalities. In order to increase the sensitivity of the test, the requirement for overall test significance has to be relaxed and individual cells considered in the analysis. This potentially limits the validity of the proposed approach and as a cell could have a larger than expected number of observations purely by chance. The findings also put some doubt on the statistical validity of the original t-test approach as it also uses the deviation from the means (in the form of standard deviation) rather than statistical significance, as a tool for assessing the strength of individual preferences. The causes of the discrepancies between the two scorings techniques need to be examined further, before recommendations can be made on which approach is better suited to identifying the strength of individual preference for a information input modality.

References

[1] N. D. Fleming and C. Mills, "Not another inventory, rather a catalyst for reflection," 1992.

[2] K. R. Scherer, "What are emotions? And how can they be measured?," *Soc Sci Inform*, vol. 44, no. 4, pp. 695–729, Dec. 2005.

[3] W. L. Leite, M. Svinicki, and Y. Shi, "Attempted Validation of the Scores of the VARK: Learning Styles Inventory With Multitrait-Multimethod Confirmatory Factor Analysis Models," *Educ Psychol Meas*, vol. 70, no. 2, pp. 323–339, Mar. 2010.

[4] J. A. Krosnick, "Response strategies for coping with the cognitive demands of attitude measures in surveys," *Appl Cognitive Psych*, vol. 5, no. 3, pp. 213–236, May 1991.

[5] J. M. Batista-Foguet, B. Ferrer-Rosell, R. Serlavós, G. Coenders, and R. E. Boyatzis, "An Alternative Approach to Analyze Ipsative Data. Revisiting Experiential Learning Theory," *Front. Psychol.*, vol. 6, pp. 575–10, Nov. 2015.

[6] A. Brown and A. Maydeu-Olivares, "Item Response Modeling of Forced-Choice Questionnaires," *Educ Psychol Meas*, vol. 71, no. 3, pp. 460–502, May 2011.

[7] J. F. Salgado, N. Anderson, and G. Tauriz, "The validity of ipsative and quasi-ipsative forced-choice personality inventories for different occupational groups: A comprehensive meta-analysis," *J Occup Organ Psychol*, vol. 88, no. 4, pp. 797–834, Dec. 2014.

[8] N. D. Fleming, "I'm different; not dumb Modes of presentation (V.A.R.K.) in the tertiary classroom," presented at the Research and Development in Higher Education, Proceedings of the 1995 Annual Conference of the Higher Education and Research Development Society of Australasia, 1995, vol. 18, pp. 308–313.

Journal of Physics: Conference Series **772** (2016) 012014 doi:10.1088/1742-6596/772/1/012014

Multi-scale biomedical systems: measurement challenges

R Summers
Professor of Information Engineering
Loughborough University, Loughborough, UK, LE11 3TU.

E-mail: r.summers@lboro.ac.uk

Abstract. Multi-scale biomedical systems are those that represent interactions in materials, sensors, and systems from a holistic perspective. It is possible to view such multi-scale activity using measurement of spatial scale or time scale, though in this paper only the former is considered. The biomedical application paradigm comprises interactions that range from quantum biological phenomena at scales of 10-12 for one individual to epidemiological studies of disease spread in populations that in a pandemic lead to measurement at a scale of 10+7. It is clear that there are measurement challenges at either end of this spatial scale, but those challenges that relate to the use of new technologies that deal with big data and health service delivery at the point of care are also considered. The measurement challenges lead to the use, in many cases, of model-based measurement and the adoption of virtual engineering. It is these measurement challenges that will be uncovered in this paper.

Introduction

The aim of this paper is to uncover the measurement challenges that arise from a systemic appreciation of a disease process that affects both an individual and the population within which that individual belongs. By drawing a systems boundary that encompasses both individuals and populations is possible via a multi-scale approach to physiology and epidemiology. This approach can be defined as the consideration of physiological systems that possess measurable characteristics that can be exhibited at multiple spatial or time scales simultaneously. For this paper only multiple spatial scales are considered; and to demonstrate the variety of measurement issues and their ensuing complexity a snap-shot is taken at either end of the spatial scale under consideration (quantum level phenomena within an individual to epidemiological studies of populations) as well as an application-based intervention at the limit of the whole body level (telemedicine). Quantum biology can be defined as the purposeful use of quantum mechanics applied to the system of interest in order that some sort of biological advantage can be gained. Telemedicine at its most simple is the delivery of care at a distance using a digital network to connect and integrate component parts. Finally, epidemiology refers to measurement of some defined individual outcome within a specified population. Measurement challenges are likely to be different across the spatial scale, it is the purpose of this paper to uncover the nature of these differences as well as any similarities.

Multi-scale biomedical systems

The traditional view of biology, physiology and epidemiology is to gain an understanding of a specified process via a reductionist approach. This systematic mode of thought and analysis is common place in engineering and science. However, there is evidence accumulating that a more systemic appreciation may bring dividends when trying to understand problems that surpass one level of scale simultaneously. Perhaps the current best example of achievement in this field is the physiome project which collects and curates multi-scale physiological models from research groups throughout the world. An early description of this project demonstrates the model-based philosophy that still prevails [1]. The vast majority of the current work provides details of models that apply to structure and function of specified systems in the adult. To extend the application base as well as to explore measurement at the boundaries of current knowledge and understanding, a multi-disciplinary international team was brought together to investigate the utility of multi-scale models for cardiac morphogenesis, and more specifically epithelial to mesenchymal transition (EMT) of cells that form the internal structure of the heart [2, 3]. EMT will continue as a system of interest, as it has been shown that it also plays a role in wound healing and tumour growth.

Quantum biology

It is often quoted that should the 20th Century be labelled the 'physicists' century, then the 21st Century has potential to be the 'biologists' century. Perhaps no one topic fits that bill other than quantum biology – a discipline that is looking for a purpose. There are many missing pieces of the jigsaw that would allow a complete picture to be elucidated, held back by the lack of a formalism to describe the state space of biological systems and their dynamics in quantum theoretical terms. Thus the discipline is immature mainly due to the difficulty of controlling, measuring and modelling quantum level systems. However, measurement challenges can still be discerned and relate to: adequacy of language used to describe interventions, the exact system information needed, the lack of experimental design, as well as the completeness (or lack of it) of the theory that underpins the measurement. There still remains a gap between the disciplines of quantum physics and the physical principles of quantum biology but exploring this gap has potential to reap rewards in terms of yielding new and interesting results. In roads have already been defined through the study of quantum biology in photosynthesis [4], magneto-reception [5], olfaction [6], vision [7], and early reviews of the discipline [e.g. 8].

Telemedicine

The increasing burden of disease caused by the ageing population coupled with ever-increasing costs of hospital-based healthcare have created a perfect storm out of which has emerged the need for telemedicine-based interventions that enable swifter hospital discharge and delivery of care in the community. To a large extent telemedicine relies on patient engagement with self-management and though the measurement challenges are not limited to patient-technology interactions, there are still some that can be alleviated by better education and training opportunities. Some of the issues relate to: time of measurement, missed measurement, proxy measurement, incorrectly inputed data. As an example in the last category, consider the routine measurement of blood pressure, say by an automatic sphygmomanometer. The cuff may be incorrectly placed anatomically; the cuff may be placed on top of a shirt/blouse; the sphygmomanometer itself may need calibrating, or the traceability chain may be broken. The full paper will draw upon experiences from EU Doc@ Home

Epidemiological studies

The major measures of disease in a population are related to prevalence and incidence, where the former measures existing cases of disease (or other outcome) and the latter measures new cases of disease within a specified population. In fact there are two measures of both in common practice. Point prevalence is the proportion of individuals in a defined population that possess the health outcome of interest at a specified point in time; whereas period prevalence is similar though measured over a specified time period. As an example of point prevalence consider a UK General Practitioners Surgery list of 20,000 outpatients on 1st August 2016, should 400 of them have diabetes the point prevalence of diabetes within the given population is 400/20,000 which is 0.02 or 2%.So although prevalence in this simple example provides an example of disease burden at the given GP surgery, it assumes that diagnostic accuracy is 100% with no Type I or Type II errors. The two main measures of incidence are termed incidence risk and incidence rate. Incidence risk is the more straightforward of the two to adopt and is the proportion of individuals in a population that were initially free from disease but who have been contaminated within a specified time period. This measure assumes that there are no naturally immune individuals in the given population and, of course, individuals may die within the study period, move away or indeed move in to the area of interest. It is incidence rate that is perhaps the more purposeful of the two measures and tries to take into account some of the measurement issues associated with incidence risk by using the sum of the time that each individual has remained at risk of developing the disease (or other outcome) of interest. To adopt incidence rate there is thus the need to calculate a factor 'person-time at risk' which itself has several sources of measurement error. This factor is defined as the sum of each individual's time at risk over a given time period (usually months or years). Should the individual develop the disease/outcome, die, migrate, (s)he is no longer at risk and therefore cease to contribute to the factor. Clearly tracking of each individual in the population is extremely important to obtain a true measure of incidence rate. Poor study management can cause an underestimate of the true incidence rate. Prevalence and incidence are related by the following equation:

$$P = ID$$

Where P is Prevalence; I is Incidence rate; and D is the average duration of the disease.

In most epidemiological studies it is rare to use the complete population of a geographical area as the given population, instead a population sample is used. The population sample should be a representative of the whole, but inter-comparisons between outcomes measured from different samples from the same population have shown that this is not the case. Differences occur due chance variations in each of the samples, but the crux of the issue is that it is not known whether the variation causes an overestimation or an underestimation of the true value. Statistical tests are available that measure this chance effect and are related to 'standard error' measures. When comparing results within a population sample, the standard error can cause Type I or Type II errors: rejecting the null hypothesis when it is true (true negatives) and accepting the null hypothesis when it is false (false positive).. Study design is aimed at reducing the sampling error which can be accomplished by allowing the population sample to be as large as possible. There are also measurement errors associated with the research methods used within the study design: lack of or incomplete data, bias, confounding variables, temporal sequence between having a disease or not may be fuzzified (e.g. by having a susceptible population), and so on.

Conclusions

This paper has considered measurement issues and challenges across the spatial scale spectrum and it is clear from the snap-shot approach used that most challenges are unique to a particular level of analysis. However there are also similarities – for each there is a gap between conceptualisation and realisation. Another reason for choosing the three areas is the difference in maturity of the measurement methods associated with systemic intervention. Epidemiology has been with us the longest so it is not unusual to see the bank of well-validated statistical methods associated with reducing the standard error. If leading edge telemedicine applications are ignored, it has been available since the 1990s. It can be argued that further methods need to be adopted before acceptance by the general public. Should we ignore Schrödinger [9] and the like who worked individually at different points in the twentieth century, then quantum biology is the newest sub-discipline. There remains much to do to gain acceptance within the scientific community before exposure to the general population. It is quantum biology where the leading edge now lies, perhaps waiting for novel visualisation methods to allow accelerated development of its purposeful use in biology and medicine.

References

[1] Hunter P J, Robbins P and Noble D 2002 The IUPS human physiome project *Eur. J. Physiol.* **445** (1) 1-9.

[2] Abdulla T, Luna-Zurita L, de la Pompa J L, Schleich J M and Summers R 2013 Epithelial to mesenchymal transition – the roles of cell morphology, labile adhesion and junctional coupling *Comput. Meth. Prog. Biomed.* **111** (2) 435-46

[3] Schleich J M, Abdulla T, Summers R and Houyel L 2013 An overview of cardiac morphogenesis *Arch. Cardiovasc. Dis.* **106** (11) 612-23

[4] Sarovar M, Ishizaki A, Fleming G R and Whaley K B 2010 Quantum entanglement in photosynthetic light-harvesting complexes *Nature Phys* **6** 462-7

[5] Ritz T 2011 Quantum effects in biology: bird navigation *Proc Chem* **3** 262-75

[6] Brookes J C, Hartoutsiou F, Horsfield A P and Stoneham A M 2007 Could humans recognize odor by photon-assisted tunnelling? *Phys. Rev. Lett.* **98** 038101

[7] Polli D et al. 2010 Conical intersection dynamics of the primary photoisomerization event in vision *Nature* **467** 440-3

[8] Lambert N, Chen Y N, Cheng Y C, Li C M, Chen G Y and Nori F 2013 Quantum biology *Nature Phys* **9** 10-18

[9] Schrödinger E 1944 *What is life?* (Cambridge: Cambridge University Press)

Development of radar-based system for monitoring of frail home-dwelling persons: A healthcare perspective

Tobba T. Sudmann[1,2], Ingebjørg T. Børsheim[1,3], Knut Øvsthus[1,4], Tomasz Ciamulski[1,4], Andrzej Miękina[6], Jakub Wagner[6], Paweł Mazurek[6], Frode F. Jacobsen[1,5]

[1] Centre for Care Research – Western Norway, Bergen University College, P.O. Box 7030, N-5020 Bergen, Norway

[2] Department of Social Science and Social Education, Bergen University College, P.O. Box 7030, N-5020 Bergen, Norway

[3] Department of Occupational Therapy, Physical Therapy and Radiography, Bergen University College, P.O. Box 7030, N-5020 Bergen, Norway

[4] Department of electrical engineering, Bergen University College, P.O. Box 7030, N-5020 Bergen, Norway

[5] VID Specialized University, P.O. Box 184 Vindern, N-0319 Oslo, Norway

[6] Warsaw University of Technology, Faculty of Electronics and Information Technology, Institute of Radioelectronics and Multimedia Technology, Nowowiejska 15/19, 00-665 Warsaw, Poland

Abstract. This interdisciplinary project aims to develop and assess the functional potential of radar technology in the care services. The project mainly has an exploratory character where the technological and functional potential of impulse-radar sensor are tested out in monitoring of elderly and disabled people living in their own home. Designing a non-invasive system for monitoring of movements of frail persons living at home is the main goal, with the intent of assessing health and functional status through monitoring of activities of daily life (ADL) and detecting potentially dangerous situations, not the least related to a long lie following falls.

1. Introduction: Development of novel radar technology in home care services

The task of this Polish-Norwegian project is the development of novel technology for care services; specifically – designing a non-invasive and non-intrusive radar-based system for monitoring of movements of frail persons living at home is the main goal. Monitoring of activities of daily life (ADL) and gait characteristics provides vital information regarding health and functional status; such information may provide early warning signs with regard to potential dangerous falls.

We are presently viewing an increased need for complex care services due to the longevity of people, and a rising number of elderly who will suffer from several chronic conditions, disability or frailty. Today we observe a marked tendency towards compression of morbidity, *i.e.* more people live longer and an increasing number of them are disabled or frail only a short time at the end of their lives (from a few weeks to a few years) [1]. Falls among elderly people are one of the main causes of their hospitalization [2-10]. The number of deaths globally caused by fall events was around 391,000 in 2003, where approximately 40% of the falls were from people over 70 years of age [11,12]. A recent WHO publication [13] suggests that future health and care needs should be met by strengthening health promotion and community-based services, and by acknowledging family and informal home-based care. These ends can be reached by, *e.g.*, facilitating innovation, including employment of ambient technology. The present project deals with development of the telecare part of ambient technology. We aim at developing a system, which is less intrusive than existing vision-based or body-worn devices. Our paper will convey some preliminary experiences regarding opportunities and challenges in this interdisciplinary project.

This paper reports on experiences, with regard to designing a non-intrusive and non-invasive system for fall detection, and preliminary findings in the project. The lessons learned from the interdisciplinary teamwork are likely to be applicable to several technology development projects.

2. Methods

The design of radar modules employed is based on impulse-radar sensors NVA series 6000 (manufactured by Novelda AS, https://www.novelda.no/content/radar-ics) being fully integrated nanoscale radar transceivers, designed for low-power applications. In order to obtain the desired parameters of the developed radar modules, two separate antennas have been designed taking into account the limitation concerning the radiation power in the target frequency band (6.5–8GHz) and characteristics of the receiver and of the transmitter. The radar sensors are connected with an amplifier board and the antennas by means of low-loss semi-rigid coaxial cables and high-grade SMA connectors. The single-board computer is connected to a compact PCB that interconnects its USB ports and routes them to the sensor's USB-to-SPI converter. All the electronic elements of the Radar Module are contained within an aluminum box whose dimensions are 120 mm × 120 mm × 57 mm [14]. The observational angle of each radar is 90 degrees. The radars are capable of measuring the distance in the range from 0.1 to 6 meters up to 15 times per second. One of the radar modules is shown in figure 1.

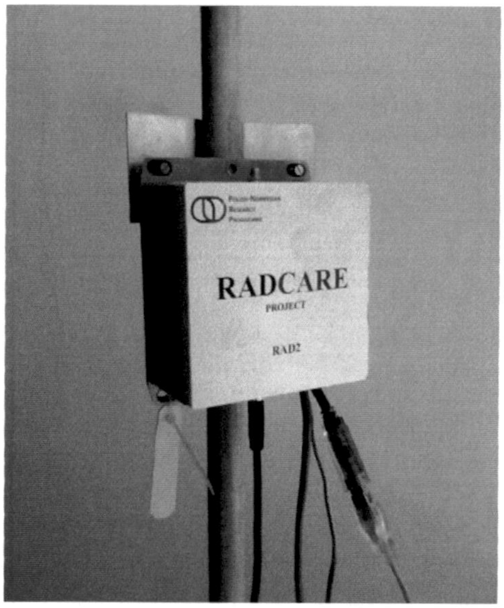

Figure 1. An exemplary RADCARE radar module.

In the first phase of experiments, a set of fall scenarios was designed in collaboration with two physiotherapists and one occupational therapist in Bergen, Norway. Because the RADCARE radar

Journal of Physics: Conference Series **772** (2016) 012015 doi:10.1088/1742-6596/772/1/012015

sensors were not yet available, the falls were performed and recorded using APDM Opal sensors [15], synchronized and controlled using a wireless communication system. These sensors provided us with three-dimensional accelerometric data, as well as additional data from gyroscopes, magnetometers and temperature sensors. The sensors were attached to the body at different levels: hip, shoulders and head. Fall experiments were performed in cooperation with Warsaw University of Technology (WUT) which provided technical requirements. The data acquired in this set of experiments were sent from Bergen University College (BUC) to WUT and used for testing signal processing algorithms before first data from RADCARE radar sensors would be available.

The second phase of fall-related experiments was also set up in collaboration with the above mentioned two physiotherapists and one occupational therapist, this time in the movement laboratory at BUC, using Qualisys movement capture system [16] based on a set of very sensitive infrared cameras which can observe markers reflecting the infrared radiation. The test person was marked with 20 markers in frontal, sagittal and transversal planes (bilateral on ankles, knees, hips, shoulders, elbows, wrists, forehead, ears, back head, and the back). We had 7 motion capture cameras, and an area of ca. 3 m × 3 m covered with multiple layers of thick Airex mats. We explored falls based on typical ADL (Activities of Daily Life) situations and known fall risk situations, like stumbling (*e.g.* related to doorsills and carpets) and sitting to standing movements (*e.g.* from bed or toilet). This broad focus covers falls due to intrinsic (*e.g.* low blood pressure), extrinsic (*e.g.* walking sticks) and environmental causes (*e.g.* insufficient light).

We also assessed the possibility of using the RADCARE technology for the indication of the physical activity of the monitored person. The third phase of experiments involved the use of a system for estimating the two-dimensional position of monitored persons, based on two RADCARE radar modules developed at WUT. Some experiments were performed in laboratory conditions in order to estimate the achievable measurement uncertainty. In these experiments, marks – distant from each other by 1 m – have been placed on the floor at 21 points in the observed area, which enabled us to walk along predefined trajectories. Walking with known speed has been assured by making half-meter steps (according to the marks on the floor) in equal time intervals, signalled by a metronome. The measurement setup used in these experiments is presented in figure 2.

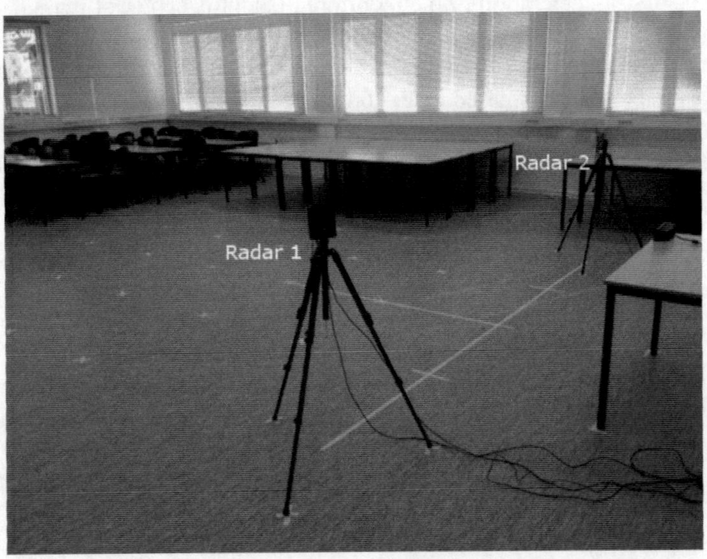

Figure 2. The measurement setup in the laboratory experiments involving a radar-based system for measuring the two-dimensional position of monitored person.

In other experiments, data representative of 10 different movement scenarios were acquired by means of the radar-based system, *viz.* sit to stand, stand to sit, walking, turning, lying down, rising from floor, as well as sitting activities as drinking from a cup, doing ones hair and tying/untying shoe laces.

There are also ongoing experiments which involve the monitoring of an elderly person in her household.

3. Results and discussion

Since the character of the project was mainly exploratory, the capabilities of radar technology, regarding the monitoring of elderly and disabled persons, were tested out, taking into account the present and future needs related to assistive care. It turned out that such objective requires an interdisciplinary approach in various aspects. The relevance and reliability of the developed system depends as much on health sciences insight and healthcare personnel experiences as on technology knowledge and experience. The input from the health sciences proved to be important for identification of significant movements and vital signs in daily life activities, and for developing promising areas for observation and measurements. The most promising results have derived from mapping the capability and feasibility of the RADCARE technology to infer information that acts as indicators of general health status and wellbeing. Measurements of daily life movements in the test home have proven to have clinical relevance, and to provide telecare support and assistance for living well and longer at home.

More specifically, it was possible to detect presence and length of stay at selected places at specific points of time (also behind non-metal walls), and distance covered in monitored space. It was also possible to measure the amount of time spent by a person without any motion, which could indicate the occurrence of a dangerous incidence, such as a fall. Such data can be used for activating alarms if motion is not detected in a selected interval of time, *e.g.* one hour during daytime.

It was also proved that the monitoring system based on two radar sensors can provide estimates of several quantities useful for the healthcare staff, such as:

- average gait speed,
- movement direction,
- the travelled distance,
- acceleration and de-acceleration of the monitored person's center of mass.

Although the movement direction is not an informative quantity in itself, it provides data that can be used to infer vital information about the patient's ability to plan, to remember, to find and replace utensils, clothes or props, or indicate a tendency to deliria at specific times. Information on meals and medication and sleeping habits can be used as a backdrop when interpreting movement data.

The estimates of the travelled distance can be summarised for a selected period, such as the last eight hours or the last week. The visualisation of the distance travelled each day during a longer period may help the medical or healthcare staff to detect a long-term trend indicating an improvement or deterioration of health.

Acceleration and de-acceleration related to activities like starting to walk, stopping, and changing the direction, provide valuable information about the monitored person's balance, postural and motor control.

It proved possible to map movement in different positions (lying, sitting, upright position) and quality of gait (rhythm, sway), all of which are vital information for healthcare personnel. We were, however, not able to identify position and movements of arms and legs.

With regard to falls, the project revealed that the most important issue does not appear to be the trajectory of a fall and hence fall detection, but to differentiate between injurious and non-injurious factual falls. We are presently exploring to which extent the RADCARE technology can contribute to making this most important distinction.

The experiments challenged the key question of what is a fall, how do people fall, what happens during fall, and what precedes a fall. The technology also posed some challenges when it came to production of artefacts and markers falling off the test person. Our focus in the project has hence increasingly been directed towards identification of important components of a fall:

1. speed of movement in a horizontal and vertical planes,
2. positive and negative acceleration,
3. the long lie, *i.e.* the time spent motionless from the moment of impact until the time the person is helped out; the long lie dramatically increases morbidity and mortality.

RADCARE technology has not so far been capable of delivering data that can be used in fine-tuned movement analysis, or data that are refined enough to calculate the relationship between base of support and center of mass, or detecting the amount for compensating movements during everyday living – those data could enable the healthcare personnel to recognize person's reduced balance and increased risk of falling. Nevertheless, the information on gait, provided by the developed system, *e.g.* its speed, cadence, acceleration, and variability, provides the healthcare researchers with valuable information for predicting the functional state and health of the person and hence – the risk of falling [17-20].

4. Conclusions

Interdisciplinary collaboration, where insight from health and social scientists and practitioners is as important as technical knowledge, was proved to be crucial in the RADCARE project – aimed at the development of novel technology for the healthcare sector. Based on the experience we gained when designing a system for detecting and indentifying body movements and ADL of a home-dwelling elderly who may suffer from several chronic conditions, disability or frailty, we were able to define challenges, opportunities and important topics. The wider context of falls in homes needs to be taken into account, with regard to intrinsic, extrinsic and environmental factors.

Our experiments provided valuable data related to human movements, vital for assessing the general health and functional status of a person and, therefore, the risk of falling. The RADCARE technology also seems promising for detecting the most dangerous types of situations, *i.e.* falls followed by a long lie. More effort will be put into this work, and further experiments will be carried out to enable the detection of the potentially very harmful long lie. From our point of view, it is feasible to use a radar-based monitoring system – alone or in combination with other non-invasive monitoring techniques – for preventing falls.

Acknowledgments

This work has been supported by EEA Grants – Norway Grants financing the project PL12-0001 (http://eeagrants.org/project-portal/project/PL12-0001). The involved institutions are the Warsaw Technological University (WUT), with Prof. Wieslaw Wieniecki as principal investigator, and, The Center of Care Research – Western Norway at Bergen University College, with Prof. Knut Øvsthus as principal investigator, Associate Prof. Dr. Tobba Sudmann, Assistant Prof. Ingebjørg Børsheim and Dr. Tomasz Ciamulski as co-investigators, and Prof. Frode F. Jacobsen as project leader.

References

[1] Andersen S L, Sebastiani P, Dworkis D A, Feldman L and Perls T T 2012 *The Journals of Gerontology Series A: Biological Sciences and Medical Sciences* **67** 395-405
[2] Bergland A, Jarnlo G B and Laake K 2003 *Aging Clinical and Experimental Research* **15** 43-50.
[3] Clemson L. Kendig H, Mackenzie L and Browning C 2014 *Journal of Aging and Health* **27** 239-256.
[4] Lord R 2007 *Falls in Older People: Risk Factors and Strategies for Prevention* 2007 (Cambridge: Cambridge University Press)
[5] Rubenstein Z 2006 *Age and Ageing* **35** ii37-ii41.
[6] Scheffer A C, Schuurmans M J, Van Dijk N, Van der Hooft T and De Rooij S E 2008 *Age and Ageing* **37** 19-24.
[7] Stevens J A, Mahoney J E and Ehrenreich H 2014 *Inj Epidemiol* **1** doi: 10.1186/2197-1714-1-5.

[8] Tideiksaar R 2010 *Falls in Older People: Prevention and Management* (Baltimore: Health Professions Press)
[9] Tinetti M E, Gordon C, Sogolow E, Lapin P and Bradley E H 2006 *The Gerontologist* **46** 717-725.
[10] Rashya R and Sindhuja M 2015 *International Journal of Research in Engineering and Science (IJRES)* **3** 50-57.
[11] World Health Organization, *WHO Global Report on Falls Prevention in Older Age* 2008 (Geneva: World Health Organization)
[12] Abbate S, Avvenuti M, Corsini P, Vecchio A and Light J 2010 *Wireless Sensor Networks: Application-Centric Design* Eds. Merret G V and Y. K. Tan Y K (Rijeka, Croatia: Tech) 147-166.
[13] World Health Organization (WHO) 2015 *World report on ageing and health.* (Luxembourg: WHO)
[14] Morawski R Z, Yashchyshyn Y, Piórek M, Jacobsen F F, Øvsthus K and Winiecki W 2015 *Przegląd Telekomunikacyjny & Wiadomości Telekomunikacyjne* **18** 598-602
[15] *OPAL Wearable Sensors*, 29 June 2015, [Online], Available at http://apdm.com/Wearable-Sensors/Opal
[16] *Qualisys Motion Capture* [Online] Accessed 29 June 2015, Available at http://www.qualisys.com/
[17] Abellan Van Kan G, Rolland Y, Andrieu S, Bauer J, Beauchet O and Bonnefoy M 2009 *The Journal of Nutrition, Health & Aging* **13** 881-889
[18] Brodie M, Lord S, Coppens M, Annegarn J, and Delbaere K 2015 *IEEE Transactions on Biomedical Engineering* **62** 2588-94
[19] Hausdorff J M and Buchman A S 2013 *The Journals of Gerontology Series A: Biological Sciences and Medical Sciences* **68** 409-411.
[20] Ijmker T and Lamoth C J 2012 *Gait & posture* **35** 126-130

Quality by Design approach in the development of a magnetic transducer for biomedical measurements: preliminary results on Design Space configuration

D R Louzada[1], E C Monteiro[1], A M Rivero[1], L G S Fortaleza[1], C R H Barbosa[1], D R T Velázquez[1], E C Silva[2], L A P Gusmão[2]

[1] Postgraduate Programme in Metrology, Pontifical Catholic University of Rio de Janeiro, PUC-Rio, Brazil.

[2] Department of Electrical Engineering, Pontifical Catholic University of Rio de Janeiro, PUC-Rio, Brazil.

daniel_louzada@esp.puc-rio.br

Abstract. The concept of Quality by Design (QbD) has been widely used by the pharmaceutical industry since 2004, changing the focus on inspections to embrace greater control in the manufacturing process. This new approach contributed to lower production costs and higher quality of medicines. This work discusses the first steps towards the implementation of a QbD approach aimed at ensuring the quality of a biomedical device under development. This device will be applied as a non-magnetic metallic foreign body localization system for surgical removal guidance. It should be highlighted that a new approach was used to define the Design Space, based on fuzzy logic rather than the typical statistical techniques.

1. Introduction

Removal of foreign bodies is a common challenge in the medical practice. The widely available localization procedures are radiography, computed tomography and radioscopy [1, 2]. However, such techniques do not give sufficient information and are inefficient to localize small objects. In addition, they can cause excessive exposure of the patient and surgical staff to ionizing radiation [1]. An even greater challenge is posed by foreign bodies consisting of firearms projectiles, since they are usually made of materials that do not possess a remnant magnetic field [3]. In this case, the induction of eddy currents in the metallic object by a primary alternating magnetic field source generates a secondary magnetic field that can be measured by high sensitivity magnetic sensors [3-5].

The Laboratory of Biometrology (LaBioMet) at PUC-Rio conducts research in the area of non-invasive clinical diagnosis [1-3, 5-9], and is currently developing a biomedical device aimed at locating non-magnetic metallic foreign bodies and guiding their surgical removal. This device is based on a high sensitivity magnetic transducer, since the secondary magnetic fields are extremely weak [3, 5]. The device design is being guided by some major aspects relevant for biomedical technology developments, such as: high sensitivity, low cost of fabrication and operation, low complexity, safety, portability, among others [10, 11].

Quality by Design (QbD) is a tool that has been successfully and widely used by the pharmaceutical and biotechnology industry since 2004, when the Food and Drug Administration (FDA) recommended a new method for quality assurance [12]. According to the International

Conference on Harmonisation of Technical Requirements for Registration of Pharmaceuticals for Human Use (ICH) [13], the QbD implementation requires identification of a set of information associated with the development of the product, including:

- Target Product Profile (TPP) - characteristics that guarantee the quality of a product and ensure its safety and efficiency. It constitutes a set of attributes that indicates the purpose of the product, on which are established development strategies. The pharmaceutical industry associates TPP with aspects such as pharmacology, indications and method of use; and relating to chemical properties, as solubility and stability.
- Critical Quality Attributes (CQA) – physical, chemical or biological properties, which must operate within a certain range or distribution so as to ensure a pre-defined product quality.
- Critical Process Parameters (CPP) - process parameters whose variability has an impact on any critical quality attribute and, therefore, should be monitored or controlled to ensure the desired quality.
- Design Space - multidimensional space assigning a combination of the material's attributes and/or parameters of the process to ensure quality. It is the best set of parameters, determined statistically, which enables the accommodation of any change in the CPPs.
- Control Strategy - a planned set of controls, derived from current product and process understanding that ensure process performance and product quality.

The QbD concept improves the scientific understanding of the entire production chain of a given product, and consequently brings light to the critical attributes for the quality of the product [14-20]. In that way, it enables the development of a set of control strategies based on articulation bands and operating limits [19]. Thus, the emphasis is no longer in control inspections because, as mentioned in the ICH Q8 guidance, "quality cannot be tested into products; it should be built-in or should be by design" [12, 13]. As a consequence, lower production costs and increasing quality are among the expected benefits. Recent studies are pointing out new solutions as alternative to statistical techniques conventionally used in the QbD concept by means of computing methods for pharmaceutical industry [19].

Steps toward another line of application of QbD have been addressed by recent works, that discuss the possibility of adapting the QbD approach in order to apply it in the development of biomedical technologies [11, 13, 20, 21].

In this work, QbD is adapted in order to be applied in the development of a health technology for non-magnetic metallic foreign bodies localization, by means of a high sensitivity and low cost magnetic sensor, based on giant magnetoimpedance (GMI). Aiming at implementing this quality tool for the biomedical device under development, their specific parameters of TPP, CQA and CPP were identified. Moreover, it should be noticed that the present work proposes a new approach to configure the Design Space, based on a fuzzy inference system (FIS) framework, rather than the conventional statistical approaches.

2. Methods

Aiming at meeting the metrological priorities, considering the biomedical application of the measuring device under development [10, 11], the Target Product Profile (TPP) of QbD, associated to the functionalities of the technology, was defined. Based on the TPP and on risk analysis techniques, the Critical Quality Attributes (CQA) were specified, consisting of the technology characteristics which need to be controlled during the assembly process, in order to guarantee its quality. Likewise, the Critical Process Parameters (CPP) were outlined, consisting of the parameters that have significant influence on the critical quality attributes of the device. These three earliest stages of QbD were determined with the support of a brainstorming activity performed by a multidisciplinary team.

In order to outline the next step, which is the definition of the Design Space, the input ranges of the CPPs were defined. These ranges were established aiming at ensure suitable operational values for all CQAs.

On the contrary of the huge databases available in the pharmaceutical industry, there is a lack of such databases for biomedical technologies, precluding the use of the statistical methods conventionally employed in that sector. Consequently, the adaptation of the conventional QbD method was necessary. Therefore, the Design Space configuration was formulated by a new approach based on a fuzzy inference system, enabling the analysis of complex systems without the need for sophisticated statistical models. Such inference is based on the interaction of linguistic variables and membership functions, which may be defined by experts brainstorm.

Aiming at exemplifying the process for Design Space configuration, the developed algorithm was applied to characterize the relationship of a set of two CPPs and two CQAs. The estimated optimal operating ranges for the two selected CPPs (amplitude of the AC electric current passing through the solenoid that generates the primary magnetic flux density, CPP_{B2}; and its frequency, CPP_{A2}) with respect to two CQAs (transducer sensitivity, CQA_3; and secondary magnetic flux density, CQA_7). The linguistic variables adopted represent the level of suitability of each process parameter value (Unsuitable, Acceptable and Ideal) to the adequacy of each defined CQA to the desired quality of the device's TPPs. Furthermore, fuzzy membership functions were employed to represent how much each operating value refers to the linguistic variables. The fuzzy inference for the two CQAs generated two maps of operating ranges. In this way, the Design Space can be obtained by intersecting the best performance area for the two CQA maps.

3. Results

The Target Product Profile (TPP) was defined according to the health technology features that meet the associated surgical demands of foreign body detection and localization. The next step consisted in determining the Critical Quality Attributes (CQA) corresponding to the critical technology features that need to be evaluated throughout the device development and manufacturing process. With the CQA data, risk analysis, as well as brainstorming, the Critical Process Parameters (CPP) related with each CQA were determined. Table 1 presents the resulting features associated with these three early stages of QbD (TPP, CQA and CPP) applied to the developing biomedical system under development.

Based on TPP, CQA and CPP indicators (Table 1), a risk analysis between each TPP and CQA, as well as between each CQA and CPP, indicated the degree of relevance between each other (classified as High, Medium and Low).

As stated in the previous section, the Design Space configuration was adapted for biomedical technology applications by incorporating a fuzzy system. For assessing the device performance according to Design Space, it is proposed the adoption, among the QbD requirements, of metrological traceability of the measurement results of the defined CPP parameters, including its adequacy to a specified target measurement uncertainty.

In order to illustrate this adaptation, two critical attributes associated to the technology, the transducer sensitivity (CQA_3) and the secondary magnetic flux density (CQA_7), were studied in relation to two critical parameters, the frequency of the primary magnetic flux density (CPP_{A2}) and the current intensity (CPP_{B2}) passing through the solenoid that generates this primary magnetic source. The definition of CPP ranges of values for process parameters were obtained by means of theoretical models, simulations or experimental results. For instance, the relation of the secondary magnetic flux density (CQA_7) with the frequency of the primary magnetic field (CPP_{A2}) can be established in such way that the magnetic flux density increases with the frequency until about 100 kHz, when it saturates around a maximum value [2]. Thus, the range between 0 and 100 Hz presents lower performance for CQA_7, while the acceptable range between 100 Hz and 10 kHz evolves into a better classification, and frequencies above 10 kHz can be considered ideal, being predominantly of higher performance, as shown in figure 1b.

The optimal ranges for the values of frequency of the primary magnetic flux density (CPP_{A2}) and the current intensity passing through the solenoid (CPP_{B2}), considering the two critical attributes of "transducer sensitivity" and "secondary magnetic flux density", CQA_3 and CQA_7 respectively, are represented in figures 1a and 1b. Although for CQA_7 a wide area of the surface plot corresponds to ideal performance, for CQA_3, higher performance is obtained only within a limited range of CPP values.

Table 1. TPP, CQA and CPP indicators for the system under development, for non-magnetic metallic foreign bodies localization in humans.

Code	Target Product Profile (TPP)	Code	Critical Quality Attributes (CQA)	Code	Critical Process Parameters (CPP)
A	High sensibility magnetic transducer	1	Spatial Resolution	A2	Frequency of the primary field
B	Location of foreign body (non-magnetic)			B2	Current intensity
		2	Operation at ambient temperature	C2	Current intensity (circuit)
				D2	Circuit's supply voltage
C	High accuracy	3	Transducer sensitivity	E2	Solenoid dimensions
D	Low measurement uncertainty			F2	GMI supply voltage
		4	Gradiometric configuration	G2	Oscillator supply voltage
E	Foreign body location at different distances	5	Signal to noise ratio	H2	Supply Voltage (Conditioner)
F	Environmental sustainability			I2	AC in GMI sensor
G	Safety	6	Primary magnetic flux	J2	AC frequency of the current in the sensor
H	Low complexity manufacturing	7	Secondary magnetic flux	L2	Sensor size
I	Low manufacturing cost			M2	Homogeneity of the sensor
J	Harmlessness	8	Resolution time	N2	Voltage Stability
				O2	Signal amplification
L	Low operating cost	9	Circuit characteristics (size)	P2	Relative position between the sensors and solenoid
M	Reduced complexity of operation			Q2	Circuit noise
N	Portability	10	Circuit characteristics (components)	R2	Filter
				S2	Processing Software
O	Low maintenance complexity (calibration included)	11	Sample characteristics sensor	T2	Acquisition software
				U2	Frequency of the primary field
P	Low cost of maintenance (calibration included)	12	Signal processing	V2	Current intensity

The Design Space obtained by combining both CQA requirements, as shown in figures 2a and 2b, indicates that the highest performance is obtained for frequencies between 7 and 9 kHz for the primary magnetic field, with an AC current passing through the solenoid between 0.6 and 1.8 mA.

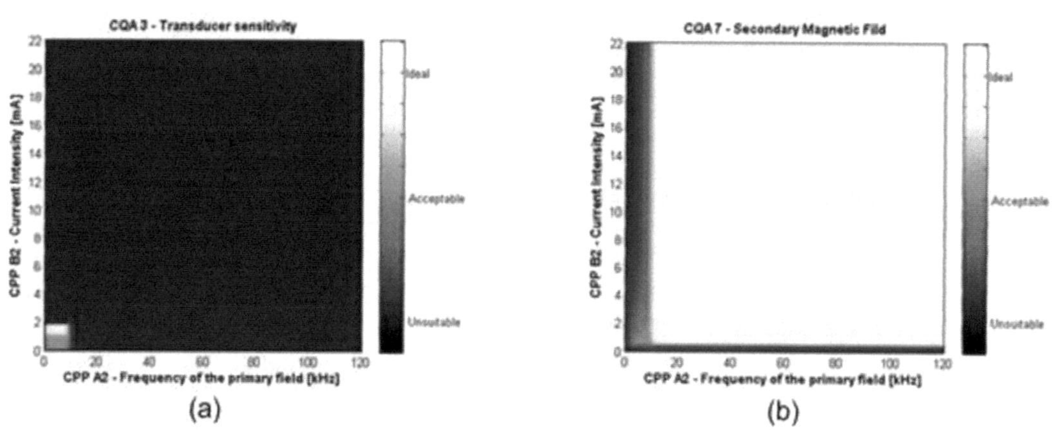

Figure 1. Surface plot of the FIS of CPP_{A2} and CPP_{B2} in respect of (a) CQA_3 and (b) CQA_7.

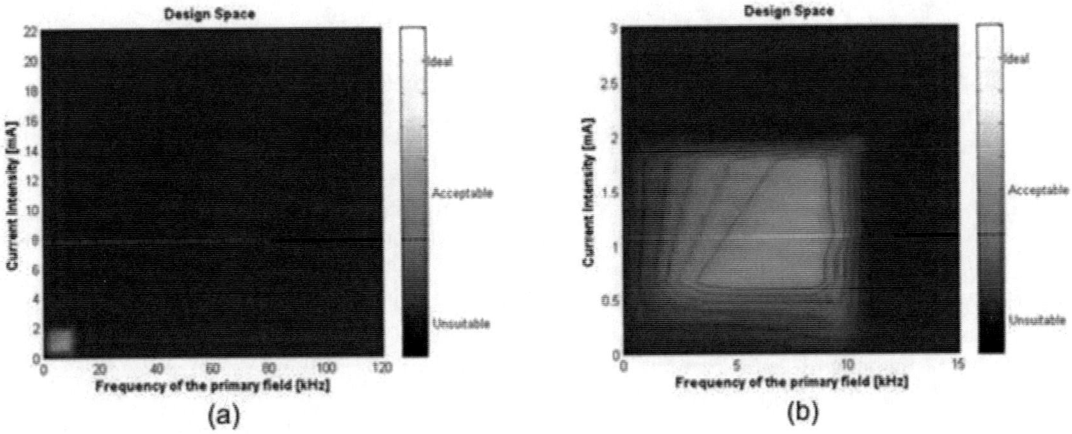

Figure 2. Design Space of the CPP_{A2} and CPP_{B2}, considering the combined requirements of CQA_3 and CQA_7. In (a), the full scale is presented and, in (b), the results are zoomed, highlighting the CPP range of higher performance.

By analyzing the zoomed surface plot of the Design Space (figure 2b), the possibility of applying frequencies between 3 and 9 kHz is evidenced, provided that the maximum allowed current intensity is reduced as frequency decreases.

4. Conclusion

In the present work, the modern quality tool, Quality by Design (QbD), that has been applied since 2004 in the pharmaceutical and biotechnology industries is adapted for application in the development of a biomedical technology aimed at non-magnetic metallic foreign body localization.

The basic stages of the QbD methodology were adapted in order to define the health technology features that meet the associated surgical demands of foreign body detection and localization, including the Target Product Profile (TPP), the Critical Quality Attributes (CQA) and the Critical Process Parameters (CPP). Considering that the lack of databases for biomedical technologies precludes the use of the statistical methods conventionally employed by QbD in the pharmaceutical industry, the Design Space configuration becomes a challenge to be overcome. Therefore, a recently proposed QbD-fuzzy system was adapted to the biomedical localization system in order to designate the range of CPP values associated with the highest performance of the device under development [21]. It is proposed the adoption of the requirement of metrological traceability and adequacy to a target uncertainty in the CPP values assessment. These aspects are not taken into consideration in the classical approach of QbD. In order to illustrate the possible information obtained by outlining the Design Space, two critical attributes associated to the technology (CQA_3 and CQA_7) were studied in relation to two CPPs (CPP_{A2} and CPP_{B2}). As a result, the optimal ranges for the values of frequency of the primary magnetic flux density and the current intensity passing through the solenoid that generates this primary magnetic field, considering the two critical attributes of "transducer sensitivity" and "secondary magnetic flux density", CQA_3 and CQA_7 respectively, were represented by a Design Space area. The obtained results indicate that the best possible performance combining both CQA requirements are obtained for frequencies between 7 and 9 kHz and AC current intensities between 0.6 and 1.8 mA. According to the Design Space configuration, it is also possible to apply frequencies between 3 and 7 kHz, as long as the maximum allowed current intensity is reduced as the frequency decreases.

The obtained Design Space configuration, for the selected CQAs and CPPs associated to the non-magnetic metallic foreign body localization system, points out the potential contribution of QbD for the scientific understanding of the processes and, consequently, for ensuring proper and safe performance of biomedical technologies under development.

References

[1] Monteiro E C, Barbosa C H, Lima E A, Ribeiro, P C and Boechat P 2000. Locating steel needles in the human body using a SQUID magnetometer. *Phys. Med. Biol*, **45(8)**, 2389.
[2] Barbosa C H, Monteiro E C, Lima E A, Santos S F, Cavalcanti E G and Ribeiro P C 2001. Improvement of a technique for localization of steel needles in humans using a SQUID magnetometer. Applied Superconductivity, *IEEE Transactions on*, **11**, 677-680.
[3] Barbosa C H 2004. Localization of firearm projectiles in the human body using a superconducting quantum interference device magnetometer: A theoretical study. *Rev. Sci. Instrum.*, **75(6)**, 2098-2106.
[4] Smythe W R 1989 *Static and dynamic electricity* 3rd Ed., (USA), Hemisphere Publ. Corp.
[5] Fortaleza L G S, Barbosa C R H, Monteiro E C, Siva E C and Gusmão L A P 2015. Detecting Non-Magnetic Metallic Foreign Bodies by GMR Sensors through the Use of Eddy Currents. *In: 8º Congresso Brasileiro de Metrologia*, 1-4.
[6] Louzada D R, Monteiro E C, Gusmão L A P and Barbosa C H 2007. Medição não-invasiva de ondas de pulso arterial utilizando transdutor de pressão MIG. *In IFMBE proceedings Springer Berlin Heidelberg*, 436-439.
[7] Pompéia F, Gusmão L A P, Barbosa C R H, Monteiro E C, Gonçalves L and Machado F L A 2008. Ring shaped magnetic field transducer based on the GMI effect. *Meas Sci.Tech.*, **19(2)** 025801.
[8] Silva E C, Gusmão L A P, Barbosa C H and Monteiro E C 2013. Progress toward a hundredfold enhancement in the impedance phase sensitivity of gmi magnetic sensors aiming at biomagnetic measurements. *InV IFMBE proceedings, Springer Berlin Heidelberg*, 742-745.
[9] Silva E C, Gusmão L A P, Barbosa, C H, Monteiro E C and Machado F L A 2011. High sensitivity giant magnetoimpedance (GMI) magnetic transducer: magnitude versus phase sensing. *Measurement Sci. and Tech.*, **22(3)**, 035204.
[10] Monteiro E C 2007. Biometrologia: confiabilidade nas biomedições e repercussões éticas. *Metrologia e Instrumentação*, **6**, 6-12.
[11] Monteiro E C and Leon L F 2015. Metrological Reliability of Medical Devices. *Journal of Phys. Conf. Series*, **588**, 012032.
[12] Us food and drug adminsration et al, 2015. US Department of Health and Human Services.
[13] ICH Harmonised Tripartite. Pharmaceutical development. Q8 (2R). 2009. *Int. conf. on harmonisation of tech. Requiremens for registration of pharmaceutical for human use.*
[14] Veljanovki I 2011. Integrating quality by design (QbD) in medical device manufacturing– concept, benefits, and challenges. White Paper. *Stevens Inst. of Tech.* 1-9.
[15] Lionberger R A et al 2008. Quality by design: concepts for ANDAs. *The AAPS journal*, **10(2)**, 268-276.
[16] Nadpara N P et al. 2012. Quality by design (QbD): A complete review. *Int. journal of pharmaceutical. Scie. Rev. and research.*
[17] Rathore A S 2009. Roadmap for implementation of quality by design (QbD) for biotechnology products. *Trends in biotechnology*, **27(9)**, 546-553.
[18] Lawrence X Y 2008. Pharmaceutical quality by design: product and process development, understanding, and control. *Pharmaceutical research*, **25(4)** 781-791
[19] Birle S, Hussein M A and Becker T 2013. Fuzzy logic control and soft sensing applications in food and beverage processes. *Food Control*, **29(1)** 254-269.
[20] Rivero A M, Monteiro E C, Leite D S, Tannenbaum F S, Pinto M A G and Leite K T F 2015. Implementing First Stages of Quality by Design Approach in The Development of an Assistive Robotics Technology. *In: XXI IMEKO World Congress*, 1295-1299.
[21] Rivero A M 2015. New Quality by Design – Fuzzy Method Applied in the Development of Biomedical Techonology. Masters dissertation, PUC-Rio.

Evaluation applications of instrument calibration research findings in psychology for very small samples

W P Fisher Jr[1], **P Petry**[2]

[1]Research Associate, BEAR Center, Graduate School of Education, University of California, Berkeley, CA USA
[2]Principal, PaulaPetry.com
Miami, FL USA
E-mail: wfisher@berkeley.edu

Abstract. Many published research studies document item calibration invariance across samples using Rasch's probabilistic models for measurement. A new approach to outcomes evaluation for very small samples was employed for two workshop series focused on stress reduction and joyful living conducted for health system employees and caregivers since 2012. Rasch-calibrated self-report instruments measuring depression, anxiety and stress, and the joyful living effects of mindfulness behaviors were identified in peer-reviewed journal articles. Items from one instrument were modified for use with a US population, other items were simplified, and some new items were written. Participants provided ratings of their depression, anxiety and stress, and the effects of their mindfulness behaviors before and after each workshop series. The numbers of participants providing both pre- and post-workshop data were low (16 and 14). Analysis of these small data sets produce results showing that, with some exceptions, the item hierarchies defining the constructs retained the same invariant profiles they had exhibited in the published research (correlations (not disattenuated) range from 0.85 to 0.96). In addition, comparisons of the pre- and post-workshop measures for the three constructs showed substantively and statistically significant changes. Implications for program evaluation comparisons, quality improvement efforts, and the organization of communications concerning outcomes in clinical fields are explored.

1. Introduction

An extensive body of published research documents the methods, models and processes by which item calibration invariance across samples can be established [1-3]. Application of these kinds of calibration research results in workshop evaluation contexts may provide previously unavailable advantages for demonstrating value obtained. Workshop providers and clinical practitioners may find new degrees of scientific credibility in being able to demonstrate repeatable and reproducible effect sizes expressed in comparable units, and as a result of establishing baselines for their own internal benchmarking and improvement activities. Workshop sponsors and funding organizations will be better able to separate successful and unsuccessful programs, enabling them to more clearly define returns on investments. Workshop participants will have the means to assess their status on any given construct at any time, to better maintain momentum on the therapeutic trajectory, and may also become informed enough about

[1] To whom any correspondence should be addressed.
[2] To whom any correspondence should be addressed.

measurement to demand comparable outcome measures from competing workshop providers. Future research will focus on developing practical scaffolding from the therapeutic progressions defined by each scale's invariant item hierarchy.

Participants participated in workshop programs designed to build awareness of the impact of misgivings, victim/bully/rescue cycles, and conscious engagement on their lives. The programs included leadership coaching sessions presented in a positive, uplifting context, to open participants' perspectives to harmony and flow. Participants kept journals recording their observations of how misgivings and rescue dynamics impact their lives, and about the approaches they used to step into conscious engagement with aspects of life they had previously taken for granted.

The data analyzed here were originally collected with the intention of evaluating the effectiveness of a stress-reduction workshop, which was not funded by federal research money. The original evaluation was completed to the satisfaction of the sponsoring organizations. In the course of examining the data for the purposes of the providing the evaluation, some unexpected results emerged. These pre-existing data gathered for the evaluation were stripped of all identifying information, such that there was no responses could be linked back to the persons originally providing them through a key to a coding system or any other means. There being no interaction at all between the research analyst and any individual workshop participants, and as no identifiable private information was involved, this project did not require review by any committee or office for the protection of human subjects. (For more information, see, for instance, http://cphs.berkeley.edu/secondarydata.pdf.)

2. Method

Typical workshop evaluations ask participants to rate the quality and organization of the presentations, the presenters' skills and knowledge, the efficiency of the registration process, and the comfort and cleanliness of the facility. Results are usually compiled by calculating percentages of responses indicating various levels of agreement with each item. Measures of the actual constructs intended to be impacted are not commonly obtained.

The Joyful Living Workshop evaluations, in contrast, were designed and conducted with the goals of establishing measures exhibiting consistent interpretability and meaningfulness, building on past research, and measuring change with enough sensitivity and power to obtain meaningful and statistically significant differences in individual comparisons of the pre- and post-workshop measures.

3. Samples

The Joyful Living Workshop program is designed to expand consciousness as to a productive sense of personal identity and to help shift away from misgivings, victim/bully/rescue cycles, and toward conscious engagement in everyday life. Several one-to-two day, and two-and-a-half day workshops were held over three-month periods in 2012 and 2014. Sixteen participants provided pre- and post-workshop measures on all three constructs in 2012, and 14 did in 2014.

In 2012, there were 19 total workshop participants, and 12 of the 15 who provided demographic information were employees of a Miami (FL) community resource center for parents of children with disabilities. Ages ranged from 20s to 60s, with the mode and median in the 40s. All 15 respondents identified themselves as Hispanic, 13 of the 15 were female, and education levels ranged from high school to doctorate, with the mode and median at completed college. The 2014 participants were of similar demographic composition.

4. Instruments

Rasch-calibrated instruments measuring the intended outcomes of the workshop (reduced depression, anxiety and stress, and enhanced joy for living as an effect of mindfulness practice) were identified in the research literature. These instruments were the 21-item Depression, Anxiety and Stress Scales (DASS) [4-6] and the Solloway Mindfulness Scale (SMS) [7,8].

The calibrations of these instruments on data from samples of persons belonging to the same general population as the workshop participants provided a reasonable degree of confidence in the possibility that the item hierarchies could be reproduced in the workshop data. Items were selected for retention, modification, or addition with the goal of achieving high ratios of variation to error (reliability coefficients) in the measures.

5. Results

Correlations (not disattenuated) of the two sets of small-sample workshop item calibrations with the published values for the three scales range from 0.85 to 0.99. In addition, pre- and post-workshop item calibrations showed the same patterns of invariance (correlations (not disattenuated) range from 0.86 to 0.92). Finally, comparisons of the pre- and post-workshop measures for the three constructs showed substantively and statistically significant changes.

6. Discussion

There were five primary outcomes. Each of these is a significant innovation in its own right, and will be found in other workshop evaluations only rarely, if at all.

- First, three well defined scales were constructed by adapting previously researched instruments.

  Long experience with instruments of these kinds guided identification of tools likely to be both useful and well-based in evidence and strong theory. That experience also indicated what kinds of changes might prove workable.

  The vast majority of workshop evaluations are not based in previously researched instruments.

- Second, each of the adapted scales maintained the scientific qualities expected of it.

  The new small-sample calibrations retained their expected values when compared to their original published values (see the figure below).

  The pre-workshop scale values were invariant when compared with the post-workshop scales.

- Third, linear measures with known uncertainty values and internal consistency (model fit) indices were constructed.

- Fourth, measurement precision was sufficient to distinguish pre- and post-workshop measures to statistically significant degrees. (See Table 1.1.)

- Fifth, the substantive significance of the differences between the pre- and post-workshop measures is evident in the content of the survey items positioned on the scales between the two measures. (See Table 1.1.)

Generalized application of these results should anchor the items in each scale at established scale values, or could be converted to more intuitively accessible scale ranges, if connections to published values were maintained.

7. Conclusion

Application of Rasch instrument calibration research results in workshop evaluation contexts may provide previously unavailable advantages for demonstrating the value of results. Most workshop evaluations do not maintain instrument comparability with published calibrations, nor do they experimentally establish a basis for comparing pre- and post-workshop measures. Workshop providers and clinical practitioners may find new degrees of scientific credibility in being able to demonstrate repeatable and reproducible effects, and as a result of establishing baselines for their own internal benchmarking and improvement activities. Workshop sponsors and funding organizations will be better able to separate successful and unsuccessful programs, enabling them to more clearly define returns on investments. Workshop participants will have the means to assess their status on any given construct at any time, to better maintain momentum on the therapeutic trajectory, and may also become informed enough about measurement to demand comparable outcome measures from competing workshop providers. Future research will focus on developing practical scaffolding from the therapeutic progressions defined by each scale's invariant item hierarchy.

8. References

[1] Fisher W P Jr 1997 *Physical Med & Rehabil State Art Revs* **11** 357-373
[2] Fisher W P Jr, Harvey R F, & Kilgore K M 1995 *NeuroRehabil* **5** 3-25
[3] Fisher W P Jr 1999 *J Louisiana St Med Soc* **151** 566-578
[4] Antony M M, Bieling P J, Cox B J, Enns M W, Swinson R P 1998 *Psychol Assessment* **10** 176-181
[5] Lovibond S H, Lovibond P F 1995 *Manual for the Depression Anxiety Stress Scales. 2nd edition* Sydney Psychology Foundation of Australia
[6] Shea T L, Tennant A, & Pallant J F 2009 *BMC Psychiatry* **9** [doi:10.1186/1471-244X-9-21]
[7] Solloway S, & Fisher W P Jr 2007 *International J Transpersonal Studies* **26** 58-81
[8] Solloway S, & Fisher W P Jr 2007 *J Appl Meas* **8** 359-372

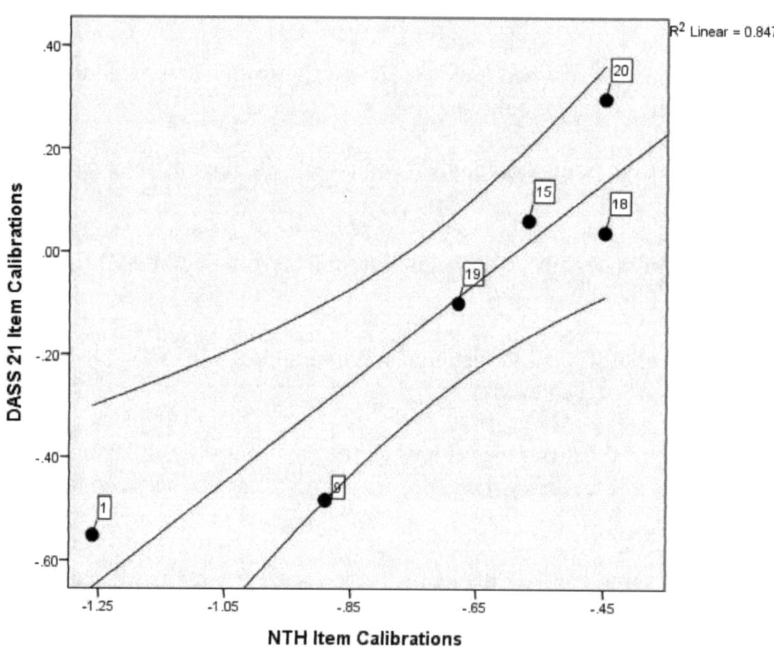

Journal of Physics: Conference Series **772** (2016) 012017 doi:10.1088/1742-6596/772/1/012017

```
TABLE 1.1 Joyful Living Pre&Post-Workshop Eval    Joy2.out.txt  Oct  7 16:35 2012
INPUT: 52 PERSON  73 ITEM  REPORTED: 35 PERSON  34 ITEM  4 CATS  WINSTEPS 3.72.0
--------------------------------------------------------------------------------

         MAP OF PERSON AND ITEM
MEASURE                                    |                                 MEASURE
  <more> ---------------------- PERSON  -+- ITEM     ----------------- <rare>
                                          +  I see ordinary things as extraordinary.
                                          |
                                          |
                                          |
                                          |  I have felt peaceful.
                                          |
                                          +
                                          |
EXPECTED RATINGS ON ITEMS AT THIS LEVEL   |  I feel free to choose what my life will be like.
       PREWORKSHOP: 20% S OR VS AGREE     |  I have self-control.
      POSTWORKSHOP: 45% S OR VS AGREE     |
                                          |
                                          +
  AVERAGE EXPECTED RATING HERE IS 2.4     |
      AVERAGE POSTWORKSHOP MEASURE ---->  |  There's been a lot to look forward to.
                        ^                 |  I enjoyed being lighthearted.
                        ^                 |
                        ^                 |
                        ^                 +
                                          |
                        A                 |  I've affirmed my value to others.
                                          |
                        V                 |
                                          |  I am aware that other people appreciate who I am.
                        E                 +
                                          |
                        R                 |  I experience life in new ways.
                                          |
                        A                 |
                                          |
                        G                 +
                                          |
                        E                 |
                                          |  Unexpected things brought me joy.
                                          |
                                          |
                        G                 +
                                          |
                        A                 |
                                          |
                        I                 |  I enjoy discovering amazing things in nature.
                                          |
                        N                 +  I breathed freely.
                        ^                 |
                        ^                 |
                        ^                 |
                        ^                 |
                        ^                 +
                        ^                 |
                        ^                 |
     AVERAGE PREWORKSHOP MEASURE ---->    |
  AVERAGE EXPECTED RATING HERE IS 2.1     |
EXPECTED RATINGS ON ITEMS AT THIS LEVEL   |  Life seems full of meaning.
       PREWORKSHOP: 55% S OR VS AGREE     +  I find joy in ordinary things.
      POSTWORKSHOP: 70% S OR VS AGREE     |  I see how things change from moment to moment.
                                          |
                                          |
  <less> ---------------------- PERSON  -+- ITEM     ----------------- <frequent>
```

Journal of Physics: Conference Series **772** (2016) 012018 doi:10.1088/1742-6596/772/1/012018

Forensic Metrology: Its Importance and Evolution in the United States

Ted Vosk, JD
Attorney at Law/Consultant
8105 NE 140th Pl, Kirkland WA 98034

Email: tvosk@comcast.net

Abstract. Forensic measurements play a significant role in the U.S. criminal justice system. Guilt or innocence, or the severity of a sentence, may depend upon the results of such measurements. Until recently, however, forensic disciplines were largely unaware of the field of metrology. Accordingly, proper measurement practices were often, and widely, neglected. These include failure to adopt proper calibration techniques, establish the traceability of results and determine measurement uncertainty. These failures undermine confidence in verdicts based upon forensic measurements. Over the past decade, though, the forensic sciences have been introduced to metrology and its principles leading to more reliable measurement practices. The impetus for this change was driven by many forces. Pressure came initially from criminal defense lawyers challenging metrologically unsound practices and results relied upon by government prosecutions. Litigation in the State of Washington led this movement spurring action by attorneys in other jurisdictions and eventually reform in the measurement practices of forensic labs around the country. Since then, the greater scientific community, other forensic scientists and even prosecutors have joined the fight. This paper describes the fight to improve the quality of justice by the application of metrological principles and the evolution of the field of forensic metrology.

1. Beginnings

Forensic measurements play a prominent role in the criminal justice system. They are relied upon to investigate and prove charges ranging from murder to simple traffic offenses. Some crimes and punishments, for example driving with a breath alcohol concentration (BrAC) in excess of a prohibited level, even require forensic measurements to be established. As ubiquitous as forensic measurements are, though, historically they have often been poorly understood and misrepresented by forensic and legal professionals alike. This undermines the ability of fact finders to properly weigh such evidence and the public's confidence in verdicts based upon it.

The fight to change this state of affairs through the imposition of a complete and coherent metrological framework upon forensic measurements was taken up by attorneys in Washington State defending individuals charged with the crime of Driving Under the Influence of Alcohol and Drugs (DUI). DUI defense was a natural place for such a movement to take hold as almost all DUI prosecutions include the results of forensics measurements as evidence. These include breath and blood alcohol test results as well as measurements of the amount or concentration of drugs in an individual's blood. The lion's share of the litigation undertaken in this cause centered on BrAC tests as they are by far the most frequent forensic measurement performed.

Breath alcohol tests are used in the investigation and prosecution of DUI throughout the United States. Their purpose is to measure either an individual's breath or blood alcohol concentration. An important element of many breath test machines is a simulator. A simulator is a device that is filled with an aqueous-alcohol solution, referred to as a simulator solution. The simulator heats the solution to a specified temperature producing a vapor of a specified alcohol concentration. This vapor is used to both calibrate breath test instruments as well as to check their accuracy during the performance of a breath test. In order to ensure that the vapor created is the correct concentration, a thermometer is used to monitor the temperature of the solution it is a product of.

In early 2001, the Washington State Toxicology Lab claimed that the accuracy of the temperatures reported for the solutions were determined by a specified "margin of error" attributed to the thermometers used to measure them. An investigation by the defense turned up several problems with the State's claim, however. First, the uncertainty of the temperatures measured by the thermometers used was found to be significantly greater than the margin of error that had been claimed. Second, the thermometers themselves were not being used in a manner consistent with their validation rendering any values reported unreliable even had the claimed margin of error been correct. After a day-long hearing wherein these issues were addressed, a Court suppressed the breath test results. [1]

The victory wasn't about "just trying to get another guilty person off" as is so often lamented by critics, though. It was about preventing the government from using flawed science to deprive Citizens of their liberty. Every one of us is innocent until proven guilty. That is one of the safeguards against tyranny provided by our Constitution. When the government tells a judge or jury that science supports claims that it does not, it is tantamount to committing a fraud against our system of Justice. It doesn't matter whether the deception is purposeful or not because the result is the same: a Citizen's liberty is imperilled by a falsehood.

The government sought to fix the problems that had been encountered in the use of the thermometers by requiring that the temperatures measured by such thermometers must, going forward, be traceable to standards maintained by NIST. Unfortunately, the government neither knew the definition of traceability nor understood how to establish it. The Defense Bar put an end to this bad government science by showing what was required to establish the traceability of the temperature reported by these simulator thermometers. The Washington State Supreme Court suppressed breath tests state-wide in the first published decision explicitly recognizing metrology in a forensic context. In doing so it enunciated a clear and common sense principle:

> "If the citizens of the State of Washington are to have any confidence in the breath testing program, that program has to have some credence in the scientific community as a whole." [2]

2. Not Just Bad Science, Bad Tribunals

But bad government science doesn't necessarily arise from bad government scientists. Nor is the desire to ensure that science is used correctly to discover truth in the courtroom confined to defense attorneys. Forensic scientists, prosecutors and judges have sought the same goals.

In 2004, the head of the Washington State Breath Test Program helped the defense to keep the government from administratively suspending a women's driver's license based upon a misleading test result. The woman had submitted to a breath test that yielded duplicate results both in excess of the legal limit. Through the State expert, the defense was able to show that the uncertainty associated with the results proved that there was actually a 56.75% probability that her true BrAC was less than the legal limit.

The bad government science in this case was not that done by forensic scientists. To the contrary, it was one of the State's top forensic scientists who used metrology to establish that this motorist had most likely *not* violated the law. Rather, it was what government officials did with otherwise good science that rendered it bad. Despite the testimony provided by the scientist in charge of the State's breath test program, the Washington Department of Licensing chose to ignore what science said about the conclusions these results supported, and suspended this Citizen's license. It was only on appeal that a court recognized the absurdity of the ruling and reversed the suspension. [3]

3. Metrology: A Tool for the Critical Analysis of All Forensic Measurements

Measurements involving the determination of a person's breath or blood alcohol concentration are quite common because the crime of DUI is defined by the results of these measurements. They are by no means the only type of forensic measurements, though. Determining the weight of seized drugs using a scale; the speed of a motor vehicle using a radar; the angle at which a bullet entered a wall using a protractor; and even the distance between a drug transaction and a school using a measuring wheel; these are just a few of the many types of forensic measurements that are performed. The same underlying metrological principles that allowed for the analysis of the breath alcohol measurements discussed above apply to every other forensic measurement as well.

This leads to an astonishing conclusion. Since the science of metrology underlies all measurements, its principles provide a basic framework for the critical evaluation of *all* measurements, regardless of the field they arise out of. Accordingly, given a familiarity with metrology, scientists and police officers can better perform and communicate the results of the forensic measurements they perform; Lawyers can better understand, present and cross-examine the results of forensic measurements intended to be used as evidence; Judges will be better able to subject testimony or evidence based on forensic measurements to the appropriate gatekeeping analysis; and each of these participants will be better prepared to play their role in ensuring that the misuse of science doesn't undermine the search for truth in the courtroom.

4. Reform Through Litigation

This was the idea attorneys had in mind when, in the Summer of 2007, a forensic scientist within the Washington State Toxicology Lab was discovered committing perjury, claiming to have performed measurements that she had not. Upon further investigation, though, the defense discovered that the Lab's problems went far deeper than perjury. The Lab's process for creating simulator solutions for the calibration and checking of breath test machines was in a state of disarray. Failures to validate procedures, follow approved protocols, adhere to scientifically accepted consensus standards, properly calibrate or maintain equipment, and even to simply check the correctness of results and calculations were endemic.

In a private memo to Washington's Governor, the State Toxicologist explained that the measurement procedures in question "had been in place for over twenty years and had gone unchallenged, leading to complacency." What allowed the defense to find what others had missed over the years was, again, metrology. Viewed through the appropriate metrological framework, it became clear that what complacency had led to was the systemic failure of the Lab to adhere to fundamental scientific requirements for the acquisition of reliable measurement results. After a seven-day hearing that included testimony from nine experts, declarations from five others, as well as one hundred and sixty-one exhibits, a panel of three judges issued a thirty-page ruling suppressing all breath test results until the Lab fixed the problem's identified. [4]

Under new leadership, the Lab subsequently used the same metrological framework to fix its problems that had been used to discover them. It did so by implementing fundamental metrological principles and practices and obtaining accreditation under ISO 17025, the international standard that embodies these principles and practices. Because of this, the Washington State Toxicology Lab has one the best Breath Test Calibration programs in the country. The same metrological principles that can be such effective tools in the hands of legal professionals can be even more powerful when employed by competent forensic scientists.

Criminal defense attorneys in other states, including Michigan, California, Minnesota, New Mexico, Pennsylvania and Arizona, have since begun to follow Washington's lead.

5. The National Academy of Sciences

It was at about this time, in February of 2009, that the National Academy of Sciences released a report on the state of forensic science in America. [5] The Report was very critical of the practices engaged in by many of the forensic sciences identifying problems in the areas of method validation, adherence to appropriate practices as evidenced by consensus standards, the determination and reporting of measurement uncertainty and many others. A majority of the scientific issues identified by the Report, though, are those that metrology addresses. What had been done in Washington State with respect to forensic measurement was to not only beat the Academy to the punch in the discovery of these issues, but also to the identification of the appropriate framework for their solution.

In the forensics community, metrology is now being relied upon to address many of the issues identified by the National Academy of Sciences. Its principles are helping to improve how forensic measurements are developed, performed and reported. Accreditation and adherence to international scientific standards are restoring confidence that forensic measurements comply with the same rigorous methodology followed in other sciences. And it is providing a common language for all those engaged in making or relying upon forensic measurements to communicate about them regardless of application.

6. A Growing Movement

Despite having had it's breath test calibration program accredited, by the Summer of 2009, the Washington State Toxicologist had decided that the uncertainty associated with breath test results obtained in Washington would neither be determined nor provided to Citizen's who were being prosecuted on the basis of those results. Accordingly, Washington defense lawyers engaged the fight again. Their argument was that breath test results could not be properly weighed by fact finders absent their uncertainty. In other words, judges and juries could only understand the conclusions supported by a result if they were provided the uncertainty that revealed the range of values that could reasonably be attributed to an individual's BrAC based on that result.

The Defense drove home it's point during cross-examination of the State's primary expert. The witness was handed a breath alcohol test ticket containing results of 0.081 g/210L and 0.080 g/210L. Assuming proper quality assurance procedures and testing protocols were followed, all parties agreed that these were the results of an "accurate and reliable" test. The expert was then asked, given these results, could he state beyond a reasonable doubt that this individual's BrAC exceeded a 0.080 g/210L (the *per se* limit in Washington State). The expert responded: "I would have to say yes based on these results here."

Similar evidence and testimony, concerning a range of forensic measurements, is introduced in courtrooms around the country every day. And based on such evidence and testimony, citizens accused of all manner of crimes plead or are found guilty. The problem is that an accurate and reliable test doesn't necessarily mean what most lay factfinders presume. In fact, as the Court later noted, even an expert may be misled by what is deemed an accurate and reliable result. This was demonstrated by subsequently providing the expert with the breath test result's uncertainty and asking whether he still believed that these results supported the conclusion that this individual's BrAC exceeded a 0.080 g/210L that beyond a reasonable doubt. His response was no.

The expert conceded that, despite the fact that the test was accurate and reliable, based on the result's uncertainty there was actually a 44% likelihood that this individual's BrAC was below 0.080 g/210L! Far more than a reasonable doubt, these "accurate and reliable" test results barely established the conclusion as more likely than not!

After a five-day hearing that included testimony from four experts as well as 93 exhibits, a Panel of three judges issued a thirty-page order explaining that breath test results would henceforth be inadmissible unless they were accompanied by their uncertainty. [6]

The rulings from this and other Washington cases concerning measurement uncertainty garnered nationwide attention. Lawyers, judges, forensic scientists and scholars from around the country began discussing and writing about the importance of providing a measured result's uncertainty when the result will be relied upon as evidence at trial. [7] Thomas Bohan, former president of the American Academy of Forensic Sciences, declared it to be "a landmark decision, engendering a huge advance toward rationality in our justice system and a victory for both forensic science and the pursuit of truth." [8] The battle was subsequently taken up by defense attorneys in several other State and Federal Courts, and continues to spread as of the time of this writing.

Unfortunately, in a decision that defies reason, the Washington State Court of Appeals decided that it understood science better than scientists. It found that: 1) Proper science does not require that the uncertainty of measured results be either determined or provided to the users of those results; and 2) despite the State's top expert having been fooled by breath test results that were not accompanied by their uncertainty, that measured results were best understood without their uncertainty. [9] Given the embarrassment suffered by the State Toxicology Lab during the lower Court proceedings, however, it now "voluntarily" determines and provides the uncertainty of the results of all breath tests performed within the State of Washington.

7. The Prosecution Begins to See the Light

It's not just defense counsel who are concerned about the effects of bad and misleading forensic evidence in the courtroom, though. In a 2013 paper published in the Santa Clara Law Review, a California prosecutor provided the rational for why all those advocating on behalf of the state should also be fighting for the imposition of metrological requirements on state administered breath and blood tests. In fact, after a trial court denied a defense motion to require the reporting of uncertainty and traceability with blood test results, he worked with that jurisdiction's lab to make sure that this was done for all future test results despite the court's ruling. And he subsequently worked to make this a mandatory regulatory requirement. Why? Because he wants to ensure that the science presented by the state in court is "the best science regardless of what the law requires." [10]

8. Back to Fundamentals: The Measurand

The latest battles in this saga concern the most fundamental aspect of the measurement process: identification of the measurand. The problem at issue is identification of measurands in statutes by scientifically unsophisticated legislators that are either so significantly under-defined, or vary by jurisdiction in such a way, that they create confusion for both forensic science and legal professionals.

An example of the first type of problem arises under controlled substance statutes that prohibit certain substances and their "analogues." An analogue is typically defined as having a chemical structure "substantially similar" to the named controlled substance. What constitutes substantially similar, however, is a matter of opinion on which forensic scientists in the same laboratories often disagree on. The second type of issue arises primarily in the DUI context where statutes of different jurisdictions specify distinct measurands for a common breath alcohol test. [11] Depending upon the measurand identified, the result of a breath test means something very different between these jurisdictions and the uncertainty of the results varies significantly as well.

Statutes suffering from such infirmities create a confusing landscape. Given the role of lawmakers, as opposed to scientists, in the specification of a test's measurand, though, they are representative of the type of problems that arise at the intersection of law and science. With a lack of scientific sophistication, many lawmakers fail to realize the necessity of, or required detail concerning, the measurand in a measurement. In fact, many have never even heard the term measurand before.

9. Conclusion

The truth about any scientific measurement is that it can never reveal what a quantity's true value is. The power of metrology lies in the fact that it provides the framework by which we can determine what conclusions about that value are supported by measured results. It tells us how to develop and perform measurements so that high quality information can be obtained. It helps us to understand what our results mean and represent. And finally, it provides the rules that guide our inferences from measured results to the conclusions they support. Whether you are a prosecutor or defense attorney, judge or forensic scientist, or even a law enforcement officer who performs measurements as part of investigations in the field, forensic metrology provides a powerful tool for determining the truth when forensic measurements are relied upon. Forensic science, legal practice and justice itself are improved by a familiarity with the principles of forensic metrology. [12]

References

[1] *City of Bellevue v. Tinoco*, No. BC 126146 (King Co. Dist. Ct. WA 09/11/2001)
[2] *City of Seattle v. Clark-Munoz*, 93 P.3d 141 (Wash. 2004).
[3] *Herrmann v. Dept. of Licensing*, No. 04-2-18602-1 SEA (King Co. Sup. Ct. WA 02/04/2005)
[4] *State v. Ahmach*, No. C00627921 (King Co. Dist. Ct. WA 01/30/2008)
[5] Nat'l Research Council, Nat'l Academy of Sciences, Strengthening Forensic Science in the United States: A Path Forward (2009).
[6] *State v. Fausto*, No. C076949 (King Co. Dist. Ct. WA 09/20/2010)
[7] Imwinkelried E. The Importance of Forensic Metrology in Preventing Miscarriages of Justice: Intellectual Honesty About the Uncertainty of Measurement in Scientific Analysis. *John Marshall Law Journal* 2014; VII(2): 331-370.
[8] Ted Vosk, *Trial by Numbers: Uncertainty in the Quest for Truth and Justice*, The NACDL Champion, Nov. 2010, at 48, 54.
[9] *State v. King County Dist. Court West Div.*, 307 P.3d 765 (Wash. App. 2013)
[10] Christopher Boscia, *Strengthening Forensic Alcohol Analysis in California DUI Cases: A Prosecutor's Perspective* 53 Santa Clara L. Rev. 733, 764-765 (2013).
[11] Ted Vosk et al, *The Measurand Problem in Breath Alcohol Testing*, 59 J Forensic Sci 811–815 (2014)
[12] Ted Vosk & Ashley Emery, Forensic Metrology: Scientific Measurement and Inference for Lawyers, Judges, and Criminalists (CRC Press 2014).

Calibrating Communication Competencies

Donna Surges Tatum

CEO, Meaningful Measurement, Inc. 3170 N. Sheridan Rd. #1008 Chicago, IL 60657

donnatatum@meaningfulmeasurement.com

Abstract. The Many-faceted Rasch measurement model is used in the creation of a diagnostic instrument by which communication competencies can be calibrated, the severity of observers/raters can be determined, the ability of speakers measured, and comparisons made between various groups.

Keywords: Rasch Measurement, Communication Competency, Multifaceted Models

1. Communication Competence

Competence is the application of knowledge, skills and abilities. It is the amount and quality of the things a person can do. Success in most areas of life are dependent upon good communication skills.

Aristotle wrote the *Rhetoric* [9] as a practical guide for those who needed to be persuasive in everyday life. Nearly twenty-five hundred years later, we are still using his ideas and concepts. Many high schools, colleges and universities require the study of discourse, both oral and written. Communication is one of the most important competencies required for almost any job. This is borne out in many surveys through the years. A study from 1975 by Magill, et al. shows that of all the courses taken, college graduates considered public speaking one of the most important and useful in later life. [10] In 2001 a *Wall Street Journal* report stated that 90% of recruiters said they consider communication and interpersonal skills very important [11]. The six core competency skills for physicians include interpersonal and communication skills [12-13]. The 2008 Society of Human Resource Management (SHRM) Leadership Competencies includes effective communication [14]. A quick internet search for job competencies produces many articles which almost always list communication, including such skills as oral, written, teamwork, and the ability to influence/persuade [15-18].

2. Instrument Development

In 1991 an investigation of public speaking evaluation produced a pilot diagnostic instrument with which to measure speeches.19] Aristotle's principles contained in *The Rhetoric* [9] and elaborated by Monroe in the Motivated Sequence [20] are used as the framework for defining the competencies required to give a persuasive presentation. Hundreds of communication books are published every year. Although the terminology may not be used, most of them contain much of Aristotle's practical advice on discourse.

Aristotle considered rhetoric a neutral tool to be used to analyze a situation and then artfully present a position. He clearly differentiated between "inartistic" are "artistic" proofs. (I_{2-9}) Inartistic proofs are the facts of the case which must be arranged according to the precepts of the artistic proofs; ethos, logos and pathos. These sources of belief must all be present in order for a persuasive transaction to occur.

Ethos is personal credibility. There are two types of ethos: ascribed and earned Ascribed ethos is what others think prior to the beginning of the communication. It is the halo effect of reputation, profession, organization represented, or perhaps just the color of a tie. Although a person may start with a certain amount of credibility, it must be earned every time that person speaks, for ethos is ephemeral

Journal of Physics: Conference Series **772** (2016) 012019 doi:10.1088/1742-6596/772/1/012019

and can be lost in an instant. Competencies which contribute to credibility are sincerity, mastery of the subject, demeanor, commitment and fairness. Ethos is also enhanced by good delivery skills skill such as eye contact, posture, fluency, enthusiasm, smooth and complete gestures, authoritativeness, articulation and body movement

Logos is Greek for "word" so this form of proof is related to language usage and message construction. Competencies include using originality of expression and striking statements, language that is precise and appropriate for the situation, descriptive, vivid and colorful. Message construction includes a well-organized structure, smooth flow, variety of transitions and unusual tactics. The topic must be interesting, manageable and fulfill a purpose. The problem must be expressed clearly, the solution must be clear, concise, practicable and contain novel ideas. Evidence should be appropriate and logical, with a variety of supporting material and understandable numerical information.

Pathos are the emotional appeals inducing states of feeling in the listener. The speaker must have expressive eye contact, identify with the audience and create common ground, demonstrate the importance of the topic, focus on the advantages to the listeners, generate appropriate responses, and have a compelling style, riveting the listener. When appropriate, humor can illustrate key points and be an effective persuasive strategy. Above all, the speaker must create a strong need in the listener to listen, then relieve the tension by effectively resolving cognitive dissonance.

The Motivated Sequence was first published in 1935 by Alan H. Monroe in the classic textbook, *Principles and Types of Speech*. It has been in continuous publication and is now in its 18th edition. This plan to organize and present thoughts in a persuasive manner consists of five chronological steps. It psychologically follows the stages by which people learn and become motivated to adopt an attitude or take action. Briefly it involves: 1) Attention Step: Get and maintain the attention of the listeners. 2) Need Step: Demonstrate the need; pose a problem; create cognitive dissonance. 3) Satisfaction Step: Fulfill the need; present the solution; give the plan; relieve cognitive dissonance. 4) Visualization Step: Project the results of what will happen if the solution is or is not adopted by involving the listeners in picturing the future. 5) Action Step: Request specific action by telling the listeners exactly what is wanted for them to do, what they should do, or what they can do. When implemented, this method gives the speaker a structure with which to help the listeners persuade themselves. It encompasses the artistic proofs Aristotle says are necessary to have a successful persuasive transaction.

From 1992 – 2000 this author experimented with many permutations of the pilot diagnostic instrument. Some versions were shorter, or new items were added or exchanged with existing items. Ratings scales varied as well, using three, four, five or six categories. In 2000 the instrument was finalized with some revisions to the original 52 pilot items using a six-point rating scale.

3. Method

Some research situations require direct observations which are then recorded by a rater/judge who uses a rubric to assess the activities under investigation. Pendrill states "The Rasch approach...is not simply a mathematical or statistical approach, but instead [is] a specifically metrological approach to human-based measurement." [10, 26] The multi-faceted extension of the Rasch model solves the problem of subjective ratings by conditioning the raw scores into objective measures. One of the problems with rater mediated data is that people do not always view things in the same way and may have different levels of severity when rating the same phenomenon. One person's rating of "3" could very well be harder to attain than another person's "4" on the rating scale. Although inter-rater reliability is an important goal, the most essential issue is *intra*-rater consistency. When raters are internally consistent, then their relative severity levels can be accounted for in the results when using the multi-faceted model.

The FACETS computer program version 3.8.0 [7] was used to analyze the data. Additional analyses were performed using SPSS 18. The following model was employed:

$$\log\left(\frac{P_{nmjgik}}{P_{nmjgik}-1}\right) = B_n - C_j - D_{gi} - F_{gk}$$

B_n $n = 1 - 1070$ (speakers/speeches)

C_j	$j = 1 - 773$	(judges)
D_{gi}	$g = 1, i = 1 - 52$	(items)
F_{gk}	$g = 1, k = 1 - 6$	(6-point rating scale)

4. Results

Data for this study consists of people who participated in a class or workshop and gave at least one persuasive presentation from 2000 to 2015. These presentations were evaluated by instructors, fellow students and independent raters. There were 1,070 presentations given by 739 speakers, rated by 773 raters who evaluated at least 3 presentations using 9,218 of the finalized rating forms. Multifaceted models allow speakers, raters and items to all be along the same line of inquiry. The origin of the scales was changed from the default to a scale centered on 10.00. Table 1 shows the summary statistics, which indicate a satisfactory overall fit of the data to the model.

Table 1 Summary Statistics

Facet	N	Average	Measure	Model SE	SD	Infit MnSq	Outfit MnSq	Pt Bis	Sep	Rel
Items	52	4.46	10.00	0.01	0.41	1.00	1.01	0.44	29.30	1.00
Raters	776	4.46	10.00	0.08	0.80	1.05	1.04	0.30	9.57	0.99
Speeches	1070	4.48	11.53	0.07	0.78	0.96	0.96	0.33	9.82	0.99

One of the first things to check is whether the rating scale is working properly [21]. The probability curves for the six-point scale show all category measures and thresholds are ordered, expected is close to observed, and the mean squares are well behaved. Figure 1 illustrates the importance of modeling each identifiable facet affecting the measures, which accounts for rater differences, rather than simply assuming raw score averages, for assessment purposes. Of course, the measures are drawn from the raw scores so there is a high correlation between measures and scores.

However, in some cases there is a substantial disparity. One hundred presentations were randomly selected from the data. Thirty presentations were then randomly chosen from that group to create a scatterplot of raw scores and measures. As one can see, some presentations do not have raw scores and measures that lineup. For instance, presentations 1, 21, 28 and 10 all have an average raw score of 4.01. Yet, when the difficulty of the items and the severity of the raters are taken into account, the measures are quite different: 9.91, 10.26, 10.66 and 11.15 respectively. Another example shows two presentations, 11 and 28, with the same measure of 10.66, yet the raw scores are 4.72 and 4.01.

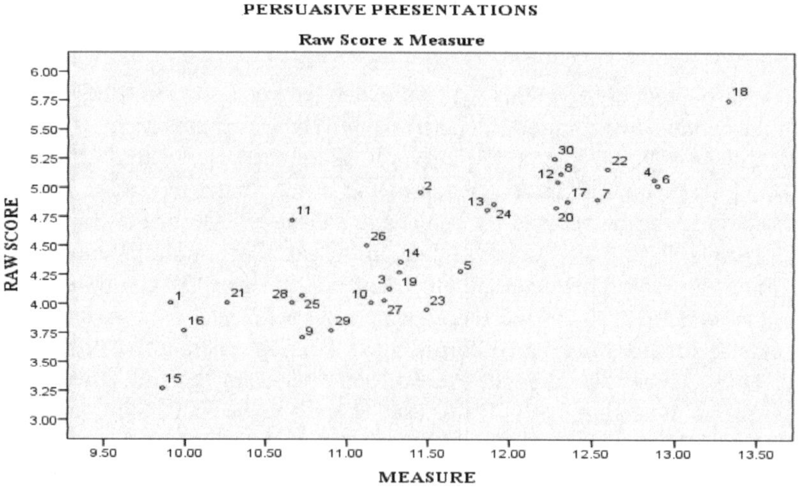

Figure 1 Persuasive Presentations: Raw Scores x Measures

The construct for persuasive presentation communication competencies is stable. The diagnostic tool to evaluate speeches and to calibrate the competencies necessary to be effective was first developed and piloted in 1991. Two hundred seventy-nine rating forms for 31 speeches were filled out by the two instructors and 14 class members using a 4-point scale. Those item measures were compared to the measures from the current analysis with data from 2000 – 2015; about 35 times more data than the pilot. Figure 2 is a graphic representation of the .918 correlation of the 52 competencies from the two different data sets.

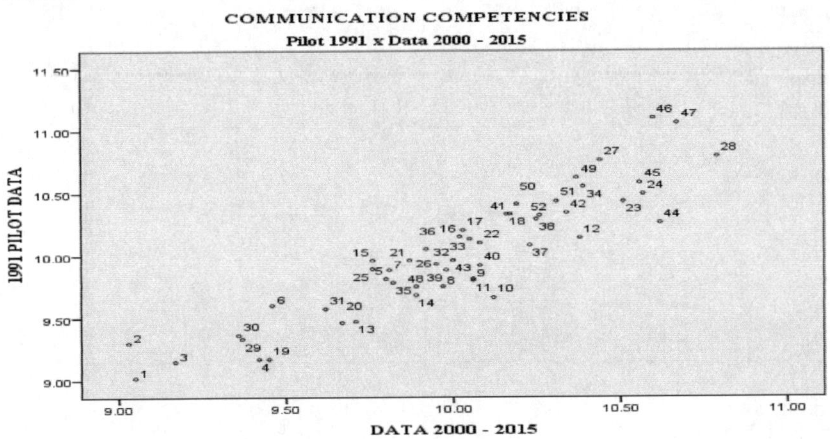

Figure 2 Communication Competencies Pilot 1991 x Data 2000-2015

An examination of rater behaviour shows the majority are internally consistent. Examining the various fit statistics reveals that about 85% of the raters maintain a frame of reference that makes their judgements useful, consistent and reliable.

The average ability measure of all 1,070 presentations is 11.53, about a logit and a half above the theoretical person of average ability. This makes sense because all participants are college graduates, many with advanced degrees, who are a cross section of mid-career urban professionals such as attorneys, engineers, accountants, sales and marketing experts. The goal of any course is demonstrable improvement. Two hundred seventy-seven participants gave two presentations. The mean measure of the first speech is 11.25, and the second is 11.82; a gain of .57, or more than half a logit. A paired samples t-test statistic of -14.43 (.000significance) confirms the fact this progress is not due to chance.

The competency items covered the range of the variable without significant misfit. All items had both infit and outfit mean squares between .72 and 1.30; point biserial correlation was .51 to .37 with only four items below .40; the estimated discrimination ranged from 1.29 to .67 with only 2 items less than .80. Thus this data set from the persuasive presentation evaluation instrument is proven valid and reliable.

5. Discussion

A map of communication competencies for persuasive presentations is a useful tool for both instructor and learner. Understanding the scientific structure of persuasive presentations allows rhetorical choices and decisions to be based on the facts of measurement. Figure 3, The Communication Competency Map, [22] [23] shows the hierarchy of proficiencies required to give a good presentation. It helps with understanding the construction of an effective presentation, and to identify the items which transform a competent speech into an excellent speech. Assessment is appropriately targets, based on educational level, professional objective and personal goals.

When a student is given diagnostic feedback with the overall and subscale measures for a persuasive presentation, it is a personalized guide for improvement. The distribution graph helps the student acquire

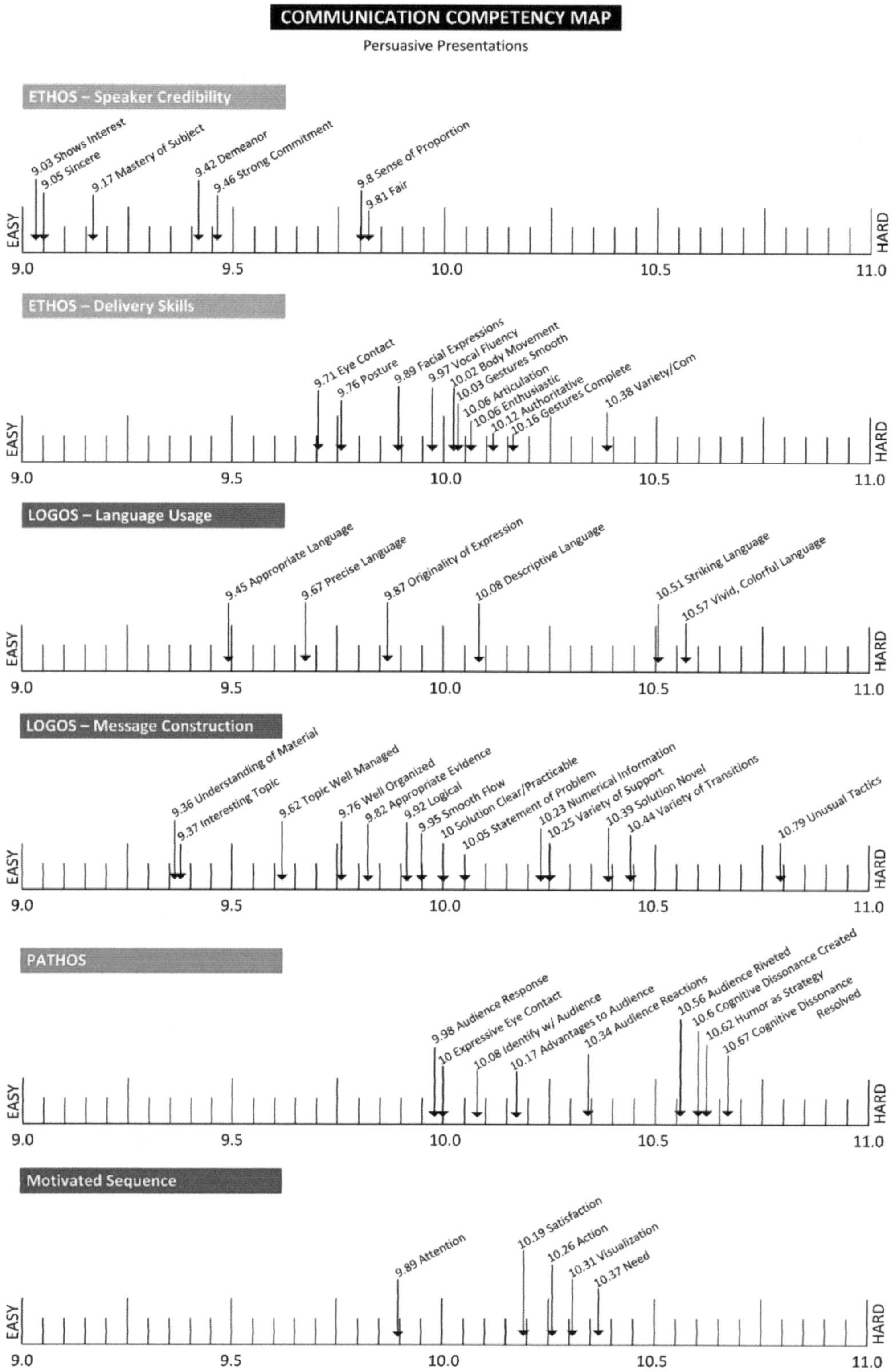

Figure 3. Communication Competency Map

Journal of Physics: Conference Series **772** (2016) 012019 doi:10.1088/1742-6596/772/1/012019

a frame of reference for his or her performance. This information contributes to the gain in knowledge, skills and abilities required to give a successful presentation. Many of the competencies necessary in persuasive presentation are transferable to other types of communication. They can be used to assess interpersonal skills, teamwork, leadership, professionalism or other needs. Currently there is a study being conducted measuring physician patient interaction. Much remains to be done to fully realize the value of calibrating communication competencies to a linear unit of comparison that takes rater variation into account. "Without access to recognized metrological standards, it will be difficult to judge objectively the relative levels of quality of products and services appraised by humans.... Fortunately, it turns out that the Rasch approach, with its explicit separation of person and item attribute estimation, is well suited for introducing metrological traceability to human-based measurement." [9 $_{28}$]The value of having a calibrated diagnostic tool lies in its application. It allows researchers to be able to objectively compare results from time to time and group to group. It allows standards to be set for required proficiency at various job and professional levels. The communication of communication competence transcends all disciplines, for it is fundamental to successfully create understanding.

References

[1] G Rasch 1960 Probabilistic models for some intelligence and attainment tests (Reprint, with Foreword and Afterword by B. D. Wright, Chicago: University of Chicago Press, 1980) (Copenhagen, Denmark: Danmarks Paedogogiske Institut)
[2] B D Wright and M H Stone 1979 *Best test design: Rasch measurement* (Chicago, Illinois: MESA Press)
[3] B D Wright and G N Masters 1982 *Rating scale analysis: Rasch measurement* (Chicago, Illinois: MESA Press)
[4] E B Andersen 1973 *British J Math Stat Psychol* **26** 31-44
[5] D Andrich 1978 Psychometrika **43** 561-573
[6] G H Fischer 1973 *Acta Psychologica* **37** 359-374
[7] J M Linacre 1989 *Many-facet Rasch measurement* (Chicago, Illinois: MESA Press)
[8] L Pendrill 2014 Man as a measurement instrument [Special Feature]. *NCSLi Measure: The Journal of Measurement Science* **9** 22-33
[9] Aristotle *Rhetoric (*Trans. W.Rhys Roberts 1954. New York: Modern Library)
[10] B Magill, et al. *Am Business Communic Assoc Bullet* Vol? 31-33
[11] http://online.wsj.com/documents/topschools2001.htm
[12] http://www.abms.org/board-certification/a-trusted-credential/based-on-core-competencies/
[13] https://www.aamc.org/initiatives/admissionsinitiative/competencies/
[14]https://www.shrm.org/research/articles/articles/pages/leadershipcompetencies.aspx
[15] http://money.usnews.com/money/blogs/outside-voices-careers/2014/12/10/the-top-10-skills-you-need-to-be-successful
[16] https://targetjobs.co.uk/careers-advice/career-planning/273051-the-top-10-skills-thatll-get-you-a-job-when-you-graduate
[17] http://youremployment.biz/competency/examples-of-competencies/
[18].E Riggio April 27, 2014 The Top 10 Leadership Competencies https://www.psychologytoday.com/blog/cutting-edge-leadership/201404/the-top-10-leadership-competencies
[18] http://www.best-job-interview.com/12-core-competencies.html
[19] D. Surges Tatum 1991 *A measurement system for speech evaluation* (University of Chicago: unpublished dissertation)
[20] A H Monroe 1935 *Principles and types of speech* (New York: Scott, Foresman, and Co.)
[21] Linacre 1999 *J Outcome Meas* **3** 103-122
[22] D Surges Tatum 2001 *Persuasive communication: business & professional speaking* (Chicago, Illinois: Meaningful Measurement)

Journal of Physics: Conference Series **772** (2016) 012020 doi:10.1088/1742-6596/772/1/012020

Linear units improve articulation between social and physical constructs: An example from caregiver parameterization for children supported by complex medical technologies

N Bezruczko[1], T Stanley[2], M Battle[3] and C Latty[3]

[1]The Chicago School of Professional Psychology, Chicago, USA
[2] Indiana University Health, Indianapolis, USA
[3]Riley Hospital at Indiana University Health, Indianapolis, USA

E-mail: nbezruczko@msn.com

Abstract. Despite broad sweeping pronouncements by international research organizations that social sciences are being integrated into global research programs, little attention has been directed toward obstacles blocking productive collaborations. In particular, social sciences routinely implement nonlinear, ordinal measures, which fundamentally inhibit integration with overarching scientific paradigms. The widely promoted general linear model in contemporary social science methods is largely based on untransformed scores and ratings, which are neither objective nor linear. This issue has historically separated physical and social sciences, which this report now asserts is unnecessary. In this research, nonlinear, subjective caregiver ratings of confidence to care for children supported by complex, medical technologies were transformed to an objective scale defined by logits (N=70). Transparent linear units from this transformation provided foundational insights into measurement properties of a social-humanistic caregiving construct, which clarified physical and social caregiver implications. Parameterized items and ratings were also subjected to multivariate hierarchical analysis, then decomposed to demonstrate theoretical coherence (R^2 >.50), which provided further support for convergence of mathematical parameterization, physical expectations, and a social-humanistic construct. These results present substantial support for improving integration of social sciences with contemporary scientific research programs by emphasizing construction of common variables with objective, linear units.

1. Introduction

Since the 1980s, organizations such as the National Academy of Sciences, US Global Change Research Program, International Social Science Council, and International Geosphere-Biosphere Program among others have emphasized social and physical science collaboration to address international research problems. Growing recognition that social forces are a dominant contributor to global warming, oceanic pollution, and international terrorism emphasizes mobilization of social and physical science collaboration to solve international problems. Despite early disappointments, progress toward collaboration continues and today several initiatives such as the Intergovernmental Platform on Biodiversity and Ecosystem Services, as well as ongoing basic and applied science programs integrate physical and social sciences [1, 2].

2. Problem

Despite broad consensus about collaboration benefits to scientific problem solving, and enormous strides toward establishing organizational infrastructure, social scientists express deep frustration with institutionalized organizational barriers, as well as entrenched social attitudes against physical and social science integration [3]. In general, social and physical collaborations tend to display asymmetrical relations. Not surprisingly, social scientists are calling for more meaningful collaboration with physical science initiatives and specifically greater access to decision making and research policy [4].

While these complaints are certainly legitimate, social scientists demonstrate little recognition of their own dysfunctional role in blocking advances toward meaningful collaboration. The social sciences typically maintain incommensurable paradigms, divergent epistemologies, as well as incompatible ontologies, which has led to their marginalization in scientific policy and governance. At a practical level, lack of a common language or complementary methodology creates enormous barriers to collaboration with physical sciences. Not surprisingly, the social sciences are commonly viewed as intellectually provincial outposts on the outskirts of scientific knowledge. A goal of this research is to show this general bifurcation into social and physical sciences is no longer necessary.

3. Objective

Genuine efforts to improve physical and social science collaboration have turned to mutual benefits derived from shared empirical methods. For example, correlational statistical methods originally developed by the social sciences, now provide a venue for convergence between otherwise disparate social and physical research traditions. Specifically, statistical correlation provides a common language and concepts between social and physical scientists, a foundation for collaboration. Unfortunately, statistical correlation also exposes the weak measurement conventions of contemporary social science research, which erode and undermine common goals.

Unlike the theoretical architecture of physical science, which is built on a foundation of objective, linear measures, social sciences remain fixated on ordinal raw scores and ratings. Broadly across the social sciences, researchers are deeply committed to measurement with "assumed" rather than objective properties, which is simply anathema to physical science collaboration. The function of objective, international measurement standards in physical science makes folly of conventional social science measurement practices. Sensitivity among social scientists to weak measurement has led to important advances such as generalized linear models (GLM) [5], but support for them is restricted, and social researchers typically find GLM too complicated for ordinary applications.

An objective of present research was to demonstrate a simple linear probabilistic measurement model, which addresses social science measurement limitations. This demonstration establishes a common language between social and physical constructs and, implemented broadly, would improve possibilities of effective collaboration.

Results from present, ongoing research demonstrates simple mathematical parameterization with a Rasch model, which clarifies axiomatic foundations of a real number scale. Measuring units in this framework were shown to have Euclidian properties necessary to represent both physical and social observations in a common metrological framework. Then hierarchical statistical methods examined theoretical coherence of a psychophysical construct in this context. A practical effect here is to render the long standing bifurcation into social and physical sciences certainly less meaningful, and, arguably, unnecessary.

More specifically, an experiment was conducted that compared subjective self-confidence ratings of mothers caring for children supported by complex medical technologies. One group received specially designed simulation training, while another group participated in conventional, instructor-led hospital-based training. A functional caregiving construct representing mothers' confidence was operationally defined with physical tasks, which were intended to manifest a quantitative performance gradient. Those mothers more confident about caring for their children supported by medical technologies were expected to demonstrate significantly higher confidence on the measurement construct, and overall probabilistic measures were compared between training groups.

4. Methods

4.1. Sample

Participants consisted of all caregivers enrolled in supervised training at Riley Children's Hospital at Indiana University Health (N ~ 70). While infant tracheostomy patients ranged from one month to 17 years, primary caregivers were typically parents. All participants were from American Midwestern states and predominantly with English language background.

4.1. Data

Both competency and confidence observations using standard rating scales were collected. Caregiver competency was scored on several tasks central to hospital-led training. For example, performance tasks such as "Alleviate respiratory distress for a child with a tracheostomy" and "Perform CPR on a child with a tracheostomy" consisted of several discrete steps, which were scored dichotomously (0/1). Caregivers reported self-confidence ratings using items with rating scales (1 = none, 2 = A little, 3 = somewhat, 4 = A lot, and 5 = completely).

4.2. Analysis

Competency and confidence ratings were transformed to linear measures with a Rasch model for rating scales and results were reported in both raw form (scores and ratings) and linear logits (log odds units). Instrument reliabilities of all psychometric measures exceeded conventional standards, that is, alpha >= .80.

4.2.1. Measurement model. Raw rating responses were first transformed to linear units (log-odds) with Winsteps software for Rasch measurement [6], then differences between item difficulties and caregiver's confidence were modeled with one parameter logistic function, which established a common real number scale for items and caregivers. The mathematical model for the raw data transformation is:

$$\Pi_{nix} = \frac{\exp \sum_{j=0}^{x} [\beta_n - (\delta_i + \tau_j)]}{\sum_{k=0}^{m} \exp \sum_{j=0}^{k} [\beta_n - (\delta_i + \tau_j)]} \qquad (1)$$

Where β = caregiver's confidence, δ = task difficulty, and τ = rating scale thresholds. Π_{nix} is the probability that any item δ_i will be rated X by caregiver β_n where X takes a value from a fixed range (j = 1, 2, 3, 4, and 5), m = number of rating steps for an item, and k = ith step.

Dichotomous competency scores were transformed with a Rasch model for achievement scales [7].

4.2.2. Measurement properties. Transformation of caregiver self-confidence ratings to linear measures were evaluated with appropriate separation indices and fit statistics, as well as conventional psychometric reliability. A comparable evaluation was conducted with competency scores.

4.3 Conventional statistical analyses. Competency and Confidence were statistically compared in a 2 X 2 within group MANCOVA with repeated measures, which controlled for age, language, and gender covariates. Model assumptions concerning normality, linearity, additivity, and independence were examined. Standardized effect sizes were computed to represent group differences. Contributions

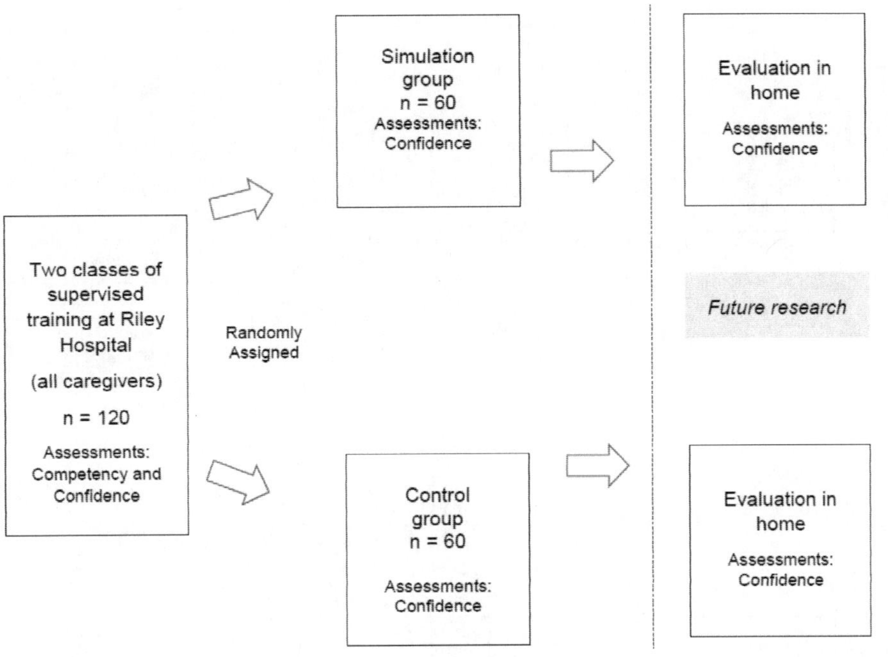

Figure 1. Functional caregiving experiment to validate physical implications of simulation training.

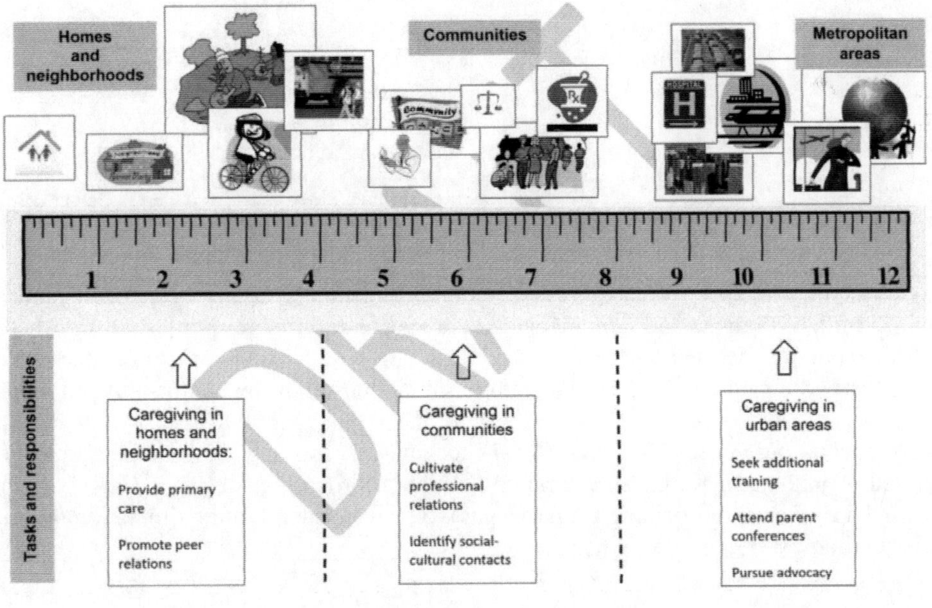

Figure 2. Physical caregiving challenges increase as caregivers and children advance from homes into neighbourhoods, communities, and urban areas. Social-cultural context defines specific skills caregivers need at each level. Measurement on a linear scale clarifies probabilistic differences in caregiving difficulty as complexity increases across the continuum.

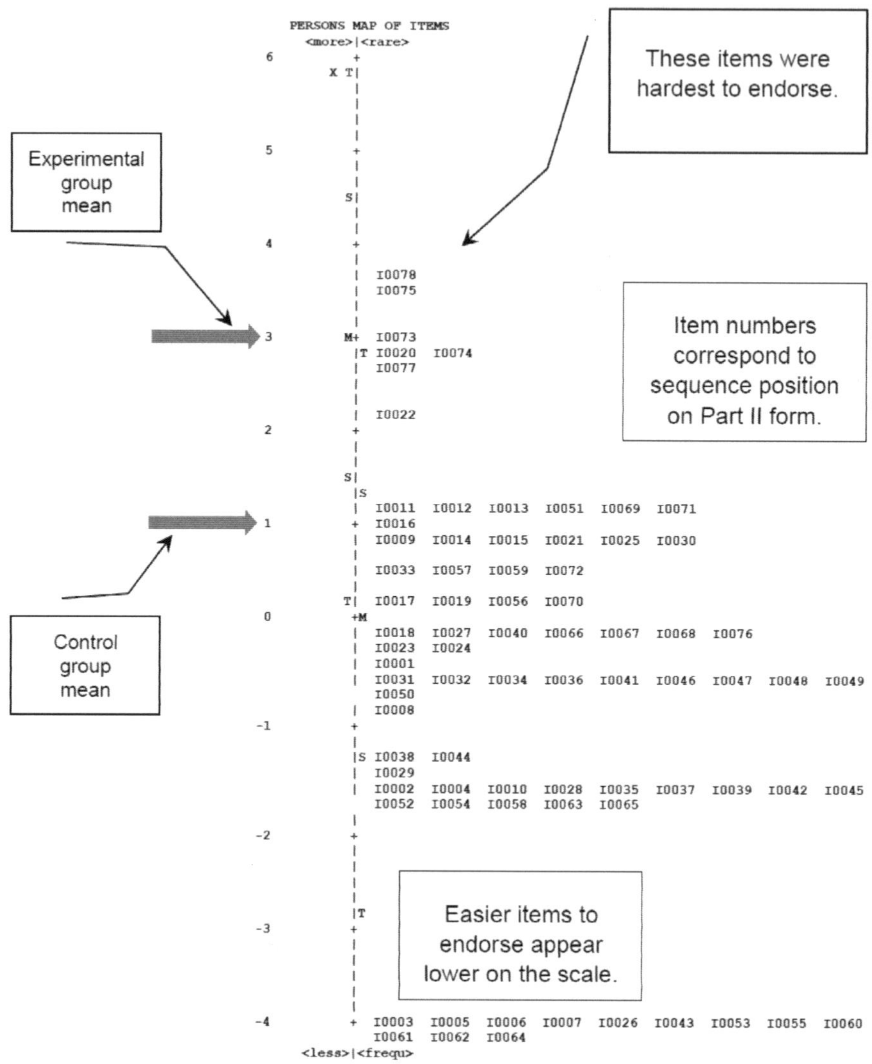

Figure 3. Parameterization of caregiver self-confidence ratings in an objective linear framework. Mothers's confidence ratings and item difficulties are represented here in a framework that describes probabilistic relations between perceived task difficulties and caregiver confidence. Results are presented after participation in medical technology simulation versus instructor-only led training.

to caregiver self-confidence after traditional hospital-led training versus experimental training (hospital-led + simulation) were also examined with the following model: Y = Caregiver confidence, X_1 = Confidence after initial training (scale values), X_2 = Confidence after simulation training (coded for simulation and control groups), when:

$$Y = a + b_1 X_1 + b_2 X_2 + e \quad (2)$$

Then, significant change in R^2 between caregiver self-confidence attributable to training + simulation confidence, when controlling for age, language, and gender compared to hospital training only clarified effectiveness strength of simulation training. Figure 1 presents the study design.

5. Results

Figure 2 presents an idealization of qualitative physical performance goals in an objective scientific framework. Figure 3 presents their parameterization in a common framework for caregivers' self-confidence and perceived physical task difficulties. Both mothers' confidence ratings and physical task difficulties defined a caregiving construct that was mathematically transformed to common logit units. Therefore, objective probabilistic relations can now be expressed about mothers' confidence to complete specific caregiving tasks. Likewise, caregiver confidence measures can be aggregated by groups to establish effect sizes. Moreover, ongoing observations are incrementally increasing the capacity to investigate the dynamics underling mothers' confidence and better understand the conditions that promote mothers' confidence to implement medical technologies in their homes.

As expected, caregiving tasks defined a coherent empirical variable, which described mothers' confidence to care for children with medical technology. Moreover, the physical task hierarchy conformed to predictions concerning qualitative caregiver shifts across a humanistic construct.

6. Discussion

A goal of this research was to demonstrate that objective linear units contribute to transparent communication among disciplines in scientific research programs, which should improve understanding and facilitate problem solving in collaborations. Results presented here show human self-perceptions parameterized on probabilistic linear scales corresponded directly to physical performance expectations, both defined by objective thresholds also expressed in linear units. This framework mediated parallel social and physical research paradigms, which mitigate those differences that typically stymie collaboration. An important outcome here was empirical analysis meaningful to both physical and social perspectives. These results suggest the traditional bifurcation of physical and social observations need not have central importance to scientific problem solving.

Finally, this facilitation of research between social and physical sciences could logically have been accomplished with alternative mathematical approaches to metrology. However, the measurement system implemented here, arguably provided an efficiency, convenience, and transparency currently not widely prevalent in the social sciences.

6.1. Implications

First, these results show simple linear measurement methods are already available for representing social and physical observations in common frameworks with objective linear properties. Common thresholds on this framework can now be articulated between social-behavior and physical science targets. In addition, the strong properties of this framework support meaningful and convenient measurement of change, which is typically only assumed in weaker social science paradigms. These advances have broad implications across health and biological sciences where integration of social constructs is desperately needed. An integration of physical and social methods is already documented in numerous laboratory studies and broader applications should be considered of those advances.

7. References

[1] Mooney H A, Duraiappah A, and Larigauderie, A 2013 Evolution of natural and social science interactions in global change research programs. *PNAS* **110** 3665

[2] Weaver C P, Mooney S, Allen D, Beller-Simms N, Fish T, Grambsch A E, and Langner L 2014 From global change science to action with social sciences. *Nature Climate Change* **4** 656

[3] Viseu, A 2015. Integration of social science into research is crucial. *Nature* **525** 291

[4] Sovacool B K, Ryan, S E, Stern, P C, Janda K, Rochlin G, Spreng D, and Lutzenhiser, L 2015 Integrating social science in energy research. *Energy Research & Social Science* **6** 95

[5] McCullagh P and Nelder J A 1989 *Generalized Linear Models* (Boca Rotan, FL: Taylor)

[6] Linacre J M 2015 Winsteps software Version 3.9 (Beverton, OR: Winsteps.com)

[7] Wright B D and Stone M H 1979 *Best Test Design* (Chicago: MESA Press)

[8] Wright B D and Masters F N 1982 *Rating Scale Analysis* (Chicago: MESA Press)

Journal of Physics: Conference Series **772** (2016) 012021 doi:10.1088/1742-6596/772/1/012021

Expressing Academic Growth on a Conjoint Developmental Scale with General Objectivity

G L Williamson

Senior Research Scientist, MetaMetrics, 1000 Park Forty Plaza Drive, Suite 120, Durham, NC 27713, USA

E-mail: gwilliamson@lexile.com

Abstract. Individual growth curves yield insights about growth that are not available from any other methodology; and, developmental scales based on conjoint measurement models provide unique interpretive advantages for investigations of academic growth. Benefits are demonstrated in three examples. First, a series of fifteen statewide reading growth curves is annotated with historical policy actions related to assessment, accountability and early interventions. Second, a common measurement framework simultaneously addresses five interpretive perspectives—student reading growth; achievement level standards; K-12 text complexity standards; postsecondary reading demands; and, occupational reading demands. Third, incremental velocity norms are introduced for average reading growth based on a parametric mathematical model for individual growth curves.

1. Perspective

There are five essential properties for the optimal measurement of student growth. The measurement scale must be *unidimensional*; *continuous*; *developmental*; *equal-interval*; and, must have *invariance of scale unit*. The first four properties can be achieved through item response theory [1]. Invariance of scale unit can be attained when a Rasch-based measurement scale is anchored by means of a construct specification theory. See [2] for a detailed discussion of these principles.

As of this writing, there are only two conjoint measurement scales for reading ability that have all five properties mentioned above. One of them, The Lexile® Framework for Reading [3], was adopted in North Carolina (NC) around two decades ago as a supplemental scale for its assessment program. This has yielded unprecedented interpretations of reading growth featured in the Results section.

2. Approach

The "data" for the examples were drawn from multiple traditions of previous research. One tradition produced distributions of text-complexity measures for reading materials characteristic of the public schools [4] and postsecondary life in the USA [5, 6]. Another tradition produced reading achievement measures for individuals who attended public schools in the state of North Carolina [7, 8, 9]. The third tradition of research consisted of multilevel growth models applied to longitudinal data [2, 10].

3. Results

3.1 Student growth with historical policy annotations

Aggregate growth curves for 15 successive panels of students are featured in Figure 1 using a common scale for reading achievement. The growth curves are ordered chronologically from left (earlier) to right (more recent). Each growth curve spans Grades 3-8 in different years, noted along the horizontal axis. Each panel consists of students who progressed from grade to grade without repeating a grade and who had complete data across all six occasions.

In addition to the displayed growth curves, Figure 1 is annotated schematically with historical policy actions that took place during the time frame (1993-2014). There are three categories of operational policy initiatives: a) assessment editions; b) accountability programs; and, c) early intervention reading initiatives. A few comments about the three categories will facilitate understanding the subsequent discussion of results.

There have been four editions of NC End-of-Grade (EOG) reading assessments administered statewide in Grades 3-8—Edition 1 (1993-2002); Edition 2 (2003-2007); Edition 3 (2008-2012); and a fourth edition, called NC READY EOG (2013-present). These editions are annotated across the top of Figure 1. Each edition had its own unique vertical scale. Three of the four scales were directly linked to the Lexile scale via symmetric linking functions. However, Edition 2 was not directly linked to the Lexile scale, though it was linked to Edition 1 (equi-percentile method) and Edition 3 (Stocking-Lord method). These linking studies provided the means to translate all scores to the Lexile scale.

Six waves of accountability initiatives occurred during the time frame and are annotated from left to right in the mid-section of Figure 1. From 1993-1996 institutional accountability was focused on cross-sectional analyses of school district average performance (Senate Bill 2). In school year 1995-96, the state piloted a new accountability system in ten local education agencies using the school building as the unit of analysis for institutional accountability. The program, dubbed the *ABCs*, was first implemented statewide in elementary schools in 1996-97 and additionally in high schools the following year. The signal feature of the ABCs was that unique growth expectations were set for each school and growth was based on the performance of the same students over time. There were two generations of the ABCs, distinguished by different methods (gain-score versus status-projection) for calculating growth. The current generation of school-level accountability, NC READY, was initiated in 2012-13. Under NC READY, the Education Value-Added Assessment System (EVAAS) provides school value-added growth results. Finally, individual accountability became effective for students in Grades 3 and 8 in 2002. Student accountability standards required that individual students must score at or above a grade-level cut score in order to be promoted to the subsequent grade.

Two early-intervention initiatives were implemented during the time frame for Figure 1. They are annotated just above the horizontal axis. In 1993, SMART START was enacted to provide educational resources for four-year-olds to better prepare them to enter Kindergarten at age five. SMART START was gradually implemented. By 1998, it had been deployed in 55 of the state's 100 counties. By 2001, SMART START was fully deployed statewide. Also in 2001, a second early intervention program called *More at Four* was added to provide greater support for four-year-olds.

Each growth curve in Figure 1 visually manifests three features of growth—magnitude, velocity and acceleration. The overall elevation of each curve in relation to the vertical axis reflects achievement status on each occasion (i.e., the magnitude of growth). The slope of a tangent to a curve quantifies the instantaneous velocity of growth at a particular time. The curvature of each trajectory connotes the acceleration of growth. The general impression is that growth was fairly consistent from panel to panel, characterized by positive initial velocity and deceleration over time. Furthermore, there was systemic improvement in the magnitude of growth over time as evidenced by the general rise of the curves moving from left to right in chronological sequence. Other insights are revealed by considering the three categories of policy initiatives.

In relation to changing assessment editions, an important insight is that the general shape of the growth curves is robust and consistent across time, even though four successive editions of assessments were used. Capturing this consistency of growth is possible because all four editions of reading tests were calibrated to the same (Lexile) scale.

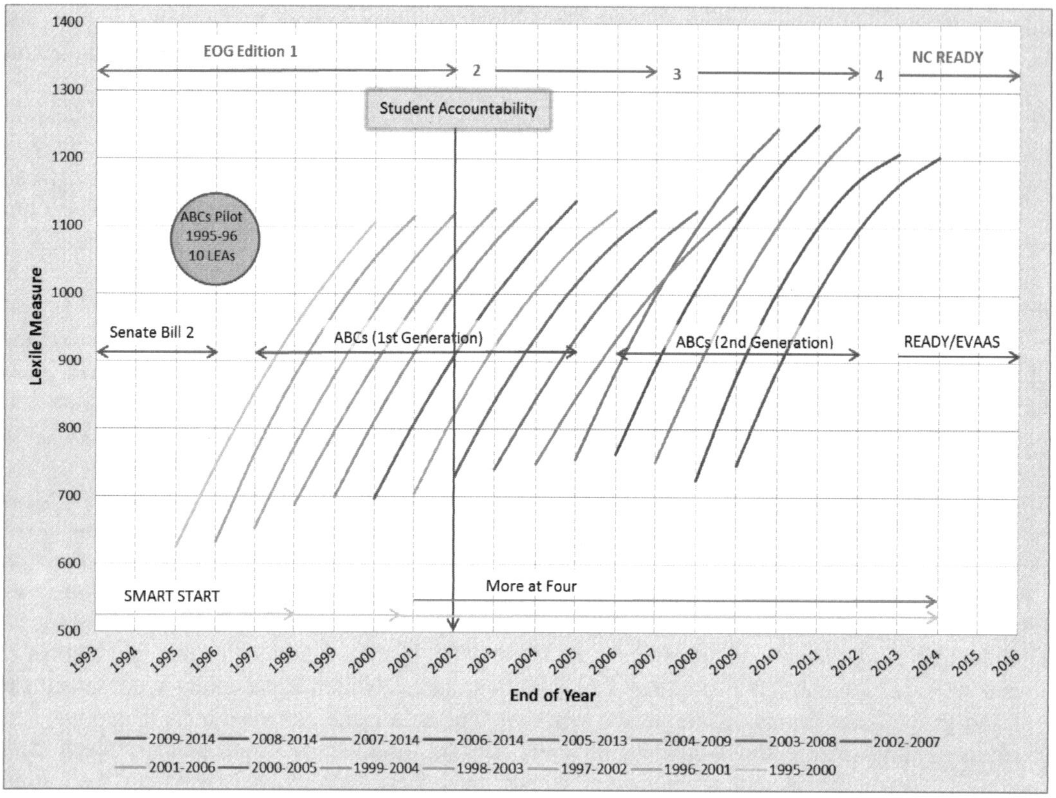

Figure 1. North Carolina (NC) growth curves (n = 1,012,118) annotated with assessment, accountability and early intervention initiatives

When growth curves and policy initiatives are viewed together in Figure 1, there are interesting correspondences that suggest how policy actions *may* have influenced changes in growth patterns over time. Although, I offer some thoughts on possible causal hypotheses, they are limited to North Carolina accountability and early intervention initiatives. For this reason, I must emphasize that what I present is not a demonstration of cause and effect. Rather, I am merely illustrating how one might identify potential causal hypotheses that could be relevant to interpreting growth and performance.

Let the reader scan Figure 1 from left to right and notice the beginning points of the growth curves. There seems to be an uptick in Grade 3 performance in 1997 and initial performance continues to rise for several years before leveling off around 2000. This uptick in the starting point for growth is interesting because 1997 was the first year that ABCs growth expectations applied to all elementary schools and it is also the year when the first students to experience SMART START reached Grade 3.

There seems to be another uptick in initial status of growth in spring 2002 and again several years when initial status continued to rise. This is interesting because student accountability standards first applied to Grade 3 in 2002. Also, the panel that began in 2002 would have contained larger numbers of students who benefited from SMART START because the four-year-olds from 1998 (when 55 of 100 counties had the program) would have reached the end of Grade 3 in spring 2002.

Now, observe the end-points of the growth curves (i.e., student performance at the end of Grade 8) for the first six panels. Eighth-grade performance was gently rising for the first three panels, possibly due to the ABCs. Then eighth grade performance appears to increase more rapidly across the next three panels after the introduction of student accountability standards for eighth-graders in 2002.

Eighth-grade achievement peaked in 2004 and 2005. The subsequent four panels exhibited lower Grade 8 performance due to increased deceleration of growth. This was likely a compound result of slowly declining reading growth in middle schools (particularly sixth grade) and the subsequent adoption of revised growth standards for the ABCs (2nd generation) effective in 2006.

Other interesting scenarios can be constructed for Figure 1. Due to space limitations, I mention only one more—the fact that the last five growth curves seem to be different from the first ten growth curves, primarily in terms of their overall length. This can be traced, in part, to a methodological artifact. For the first ten panels, Edition 2 scores were translated to the Lexile scale after applying the equi-percentile link between Edition 2 and Edition 1. For the last five panels, Edition 2 scores were translated to the Lexile scale after applying the Stocking-Lord link between Edition 2 and Edition 3. Though this may not be the only explanation, it does account for some of the observed difference in the length of the growth curves. Whether a methodological artifact or some combination of factors is ultimately the explanation, one point cannot be ignored. Placing all scores on a common scale may help identify anomalies as well as intentional changes in the features of growth.

3.2 Five perspectives for growth

In Figure 2, I display the earliest growth curve from the previous example and a recent growth curve spanning Grades 3-11 along with four additional background contexts: a) student reading achievement level standards; b) K-12 text-complexity standards; c) postsecondary text complexity; and, d) reading demands associated with specific occupations. Figure 2 originally appeared in *The Education Standard* [11] with detailed discussion; the authors used it to illustrate how states can assure that student growth is well-aligned with achievement standards and postsecondary expectations associated with college and career readiness. I incorporate the figure here to illustrate how conjoint measurement provides unique insights and interpretive perspectives for student growth and academic performance.

The rectangle located in the middle of Figure 2 represents the interquartile range of postsecondary text complexity and the diamond inside the rectangle represents the median [5]. The dots on the right-hand side of Figure 2 represent the median text complexity of reading materials associated with 59 *Bright Outlook Occupations* [12]. The dashed line segments on the left-hand side of the figure represent text complexity standards widely adopted by states for use in Grades K-12. Finally, the hash marks at Grades 3-8 represent rising student achievement standards adopted in North Carolina.

Looking at Figure 2, it is immediately obvious that the state raised achievement standards substantially over twenty years. As standards were raised they were increasingly aligned with the text requirements associated with postsecondary contexts. Most importantly, the average student growth curves also rose over time, keeping track with increasing policy and postsecondary reading requirements. These disparate perspectives are unified by the use of a common scale with conjoint measurement of persons and texts.

3.3 Incremental velocity norms

For my final example, I reproduce a table which appears in [13]. I used a parametric growth model to construct incremental velocity norms for academic growth from the end of Grade 3 to the end of Grade 11. The results are displayed in Table 1. Along the diagonal, the table contains the spring-to-spring gains in reading achievement between adjacent grades. Thus a fifth-grade teacher who wishes to know how much growth is typical during Grade 5 only needs to refer to the entry at the juncture of the row for Grade 4 and the column for Grade 5 to discover the answer—specifically, 122L. The table also contains gains between any pair of grades in the time frame. So for example a middle school principal serving Grades 6-8 can ascertain that the amount of growth exhibited by students during those three years was 260L.

Cross-sectional status norms have been the standard in education for the last 100 years. Table 1 provides growth norms based on longitudinal panel data—the same students are reflected across the entire time frame from the end of Grade 3 to the end of Grade 11. Table 1 thus fills a void for educators who wish to have defensible growth standards as a reference for interpreting aggregate student growth over time. Table 1 provides estimates of the amount of developmental reading growth between any pair of years spanned by the normative growth curve.

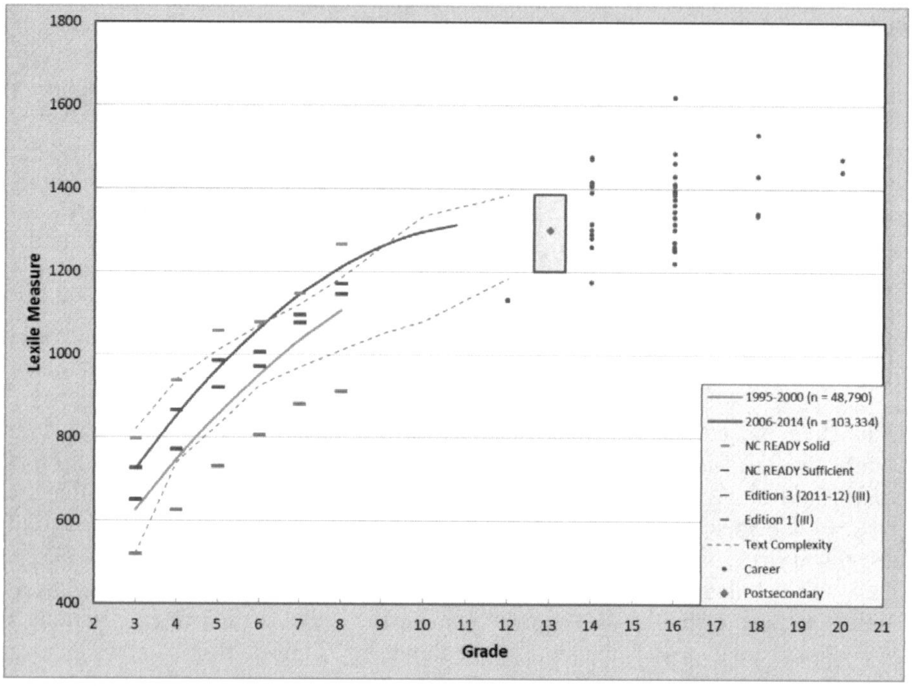

Figure 2. North Carolina reading growth and achievement standards (1995-2014) relative to K-12 text-complexity standards, postsecondary text requirements and text complexity for 59 Bright Outlook Occupations. Adapted from "Aligning the K-12 academic journey with a postsecondary destination," by M Smith and G L Williamson 2016 *The State Education Standard* **16**(1) p 16. Copyright 2016 by the National Association of State Boards of Education (NASBE).

Table 1
Incremental Velocity Norms for Average Reading Growth Denominated in Lexile Scale Units

	Student Achievement Estimated from the Average Reading Growth Curve								
	664L	803L	925L	1029L	1116L	1185L	1237L	1271L	1288L
Grade	3	4	5	6	7	8	9	10	11
3		139L	261L	365L	452L	521L	573L	607L	624L
4			122L	226L	313L	382L	434L	468L	485L
5				104L	191L	260L	312L	346L	363L
6					87L	156L	208L	242L	259L
7						69L	121L	155L	172L
8							52L	86L	103L
9								34L	51L
10									17L

Note. Based on average reading growth for North Carolina students (n = 101,610), 2002-2010. The fitted model is given by the equation: $E(L|T) = 663.8 + 148.0\ T - 8.7\ T^2$ (where T = Grade - 3). Velocity increments for adjacent grades (i.e., year-to-year gains) are shown in the shaded diagonal. Adapted from "Novel interpretations of academic growth," by G L Williamson 2016 [Manuscript submitted for publication]. Copyright 2016 by G L Williamson.

Journal of Physics: Conference Series **772** (2016) 012021 doi:10.1088/1742-6596/772/1/012021

4. Summary
The three examples presented here serve to illustrate the power of educational measurement scales that are designed to measure student growth with general objectivity. Developmental analyses of growth benefit from maintaining such scales over long periods of time because growth norms are improved and relationships between features of growth and policy contexts become more apparent. Additionally, with conjoint measurement, student reading growth and performance are easily interpreted in relation to the complexity of texts that students read.

References

[1] Yen W M and Fitzpatrick A R 2006 *Item response theory* (Educational measurement 4th ed) ed R L Brennan (Westport, CT: American Council on Education and Praeger Publishers)

[2] Williamson G L 2015 Measuring academic growth contextualizes text complexity *Pensamiento Educativo: Revista de Investigación Educacional Latinoamericana* **52**(2) pp 98-118.

[3] Stenner A J, Burdick H, Sanford E and Burdick D S 2007 *The Lexile Framework for Reading technical report* (Durham, NC: MetaMetrics) Retrieved from http://www.lexile.com/research/9/

[4] Williamson G L, Koons H, Sandvik T and Sanford-Moore E 2012 *The text complexity continuum in grades 1-12* Research Brief (Durham, NC: MetaMetrics)

[5] Stenner A J, Sanford-Moore E and Williamson G L 2012 *The Lexile® framework for reading quantifies the reading ability needed for "College & Career Readiness"* Research Brief (Durham, NC: MetaMetrics)

[6] Williamson G L 2008 A text readability continuum for postsecondary readiness *J. Advanced Academics* **19**(4) pp 602-632 doi: 10.4219/jaa-2008-832

[7] Bazemore M and Van Dyk P B 2004 *North Carolina reading comprehension tests: Technical report* (2nd ed) (Citable draft) (Raleigh, NC: North Carolina Department of Public Instruction)

[8] North Carolina Department of Public Instruction 2009 *North Carolina Reading Comprehension Tests: Technical report* (3rd ed) (Raleigh, NC: Author) Retrieved from http://www.ncpublicschools.org/docs/accountability/testing/reports/eogreadingtechman3.pdf

[9] Sanford E E 1996 North Carolina end-of-grade tests technical report # *1: Reading comprehension, mathematics* (Raleigh, NC: Department of Public Instruction)

[10] MetaMetrics 2011 *NC EOC feasibility studies research & development report* (Durham, NC: Author)

[11] Smith M and Williamson G L 2016 Aligning the K-12 academic journey with a postsecondary destination *The State Education Standard,* **16**(1) pp 14-18

[12] Williamson G L and Baker R F 2013 *Enriching the concept of career preparedness by examining text complexity associated with Bright Outlook Occupations* Research Brief (Durham, NC: MetaMetrics)

[13] Williamson G L 2016 Novel interpretations of academic growth [Manuscript submitted for publication]

Journal of Physics: Conference Series **772** (2016) 012022 doi:10.1088/1742-6596/772/1/012022

A large-scale, long-term study of scale drift: The micro view and the macro view

W He,[1] S Li,[1] G G Kingsbury[2]
[1]NWEA, 121 NW Everett St, Portland, OR 97209
[2]Independent consultant, 6134 SE Lincoln, Portland, OR 97215

E-mail: wei.he@nwea.org

Abstract. The development of measurement scales for use across years and grades in educational settings provides unique challenges, as instructional approaches, instructional materials, and content standards all change periodically. This study examined the measurement stability of a set of Rasch measurement scales that have been in place for almost 40 years. In order to investigate the stability of these scales, item responses were collected from a large set of students who took operational adaptive tests using items calibrated to the measurement scales. For the four scales that were examined, item samples ranged from 2183 to 7923 items. Each item was administered to at least 500 students in each grade level, resulting in approximately 3000 responses per item. Stability was examined at the micro level analysing change in item parameter estimates that have occurred since the items were first calibrated. It was also examined at the macro level, involving groups of items and overall test scores for students. Results indicated that individual items had changes in their parameter estimates, which require further analysis and possible recalibration. At the same time, the results at the total score level indicate substantial stability in the measurement scales over the span of their use.

1. Introduction

Developing and maintaining vertical scales for the measurement of educational achievement is a complex task. Beyond the psychometric issues, the educational setting brings its own set of challenges to accurate measurement. Among these are the following:

- From grade to grade, the content that a student is learning may change noticably, which will change the questions that it is appropriate to administer.
- From school to school, the amount and order of content changes noticably, even for students in the same grade.
- From year to year, schools may change their curriculum, instructional approaches, and instructional materials.
- As they grow, students change in their willingness to engage in assessments, and their interest in the subject being measured.
- As time passes, federal and state regulations change the requirements placed on the schools and their students.

The nature of education is an ecosystem that is constantly changing, so creating a measurement scale that measures student achievement in a consistent fashion is an interesting challenge.

In the late 1970s, the Northwest Evaluation association (NWEA) began to develop a set of measurement scales to examine student achievement in mathematics and reading. The initial work to develop these scales attached items to a Rasch measurement scale using the four-square linking design developed by George Ingebo [1] and explicated by Ben Wright [2]. The purpose of the initial development was to create item banks that could be used by multiple school districts to create assessments that matched their curriculum. The original items were associated with content from grades 3 to 8, and all assessments were administered in paper form to students in the pacific northwest.

Since that time, NWEA has expanded and updated the item banks, extended the range of the scales to measure students from kindergarten to high school, added new item styles, added a host of accommodations, moved to adaptive testing, moved to electronic delivery of all tests, and expanded to work with students across the United States, Canada, and a number of schools in other countries. Given these changes, it makes sense to examine the stability of the measurement scales, to assure that a student with a particular level of achievement would obtain approximately the same score today that they would have obtained several decades ago. A previous study by Kingsbury and Wise [3] indicated that the NWEA measurement scales in reading and mathematics showed little drift in their first twenty years of use. The current study expands on that earlier study by adding fifteen years of data and adding the language usage scale and the general science scale.

2. Methodology

A total number of 26,059 items were used in this study. These items were originally calibrated between January, 1980 and January, 2014. The items were recalibrated using responses gathered during January 1st, 2014 and November 30th, 2015. To perform the recalibration, at least 1,000 responses were collected for each item in an operational test. On average, items in each item pool contained an average of 2,953 student responses (with at least 500 responses in each grade). Since the operational tests were adaptive, items differed in their frequency of use and also in the achievement levels of students to whom they were administered. Table 1 presents the sample size of the item pool for each subject as well as the average number of total student responses per item used for the recalibration.

Table 1. Description of Item Pools and Student Responses

Scale	Items	Average number of responses per item
Reading	7776	2949
Math	10106	2941
Language Usage	5357	2956
General Science	2820	2967

2.1. Analysis: The micro view

One approach to examining scale drift is to identify items associated with the scale that change their difficulties over time. In an educational setting, this type of change might occur for any number of reasons, including change in instructional emphasis, change in common usage of words, or change in the underlying scale. This examination of item parameter stability serves as the micro view of scale stability. Consistent item calibrations allow us to conclude that the scale is remaining stable, but changes in item parameter estimates may spring from a variety of sources.

This study used the robust Z statistic developed by Huynh & Rawls [4] that originates from robust statistical procedure to detect unstable items. Z-scores are expressed in terms of standard deviations from the mean. As a result, z-scores have a distribution with a mean of 0 and a standard deviation of 1. Z score is calculated by

$$z = \frac{X - \mu}{\sigma}$$

where X is an individual score, μ is the mean, and σ is the standard deviation.

In order to compute the robust z, the mean is replaced by the median and the standard deviation is replaced by 0.74 times the interquartile range (IQR). The quantity 0.74*IQR is used to match the standard deviation of the normal distribution.

The robust statistic for each item is the ratio $z = (D - Md) / (0.74 \times IQR)$, where D is the difference between original b-parameter and the newly calibrated b-parameter, Md is the median, and IQR is the interquartile range for the differences. α in each direction was set at 5%, and the critical value is $z^*=\pm1.645$, correspondingly. All items with a robust z smaller than the absolute value of z* in absolute value were regarded as stable. Otherwise, items were flagged as drifting. This approach should identify approximately 10% of items as drifting if the null hypothesis is true. This allows the identification of many items for review, ensuring that any item with noticeable drift can be examined by content experts.

2.2. Analysis: The macro view

An additional approach to examining drift is to look at the impact of parameter estimate changes on the actual outcome measures, the total test scores for the students. This macro view accepts that individual items may drift, but asks whether this drift has an impact on students' scores and the educational decisions made with them.

Since the tests used in this analysis were adaptive, the impact of drift on total test scores will vary from student to student. In order to capture the overall impact and the variance of that impact, a set of 1,000 20-item adaptive test events were simulated using the original calibrations to select items from the item pool for General Science. General Science was selected for the macro analysis since it had the greatest percentage of items flagged for drift during the micro analysis (see below). Simulated responses were based on the new calibrations. Each test event was then scored twice, once using the original parameter estimates, and one using the new parameter estimates. These two sets of scores were then compared to identify how the change in calibration estimates would impact student scores and subsequent decisions in the classroom.

3. Results

3.1. Micro results

The existing item difficulty parameters in the bank with those from the new calibration were correlated with each other. As Table 2 indicates, the correlation coefficients $r_{old, new}$ ranged from .97 and .99 for the different scales. Using the Robust Z method (α=10%), 13-24% of the items were flagged for parameter drift for the different scales. These percentages are larger than the 10% expected under the null hypothesis, and they indicate that some amount of item drift is occurring in the items associated with each measurement scale.

While Table 2 indicates that a greater than expected proportion of the items are flagged for drift, it also indicates that the correlations between the original and new calibrations are extremely high for each of the measurement scales. This suggests that the ordering of the item difficulties is quite similar to that identified by the original calibrations.

Table 2. Item Parameter Drift Detection

Item Pool	N of items	N_flagged	% flagged	$r_{old, new}$
Reading	7776	1191	15%	0.98
Math	10106	1958	19%	0.99
Language Usage	5357	693	13%	0.97
General Science	2820	681	24%	0.97

IOP Publishing
doi:10.1088/1742-6596/772/1/012022

On average, approximately 19% of mathematics items flagged for drift. Reading items tended to be more stable than math and general science items. On average, approximately 15% of reading items were flagged as drifting, while approximately 19% of math items and 24% of general science items were flagged respectively. Since the general science items show the highest percentage of items flagged as drifting, this scale will be used as the example for further analysis.

Figure 1 shows the original calibrations for each of the general science items plotted against the new calibrations for the same items. It can be seen that the vast majority of the items have new calibrations that are quite similar to their original calibrations. Over 2000 points are on the graph, most of which have new and original calibrations that differ by two or fewer scale score points (or less than .2 logits). It is also clear that a small number of items have a new calibration that differs substantially from the original calibration (several by 2.0 logits or more). If the authors of the scale do not wish to recalibrate all of the items at once (which might lead to disruptions of trend information) they can use this figure to triage the drift to allow sequential review and recalibration.

Figure 1: Original item difficulty estimates (horizontal axis) and new item difficulty estimates (vertical axis) for general science items

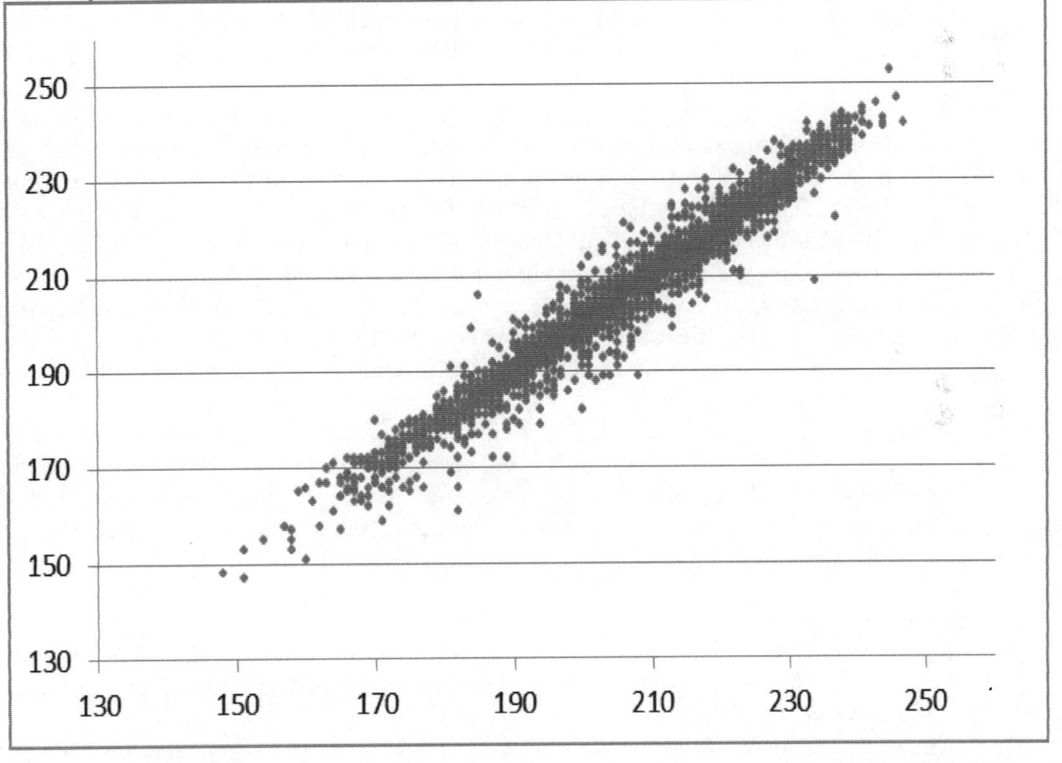

3.2 Macro Results

Since the Micro analysis indicated that the scale associated with general science had the highest percentage of items indicating parameter drift, this scale was chosen for the Macro analysis. For this scale, 24% of the items were flagged as drifting, so it is useful to ask how the drifting items influence overall test scores for students.

The simulation of 1000 adaptive test events indicated the following:

- The average difference between the scores based on the original calibrations and the scores based on the new calibrations was -0.011 logits.

- The average absolute difference between the scores based on original and new calibrations was 0.061 logits.
- The lowest difference between the scores based on original and new calibrations was -0.28 logits.
- The greatest difference between the scores based on original and new calibrations was 0.26 logits.

It is useful to note that in this simulated sample, the standard deviation of general science test scores was 1.73 logits.

4. Conclusions

The use of the micro and macro analyses allow us to look at the stability of measurement scales from different perspectives. The micro analysis allows us to look at individual items and sets of items, to determine whether they continue to measure the construct of interest in the same manner over a period of time. This allows us to identify whether processes for mid-course corrections need to be put into place for specific items in order to maintain accurate measurement. This is the part of the measurement words in which psychometricians tend to spend most of their time, and decisions here are crucial for the ongoing health of a measurement scale. In this study, the micro analysis indicates that up to 24% items should be further reviewed, to determine whether calibration values should be adjusted.

The macro analysis allows us to observe the impact that changes at the micro level have on decisions made using total test scores with groups of students. Since students see a sampling of items in any adaptive test, it is likely that they will see mostly items that haven't drifted, as well as a few items that have drifted, becoming more difficult or less difficult. It is clear from the simulation that the impact of the drifting items is quite small. The greatest differences that were observed for total test scores were less than .3 logits. This is equivalent to 3 points on the reported score scale, or less than a fifth of a standard deviation in student achievement. The average difference in scores was less than one point on the reported score scale (smaller than the smallest difference reported).

Two of the common uses of scores from this test are instructional grouping and identification of instructional need. Each of these score uses involves dividing a class into two to five instructional groupings. While these groupings are critical to good instruction, it is very unlikely that any students would be misplaced due to item drift.

This study provides strong evidence that these vertical scales in education need periodic review to identify items that may be drifting in difficulty, and need a process to deal with items that are identified. It also provides strong evidence that the drift that has occurred in the general science scale (the highest level of drift observed) does not have a substantial impact on the total test scores for students or on the decisions made from these scores.

5. References

[1] Ingebo G 1997 *Probability in the measure of achievement* (Chicago, IL: MESA Press)
[2] Wright B 1977 Solving measurement problems with the Rasch model *Journal of Educational Measurement* **14** 97-116
[3] Kingsbury G and Wise S 2011 Creating a K-12 Adaptive Test: Examining the stability of item parameter estimates and measurement scales. *Journal of Applied Testing Technology*
[4] Huynh H and Rawls A 2009 A comparison between robust z and 0.3-logit difference procedures in assessing stability of linking items for the Rasch model *Applications of Rasch Measurement in Criterion-Referenced Testing: Practice Analysis to Score Reporting* ed E V Smith Jr and G E Stone (Maple Grove, MN: JAM Press)

Use of measurement theory for operationalization and quantification of psychological constructs in systems dynamics modelling

Elena Fitkov-Norris and Ara Yeghiazarian

Kingston Business School, Kingston University, KT2 7LB, England

E.Fitkov-Norris@kingston.ac.uk

Abstract. The analytical tools available to social scientists have traditionally been adapted from tools originally designed for analysis of natural science phenomena. This article discusses the applicability of systems dynamics – a qualitative based modelling approach, as a possible analysis and simulation tool that bridges the gap between social and natural sciences. After a brief overview of the systems dynamics modelling methodology, the advantages as well as limiting factors of systems dynamics to the potential applications in the field of social sciences and human interactions are discussed. The issues arise with regards to operationalization and quantification of latent constructs at the simulation building stage of the systems dynamics methodology and measurement theory is proposed as a ready and waiting solution to the problem of dynamic model calibration, with a view of improving simulation model reliability and validity and encouraging the development of standardised, modular system dynamics models that can be used in social science research.

1. Introduction

The analysis tools available to social scientists have traditionally been developed to study and analyse physical phenomena and have been designed to meet the requirements of disciplines, which gather ratio or interval data such as physics and engineering. These tools have subsequently been adapted for application to the softer human oriented sciences such as psychology and education. Due to the different measurement approaches adopted by social scientists, who predominantly gather ordinal or nominal data, the tools are routinely applied to data without checking that it meets the assumptions required by a particular method and these issues have attracted some attention [1]. A few techniques have been developed specifically to handle qualitative concepts and these include: qualitative reasoning [2], soft systems methodology [3], causal mapping and systems dynamics [4]. These powerful qualitative approaches allow researchers to create dynamic models able to cope with uncertainty and incomplete information thought the use of latent constructs as the main unit of analysis. Not only does this allow for a flexible and relatively easy structuring of the modelled event but the approach enables direct and intuitive interpretation of the results [2].

A particular advantage of the qualitative approaches mentioned above is the use of feedback loops to represent the dependence and complex interrelations between variables as a process that develops over time, rather than a static snapshot of the system at just one point in time. This representation naturally fits in with social science phenomena which consist of a number of components from different phenomenological domains (sensory, behavioural, attitudinal) that interact and it is these interactions within a particular time frame, coupled with the structural dependence between the components, that give rise to a range of possible outcomes [5]. However, the dynamic nature of social phenomena has been generally ignored by social science researchers who traditionally take a reductionist approach in their studies [5] by considering as static snapshot of the system and just one particular point of time.

A notable attempt to move away from the reductionist approach is the work by development psychologist Esther Thelen [6] who adopted a Dynamic System Theory (DST) approach to her studies. Dynamic system theory evolved from classical mechanical engineering and uses sets of differential or difference equations to model the interactions between system variables over time and the system's output is determined by a mixture of past inputs and the output changes over time if the system is not in a state of equilibrium [7]. The dynamic systems approach was designed to handle problems from exact sciences and its approach is still predominantly reductionist as the objective is to build a mathematical model of the system whose behaviour can be examined under different initial conditions, although it has the capacity to handle feedback by using classic control theory tools.

An alternative methodology, developed in the mid 1950s by Jay Forrester for modelling analysing social science problems is system dynamics [4]. The objectives of system dynamics are: "... studying and managing complex feedback systems, in managerial, organizational and socioeconomic context..." [7, p. 5] and this is achieved by a set of simple rules for representing interrelationships between latent and observable variables to model complex dynamical entities. The system dynamics modelling methodology as identified by Sterman [8] is carried out in seven steps: the first step identifies a problem of interest to the researcher by specifying the endogenous and exogenous variables to be included in the model, followed by the statement of a tentative dynamic hypotheses that make a tentative assumption about the causes of the problem to be addressed. The researcher then builds a representation of the dynamic system using causal loop diagrams (CLD), which are used to examine the interrelationship between the different variables in the system. The interaction between different types of feedback loops, positive and balancing as well as inherent delays in the system give rise to a range of different behaviours, such as exponential growth, s-curve and oscillation, that can be exhibited by the system. The causal loop diagrams are then converted into a computer simulation using stocks and flows, which can be used to simulate a range of different scenarios in order to identify the best solution to the problem. The modelling methodology's final step is the implementation of the identified optimal strategy into practice.

System dynamics provides a natural and easy way to identify relationships between variables at the causal loop diagram stage by describing the sign of the correlation/association between two variables, rather than specifying the magnitude of the relationship. So if two variables are positively correlated an increase in one variable will result in an increase in the other, whereas negative relationships will be reflected by an increase in one variable, leading to a decrease in the other (all else being equal), however, the rates of change will not be specified at this stage of the modelling process. For example, as shown by Lizeo [9] in considering the impact of student willingness for working in groups on the group performance, we can describe the interrelationship between four of the variables in the system (willingness to work in groups, number group activities, performance and complacency) as follows: As the students willingness to work in groups increases, so would the number of learning oriented activities they undertake, which in turn will increase the group performance. As group performance increases, we can expect that this will also in turn increase the willingness to work in a group. This will be a positive loop (see Figure 1). However, as students' performance increases, the group complacency will also increase and as complacency increases, this will in turn lead to fewer learning oriented activities, leading to a decrease in performance. Complacency forms part of a negative feedback loop. So the performance of the group will be determined by the interaction of the positive loop of willingness to work and the negative loop of group complacency. The actual behaviour observed at any on point in time will depend on the initial conditions for the group, i.e. their complacency and willingness to work in a group. So it is to be expected that a group with high complacency and low willingness to work will perform less well (all other factors being equal) than a group with high willingness to work and low complacency, even though this analysis is carried out without reference to specific values.

This qualitative analysis is a useful tool for identifying drivers of group's performance, but in order to maximise the utility of the model, a dynamic simulation can be developed, quantifying the constructs used in the qualitative analysis. However, converting the causal loop diagrams to a dynamic simulation model poses two major challenges for the researcher, namely: operationalization of any latent constructs into variables and quantification of the measurement variables.

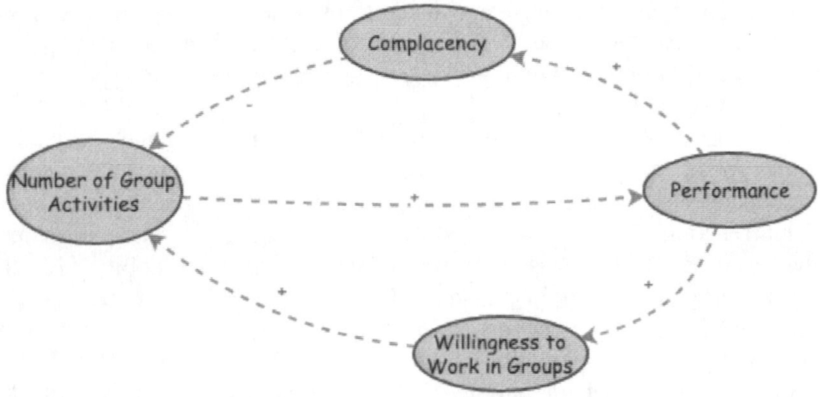

Figure 1. Causal loop diagram of factors affecting group performance

2. Challenges with dynamic simulation building

2.1. Operationalization of latent constructs.

The first issue arises as a result of the inherent difficulties of operationalizing latent scales and converting them to meaningful variables [10]. Latent constructs form an important part of psychological investigations and are used widely as tools for building frameworks representing the dimensions of latent concepts. The process is well established and involves the use of observable or manifest variables, gathered using questionnaires, and factor analysis to ascertain the underlying structure of the latent construct by identifying the manifest variables that load highly on each dimension. This is inherently a constructivist approach as it assumes that knowledge about unknown concepts is inferred from known entities (namely manifest variables) [11]. While intuitively appealing, doubts have been raised regarding the suitability of psychometric tests as a tool for operationalizing latent constructs under the positivist phenomenological umbrella as they violate the assumption of logical independence between the observed and unobservable entities [11]. Operationalizing latent constructs during the causal loop diagram stage of the systems dynamics methodology is relatively easy as systems dynamics was designed with the objective of handling easily both latent and observed variables. For example, the causal loop diagram in Figure 1 linked effortlessly the impact of complacency and group performance and allow researchers to use the variables in order to identify systemic and structural dependences in the system. However, moving onto the simulation stage of the systems dynamics methodology poses some challenges as performance is an observable variable which can be operationalized relatively easily by considering the assignment grade or the average grades for the students to date, for example, while complacency is not that easy to operationalize. The difficulties arise at a number of points in the process of converting a concept into a simulation variable, starting from the nominal definition of the concept, through to its quantification [10].

The first step of defining a latent simulation variable requires clear statements of its nominal definition, although this is not always explicitly stated in the systems dynamics methodology. Nominal definitions, normally developed by lexicographers, can encompass different aspects of the same concept. For instance, the definition of complacency in the Cambridge Online Dictionary is: "a feeling of calm satisfaction with your own abilities or situation that prevents you from trying harder" whereas the Merriam-Webster dictionary defines complacency as: "a feeling of being satisfied with how things are and not wanting to try to make them better". Although similar in defining complacency as a feeling, the two nominal definitions are not identical. The first definition suggests a necessity for calmness and is focused on the self and trying harder, while the second takes on an exogenous focus on making things better. Each definition places emphasis on a different aspect of the feeling and any operationalization of the variable will reflect the inherent assumptions taken by the researcher. Although nominal definitions are outside the remit of system dynamics, the system dynamics methodology requires researchers to be explicit about the variables included in any particular model as part of the first stage of the model building process, leading to greater transparency of the inherent assumptions made at that stage.

A further issue that affects operationalization of latent variables at the simulation stage of the systems dynamics methodology is that of generalizability. Social scientists are well aware of the multidimensional nature of latent constructs and routinely use factor analysis to build reliable and valid scales for measuring latent constructs. However, the majority of scale developed are context specific as that reduces their dimensionality and improves their reliability [12], [13]. Context-specificity ties in closely with the problem oriented nature of system dynamics but it also feeds into one of its inherent weaknesses as it encourages creation of models that are specific to the problem at hand, rather than more generic system models that could be used in a range of different contexts [7]. In a similar vein in social sciences reliability and validity are the main drivers in determining the suitability of measurement instruments such as latent constructs but the quest for improving reliability and validity also pushes researchers into greater and greater context-specificity, which in turn affects the generalizability of the research conclusions. However, while the issue of model validation has received relatively little attention in the field of systems dynamics [14], the debate of validity in the field of psychology has been on going [15], [16]. Unified approaches for testing model validity by using a network of argument based claims that can be build up to any level of complexity have been proposed [16] and this approach may offer a systematic framework for testing the validity of systems dynamics models, although some doubts have been raised with regards to the widely adopted nomological approach to validity [15]. A different approach to validity, defining it as a variation of the attitude leading to a causal variation in the measured outcome have also been proposed [15], and offer a more flexible alternative to the traditional nomological network of constructs. There seems to be further scope for developing reliability and validity measures that take into account the context-specificity of the developed model and impose a penalty for any model that is too context specific. This could be similar to the Bayesian Information Criterion for example, which tests the goodness of fit of quantitative models to the data used to build them but also imposes a penalty for loss of degrees of freedom which occurs as the number of estimated parameters and complexity of the mathematical models increases [17]. A corresponding measure in social sciences will encourage the creation of less context specific models and encourage social scientists to focus on generalizability.

2.2. Quantification of latent constructs

The second issue that a researchers face when building a system dynamics simulation is that of quantification of the latent variables [18]. The issue arises as psychological measurements are normally gathered using multidimensional instruments with nominal or ordinal scales. However a systems dynamics simulation assumes that the variables simulated are interval or ratio since their values are added, subtracted and multiplied as part of the modelling process [19]. The problem is compounded further by the fact that different latent variables could be measured on different scales. Therefore, the two aspects that need to be addressed in quantification of variables for the system dynamic simulation are:

- Converting variables' scale of measurement into one meeting the criteria of (at least) interval scale measurements.

- Defining appropriate methodologies for combining constructs with different scales with the objective of building meaningful composite measures.

Measurement theory provides the means for addressing these issues. Significant strides have been made into the development of person and item invariant measurement scales using models proposed by Guttman, Rasch and Mokken [20]. These item and person invariant scales can be combined and linked without loss of generalizability [21] and Rasch models can be used as a confirmatory tools to verify that ordinal data gathered using Likert-type scale questionnaires satisfy the equidistance constraints of interval data. The interval data can be converted into ratio scale by appropriate modification such as the one suggested by Levine [18].

Research into linking different scales using non-parametric approaches and achieving meaningful composite measures for latent variables is also on going in other disciplines such a clinical studies [22] and further development in this area can be expected as a result of growing interest in the development of meta analytic analysis tools [23]. Therefore, the research carried out in measurement theory can help overcome the limitations of systems dynamics by providing a robust set of tools for quantifying and operationalizing latent constructs and converting them into observable, ratio scale variables that can be used in dynamic simulations.

Although some researchers have questioned the assumption that "true" measurement and quantification of latent constructs is possible due to the heterogeneous nature of the intervals between the different measurement categories gathered using Likert scales [11], [24], a particular advantage of system dynamics models is their capacity to accurately represent the structural dependence of the entities within the system, without any a priori assumptions of the specific measures of the modelled variables. By focusing on the behaviour of variables over time, rather than at a particular instance in time, researchers can draw generalised conclusions about the variable behaviour. Thus system dynamics models offer an alternative structure driven analytical approach which can overcome some of the objections raised to the use of latent variables in traditional data driven analytical approaches [11].

3. Conclusion

This article examined the applicability of systems dynamics as a possible analysis and simulation tool that bridges the gap between social and natural sciences. A brief overview of the systems dynamics modelling methodology identified causal loop diagrams as a powerful tool for identifying system behavior generated by the structural dependence between the different entities, providing social scientist with an intuitive and easy to use analytical instrument. However, the simulation stage of the system dynamics methodology identified challenges with the operationalization and quantification of latent variables and measurement theory was proposed as a solution to the problem of dynamic model calibration. Of particular note are approaches for building invariant scales, which can be used to create meaningful and accurate composite measure variables that satisfy the criteria for ratio type measurement. The use of standardized variables and measurement will facilitate the development of standardized, modular system dynamics models that can be used by social scientists for model validation and in theory design.

References:

[1] J. M. Batista-Foguet, B. Ferrer-Rosell, R. Serlavós, G. Coenders, and R. E. Boyatzis, "An Alternative Approach to Analyze Ipsative Data. Revisiting Experiential Learning Theory," *Front. Psychol.*, vol. 6, pp. 575–10, Nov. 2015.

[2] L. Travé-Massuyès, L. Ironi, and P. Dague, "Mathematical Foundations of Qualitative Reasoning.," *AIM*, vol. 24, no. 4, pp. 91–106, 2004.

[3] P. B. Checkland, *Systems Thinking, Systems Practice*. John Wiley & Sons Ltd., 1998.

[4] J. W. Forrester, "System dynamics—a personal view of the first fifty years," *Syst. Dyn. Rev.*, vol. 23, no. 2, pp. 345–358, Oct. 2010.

[5] S. Salvatore, W. Tschacher, O. C. G. Gelo, and S. C. Koch, "Editorial: Dynamic systems

theory and embodiment in psychotherapy research. A new look at process and outcome," *Front. Psychol.*, vol. 6, pp. 1–3, Jul. 2015.
[6] E. Thelen, "Dynamic systems theory and the complexity of change," *Psychoanal Dialogues*, vol. 15, no. 2, pp. 255–283, 2005.
[7] L. Osipenko and J. Farr, "System Dynamics and Dynamic Systems Integration in Regulatory Environments," presented at the System Dynamics Society Conference, Oxford, 2004.
[8] J. D. Sterman, *Business Dynamics: Systems Thinking and Modeling for a Complex World.* McGraw-Hill, 2000.
[9] E. Lizeo, "A dynamic model of group learning and effectiveness," presented at the System Dynamics Society Conference, Boston, 2005.
[10] C. Jacobsen and R. Bronson, "Defining sociological concepts as variables for system dynamics modeling," *Syst. Dyn. Rev.*, vol. 3, no. 1, pp. 1–7, Dec. 1987.
[11] J. Michell, "Constructs, inferences, and mental measurement," *New Ideas Psychol*, vol. 31, no. 1, pp. 13–21, Apr. 2013.
[12] C. Robie, M. J. Schmit, A. M. Ryan, and M. J. Zickar, "Effects of Item Context Specificity on the Measurement Equivalence of a Personality Inventory," *Organ Res Methods*, vol. 3, no. 4, pp. 348–365, Oct. 2000.
[13] M. Kunter, A. Frenzel, G. Nagy, J. Baumert, and R. Pekrun, "Teacher enthusiasm: Dimensionality and context specificity," *Contemp Educ Psychol*, vol. 36, no. 4, pp. 289–301, Oct. 2011.
[14] Y. Barlas, "Formal aspects of model validity and validation in system dynamics," *Syst. Dyn. Rev.*, vol. 12, no. 3, pp. 183–210, Sep. 1996.
[15] D. Borsboom, G. J. Mellenbergh, and J. van Heerden, "The Concept of Validity.," *Psychol Rev*, vol. 111, no. 4, pp. 1061–1071, 2004.
[16] M. T. Kane, "The argument-based approach to validation," *School Psych Rev*, 2013.
[17] J. J. Dziak, D. L. Coffman, S. T. Lanza, and R. Li, "Sensitivity and specificity of information criteria," Penn State University, 12-119, 2012.
[18] R. L. Levine, "System dynamics applied to psychological and social problems," presented at the System Dynamics Society Conference, Bergen, Norway, 2000.
[19] C. Nuthmann, "Using human judgment in system dynamics models of social systems," *Syst. Dyn. Rev.*, vol. 10, no. 1, pp. 1–27, Mar. 1994.
[20] G. Engelhard Jr., "Historical Perspectives on Invariant Measurement: Guttman, Rasch, and Mokken," *Measurement (Mahwah N J)*, vol. 6, no. 3, pp. 155–189, Aug. 2008.
[21] W. P. Fisher, "Invariant thinking vs. invariant measurement," *RMT*, vol. 14, no. 4, pp. 778–781, 2001.
[22] K. M. Wittkowski, E. Lee, R. Nussbaum, F. N. Chamian, and J. G. Krueger, "Combining several ordinal measures in clinical studies.," *Statist. Med.*, vol. 23, no. 10, pp. 1579–1592, May 2004.
[23] I. K. Crombie and H. T. Davies, "What is meta-analysis," *What is*, 2009.
[24] D. Borsboom, G. J. Mellenbergh, and J. van Heerden, "The theoretical status of latent variables.," *Psychol Rev*, vol. 110, no. 2, pp. 203–219, 2003.

Journal of Physics: Conference Series **772** (2016) 012024 doi:10.1088/1742-6596/772/1/012024

The NIST *Simple Guide* for Evaluating and Expressing Measurement Uncertainty

Antonio Possolo

Statistical Engineering Division, Information Technology Laboratory, National Institute of Standards and Technology, U.S. Department of Commerce, Gaithersburg, MD

E-mail: antonio.possolo@nist.gov

Abstract. NIST has recently published guidance on the evaluation and expression of the uncertainty of NIST measurement results [1, 2], supplementing but not replacing B. N. Taylor and C. E. Kuyatt's (1994) *Guidelines for Evaluating and Expressing the Uncertainty of NIST Measurement Results* (NIST Technical Note 1297) [3], which tracks closely the *Guide to the expression of uncertainty in measurement* (GUM) [4], originally published in 1995 by the Joint Committee for Guides in Metrology of the International Bureau of Weights and Measures (BIPM).

The scope of this *Simple Guide*, however, is much broader than the scope of both NIST Technical Note 1297 and the GUM, because it attempts to address several of the uncertainty evaluation challenges that have arisen at NIST since the 1990s, for example to include molecular biology, greenhouse gases and climate science measurements, and forensic science.

The *Simple Guide* also expands the scope of those two other guidance documents by recognizing observation equations (that is, statistical models) as *bona fide* measurement models. These models are indispensable to reduce data from interlaboratory studies, to combine measurement results for the same measurand obtained by different methods, and to characterize the uncertainty of calibration and analysis functions used in the measurement of force, temperature, or composition of gas mixtures.

This presentation reviews the salient aspects of the *Simple Guide*, illustrates the use of models and methods for uncertainty evaluation not contemplated in the GUM, and also demonstrates the *NIST Uncertainty Machine* [5] and the *NIST Consensus Builder*, which are web-based applications accessible worldwide that facilitate evaluations of measurement uncertainty and the characterization of consensus values in interlaboratory studies.

References

[1] Possolo A 2015 *Simple Guide for Evaluating and Expressing the Uncertainty of NIST Measurement Results* (Gaithersburg, MD: National Institute of Standards and Technology) NIST Technical Note 1900

[2] Possolo A 2016 *Metrologia* **53** S17–S24

[3] Taylor B N and Kuyatt C E 1994 *Guidelines for Evaluating and Expressing the Uncertainty of NIST Measurement Results* (Gaithersburg, MD: National Institute of Standards and Technology) NIST Technical Note 1297 URL physics.nist.gov/Pubs/guidelines/TN1297/tn1297s.pdf

[4] Joint Committee for Guides in Metrology 2008 *Evaluation of measurement data — Guide to the expression of uncertainty in measurement* (Sèvres, France: International Bureau of Weights and Measures (BIPM)) BIPM, IEC, IFCC, ILAC, ISO, IUPAC, IUPAP and OIML, JCGM 100:2008, GUM 1995 with minor corrections URL www.bipm.org/en/publications/guides/gum.html

[5] Lafarge T and Possolo A 2015 *NCLSI Measure Journal of Measurement Science* **10** 20–27

On Trial: the Compatibility of Measurement in the Physical and Social Sciences

S J Cano[1], T Vosk[2], L R Pendrill[3], A J Stenner[4]

[1]Modus Outcomes, Spirella Building, Letchworth Garden City, SG6 4ET, UK, stefan.cano@modusoutcomes.com; [2]8105 NE 140th Pl, Kirkland, WA 98034, tvosk@comcast.net; [3]SP Technical Research Institute of Sweden, Metrology, Box 857, SE-50115 Borås, Sweden, leslie.pendrill@sp.se; [4]MetaMetrics, Inc., 1000 Park Forty Plaza Drive, Suite 120, Durham, NC 27713, USA & The University of North Carolina, Chapel Hill, North Carolinam USA, jstenner@Lexile.com.

Abstract

In this paper, we put social measurement on trial: providing two perspectives arguing why measurement in the social and in the physical sciences are incompatible and counter with two perspectives supporting compatibility. For the case 'against', we first argue that there is a lack of definition in the social sciences. Thus, while measurement in the physical sciences is supported by empirical evidence, calibrated instruments, and predictive theory that work together to test the quantitative nature of properties, measurement in the social sciences, in the main, rests on a vague, discretionary definition of measurement that places hardly any restrictions on empirical data, does not require calibrated instruments, and rarely articulates predictive theories. The second argument for the case 'against' introduces the problem associated with psychometrics, including different approaches, methodologies, criteria for success and failure, and considerations as to what counts as measurement. Making the first case 'for', we highlight practical principles for improved social measurement including units, laws, theory, and metrology. The second argument 'for' introduces the exemplar of the Lexile Framework for reading that exploits metrological principles and parallels the paths taken by, for example, thermometry. We conclude by proposing a way forward potentially applicable to both physical and social measurement, in which inferences are modelled in terms of a measurement system, where specifically the output of the instrument in response to probing the object ('entity') is a performance metric, i.e. how well the set-up performs the assessment.

1. Introduction

In the social sciences, measurements are frequently made using rating scales, questionnaires or ability tests, which are types of 'instruments', but different from the devices, such as voltmeters, used for measurement in physics and engineering. This is because the variables, or constructs, 'social measurement' seeks to capture (e.g., educational ability, psychological function, cognitive function, disability, and fatigue) are not generally quantifiable using the instrumentation familiar to physical scientists [1]. In contrast to the access to directly observable variables physicists and engineers are often presumed to have, social scientists are typically seen as inferring their measurements from manifestations of hidden variables assumed to be observable only indirectly [2]. In this paper we put social measurement on trial: providing two perspectives arguing why measurement in the social and in the physical sciences are incompatible and counter with two perspectives supporting compatibility. We conclude with a proposal of a potential way forward. It is important to flag that, we do not include discussions surrounding model-based theories (e.g., see references in [3]), "replication crisis" in the social sciences [4], or statistical models used in complex hierarchical systems, as these are beyond the scope of this article.

2. The Case 'Against' Part 1: A Lack of Definition

Just as there are many instruments[a] measuring physical properties in the physical sciences, there are many instruments purporting to measure educational, psychological and health-related variables in the

[a]Measuring instrument: "device used for making measurements, alone or in conjunction with one or more supplementary devices" (§3.1 VIM)

Published under licence by IOP Publishing Ltd

Journal of Physics: Conference Series **772** (2016) 012025 doi:10.1088/1742-6596/772/1/012025

social sciences [5]. However, there are crucial differences between the physical and social sciences in what they require from their instruments. Historically, psychological effects initially entered into measurement science in the field of psychophysics in the 19th Century in attempts to relate abstract human sensations to quantifiable physical external stimuli (such as touch pressure, and sound pitch). Psychometrics developed thereafter to include other mental attributes (such as attitude, knowledge, empathy) which are not simply responses to physical stimuli, and are termed 'latent traits' [6].

While measurement in the physical sciences is supported by empirical evidence, calibrated instruments, and predictive theory that work together to test the quantitative nature of properties, measurement in the social sciences is, in the main, made possible only by a vague, discretionary definition of measurement that places hardly any restrictions on empirical data, does not require calibrated instruments, and rarely articulates predictive theories [7]. The key requirement of a scientific theory is to make predictions that can be tested empirically and that can be relied upon to be reproducible in the future [7,8]. This applies to theories relating different variables as much as it applies to the theory that a particular variable exists as a quantitative attribute. However, in the social sciences, the proposal of a latent variable is hardly ever framed as a theory, the hypothesis of a unit amount is rarely tested, and very few predictions are made that would indicate the quantitative character of the attribute.

Despite a long history of clearly grounded and well-articulated measurement theory and practice [7], measurement in the social sciences is usually completely detached from the ontological claim of a quantitative latent variable. Instead, Stanley Smith Stevens famously proposed that measurement be defined as 'the assignment of numerals to objects or events according to rules' [9]. However, Stevens did not specify the rules explicitly but referred to permissible transformations: 'In what ways can we transform its values and still have it serve all the functions previously fulfilled?' [9; p. 680]. In practice, this reasoning results in a circularity as the scale level implies permissible transformations, which, in turn, determine the scale level [10]. In case of metric scales, numerals are interpreted as numbers that 'represent aspects of the empirical world' [9; p. 677].

The first case against the social and physical sciences being compatible is that measurement in the latter is supported by empirical evidence, calibrated instruments, and predictive theory that work together to test the quantitative nature of properties. In contrast, measurement in the social sciences, in the main, rests on a vague, discretionary definition of measurement that places hardly any restrictions on empirical data, does not require calibrated instruments, and rarely articulates predictive theories.

3. The Case 'Against' Part 2: The Problem of Psychometrics

The methodology underpinning social measurement is known as psychometrics. There are three major psychometric paradigms[b]: Classical test theory (CTT [11]), item response theory (IRT [12]), and Rasch Measurement Theory (RMT [13]). For nearly a century, the dominant paradigm has been the traditional psychometric methods that are underpinned by CTT. This theory of measurement stemmed from seminal works of scientists including Galton, Spearman, and Pearson [7]. However, the last half-century has seen the development of psychometric methods that have refocused the way we can and should measure. These methods (e.g., RMT) have significant conceptual and empirical advantages over traditional approaches [14]. However, social measurement researchers have been slow to accept, embrace, understand, and apply them in their work [7].

We propose three central challenges to social measurement and psychometric methods. The first challenge is to arbitrate the differences between the three main psychometric paradigms (i.e., CTT, IRT, RMT). Each prioritizes paradigmatically incommensurable approaches, criteria for success and failure, and considerations as to what counts as measurement. Might there be scientific, mathematical, aesthetic, moral, and practical reasons why one paradigm is superior to the others? The second challenge, and in part a sequela of the first, is to determine whether there are any practical methods for evaluating in common terms the quality of the many thousands of measurement instruments in the social sciences.

[b] It could be argued that the rarely-used, but historically important Guttman paradigm, is a fourth paradigm.

Third, irrespective of the psychometric paradigm chosen to develop or evaluate instruments, statistical analyses alone do not and cannot inform us as to what is being measured. Statistical adequacy does not guarantee valid measurement. As such, psychometric statistics can be misleading when considered in isolation, and cannot be expected to produce consistently meaningful results when considered apart from qualitative scale content evaluations.

To sum up the second case against, measurement, whether physical or social, to be compatible ought to have similar definitions (i.e., the ratio of two magnitudes of the same *thing* in which the denominator is the unit, providing for invariant comparisons) and the same broad goal (i.e., quantification of meaningful variables). Social measurement neither aspires to nor approximate physical measurement [15], probably because it is not achievable [16]. However, the role of social measurement methods in high-stakes decision making for education, psychology, and health, highlights the need for the field to advance. In particular, there is an urgent need for: 1) testable psychometric theories; 2) testable construct theories, and 3) an experimental hypothesis-testing approach. In the two cases for compatibility which follow, we pick up on these issues.

4. The Case 'For' Part 1: Practical Principles for Improved Social Measurement

Though a large number of significant contributions to more fully scientific measurement for psychological and social studies have been made over the last several decades, they have not yet cohered into an identifiable paradigm with an associated community of practice. Advances fall under four primary headings: units, laws, theory, and metrology.

4.1. Units

In physics, units are fixed at the same size along the entire range of variation, they are defined linearly as a ratio to a larger magnitude, they are invariant across instruments and samples, and reliability is typically conceived in terms of individual measures' uncertainty and precision. In the social sciences, fixed unit sizes, ratio definitions, linearity and individual precision estimates have been achieved, but are not typically a priority. Units in the social sciences usually reflect the ordinality of the underlying scale, vary nonlinearly across instruments and samples, with reliability conceived in terms of group-level correlations. Contrary to popular opinions concerning the supposedly deterministic structure of concrete entities in physical nature, Newtonian physics, elementary number theory, and arithmetic are now recognized as involving irreducible stochastic complexity [17,18]. This parallel motivation for stochastic models across the physical and social sciences notwithstanding, significant challenges must be met to improve the quality of units in the social sciences, not least of which involve improved methodological training and standards.

4.2. Laws

In his work in the 1950s on the measurement of reading ability, Rasch intentionally structured his measurement models so they would have the same form as Newton's laws of motion, following the method first definitively articulated by Maxwell [19-21]. Maxwell's method takes advantage of the fact that a large number of natural laws take the same form, with ratios or products of any two of three parameters equal to the third [4]. The Rasch Reading Law then relates reading ability to text complexity and comprehension rates in the same form as the Combined Gas Law [22].

The goal, however, is not a mere mechanical and superficial copying of parameters from one domain to another. Instead, Maxwell focuses on interactions, with the aims of 'constructing, modifying and merging source analogies in light of the constraints of the problem domain under investigation' [19]. Detailed construct theories, and not just model parameters, are needed.

4.3. Theory

"What I cannot create, I do not understand." This statement from Richard Feynman [23] frames a perspective on validity that focuses on the capacity to recreate or synthesize the construct in the laboratory. This fundamental criterion must be met before claims to understand a construct enough to make valid statements about it are warranted. As was the case with units and laws, there are marked contrasts between physics and the social sciences in the ways theory is used and developed. In physics, predictive control over constructs makes theory-based measurement highly efficient and effective, mass-produced instruments are not calibrated on data but in terms of theoretical properties, and constant improvements to theory drive down costs and enhance precision in primary associated industries.

In the social sciences, predictive control over constructs has been demonstrated [22, 24], but has had little impact. Further, widely-used instruments are almost always calibrated on data, high stakes test items are increasingly expensive, and a general lack of effective theory undercuts improvement efforts, allowing costs to constantly spiral higher in education, health care, social services, etc. When item calibrations and measures derived from them can be predicted from theory, the kind of reasoning processes and linguistic structures involved in all discourse and applicable in a wide variety of disciplines can be employed [25]. Instead of seeing measurement, modelling, and science as derived from physics, it may prove more accurate to see physics, the social sciences, and all other fields of inquiry as fundamentally rooted in the same kinds of cognitive processes.

4.4. Metrology

Scientific measurement is achieved with simplicity, elegance, and parsimony when an invariant, portable unit is defined via a mathematical law, predicted from theory to a useful degree of precision, and efficiently deployed in practical applications as a universally uniform metric traceable to a reference standard. In the social sciences, in contrast, units are not traceable to reference standards; communities of stakeholders do not set consensus unit standards, no resources are spent setting and maintaining uniform unit standards; and education, health care, and social service markets are accordingly highly inefficient due to the very high transaction costs incurred. Established successes with additive units, conformity to scientific laws, and predictive theory suggests metrology is feasible and would be highly desirable in the fields employing social measurement methods.

We would proffer that the most important and pressing issue for the social sciences is need for theory. And thus, to begin to employ model-based reasoning in social measurement we must adopt a more thoroughly experimental approach [26]. To that end, two key requirements are needed to advance our understanding of what instruments measure in the social sciences: explicit theories of the constructs; and explicit methods to test those theories. As such, instrument development should be from the ground up (i.e., clear definitions), rather than the traditional top down approach (i.e., arbitrary statistically-driven methods of grouping items). This necessitates the complementary roles of qualitative and quantitative methods to ensure that construct definitions and subsequent theory determines instrument content, and psychometric analyses are used to assess the validity of the resultant construct theories [27].

When item difficulties and by implication estimates of person measures are under control of a construct theory, it becomes possible to develop a construct specification equation [28]. This is the formal expression of the theory regarding the observed regularity in a set of observations generated by a given measurement process. Therefore, the equation articulates a theory of item-score variation and simultaneously provides the vehicle for confirmation or falsification of the theory. Construct specification equations are developed by regression analysis of item locations on selected item characteristics. They provide a test of fit between scale-generated observations and theory. In essence, the greater the proportion of variation in item location explained by the selected item characteristics, the greater the support for the proposed construct theory, the greater the evidence for scale validity, and the more meaningful the interpretation of person measures [29].

In summary, the first case for the compatibility of measurement in the social and physical sciences is based on the widely used practical principles for improved definition in social measurement's units, laws, theory, and metrology. An exemplar is presented in the second supportive case.

5. The Case 'For' Part 2: The Exemplar of Reading Measurement

Reading is a process in which text and reader act together to produce meaning. Reading provides the most advanced form of theory-referenced measurement [28, 29]. The central premise is to shift the focus from studying people, which is the usual focus in social measurement, to studying people relative to items. Five levels of evolving construct theory culminate in a causal model relating reader ability, text complexity and comprehension. The model is conjoint, enabling reader ability and text complexity to be denominated in a common unit and denoted as a numeric value typically ranging from 300L to 2000L [30].

It is rare in the social and behavioural sciences to find instrumentation that exploits metrological principles and copies the paths taken by, for example, thermometry [31]. The 30-year history of developments in reading measurement exemplifies an attempt to do just that. As such, a key feature of causal Rasch models [32] is that they take a form similar to some three variable physics equations like D = RT or F = MA. A crucial feature of such equations is the trade-off or cancelation property, which states that, for example, a change in mass can be traded-off for a change in acceleration to hold force constant or a change in rate can be cancelled by a change in time to hold distance constant. Michell [33] has argued that successful trade-offs and cancelations like those above are evidence for the quantitative status of attributes figuring in such equations. To sum up the second argument for the compatibility of measurement in the social and physical sciences, reading measurement stands as the exemplar of the evolution of measurement in social science being played out, and provides a potential template for constructs beyond 'reading ability'.

6. Conclusion: A Way Forward

As increased stringency is required, for instance in measurements in education, healthcare and many contemporary applications where human perception is key, one seeks the same kind of quality-assurance of social measurements as is established in physics and engineering [34, 35]. Specifically, this means assurance with objective metrological comparability ('traceability') and declared measurement uncertainty, since the decisions that need to be made about student ability or patient health should be comparable with known uncertainties, based on sound measurements. In seeking this stringency, we should at the same time be realistic, acknowledge limitations in quantifying social measurement, and avoid suffering from what has been called 'physics envy' [35].

In considering the philosophical foundations of social measurement, Maul et al. [36] recall various approaches, including empiricism, pragmatism and realism. The philosophical realism behind physical metrology assumes, as in physics, that there is an objective reality, which exists even when we do not perceive or have instruments to measure it. One might argue that there is seldom objective reality in what is measured in social science (e.g., the challenge of a task) without our actually perceiving or measuring it. Lacking an independent objective reality might lead to measurements in the social sciences providing no unique 'right' answer. This makes metrology in social measurement challenging, since: (i) independent reference standards used in metrology to ensure the comparability of different measurements would be difficult to establish separately from the actual measurement process, and (ii) measurement uncertainty would often be very large, since each new measurement set-up would produce definitions divergent from others.

The argument that **empirical operationalism** is irreconcilable with general scientific practice and vocabulary [38] (since each unique set of operations would be associated with distinct definitions) does not hold, since in physical and engineering metrology it is accepted[c] that 'a reference can be a measurement unit, a measurement procedure, a reference material, or a combination of such.'

7. References

[1] Thorndike E L 1904 *An introduction to the theory of mental and social measurements* (New York: The Science Press)
[2] Thurstone L L 1928 *Am J Sociol* **33** 529-554
[3] Pendrill L, Fisher W 2015 *Measurement* **71** 46–55

[c] VIM 2015, *International Vocabulary of Metrology – Basic and General Concepts and Associated Terms* , §1.1 Note 2, http://www.bipm.org/en/publications/guides/

[4] Open Science Collaboration Estimating the reproducibility of psychological science. *Science*, 349 (6251), aac4716, 2015.
[5] Kempf-Leonard K 2005 *Encyclopedia of Social Measurement* (Oxford: Elsevier)
[6] Tesio L 2003 *J Rehabil Med* **35** 105-115
[7] Wright B 1997 *Educ Meas: Issues Practice* **16** 33-52
[8] Popper K 1959 *The Logic of Scientific Discovery* (London, Routledge)
[9] Stevens S S 1946 *Science* **103** 677-680
[10] Duncan O D 1985 *Synthese* **42** 21-34
[11] Nunnally J C, Bernstein I H 1994 *Psychometric Theory*, (New York: McGraw-Hill)
[12] Lord F M, Novick M R 1968 *Statistical theories of mental test scores* (Reading, Massachusetts, Addison-Wesley)
[13] Rasch G 1960 *Probabilistic models for some intelligence and attainment tests* (Chicago, University of Chicago Press)
[14] Lord F, Novick M 1968 In: Lord F, Novick M, eds. *Statistical Theories of Mental Test Scores*, (Reading, Mass.: Addison-Wesley) pp. 13-26
[15] Michell J 1997 *Br J Psychol* **88** 355-383
[16] Michell J 1999 *Measurement in Psychology* (Cambridge, UK: Cambridge University Press)
[17] Chaitin G 1994 *International J Bifurcation Chaos* **4** 3-15
[18] Chaitin G J 2003 *The Limits of Mathematics* (London: Springer-Verlag)
[19] Nersessian N 1996 *Philos Sci* **63** 542-546
[20] Nersessian N 2002 In: Malament D, ed. *Reading Natural Philosophy* (Lasalle, Illinois: Open Court) pp. 129-166
[21] Fisher W 2010 Rasch, Maxwell's method of analogy, and the Chicago tradition. *Probabilistic Models for Measurement in Education, Psychology, Social Science and Health: Celebrating 50 years Since the Publication of Rasch's Probabilistic Models*, University of Copenhagen School of Business, FUHU Conference Centre, Copenhagen, Denmark, June 13-16, 2010.
[22] Burdick D, Stone M, Stenner A 2006 *Rasch Measurement Transactions* **20** 1059-1060
[23] Nersessian N 2008 *Creating Scientific Concepts* (Cambridge, Massachusetts: MIT Press)
[24] Stenner A, Stone M 2010 *J Appl Meas* **11** 244-252
[25] Fisher W, Stenner A 2011 *International J Multiple Research Approaches* **5** 89-103
[26] Stenner A J, Smith M, Burdick D 1983 *J Educ Meas* **20**, 305-316
[27] Stenner A J, Smith M 1982 *Percept Motor Skills* **55** 415-426
[28] Stenner A J, Burdick H, Sanford E E, Burdick D S *J Appl Meas* **7** 307-322
[29] Hawking S 2001 *The Universe in a Nutshell* (New York: Bantam Books)
[30] Stenner A, Stone M, Burdick D 2009 *Rasch Measurement Transactions* **23** 1204-1206
[31] Chang H 2007 *Inventing Temperature: Measurement and Scientific Progress* (Oxford: Oxford University Press)
[32] Stenner A, Fisher W, Stone M, Burdick D 2013 *Frontiers in Psychology*, **4** 1-14
[33] Michell J 1990 *An Introduction to the Logic of Psychological Measurement* (Hillsdale, New Jersey: Lawrence Erlbaum Associates)
[34] Pendrill L Petersson N 2016 *Measurement Science & Technology* (in press)
[35] Ruhm K 2016 *Measurement*, **79** 276-284
[36] Nelson R 2015 *Issues in Science and Technology* **31** 71-78
[37] Wilson M 2013 *Measurement* **46** 3766-3774
[38] Maul A, Torres Irribarra D, Wilson M 2016 *Measurement* **79** 311-320
[39] Bond T, Fox C 2015 *Applying the Rasch model* (New York, Lawrence Erlbaum Associates)
[40] Pendrill L 2014 *Metrologia* **51** S206

Journal of Physics: Conference Series **772** (2016) 012026 doi:10.1088/1742-6596/772/1/012026

Using artificial intelligence strategies for process-related automated inspection in the production environment

K Anding[1], P Kuritcyn[1] and D Garten[2]

[1]Ilmenau University of Technology, Gustav-Kirchhoff-Platz 2, 98693 Ilmenau, Germany
[2]GFE Schmalkalden e.V., Näherstiller Straße 10, 98574 Schmalkalden, Germany

katharina.anding@tu-ilmenau.de, petr.kuritcyn@tu-ilmenau.de, d.garten@gfe-net.de

Abstract. In this paper a new method for the automatic visual inspection of metallic surfaces is proposed by using Convolutional Neural Networks (CNN). The different combinations of network parameters were developed and tested. The obtained results of CNN were analysed and compared with the results of our previous investigations with color and texture features as input parameters for a Support Vector Machine. Advantages and disadvantages of the different classifying methods are explained.

1. Introduction

Lately, biologically inspired CNN are becoming more popular in the area of computer vision. The CNN is considered a special case of standard neural networks. They showed exceptional performance for various tasks as image recognition, object detection [1], feature extraction [2] and segmentation [3]. Since the first seminal paper [4], convolutional nets are continuously advancing their accuracy in different image classification tasks [5] and proved robustness against different lighting conditions, object positions, partial occlusions and translation. Despite many of the recent CNNs are very time-consuming in both training and testing, they have an important advantage in comparison to sophisticated methods in machine learning (like Support Vector Machine, decision trees etc.): CNNs usually do not need to use both external feature extraction and feature selection methods for image recognition tasks. Low-level image features like edges, lines, and corners are extracted out of the convolution layers. In the following layers, patterns are detected in these low-level features. Therefore, we get another level of abstraction within every layer and the concept can be classified among the group of deep learning methods [6].

In this paper, we propose a new method for the automatic visual inspection of metallic surfaces. We have presented a method for recognition of metal surfaces defects based on the analysis of drilled countersink holes in heat-treatable steel [7]. The classification was done using color and texture features as input parameters for a Support Vector Machine. Here we want to present results from our latest research considering the usage of Convolutional Neural Networks (CNN) for this task. We discriminated between the two classes: good parts (perfect surface) and bad parts (longitudinal rills and chatter marks) (see Figure 1).

Class Name	Example Image
Longitudinal Rills	
Perfect Surface	
Chatter Marks	

Figure 1. Perfect surface and defect samples

Firstly, we will describe the design of our approach, then present our results and compare with our previous investigations at the end.

2. Development of a novel cnn recognition approach for metal surfaces defects

2.1. Data preprocessing

Our dataset consists of 3 classes (see Figure 1) and around 250 RGB-images of resolution 1280x1024 pixels. That is a huge size for the input of convolutional network (it requires more than 11 million parameters only for one filter with size 3x3). We needed to downsample images and to reduce the number of color layers from 3 to 1 (grayscale). It does not change drastically the difficulty of detection, but improves the calculation speed and reduces the number of parameters in the net (16 times less for the first convolution layer). We obtained gray-scaled images of resolution 320x256 pixels after downsampling with antialiasing filter (package Pillow [8]). Two datasets were prepared: with two classes (good parts – bad parts) and with three classes (perfect surface - longitudinal rills - chatter marks).

2.2. Design of the network

Following packages were used for building our convolutional neuronal network: Theano [9], Lasagne [10] and nolearn[11]. The implementation consists of many elements of CNN: convolutional layers, nonlinearity functions, subsampling layers etc.

The structure of network was inspired by LeNet network [4]. Figure 2 shows the network structure. It consists of several types of layers: convolutional, pooling and dense layer. We tested different combinations of these elements with different parameters.

Three pooling layers perform downsampling after each convolutional layer. After all of them we placed two fully-connected dense (hidden) layers and then an output layer.

Each convolutional layer has a nonlinearity function placed directly after them. We tried several nonlinearity functions, which were successful in other investigations: rectifier linear unit (ReLU)[12], exponential linear unit (ELU)[13], parametric rectifier linear unit [14].

After output layer, we placed the softmax function [15], as a typical choice for the classification task, which generalizes the categorical probability distribution

For the parameter updates, we used the algorithm AdaMax [16]. This method worked well in our investigations and showed a good convergence and optimization properties in comparison to other stochastic optimization methods (Stochastic Gradient Descent (SGD)[17], Nesterov Momentum [18]).

All computations were performed on the single GPU-unit.

2.3. Data augmentation

The first problem we have faced in our investigations with CNN was a small size of the given dataset. Usually, neural networks require big datasets to achieve good results [5]. There are many techniques to make a dataset bigger without necessity of new experiments. The general term for that process is data augmentation. At first, we used flipping randomly chosen images. In the next stage we added also random rotation (0, 90, 180, 270 degrees), horizontal and vertical shift for a few pixels and zooming in the range between 0.9 and 1.1 of the original image size.
For training our net, we used small batches from the original dataset. These batches were augmented on the fly by an implemented generator. It generates new samples at "no cost".

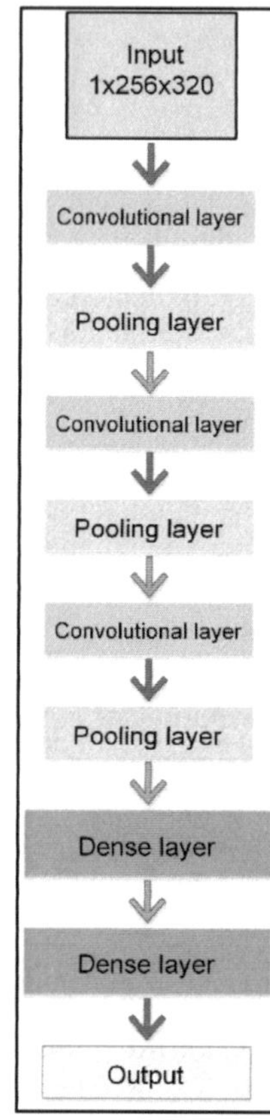

Figure 2. Structure of neuronal network for inspection of metal surfaces

2.4. Prevention of the overfitting

Usual problem when using small datasets and feedforward networks is the overfitting effect [5]. To prevent this, we applied some regularization techniques: batch normalization [19] and dropout [20]. Batch normalization is a normalization step, which fixes variances and means of layer inputs. It improves the stability and speed of the optimization process. The idea of dropout is to randomly nullify the weights of neurons from the previous layer. It prevents adaptation of neurons to specific features in images and improves the generalization ability of the network.

3. Experimental results

The results of our investigations are presented in Table 1.

There are a few most successful network configurations with the highest accuracy for the given dataset. The networks consist of a few stages with pooling layers between them. Each stage consists of one or several convolutional layers. The first number in brackets corresponds to the size of the filter, the second to the number of filters with that size. Slash sign shows stride of the filter or pooling layer. The filters on each stage are working consecutively from top to bottom, not in parallel. The networks A-D were tested on the dataset with downsampled images. The network E was tested on the dataset with full-sized grayscale images.

Table 1. Configuration and performance of several convolutional neuronal networks on the metal surface dataset

	Stage 1	Pool	Stage 2	Pool	Stage 3	Pool	Stage 4	Pool	Fully-connected layer	Total recognition rate		Nonlinearity function	Training time (sec/epoch)
										2 class	**3 class**		
A	(12, 32)	2/2 Drop. 10%	(6, 64)	2/2 Drop. 20%	(6, 128)	2/2 Drop. 30%			2x256 Drop. 30%	84.7	56.4	ELU	4.75
B	(3, 16) (3, 32)	2/2	(3, 64)	2/2 Drop. 10%					2x256 Drop. 10%	96.5	83.5	ELU	13
C	(3, 8) (3, 16)	2/2	(3, 16) (3, 32) (3, 64)	2/2	(3, 64)	2/2	(3, 128) (3, 128)	2/2 Drop. 10%	2x256 Drop. 30%	96.5	75.3	ELU	10
D	(3, 16) (3, 32)	2/2	(3, 64) (3, 128)	2/2	(3, 256)	2/2	(3, 512)	2/2 Drop. 10%	2x256 Drop. 30%	96.5	81.2	ELU	19.5
E	(11, 1)/4 (3, 16) (32, 3)	2/2	(3, 64) (3, 128)	2/2	(3, 256)	2/2	(3, 512)	2/2 Drop. 10%	2x256 Drop. 30%	95.3	75.3	ELU	21

Our first attempts showed a recognition rate of 48%. It is a decent performance and we used data augmentation and dropout to improve the results. It is worth to notice that we tried to apply batch normalization technique, but it does not improve the performance of our networks. The network A uses a relatively big filter sizes and consists of only three convolutional layers. A recognition rate of 56.4% for the 3-class dataset and 84.7% for the 2-class dataset were achieved. As showed in [21], the results can be improved by reducing the filter size and increasing the depth of network and the number of filters. The network B used two consecutive convolutional layers with small filters on the input. This stack makes possible to develop the complex features. It results in increase of the total recognition rate up to 83.5% for 3-class dataset and to 96.5% for 2-class dataset. The networks C and D use more convolutional layers. The network C consists of fewer number of filters in the first two stages that reduces the computational time, but costs a few percent of performance on the 3-class dataset (8% less in comparison to B). The network D has additional layers on stages 2, 3 and 4 in comparison to B. It results in increasing the recognition rate of objects with longitudinal rills, but also decreases the recognition rate of objects with chatter marks. In general, the network D shows a comparable accuracy with network B, but takes 50% more time for training. It also shows a good feature generalization (see Figure 3). The salience plot shows the image parts, which matter to the net.

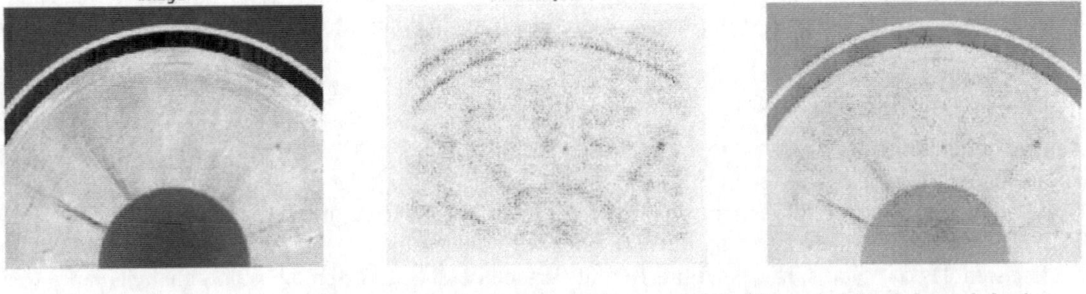

Figure 3. Salience plot of image with chatter marks (network D). (From left to right: original image, important parts of the image for classification, combination of those two images.)

The network E was designed for the dataset with full-sized images. It has the same structure as D, except the additional layer with filter size 11x11 and stride 4 on the input. This configuration was successfully used in [5] and helps to downsize the image with extracting some information in comparison to other dataset with preprocessed images. It results in a high recognition rate both on the 3-class (75.3%) and on the 2-class (95.3%) datasets. However, this did not overperform the networks B and D. The network E also requires for 7% more time than D and 38% than B.

All of our most efficient networks are using exponential linear unit as nonlinearity function after convolution layers. The tests with other nonlinearity functions (rectifier linear unit and parametric rectifier linear unit) showed no improvements of recognition rate. It can be explained with the high dependency of these methods from the initializations weight in layers.

4. Conclusions

The total recognition rate of 96.5% was achieved by using a relatively small dataset and a not fully optimized convolutional neural network. Using full sized images does not improve the classification accuracy, but it increases the computation time depending on the network architecture (not less than 5%). Using small filter sizes and deeper networks results in increasing of the classification accuracy and improving the generalization ability of the network. Applying dropout to fully connected layers helps to prevent overfitting. An optimal value of 30% in our case. Exponential linear unit as nonlinearity function and AdaMax algorithm for parameter updates showed a good synergy in our tests.

An unoptimized Support Vector Machine (SVM) reached recognition rates of about 85%. With SVM-parameter-optimization and further feature developing a total recognition rate of 99% were achieved with an SVM-classifier. The great advantages of CNN over the SVM are:

- can be efficiently implemented with field programmable gate arrays (FPGA) and graphic processing units (GPU)
- convolution kernels with integral effects can reduce image noise
- feature extraction and classification are implemented in a single process
  - ⇨ no need for time-consuming feature engineering and feature selection

A comparison between both algorithms is given in Table 2.

Table 2. Comparison between svm and convolutional neural networks.

Criteria	**svm**	**convolutional neural networks**
separate feature extraction	yes	no
separate feature selection	case-specific	no
prone to overfitting	no	with small datasets
parameter optimization needed	yes	yes

With these advantages of CNN it is possible to build classification systems in an automated way. A convolutional neural network will obtain sample images from the classes to discriminate between and a genetic algorithm will optimize the network architecture. With this optimized architecture the final network is trained.

Acknowledgments

The research project, which forms the basis of this paper, is supported by the Thuringian Ministry of Economy, Employment and Technology (TMWAT) with means from the European Social Fund (ESF). The responsibility for the content of this paper lies with the author.

The samples of metal surfaces with different surface defects, like scratches and chatter marks were produced by our project partner the society for production engineering and development (GFE e.V.) in Schmalkalden (Germany). Our special thanks goes to Dr. Daniel Garten and his colleagues for the production of metal surfaces and the image acquisition for building the dataset of metal surfaces with and without defects.

References

[1] Glorot X and Bengio Y. 2010 Understanding the difficulty of training deep feedforward neural networks *International Conference on Artificial Intelligence and Statistics* **249–256**, 2010

[2] Razavian A S, Azizpour H, Sullivan J, and Carlsson S 2014 Cnn features off-the-shelf: An astounding baseline for recogniton *CVPR 2014, DeepVision Workshop*

[3] Hariharan B, Arbel´aez P, Girshick R, and Malik J 2014 Simultaneous detection and segmentation *ECCV* **297–312**

[4] LeCun Y, Bottou L, Bengio Y, and Haffner P 1998 Gradient-based learning applied to document recognition *Proceedings of the IEEE, 86(11)* **2278–2324**

[5] Krizhevsky A, Sutskever I, Hinton G 2012 Imagenet classification with deep convolutional neural networks *Advances in Neural Information Processing Systems*

[6] Hijaz S e. a. 2016 Using Convolutional Neural Networks for Image Recognition *http://ip.cadence.com/uploads/901/cnn_wp-pdf*

[7] Garten D, Lahmann H W, Polte G and Anding K 2015 Photonic in-process quality measurements of surfaces and geometrical parameters of parts *MAZeT JENCOLOR-ZEISS SPECTROSCOPY-SENSORIK BAYERN Spectro-Net Cross-clustering Collaboration Forum, Jena http://spectronet.de/de/videos_2015/ video-photonic-in-process-quality-measurements-of-_ie2nz5z9.html*

[8] Git-Repository with source code and documentation of Pillow package *https://github.com/python-pillow/Pillow*

[9] Theano 0.8.0 documentation *http://deeplearning.net/software/theano*

[10] Git-Repository with source code and documentation of Lasagne-Framework *https://github.com/Lasagne/Lasagne*

[11] Git-Repository with source code and documentation of nolearn-Framework *https://github.com/dnouri/nolearn*

[12] Maas et al. 2013 Rectifier Nonlinearities Improve Neural Network Acoustic Models *http://web.stanford.edu/~awni/papers/relu_hybrid_icml2013_final.pdf*

[13] Clevert D A, Unterthiner T, Hochreiter S 2015 Fast and Accurate Deep Network Learning by Exponential Linear Units (ELUs) *http://arxiv.org/abs/1511.07289*

[14] He K, Zhang X et al. 2015 Delving Deep into Rectifiers: Surpassing Human-Level Performance on ImageNet Classification *http://arxiv.org/abs/1502.01852*

[15] Bishop C M 2006 Pattern Recognition and Machine Learning *Springer*

[16] Diederik K and Ba J 2014 Adam: A Method for Stochastic Optimization *arXiv preprint arXiv:1412.6980*

[17] LeCun Y, Boser B, Denker J S, Henderson D, Howard R E, Hubbard W and Jackel L D 1989 Backpropagation applied to handwritten zip code recognition *Neural computation*

[18] Nesterov Y 1983 A method of solving a convex programming problem with convergence rate O(1/k2) *In Soviet Mathematics Doklady, volume 27*, **372–376**

[19] Ioffe S and Szegedy C 2015 Batch Normalization: Accelerating Deep Network Training by Reducing Internal Covariate Shift *http://arxiv.org/abs/1502.03167*

[20] Hinton G, Srivastava N, Krizhevsky A, Sutskever I and Salakhutdinov R 2012 Improving neural networks by preventing co-adaptation of feature detector *arXiv preprint arXiv:1207.0580*

[21] He K, Sun J 2014 Convolutional Neural Networks at Constrained Time Cost *http://arxiv.org/abs/1412.1710v1*

Journal of Physics: Conference Series **772** (2016) 012027 doi:10.1088/1742-6596/772/1/012027

Instantaneous centre of rotation in human motion: measurement and computational issues

Francesco Crenna, Giovanni Battista Rossi and Alice Palazzo

Università degli Studi di Genova – DIME/MEC,
Via All'Opera Pia 15A – I 16145 Genova, ITALY

E-mail: francesco.crenna@unige.it

Abstract. The instantaneous centre of rotation plays an important role in biomechanical modelling and physical-medical interpretation of human gestures. Therefore, we consider its measurement, based on video-image acquisition and processing of human motion records. Measurement and computational aspects are discussed, including the evaluation of measurement uncertainty and the estimation of the effect of some influence quantities on the determination of the position of the instantaneous centre of rotation.

1. Introduction

Modelling of human movement is often based on a multi-bodies systems, where the bodies represent human limbs. Such bodies are linked by "joints" that either represent anatomic articulations or interactions with external entities, as in the case of foot contact with ground.

Since the motion of a rigid body is characterised, at each time instant, by an instantaneous centre of rotation (ICR), its explicit measurement may be very useful in some cases.

In our experience, we have recognised three main such cases.

One case concerns the identification of the centre of a rotational joint, for placing a marker corresponding to it. This may be relatively easy, when an external anatomical landmark is available (as the malleus for the ankle joint) or quite difficult, as in the case of hip [1]. In such cases the centre of the joint may be indirectly measured as the ICR of the relative motion [2].

Another important case, is when the relative motion in the joint is not purely rotational, as it happens, e.g., in the knee. In such cases, the ICR measurement may constitute a noteworthy improvement, in respect of assuming, as a simplifying hypothesis, a simple rotation. Such an improvement may be sensible, both from the biomechanical and the clinical standpoint [3].

Lastly, the case in which the joint represents an external contact point needs also consideration. Consider, for example, the case of hopping in place: in the landing phase, the foot tip touches ground, than it starts rotation around an ICR located somewhere in the lower toes surface, and then it moves toward the metatarsus and stabilises around it, when the foot toes lie completely on the ground. In the take-off phase, after a preliminary loading, we have the opposite movement toward foot tip [4-7].

In this paper, we consider two methods for ICR calculation, the Reuleaux technique and the rigid-body approximation, particularly convenient for on-line calculation and suited for clinic investigation [8-10]. To assess their performance, we have carried out a comparative evaluation of their performance in simulated and experimental situations. In reporting on such evaluation, we present a biomechanical application regarding ICR involved in a pedalling leg. Lastly, we summarize the results of the study and provide guidelines for uncertainty reduction and evaluation.

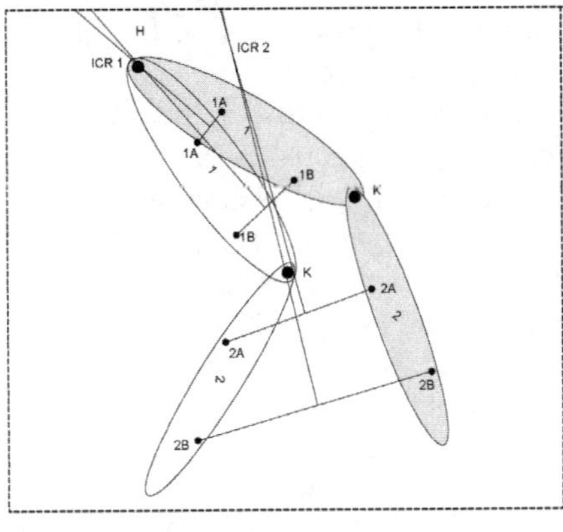

Figure 1. Lower limb scheme with two rigid bodies and hinge joints (H: hip, K: knee). Two successive positions are presented (not in scale), together with the segments connecting specific (marker) points on the segments. Perpendicular lines used to determine ICR are indicated. The ICR of the first body corresponds to H, while that of the second body results from the combination of the two rotations.

2. The instantaneous centre of rotation in biomechanics

Let's introduce briefly the concept of ICR. For the sake of simplicity, let us consider the case of planar movement of a two dimensional rigid body (Figure 1). At a specific time, ICR is the point on the plane considered around which the rigid body is rotating in that specific moment. Consequently, the ICR position generally moves in time according to rigid body motion [11-12].

This trajectory is rather intuitive in simple cases. For example consider the moving centroid for ICR corresponding to the hip when pedalling on a stationary bike. This is schematised in Figure 1 by body 1. As a first approximation, we can assume the femur (body 1) in pure rotation around the hip, fixed in a point on the saddle. Actually, from the biomechanical point of view, we could be interested in the small femur ICR movements in vertical direction, since a pelvis translation downward helps in pedalling. Beside that, hip articulation is of difficult identification on body skin, so H point position is difficult to be measured directly and an ICR measurement is required.

In biomechanics the measurement of human body kinematics is of fundamental importance: kinematic experimental data are the base for all further computation required to determine internal moments and forces and for ICR measurements also. A proper approach consists in a video-based measurement system, including a set of markers placed in specific points of interests along the body segments investigated, a video recording system and an image processing software. As a result it is possible to obtain positions, in the two or three dimensional space, of the set of points selected and individuated by markers. Measurement is affected by influence effects: camera resolution limits the spatial measurement resolution, illumination and image contrast affect accuracy in marker identification, while the frame rate corresponds to the sampling frequency. All such issues contribute to measurement uncertainty [15].

Beside that, it is worth noting that for particular gestures or in cases in which a simplified modelling is required, a planar motion reconstruction could be sufficient, for example when dealing with pedalling on a fixed bicycle. Figure 2 shows the positions of a set of markers placed on a cycling subject, when the motion is investigated in the sagittal plane.

3. ICR measurement methods

Kinematic experimental data constitute the input for ICR measurements, together with a multi-body model of the limbs under investigation. Usually positions of a cluster of markers on each segment are measured in time, in this case we will consider minimum of two markers along each segment. A few ICR measurement methods are available in literature [14], including the one deriving

Journal of Physics: Conference Series **772** (2016) 012027 doi:10.1088/1742-6596/772/1/012027

from the classical mechanics method of Reuleaux [11]. This method is relatively simple, but it sometimes present inconsistencies [6].

Let us briefly introduce it. Consider a limb constituted by two rigid bodies connected by a hinge joint, as represented in Figure 1. Two points on each segment are identified by letters A, B and we consider their positions at time t_i and t_{i-1}. The main hypothesis is that at each time the movement of a rigid body can be reconstructed as a rotation around an ICR. So we can consider the segment connecting the two successive positions of each point as the cord of circumference arch actually connecting the same points. The center of such circumference is the ICR and it is located on the line perpendicular to cord segment going through its middle point. If the body is rigid and kinematic measurement traces several points on it, it is possible to identify, in the same time interval, several perpendiculars. ICR position is determined at their intersection, as presented in Figure 1.

A possible alternative is offered by point velocity, since at each time t_i point position is measured by video system and its velocity can be evaluated by, e.g., a Savitsky-Golay interpolation [15]. In this case the ICR is located at the intersection of lines perpendicular to point velocities at each time. In the following we will consider the two alternatives, since they show important differences.

As a result, in Figure 1, body 1 ICR is located on the fixed joint (hip), while body 2 ICR position is moving due to the combination of two rotations at hip and knee. In fact, body 2 ICR will suffer from error propagation as we will demonstrate in the following.

Pure translation is another important issue since it will produce an ICR located at infinity. On the other hand a combination of linear translation and rotation will move the ICR. This is particularly interesting when linear translation is due to the behavior of the joint under investigation.

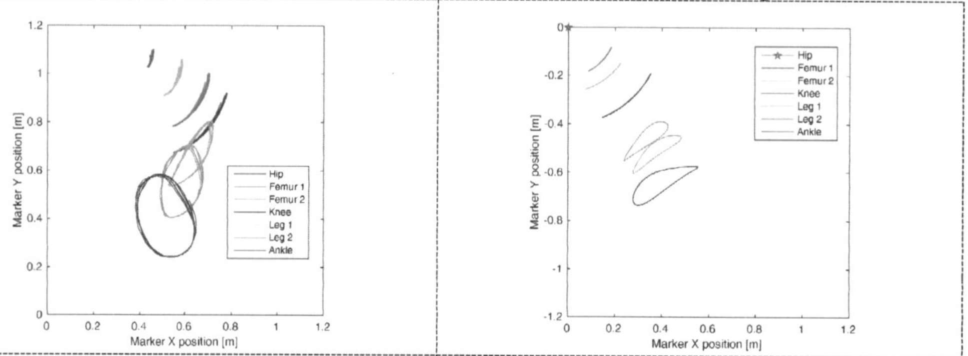

a b

Figure 2. (a) Markers trajectories for the pedalling leg experiment. 7 markers are considered: 2 on femur and lower leg respectively, plus 3 on the joints: hip, knee and ankle. (b) Rigid body assembly used for synthetic data production.

4. Experimental set up and results

Our investigation is focused on the influence of measurement noise, chord length and pure translation on ICR measurement methods presented in Section 2. For this purpose, we will use both synthetic and experimental data.

Synthetic data are useful to evaluate the processing techniques required to determine ICR under controllable reference conditions. It is possible to obtain synthetic kinematic data of a set of rigid bodies connected by pure revolute joints, with the first ICR fixed or subject to a controllable linear translation. Beside that we can vary relative angular velocities and sampling frequencies and introduce some noise if required. Figure 2b presents the situation of two bodies with 5 markers each moving with different angular velocities, while experimental data are referred to a practical biomechanical application to a pedalling leg (Figure 2a).

4.1. ICR from synthetic data

First of all the processing procedure has been validated through the use of synthetic data.

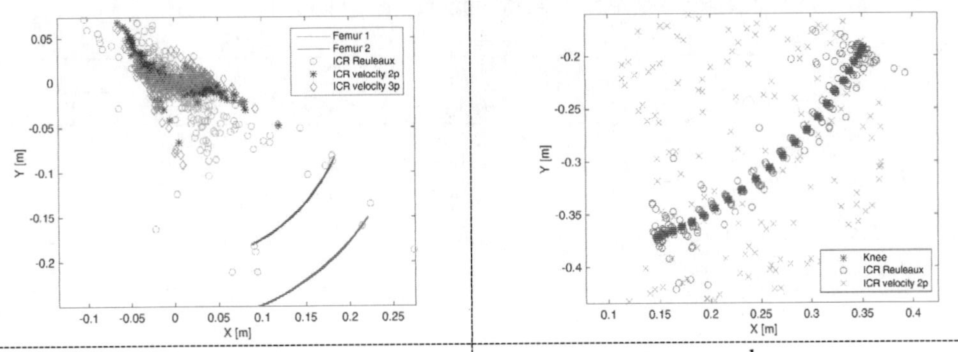

a b

Figure 3. ICR trajectories for body one (a) and two (b)

Figure 3 presents the ICR evaluated for the two bodies as presented in Figure 2a. The large ICR variability is evident in particular in the second body case. To understand the causes of such large variability, we have varied some computation parameters. ICR measurement results depend on several parameters; in particular we have investigated the following: - random Gaussian noise affecting position measurements;- length of the chord used for ICR computation; - number of points involved in the procedure. Concerning Gaussian noise we have considered a noise standard deviation of 0.5-1.5 and 3 mm. These are representative of a good experimental setup, a normal one and a difficult image processing situation. The chord length a parameter connected to rotational velocity of the segment and sampling frequency. Since we have a fixed sampling frequency, we have varied the time step between points used for ICR computation: i and i+1 or i and i+3, i and i+6 and i and i+9. Finally a minimum of two points on a rigid segment are required for computation: we have considered the possibility to use two or three points, as it may happen in practice.

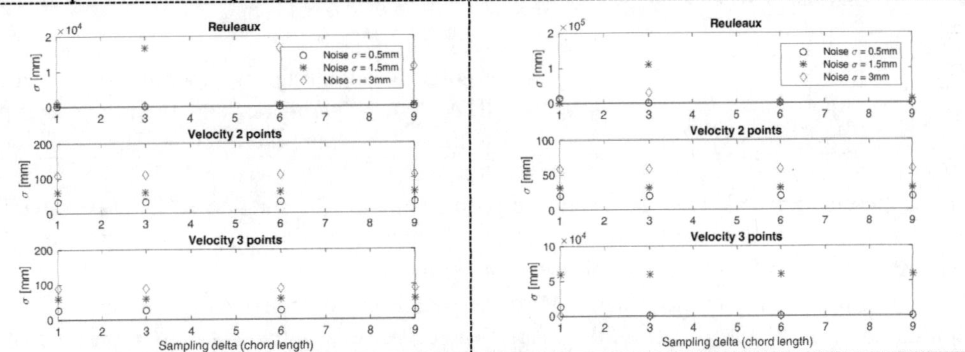

a b

Figure 4. Standard deviation of ICR position (x) for body one (a) and two (b) as a function of chord length and noise levels, for the three methods considered.

Figure 4 presents ICR results standard deviation as a function of the chord for the three noise levels considered and the three methods: Reuleaux and velocity based on two or three points. It is evident that velocity methods present more stable results when varying the chord length. As regards Reuleaux method, in this case a computation using points i and i+6 gives the lowest variability. Table 1 presents some numerical figures. It is worth noting that usually systematic deviations of mean values from nominal values are limited, while in some cases very high variability is observed. Beside that, it is evident that the uncertainties propagate toward the second body for which we have a larger variability.

Table 1. Noise effect in ICR measurements for synthetic signal at different noise levels and a mean chord length

			Hip			Knee	
			Reuleaux	**Velocity 2p**	**Velocity 3p**	**Reuleaux**	**Velocity 2p**
Noise σ=0.5mm	**σ [mm]**	X	20.2	32.9	26.0	22.6	19.9
		Y	23.6	17.9	22.3	0.5	8.9
Noise σ=1.5 mm	**σ [mm]**	Y	457	60.5	59.2	520	31.4
		X	389	83.3	75.8	1.9	9.2
Noise σ=3 mm]	**σ [mm]**	X	1650	110	88.9	3100	58.7
		Y	1540	117.7	93.6	6.4	21.5

4.2. Biomechanical application

Let's consider now the pedaling leg for which markers trajectories presented in Figure 2 are obtained at 25 Hz frame rate with a spatial resolution of about 3.5 mm (standard deviation about 2mm). In this case it is clear that the marker that was intended to be placed on the hip is misplaced along the femur since it presents some movement. For this reason ICR measurement is fundamental to identify hip position. Since the subject is cycling seated on the saddle in a first phase we can imagine a fixed ICR or hip, but results will be affected by a hip translation also due to pelvis contribution.

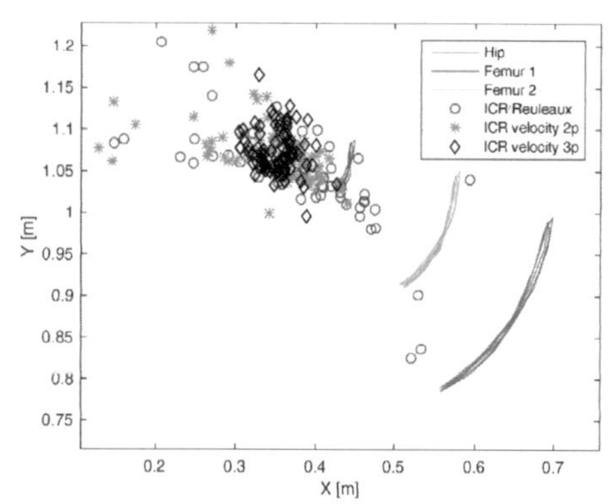

Figure 5. Pedalling leg example: trajectories for three femur landmarks and their ICR computed with the three methods considered.

Figure 5a presents femur ICR in the fixed reference frame. Considering a fixed hip we can evaluate mean value and standard deviation, which is about 35 mm. According to the synthetic case similar deviations may appear with a random noise with a sigma between 0.5 mm (σ=25 mm) and 1.5 mm (σ=65 mm), confirming a good accordance with the practical biomechanical example. Beside that a contribution due to translation is certainly present. In such cases synthetic data demonstrate by far a larger deviation (100% increment as compared with pure rotation at the same noise level), confirming difficulties in evaluating ICR when a translation contribution is present. Such problems are much more evident when dealing with the second rigid element ICR (as depicted in par 4.1). In this case, beside noise and joint translation, there is certainly a contribution due to markers motion on subject's skin during pedaling. Since all the considered methods rely on the hypothesis of rigid bodies, this seems to be the cause for the unreliable results obtained for knee ICR.

5. Conclusions

We have investigated some aspects in ICR position measurement, based on kinematic experimental data. Several issues have been addressed: noise influence, sampling or frame rate frequency, number of markers involved, presence of a translation contribution. In order to establish a controllable reference situation, synthetic data were produced by a simulator. It was possible to establish ICR measurement methods performance under controllable conditions. The velocity method has proved to be more reliable in presence of noise, in particular when three markers are involved in the procedure. When a translation is added, ICR measurement is very critical above all when dealing with the second segment in a two degrees of freedom system. We obtained some useful indication for the variability of the results and its dependence on noise and chord length or sampling rate. The velocity method has demonstrated a very good stability with both noise and chord length variations, as compared with the original Reuleaux's method. Nevertheless, when dealing with the second body in a chain, uncertainties are very large and results become unreliable in the practical application considered. In this case some further investigation is required to confirm that a violation of the rigid body hypothesis causes unpredictable ICR position deviations, and to determine if there are feasible countermeasures to limit such effects, when processing the results.

References

[1] Cappozzo A, Catani F et al. 1995 Position and orientation in space of bones during movement: anatomical frame definition and determination *Clin. Biomech.* **10** 171–178

[2] Halvorsen K, Lesser M and Lundberg A 1999 A new method for estimating the axis of rotation and the center of rotation *Journal of Biomechanics* **32** 1221-1227

[3] Galway HR and MacIntosh DL 1980 The lateral pivot shift: a symptom and sign of anterior cruciate ligament insufficiency *Clin. Orthop. Relat. Res.* **147** 45-50

[4] Frankel VH, Burnstien AH and Brookes DB 1971 Biomechanics of internal derangement of the knee. Pathomechanics as determined by analysis of the instant centres of motion. *J Bone Jt Surg* **53** 945-962

[5] Grant PG 1973 Biomechanical significance of the instantaneous centre of rotation: the human temporomandibular joint. *J Biomech* **6** 109-113

[6] Panjabi M 1979 Centers and angles of rotation of body joints: a study of errors and optimization. *J Biomech.* **12** 911-920

[7] Crenna F, Rossi GB 2015 Kinematic measurements for the characterization of the dynamics of hopping, *J. of Phys. Conf. S.* **588**

[8] Spiegelman JJ, Woo SL 1987 A rigid body method for finding centers of rotation and angular displacements of planar joint motion *Journal of Biomechanics* **20** 715-21

[9] Panjabi M, Goel VK 1982 Errors in kinematics of a planar joint: guidelines for optimal experimental design *Journal of Biomechanics* **15** 537-544

[10] Cereatti A, Camomilla V and Cappozzo A 2004 Estimation of the center of rotation: a Methodological contribution *Journal of Biomechanics* **37** 413

[11] Reuleaux F 1876 *The kinematics of machinery: outlines of a theory of machines.* London Macmillan 60-67

[12] Moorehead JD, Montgomery SC and Harve DM 2003 Instant center of rotation estimation using the Reuleaux technique and a Lateral Extrapolation technique *J. of Biomechanics* **36** 1301–1307

[13] Stagni R, Leardini A and Cappozzo A 2000 Effects of hip joint centre mislocation on gait analysis results *Journal of Biomechanics* **33** 1479-87

[14] Camomilla V, Cereatti A and Capozzo A 2006 An optimized protocol for hip joint centre determination using the functional method *Journal of Biomechanics* **39** 1096–1106

[15] Crenna F, Rossi G B and Palazzo A 2015 Measurement of human movement under metrological controlled conditions *Acta IMEKO* **4** 04-10

[16] Winter DA 2009 *Biomechanics and Motor Control of Human Movement* Wiley New York

[17] Kaufman KR and Sutherland DH 1994 *Kinematics of normal human walking in human locomotion* Williams and Wilkins Baltimore

Development of software application dedicated to impulse-radar-based system for monitoring of human movements

Andrzej Miękina[1], Jakub Wagner[1], Paweł Mazurek[1], Roman Z. Morawski[1,3], Tobba T. Sudmann[2], Ingebjørg T. Børsheim[2], Knut Øvsthus[2], Frode F. Jacobsen[2], Tomasz Ciamulski[2], Wiesław Winiecki[1]

[1] Warsaw University of Technology, Faculty of Electronics and Information Technology, Institute of Radioelectronics and Multimedia Technology, Nowowiejska 15/19, 00-665 Warsaw, Poland

[2] Bergen University College, Centre for Care Research, Mollendalsveien 6, 5020 Bergen, Norway

E-mail: r.morawski@ire.pw.edu.pl

Abstract. The importance of research on new technologies that could be employed in care services for elderly and disabled persons is highlighted. Advantages of radar sensors, when applied for non-invasive monitoring of such persons in their home environment, are indicated. A need for comprehensible visualisation of the intermediate results of measurement data processing is justified. Capability of an impulse-radar-based system to provide information, being of crucial importance for medical or healthcare personnel, are investigated. An exemplary software interface, tailored for non-technical users, is proposed, and preliminary results of impulse-radar-based monitoring of human movements are demonstrated.

1. Introduction

The life expectancy has been growing in Europe for many years, while the healthy life expectancy has been slightly diminishing since the last decade of the XXth century (*cf. http://www.healthy-life-years.eu/*). The problem of organised care for elderly people is, therefore, of growing importance. Hence the demand for research on new technologies that could be employed in monitoring systems supporting care services for such persons. The capability of those systems to detect dangerous events, such as person's fall, is of key importance [1, 2].

The systems for monitoring of elderly and disabled persons are, of course, expected not only to detect dangerous events, but also to predict those events on the basis of acquired measurement data, and therefore contribute to the prevention of such events. The analysis of gait, as well as of the itinerary and timing of activities of monitored persons, may thus contribute to prevention [2–5].

Recently, numerous attempts have been made to apply radar technology for monitoring of elderly and disabled persons (*cf.*, for example, the documents [6–9]). They are mainly motivated by the conviction that this technology may be less intrusive, less cumbersome, and less invasive with respect to the home environment than existing solutions. Another attractive feature of the radar-based systems is the possibility of the through-the-wall monitoring of human activity [10–12].

[3] To whom any correspondence should be addressed.

2. Visualisation of intermediate results of measurement data processing

It should be taken into account that the end-users of the fall detection and prevention systems, the medical and healthcare personnel, are interested only in a limited number of vital parameters, and are rather ignorant about technical details of such systems. Therefore, comprehensible visualisation of measurement data, acquired by means of various sensors, has recently attracted the attention of the research community [4, 13–17], though only with respect to well-established monitoring techniques. Depending on the applied technique, different types of information can be acquired, *e.g.*:

- if the monitoring is based on motion and pressure sensors [16, 17], then the data representative of the patient's location (*i.e.* in the occupied room) are gathered, and some inferences concerning the patient's health condition are made on the basis of the changes in his/her daily routines; *e.g.* early symptoms of depression can manifest as an increase or decrease in sleep activity, while early onset of dementia can cause the increase in wandering out of the apartment [18];
- if the monitoring is based on accelerometers and gyroscopes [19], the stress is put on the qualitative and quantitative characteristics of the patient's movements (*e.g.* number of walks per day, length of walks, and movement velocity) which can be used for identification of the changes in his/her health condition; *e.g.* as suggested in [20], walking velocity is a vital sign – it is very complicated and involves all bodily functions; to be able to walk at a normal velocity ($\geq$ 1 m/s), the overall health status must be good which means that if it deteriorates, it will be possible to observe this by a quick assessment of the person's walking manner.

Apparently, the majority of specified data can be successfully acquired in impulse-radar-based systems for monitoring of human movements. The next sections of this paper are focused on the capability of such systems to provide data being of crucial importance for medical or healthcare personnel, and on the comprehensible visualisation of those data.

3. Monitoring potential behind impulse-radar sensors

Since electromagnetic waves propagate through non-metal walls, a pair of impulse-radar sensors can localise a moving person in a monitored multi-room apartment. Therefore, a system based on such sensors can incorporate features of systems based on motion and pressure sensors; it becomes possible, *e.g.*, to create (on the daily basis) the so-called *heat maps* [4], *i.e.* diagrams depicting time spent by a monitored person in selected flat areas during a day.

An impulse-radar-based system can also incorporate some features of systems based on accelerometers and gyroscopes. Acquisition of sequences of measurement data, representative of the patient's position, followed by their numerical differentiation, enables one to quite accurately track the patient's movement, and to characterise it in terms of its direction and velocity. Moreover, it is possible to statistically summarise the movement parameters over some periods of time, *i.e.* to compute the average velocity, total distance travelled, time spent in motion, time spent motionless, *etc.* Those statistical parameters are believed to be sufficient for making useful inference on patient's health condition, and – consequently – preventing dangerous events (such as falls) based on timely identification of the health deterioration [19, 21, 22].

4. Proposed software application

Within a project aimed at the development of an impulse-radar-based system for monitoring of elderly and disabled persons, a simple software application has been designed – a kind of interface adapted to the expectations of its non-technical users.

In figure 1, the application panel designed for constant monitoring of human movements is shown. It enables one to check on-line the patient's location, his/her current movement direction and velocity, or the time spent motionless. Moreover, its user can generate a movement summary by clicking the appropriate button. An example of a movement summary is shown in figure 2; it contains the information enabling a medical caretaker to check the latest physical activities of the monitored person.

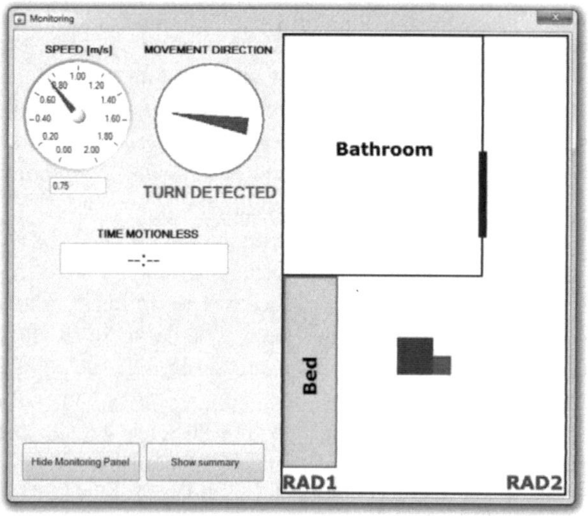

Figure 1. Monitoring panel of the application.

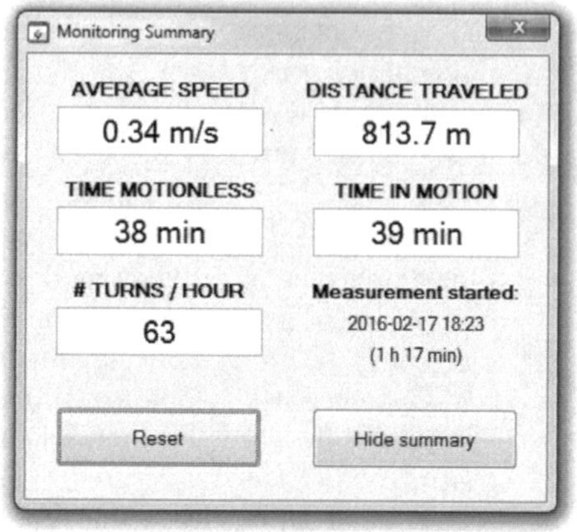

Figure 2. Example of movement summary.

The visualisation of the estimates of the patient's position and derived quantities in real time is a useful method for communication of the results of monitoring to the healthcare personnel. First, displaying the current position, together with the time passed since the last movement, enables the staff to identify a dangerous incident such as long time spent motionless after a fall. Moreover, motion detection during the night can be used to assess the quality of sleep – *e.g.* many motion events, detected during the night, may indicate sleeping problems. Next, a summary containing information on the estimates of the average movement velocity, total travelled distance, total time spent in motion and motionless, as well as the number of turns per hour may serve for inference on the patient's current health status, and on the need for its further investigation. Frequent turns, for example, may result from the patient's confusion – when he/she reminds himself/herself about something, or forgets where he/she was heading, and goes back – therefore the number of turns counted during a selected period may be an indicator of his/her mental health.

A system for monitoring of human movements, developed by the authors, may detect people behind a non-metal wall, although the accuracy of the location estimate is lower behind the wall than it is inside the room in which the radar modules are placed; however, it is sufficient for rough identification of the area. Therefore, the use of such modules allows for detecting the presence of monitored person at selected places in a household without the need to install them in each room.

After the preliminary tests in the laboratory environment, in order to acquire the real-world data, the system has been installed in the flat of one of the project participants. The radar sensors have been placed in a hallway located among other rooms – living room, kitchen, and bathroom – allowing for the acquisition of big and easily interpretable data sets. The scheme of the hallway is presented in figure 3; the symbols RAD1 and RAD2 indicate the positions of the radar modules.

In figure 3, an exemplary movement trajectory (marked with red crosses) is shown; it presents a movement from the kitchen to the living room and then to the bathroom. It can be seen that the system makes possible quite accurate tracking of the monitored person, though some small errors can also be observed. Such errors – like passing through the solid door – occur during the measurement data processing, but are mainly caused by the complexity of the measured radar signal (caused in turn by the complex structure of the human body being the subject of the monitoring).

In figure 4, an example of the movement histogram, created on the basis of the measurement data acquired during a few hours of continuous monitoring, is shown; in this figure – the warmer and more intense the colour, the more frequently the path is attended. It can be observed, that the histogram clearly indicates the paths between the rooms, with the most frequent one – between the living room and the kitchen. Moreover, as it can be seen, the monitored person quite frequently sits on the couch.

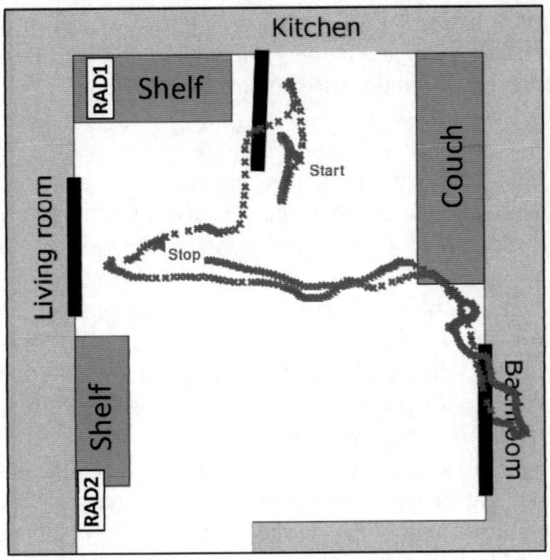

Figure 3. Example of movement trajectory.

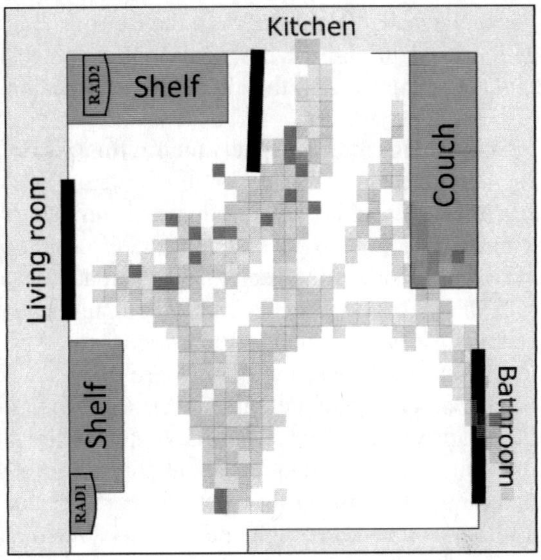

Figure 4. Example of movement histogram.

The two-dimensional location of the monitored person can be used for observing his/her presence at selected places in the household, such as the bedroom or the bathroom, at different times of day. It is then possible to determine a sequence of places visited in some equal time intervals during a number of hours or days. This type of monitoring – through the movement histograms and heat maps – facilitates *e.g.* the pre-fall, post-fall, and fall injury prevention [1]. It enables one to evaluate the variability of the walking patterns of the monitored person; moreover, those data – along with the knowledge about the patient's history and habits – can be used for designing a personalised prevention and warning system which can be tuned according to individual needs.

5. Discussion

It may be argued whether presented impulse-radar-based system is truly non-invasive, if it enables the healthcare personnel to acquire such diverse and detailed data on the monitored person. It has to be noted, though, that in order to provide a reliable monitoring of an elderly person, who is prone to fall, some compromise has to be made. From all the monitoring techniques applied today, radar sensors appear to be the least invasive with respect to monitored person's life (since they only acquire data representative of the person's location) while maximising the healthcare-related possibilities. On the contrary, there are video cameras, which obviously leave no place left for person's privacy and whose field of view is restricted only to the room they are installed in, and body-worn sensors which have to be constantly attached to the person's body and – more importantly – have to be remembered to be worn.

Another thing worth commenting on is that elderly persons aware of being monitored are easily motivated by this fact, what increases their velocity and vigour. On the one hand, more vigorous lifestyle might increase the risk for falls, but it is a mean for fall prevention, health promotion, and a facilitator for healthy behaviour on the other. Moreover, a vigorous life style increases the possibility for quick recovery after a fall, and reduces the risk of injurious falls and long lies after the impact. Therefore, another trade-off is being faced, but – ultimately – healthy ageing at home justifies a possible increase of the risk of falling.

Last but not least, some concerns, regarding the exposition of the elderly persons to the electromagnetic radiation, may be raised. In [23] it is shown that the Effective Isotropic Radiation Power (EIRP) of a single impulse-radar module nearly complies to the requirements of the ultra-wideband standards obligatory in the Europe [24]; this is sufficient for the research purposes, but must be improved in case of practical application. Nevertheless, it has to be remembered that different regulations exist, *e.g.* in the United States, causing potential limitations to the proposed system.

6. Conclusions and prospects for further work

The fall detection and prevention systems are meant to be used by the medical and healthcare personnel interested only in a limited number of vital parameters and being rather ignorant about technical details of such systems; therefore, the development of the human-system interface is not trivial and needs a multidisciplinary approach.

In this paper, the applicability of impulse-radar sensors for monitoring of human movements has been analysed and demonstrated using a prototype software application. In particular: the capability of the impulse-radar-based system to provide data needed by the medical and healthcare personnel has been investigated and preliminary results of the monitoring of human movements, visualised by means of the software interface designed for non-technical users, have been demonstrated.

Starting with information about the person's location, it is possible to derive many healthcare-related quantities being of crucial importance for medical or healthcare personnel, *viz.* the patient's current movement direction and velocity, the average movement velocity, the total travelled distance, the total time spent in motion and motionless, as well as the number of turns per hour.

There are still several issues that should be addressed in the future:

- the set of healthcare-related quantities can still be extended *e.g.* with the estimates of the person's acceleration obtained in three types of situations, *viz.* starting to walk, stopping, and changing the direction; those data can provide valuable information about the monitored person's movement dynamics, which is related to his/her health and fitness;
- the impulse-radar sensors can also be utilised to detect the breathing rhythm and heart beating of the monitored person [25]; incorporation of those functionalities would considerably enhance the quality and reliability of the monitoring;
- integration of the impulse-radar-based monitoring technique with other non-invasive techniques, *e.g.* depth sensors, can also be considered to increase the system's reliability.

Taking into account the non-intrusiveness and unobtrusiveness of impulse-radar systems, one may indicate them as promising means of reliable personalised monitoring of elderly and disabled persons in their home environment.

Acknowledgments

This work has been accomplished within the project PL12-0001 financially supported by *EEA Grants – Norway Grants* (*http://eeagrants.org/project-portal/project/PL12-0001*).

References

[1] Hamm J, Money A G, Atwal A and Paraskevopoulos I 2016 *Journal of Biomedical Informatics* **59** pp 319–45
[2] Sudmann T T, Børsheim I T and Jacobsen F F 2016 Technical Report (Bergen, Norway: Bergen University College, Centre for Care Research) pp 1–16
[3] Sudmann T T, Børsheim I T, Jacobsen F F, Øvsthus K and Ciamulski T 2016 Technical Report (Bergen, Norway: Bergen University College, Centre for Care Research) pp 1–39
[4] Baldewijns G, Debard G, van Den Broeck B, Mertens M, Karsmakers P, Croonenborghs T and Vanrumste B 2016 ed F Florez-Revuelta and A A Chaaraoui (Herts, UK: IET) pp 1–22
[5] Baldewijns G, Luca S, Vanrumste B and Croonenborghs T 2016 *BMC Medical Research Methodology* pp 16–23

[6] Morawski R Z, Yashchyshyn Y, Brzyski R, Jacobsen F and Winiecki W 2014 in *Proc. IMEKO TC-4 International Symposium* (Benevento, Italy), pp 786–91
[7] Morawski R Z, Miękina A and Bajurko P 2014 in *Proc. IMEKO Joint TC1-TC7-TC13 Symposium*), pp 1–6
[8] Su B Y, Ho K C, Rantz M and Skubic M 2015 *IEEE Transactions on Biomedical Engineering* **62** pp 865–75
[9] Jian Q, Yang J, Yu Y, Björkholm P and McKelvey T 2014 in *Proc. 8th European Conference on Antennas and Propagation* (The Hague, The Netherlands), pp 3078–81
[10] Xin L, Dengyu Q, Ye L and Huhe D 2013 in *Proc. 35th Annual International Conference of the IEEE Engineering in Medicine and Biology Society* (Osaka, Japan), pp 1013–6
[11] Yinan Y, Jian Y, McKelvey T and Stoew B 2012 *International Journal of Antennas and Propagation* **2012** pp 1–11
[12] Zetik R, Crabbe S, Krajnakc J, Peyerl P, Sachs J and Thoma R 2006 *Proceedings of SPIE – 'Sensors, and Command, Control, Communications, and Intelligence (C3I) Technologies for Homeland Security and Homeland Defense'* **6201** pp 1–12
[13] Craft M, Dobrenz B, Dornbush E, Hunter M, Morris M S and Barnes L E 2015 in *Proc. IEEE Systems and Information Engineering Design Symposium* (Charlottesville, VA, USA), pp 212–7
[14] Hassanalieragh M, Page A, Soyata T, Sharma G, Aktas† M, Mateos G, Kantarci B and Andreescu S 2015 in *Proc. IEEE International Conference on Services Computing* (New York, NY, USA), pp 285–92
[15] Mulvenna M, Carswell W, McCullagh P, Augusto J C, Zheng H, Jeffers P, Wang H and Martin S 2011 *IEEE Communications Magazine* **49** pp 110–7
[16] Munoz A, Serrano E, Villa A, Valdes M and Botia J A 2012 *Sensors* **12** pp 6282–306
[17] Robben S, Boot M, Kanis M and Krbse B 2013 in *Proc. 7th International Conference on Pervasive Computing Technologies for Healthcare and Workshops* (Venice, Italy), pp 416–9
[18] Galambos C, Skubic M, Wang S and Rantz M 2013 *Gerontechnology* **11** pp 457–68
[19] Brodie M A, Lord S R, Coppens M J, Annegarn J and Delbaere K 2015 *IEEE Transactions on Biomedical Engineering* **62** pp 2588–94
[20] Fritz S and Lusardi M 2009 *Journal of Geriatric Physical Therapy* **32** pp 2–5
[21] Maki B E 1997 *Journal of the American Geriatrics Society* **45** pp 313–20
[22] Viccaro L J, Perera S and Studenski S A 2011 *Journal of the American Geriatrics Society* **59** pp 887–92
[23] Bajurko P and Yashchyshyn Y 2015 in *Proc. 8th IEEE International Conference on Intelligent Data Acquisition and Advanced Computing Systems: Technology and Applications* (Warsaw, Poland), pp 1–4
[24] Nikookar H and Prasad R 2009 (The Netherlends: Springer Science+Business Media) Chapter 10
[25] Bajurko P and Yashchyshyn Y 2014 in *Proc. 20th International Conference on Microwaves, Radar and Wireless Communication* (Gdańsk, Poland), pp 793–6

Journal of Physics: Conference Series **772** (2016) 012029 doi:10.1088/1742-6596/772/1/012029

Measurements of non-physical quantities

Igor F Shishkin

Dr. – Professor, Principal Scientist, D.I. Mendeleyev Institute for Metrology (VNIIM), St. Petersburg, Russia

E-mail: nikhsihs@ipa-spb.com

Abstract. The aim of the paper is to suggest an approach to development of a theory of measurements for non-physical quantities. For these measurements it is not possible to ensure traceability because of their exclusive nature as substantiated by the author. This theory is presented as particularly important one for social and human sciences.

1. Introduction

The science of measurements, **METROLOGY**, is an integral part of the theory of epistemology, **GNOSEOLOGY**. It is the main instrument of obtaining accurate quantitative knowledge. It is also an instrument whereby we apply this knowledge in practice making it possible to facilitate scientific and technical progress, to promote trade relations, to increase the level of life and to improve its quality.

The science that deals with measurements of physical quantities (PQs) has a rich history and clear prospects for further development. This cannot be said about measurements of non-physical quantities (NPQs) that have nothing to do with metrological methodology. However, this gnoseology instrument is now experiencing a great demand [1, 2, and others]. If we agree with Immanuel Kant, who considered that *"in any special doctrine of nature there can be only as much proper science as there is mathematics therein"*, it appears then that all human sciences are not properly sciences, because they do not use such instruments as those that are used for NPQ measurements (in the metrological sense) and thus they are devoid of a strict quantitative basis. Let us not forget that results of researches pursued by historians, philosophers, teachers of social sciences, lawyers, sociologists, philologists, psychologists, political scientists, culture experts and a vast number of other scientists are used in decision making at all levels of governments and economy management, i.e., they touch upon fundamental interests of all people.

2. Non-physical quantities

By analogy with the definition of PQ, **by NPQ we will understand an immaterial property that, speaking in terms of quality, is common to many objects of the material and immaterial world that surround us. However, speaking in terms of quantity, this property is individual for each of these objects.**

Among the objects of our surrounding **material world** there are objects in our environment, such as: nature, urban infrastructure, household, society, family, cars, things, tools, goods, products, services, etc., that really exist in the spatio-temporal continuum. *Immaterial properties* of these objects (NPQs) are: *usefulness, beauty, necessity, importance, actuality, quality, expediency*, etc.

The **immaterial world** (noosphere[1]) includes the people's spiritual world, whose NPQs are as follows: *high-mindedness, tactfulness, giftedness, moral rectitude, vulnerability, spirituality, enthusiasm, falsity, meanness, equity* and many others.

Thus, NPQs exist only in the mind of people either as reflections of properties of the real material world (PQs), or in the form of people's own understanding. They also can be formed on the basis of insights (NPQs) of other people.

[1] Noosphere (from Greek νοῦς - mind and σφαῖρα - sphere) - sphere of mind; it is used here in the direct sense of the word without loose and broad interpretations. The term was introduced by Édouard Le Roy, professor of mathematics at the Lycée Saint-Louis in Paris, and Pierre Teilhard de Chardin, French philosopher and Jesuit priest, who together attended lectures of Vladimir Vernadsky at the Sorbonne in 1922-1923.

Just as PQs are subdivided into base and derived quantities, NPQs could be categorized as *simple* and *compound* quantities. The *simple* PQs are constituent parts of the *compound* quantities, the latter being composed of a set of simple or compound quantities of a lower level. For example, such *compound* NPQ as "*high-mindedness of a human*" includes a number of more simple NPQs (i.e. immaterial properties of that person's nature): *culture, politeness, tactfulness*. Sometimes, the *simple* NPQs constituting compound quantities are called constituent parts or components, indications, symptoms (as in psychiatry), etc.

2.1. Measure and measurement unit

"Things are to you such as they appear to you", affirmed Protagoras. All that appears to us is a fruit of our mind and belongs to the immaterial world. Insofar as the objects of such world and their *immaterial properties* are **the products of imagination** that exist only in people's mind, they cannot be measured by the same measuring instruments as those used for measurements of the PQs. Their only measuring instrument is the **HUMAN MIND**.

Protagoras believed that *"of all things the **measure** is man, of the things that are, and of the things that are not"*. Ancient Greeks proclaimed that the last and the highest gift of Gods to men is the **sense of proportion**. According to the teachings of Protagoras the world is perceived through human **senses**. Now we would rather clarify this statement by saying that we cognize the material world by means of our **sense organs** through organoleptic measurements of its quantitative characteristics (PQs), while the immaterial world is perceived through measurements of its quantitative characteristics (NPQs).

For instrumental measurements of the PQs **artefacts** can be used, such as length gauges (rulers) to measure the *length* and mass gauges (weights) to measure the *mass*. However, the noosphere has no material measures. The only **measure** here is the human **OPINION**.

It is believed that the use of **opinions** in **measurement** tasks has a big drawback: an **opinion is a subjective estimate** largely depending on people's specific features, e.g., their qualifications, erudition, experience, knowledge, personal preferences, tastes, health, etc. However, there is nothing to be surprised at: something similar happened when people used anthropometric measuring tools.

Opinions can be expressed in different forms. Therefore, for centuries, without any connection with the measurement science – metrology, have been emerging and used empirically such procedures for measurements of NPQs as decisions, surveys, sociological investigations, referenda, votes, tests, expert examinations, etc. Some of them were provided with theoretical grounds (theory of estimation, psychometry, methods of expert evaluations).

When we express our opinion about the fact that a certain immaterial property manifests itself in one case to a larger extent than in another case, we give our comparative quantitative estimate, i.e. we make a *single measurement* by an ordinal scale [3]. **Accumulating** a posteriori information enables us to proceed to a *repeated (multiple) measurement*. To do it, it is necessary to obtain an array of experimental data. In that case the **NUMBER of opinions** coinciding with the initial estimate could be a **quantitative characteristic of an NPQ**, i.e. a *result of a multiple measurement*[2], one **opinion** being the *measurement unit*.

The use of an **opinion** as the *measurement unit* having no fixed dimension means but one thing: measurements of the NPQs are performed on the basis of a DECENTRALISED realization of the measurement unit.

2.2. A way to measure NPQs

According to the second axiom of metrology, **any measurement essentially consists in empirical comparison of dimensions** [3]. As immaterial properties that exist only in people's mind have no **material** dimensions, it is possible to compare their quantitative characteristics only if they manifest themselves in some way. In this case, the **measurement** of an immaterial property boils down to a

[2] In practice a dimensionless relative index (fraction of such opinions in the general count [5]) is used as the result of multiple measurement. Its numerical values are within the interval [0; 1].

comparison of manifestations of this property, and, as a result, the question about when the property has manifested itself in a greater degree can be answered.

Nature has endowed us neither with the ability to compare quantitatively different properties with each other, nor to compare their manifestations. So, **the only way to estimate non-physical quantities is to compare their manifestations by the ordinal scale**.

2.3. Methods of NPQ measurements

When measurements are made using the ordinal scale, the comparison with a zero dimension becomes especially important. In the theory of PQ measurements, the zero dimension means the absence of this quantity, and a positive result of comparison with such a dimension is called **detection** of this physical value [3]. Manifestations of NPQs have no dimension, but the absence (i.e., 0) of any manifestation is considered as the absence (i.e., 0) of NPQs. If we have an expert **opinion** about the presence of at least one manifestation of an NPQ, it is **believed** that the NPQ has been **detected**. When the **detection method** is used for such measurements, the quantitative characteristic of an NPQ is determined by the **NUMBER of its manifestations**[2].

The second method of NPQ measurements is the **matching method**. It has two variations: **method of ranking** and **method of pairwise comparison**. When **matching method** is used, the quantitative characteristic is determined by the **NUMBER of preferences**[2] identified as "ranks" (each preference being an opinion about the "rank" to be assigned to the NPQ).

2.4. Exclusivity of NPQ measurements

Insofar as any **opinion** tends to be subjective, it is not possible to **provide its traceability to a reference**, which is the main concern of the legal metrology [4, 5]. Hence arises this question: As *"no two minds think alike"*, why is it necessary to measure non-physical quantities? The following answer springs to mind: **measurements of physical quantities (PQs) and those of non-physical quantities (NPQ) have different fields of application**.

Measurements of NPQ are:

- **a way to perceive the material world through its reflection in human minds**, which depends on their world view, creed, perception of the world, value system, priorities and a great number of other factors determining their personality. Results of measurements of the same NPQs in the material world, obtained by different people, cannot be the same IN PRINCIPLE.
- **a way to perceive the immaterial world of each person through its understanding (or lack of understanding) by other people**, acceptance or rejection of their value system and priorities, moral principles and actions, decisions, intentions and behavioral norms in the ordinary life and in emergencies. The results of such measurements obtained by different people cannot but differ from each other.

The requirement to ensure the **traceability** of PQ measurements gives way to the requirement to ensure the **exclusivity of NPQ measurements**. Not the manipulation with averaged results of NPQ measurements, but the respect of each individual person's **opinion**, taking into account his personal **opinion** when solving the most important social, political and socioeconomic tasks, are keys to the harmonious progress of civil society and to founding many social and human sciences on a strict quantitative basis.

2.5 Setting priorities

The exclusiveness of NPQ measurements can be increased by taking into account the *importance*, significance and cogency of its components. In most cases it is legislatively mandated by means of weighting factors. However, the *importance* being itself an NPQ, it can be measured as such. The iterative procedure to determine weighing factors by a commission of experts is described in details in [6]. At certain conditions, which are usually always met, it provides the values of weighing factors that

Journal of Physics: Conference Series **772** (2016) 012029 doi:10.1088/1742-6596/772/1/012029

strictly reflect correlations of importance among the objects of the expert examination once the initial data have been established by the experts.

The provision of exclusivity of measurements of *satisfaction* in different groups of users and parties interested in the quality of products, goods and services by means of weighing factors is discussed in [7].

The method of setting the priorities in accentuating the exclusiveness of different NPQs is widely used in pedagogy. In 1993 - 2011, important administrative decisions were taken on its basis in Russia [8-10]. At present, this is the main instrument to increase the exclusivity of NPQ measurements for the interest of individuals or groups and segments of the people.

2.6 Intransitive preferences

The ordinal measurement scale is characterized by the property of *transitiveness*. If the non-material properties A, B, C collated in pairs with each other are logically interrelated by proportions of preferability: B → A and C → B, then the property of transitiveness means that C → A. However, in processing the results of NPQ measurements, as in the case of PQs, gross errors (or so-called "blunders') may occur. The latter can be explained by the lack of concentration of the experts, lack of confidence or absent-mindedness, which lead to the breach of transitiveness.

For example, an expert submits to the expert commission secretary the following table of pairwise collations of five examined objects for further treatment of the results:

Table 1. Results of a pairwise collation of five examined objects

	a	**b**	**c**	**d**	**e**
a	×	↑	↑	↑	↑
b		×	←	↑	↑
c			×	←	↑
d				×	↑
e					×

The direction of preferences in Table 1 is indicated by arrows. If, when determining the ranks, the preferences marked by arrows, the objects of the expert examination will have the following ranks: $R_a = 0$; $R_b = 2$; $R_c = 2$; $R_d = 2$; $R_e = 4$. Thus, although the preferences for each pair of expert examination objects have been established, there cannot be solution in the form of ranked series, because there are *intransitive preferences*. In this case the *intransitive preferences* form a triangle (see Fig. 1), where the ranks of the expert examination objects are given in brackets.

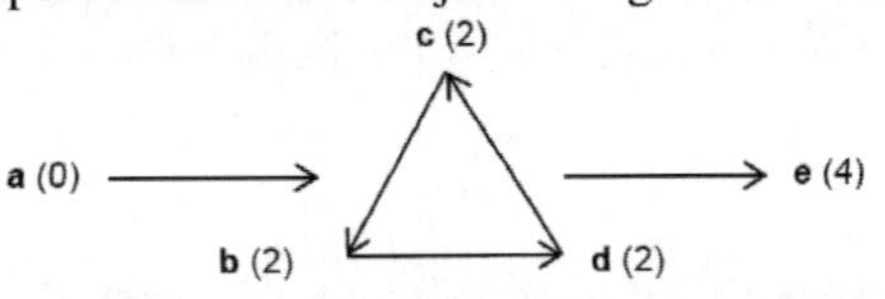

Figure 1. Results of pairwise collation of five expert examination objects

Similar ranks of certain terms of ranked series are indicative of the *intransitive preference*. They can occur everywhere: in the beginning, at the end and in the middle of the series. Intransitive references have the closed form of triangles, squares, polygons. Sometimes complex topological spaces can be present among experimental data. These spaces can include unions, absorptions, intercepts of *transitive* and *intransitive subsets*, which presupposes the possibility to use the Boolean algebra in processing the results of expert examinations. Such issues are discussed in [11].

The presence of *intransitive preferences* in the ranked series is not a singular phenomenon. The paper [9] sums up an extensive experiment performed in Russia during 10 years. From 1993 to 2003 a number of higher schools were certified according to the procedure adopted by the Russian State

Inspectorate for Certification of Educational Establishments [8]. Over 50 higher schools were certified during this campaign. The procedure allowed for taking into account opinions of pedagogical staff members about the *importance* of the factors used for certification of their educational establishments. This NPQ (*importance*) was measured by the academic councils of higher schools. In ten years of the certification campaign, *intransitive preferences* in the experimental data submitted by members of these Academic Councils for processing were detected in all cases.

It is possible to eliminate all intransitiveness by reversing one or several opinions of the expert:

$$\mathbf{a \to c \to b \to d \to e}; \qquad \mathbf{a \to b \to d \to c \to e}; \qquad \mathbf{a \to d \to c \to b \to e}.$$

Usually the expert decides himself which opinion he is ready to reverse. If he is not available, the decision is up to the expert commission.

2.7 *Quality of NPQ measurements*

In measuring an NPQ by the **detection method**, the expert either notices the presence of the NPQ, or not, i.e. he acts as an *index*. A theory that regards indexes as instruments measuring on the ordinal scale is set forth in [3].

The expert **opinion** formed as a result of the measurement carried out on the ordinal scale is the **decision**. According to the 3rd axiom of metrology, its character is random, i.e., it can be *correct* or *incorrect (wrong)*. Thus, the probability that the *quality index* of the measurement result obtained on the ordinal scale is correct appears to be an obvious quality index.

Not less frequent practical *quality index* of a decision is the probability of its being wrong. The larger is the probability of mistake, the worse is the quality of a measurement result.

Wrong decisions taken when measuring on the ordinal scale are classified into mistakes of the 1st and 2nd kinds [3]. Decreasing the probability of an error of the 1st kind (P_I) entails increasing the probability of an error of the 2nd kind (P_{II}) and vice versa. Therefore, the natural desire to minimize the errors is a contradictory one.

The decisions that in the best possible way comply with contradictory requirements are called *optimal* decisions. Ideally, the best (*optimal*) decisions would be those when the observer (operator, controller) are totally exempt from errors or, at least, when their probability (P_{err}) is minimal. Therefore, the optimization criterion $P_{err} = P_I + P_{II} = \min$ is called *ideal observer criterion*. From this condition, in the case of technical devices an *optimal detection threshold* is calculated and assigned, which thus becomes a normalized metrological characteristic of the *index* as a measuring instrument. When measuring NPQs, it is impossible to calculate, assign or verify an *optimal detection threshold*. Each expert has its own threshold and it is impossible to influence his *opinion* which is, in fact, his *decision*. However, the *reasonableness of a decision* should be confirmed by a high probability of its being correct: $P = 1 - P_{err} = 1 - P_I - P_{II}$, where the probabilities P_I и P_{II} are assigned a priori from experience.

3. Conclusions

We live in the world that does not consist exclusively of material objects. In our mind there exist notions about good and evil, justice and injustice, presumption and modesty, moral rectitude and meanness, friendliness and hostility, as well as many shades of interpersonal relations and properties of our environment. However, being manifested in larger or smaller degree, they are characterized quantitatively and, consequently, can be measured. There is no need to speak about the great importance of information exchange in the quantitative measure with respect to people's intercommunication. It is needless to mention how large could be the qualitative leap in the development of human sciences, once they have turned to be "exact" sciences in the contemporary understanding of this word, if they become founded on strict quantitative calculations. "Science begins since they start measuring. Science is inconceivable without measures" (D.I. Mendeleyev). Such qualitative leap is possible due to NPQs' measurements.

NPQs exist only in the noosphere. The material sphere and the noosphere are fundamentally different and their differences drastically affect measurements in these spheres of human activity.

1. The noosphere is individual. It is formed as a result of many years' experience, and each person has its own noosphere.
2. The instrument to measure with in the noosphere is the human (expert) mind.
3. The result of a single measurement in the noosphere is the opinion of a human (expert).
4. The result of a multiple measurement is the number of opinions[2]. In this case one opinion is equivalent to a measurement unit.
5. Opinion has no fixed dimension. This means that measurements in the noosphere are made on the basis of decentralized realization of the measurement unit.
6. The result of an NPQ measurement depends on the noosphere of the expert and it has subjective character. It is not correct to raise the question about providing traceability of measurements in this case.
7. The requirement of traceability for PQs' measurements gives way to the requirement to provide exclusiveness of NPQ measurements, i.e. the adequacy of their results to the noosphere of each individual and to the interests of different groups or segments of the people.

The exclusivity of NPQ measurements allows us to investigate the spiritual world of each individual, to understand his interests and preferences and to build productive relationships with him taking into consideration particularities of his noosphere. While average characteristic values are used in the material world when developing national, departmental, regional and sectorial programs of the economy, production scheduling programs and preparing statistical reports, the use of such average values is not possible in providing targeted assistance to the people, distributing the products, goods and services according to the needs, wishes and preferences of the consumers. Only through the study of real consumer demand and taking into consideration and respecting the opinion of every customer about the consumer properties, as well as by satisfying his personal needs, wishes and expectations, it will be possible to establish civilized market relations acceptable for all parties of this process.

Knowledge and consideration of and respect for particularities of the spiritual world of every individual in social medium that have become feasible due to the latest information technologies are tokens of the harmonious development of social relations capable of creating a conflict-free society in the foreseeable future.

References

[1] Finkelstein L 2009 *Measurement* **42** 1270–77
[2] Rossi G B 2007 *Measurement* **40** 545–62
[3] Shishkin I 2012 *Theoretical metrology. Part 1. General theory of measurements* (St. Petersburg: Piter) p 220
[4] Shishkin I 2012 *Theoretical metrology. Part 2. Ensuring traceability of measurements*: Textbook for higher schools (St. Petersburg: Piter) p 240
[5] Shishkin I 2015 Subjective measurements *J. The Econ. of Quality* **3-4** 11-12
[6] Shishkin I and Stanyakin V 1992 Qualimetry and quality management: Textbook for higher schools (Moscow: VZPI) p 256
[7] Shishkin I and Nadvidova I 2014 Measuring customers' satisfaction (St. Petersburg: National Mineral Resources University) vol 209 pp 163–165
[8] 2002 Procedure for external expert examinations to certify educational institutions of higher professional education (higher educational institutions) and their branches (Moscow: State inspection on certifying educational institutions of Russia) pp 30–35
[9] Shishkin I 2005 Measuring the quality of education and educational services *Pedagogical measurements* **1** 105-123
[10] Semenova G and Shishkin I 2011 Using quality control methods in modern pedagogy (St. Petersburg: SZTU) p 176
[11] Shishkin I 2007 On traceability of measurements carried out by expert methods *Electrika* **2** 43-48

Journal of Physics: Conference Series 772 (2016) 012030 doi:10.1088/1742-6596/772/1/012030

Cutting force measurement of electrical jigsaw by strain gauges

L Kazup
PhD Student, University of Debrecen, Department of Electrical Engineering
E-mail: kazup.laszlo@science.unideb.hu

A Varadine Szarka
Associate Professor, University of Debrecen, Department of Electrical Engineering
E-mail: angela.varadi@science.unideb.hu

Abstract. This paper describes a measuring method based on strain gauges for accurate specification of electric jigsaw's cutting force. The goal of the measurement is to provide an overall perspective about generated forces in a jigsaw's gearbox during a cutting period. The lifetime of the tool is affected by these forces primarily. This analysis is part of the research and development project aiming to develop a special linear magnetic brake for realizing automatic lifetime tests of electric jigsaws or similar handheld tools. The accurate specification of cutting force facilitates to define realistic test cycles during the automatic lifetime test. The accuracy and precision resulted by the well described cutting force characteristic and the possibility of automation provide new dimension for lifetime testing of the handheld tools with alternating movement.

1. Introduction

Thanks to the rapid progression of technology more and more new automatic test methods and devices appear in both electrical and machinery fields of industry nowadays. Due to development of modern materials and the advanced and efficient power electronic technologies the most of handheld power tools, garden tools, electric white goods of households and automotive parts can be tested by automated lifetime test methods. However, there are no fully developed test methods and constructions for testing of power tools producing fast-changing alternating movements. The major goal of our research is to develop a contactless braking method based on electromagnetics which allows automated and well reproducible lifetime-testing of these equipment. The problem is fairly difficult due to the relative high frequency of alternating movement and the complex characteristics of cutting force related to one moving period. Furthermore duration of one test cycle may be even a week. These facts require high reliability and robust structure.

The main direction of our research is the analysis of different linear brake constructions (excited by permanent magnets or electromagnets) which are capable to realize the previously mentioned test methods. The building and testing of the first prototype (including excitation coil) has been ended, the applicability of permanent magnets is in the centre of our actual research activities. Besides of the above described topic, analysis of different operational parameters related to the UUT (unit-under-test) is also an important issue. This paper presents a method for the cutting force measurement, which should result an accurate value of the most important operational parameter.

Journal of Physics: Conference Series **772** (2016) 012030 doi:10.1088/1742-6596/772/1/012030

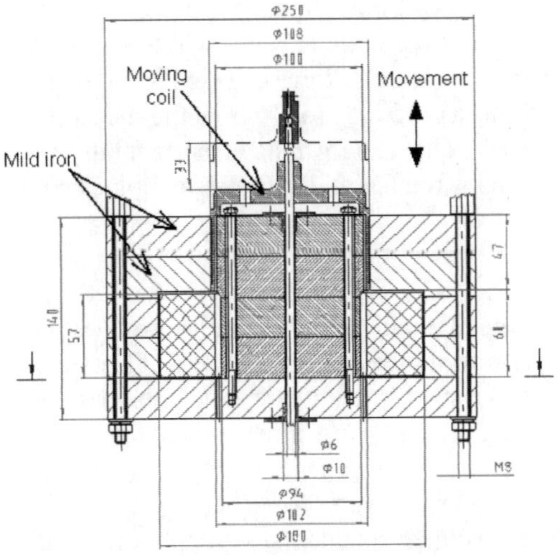

Figure 1. Prototype of dynamic linear brake for handheld tool's lifetime testing (excited soft iron stator)

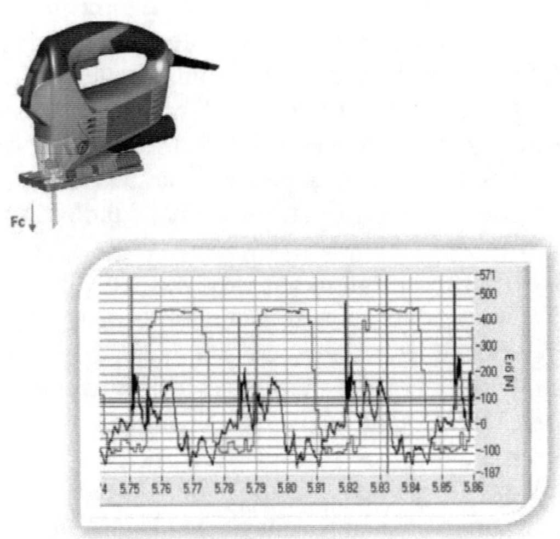

Figure 2. Estimated cutting force - Formed force between the workpiece and the footplate (measured by piezoelectric load cells)

2. Cutting force measurement with strain gauges

There are several possible measuring methods exist for determination of the cutting force of an electrical handheld jigsaw. One of the solutions uses piezoelectric load cells fixed between the workpiece and the footplate. However, the elasticity of both the footplate and the workpiece, the mass of the tool and the flexible downforce applied by the user realize a damped, complex vibration system with more degrees of freedom. The resonance frequency of this system can be close to the alternating frequency of the blade, therefore, the force measured by the load cells can be different from the real cutting force. Determination of the accurate parameter using of complex mathematical models and relationships is necessary to solve the problem.

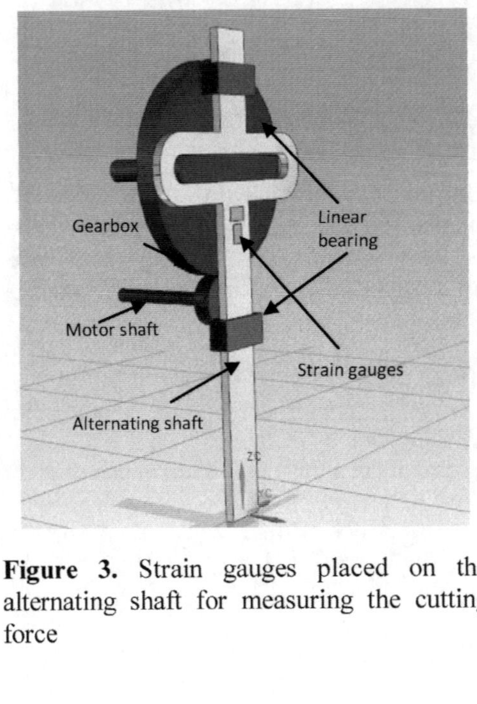

Figure 3. Strain gauges placed on the alternating shaft for measuring the cutting force

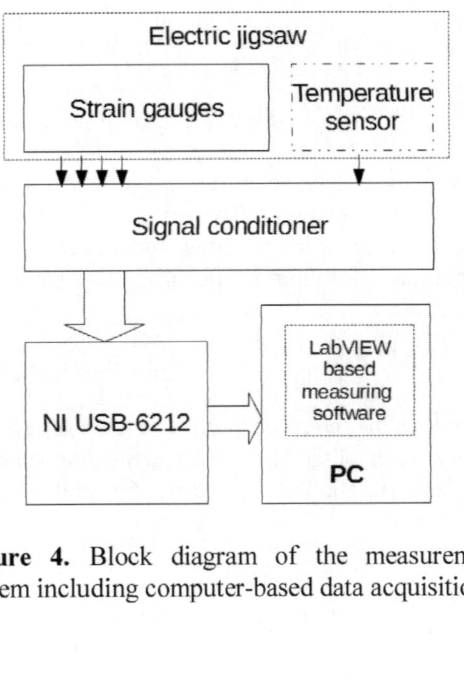

Figure 4. Block diagram of the measurement system including computer-based data acquisition

Figure 2. shows result of the cutting force measurement. Solution described in this paper measures the strain (or compression) of the alternating shaft directly. As the cutting blade is connected to the shaft, deformations measured by strain gauges are proportional to the cutting forces. The layout of the gauges is shown on Figure 3. Mounting place seen on the figure is the only one where the measuring gauges can be mounted due to the position of linear bearing. The gauges placed horizontally are responsible for the temperature compensation of the measuring circuitry and symmetrical placement provides removing of effect of shaft's deflection.

3. Structure of the measuring system

Due to the very small strain of gauges (max. 70με) a signal conditioning circuit provides the appropriate interfacing between the gauges and the computer-based measurement system. This signal conditioner handles the gauges using two Wheatstone-bridge (one for each side) and includes two pieces of AD620 type precision instrumentation amplifier. Both stages are independent from each other and provides 1000 of gain. The power supply includes a simple transformer with $24V_{AC}$ primary voltage to prevent electrical shock during testing. The self-made transformer can provide the corresponding output voltages (±12V, 5V and 24V) via linear voltage regulators to supply all circuit elements. The computer-based measurement system contains a NI USB-6212 type multifunctional data acquisition module. The data acquisition is controlled by a LabVIEW-based measuring software. The zero-adjustment of Wheatstone-bridges is also implemented by the software. The block diagram of the amplifier stage can be seen on Figure 5.

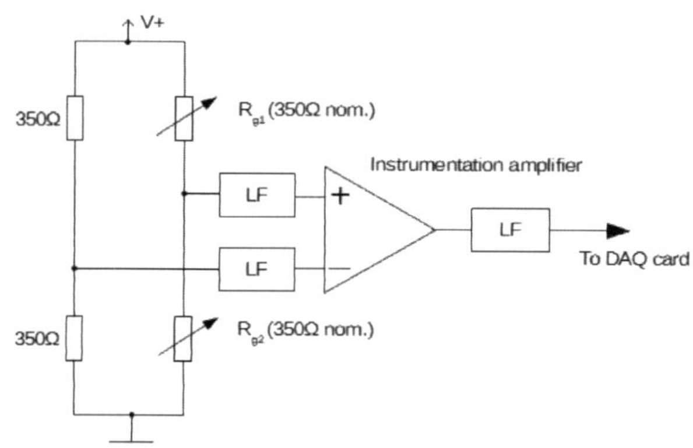

Figure 5. Simplified circuit diagram of one amplifier stage

The strain gauges used in this measurement system are constantan-based KMT-LIAS-06-1.5-350–5E-type thin film gauges. The nominal resistance is 350Ω±0.5%, the gauge factor is 2.05 ±1%. The vertical gauge is responsible for measuring the shaft's vertical strain and the horizontal one compensates the temperature drift. The strain of the shaft (ε) can be calculated by the following equation.

$$\varepsilon = \frac{-4V_r}{GF(1+2V_r)} \quad [4],$$

where V_r is the reference voltage of the bridge and GF is the gauge factor.
The mounting of strain gauges and the assembled measuring circuit are shown on Figure 6.The PCB includes a vibration sensor's driving circuit as well (see below).

Journal of Physics: Conference Series **772** (2016) 012030 doi:10.1088/1742-6596/772/1/012030

Figure 6.a. Strain gauge mounting

Figure 6.b. Assembled measuring circuit

4. Conclusions of test measurement

The first test results underlined linear relationship between the pulling/compressing force and the output signal in case of small forces. However, when the force is significantly increased, this relationship is becoming nonlinear. In case of large forces the output signal had the same polarity in both pulling and compressing period. To evaluate these results a finite element simulation has been implemented in Comsol Multiphysics 3.5. The applied vertical force was 100N. Simulated horizontal and vertical strain distributions in both "pulling" and "compressing" periods are shown on Figure 7.

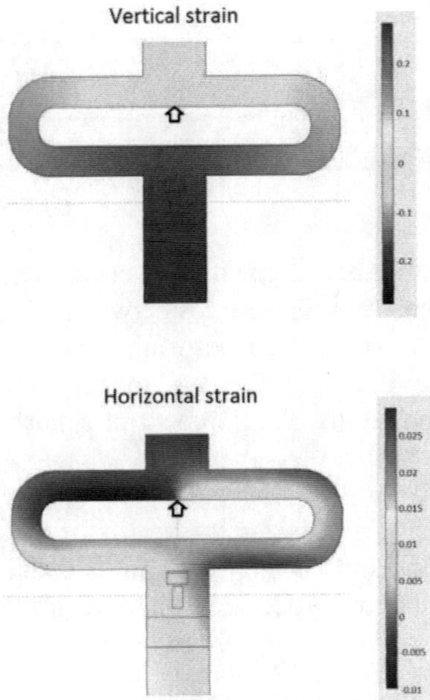

Figure 7.a. Relative strain of the alternating shaft during a "pulling" period

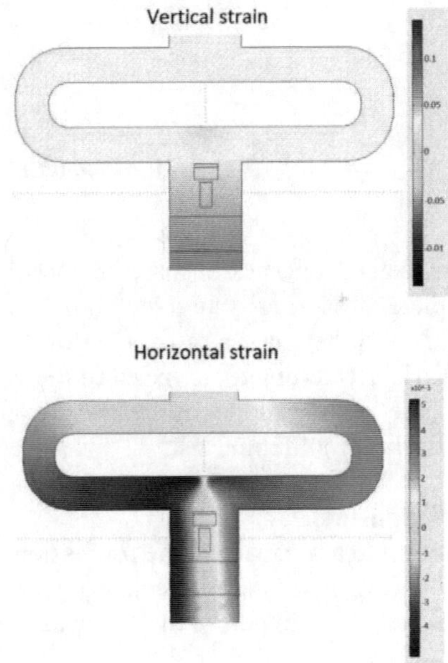

Figure 7.b. Relative strain of the alternating shaft during a "compressing" period

Geometry of the driving unit causes dominant horizontal stretching effect in the shaft during the pulling phase (about 6 times higher than the vertical stretching). In the other, pushing phase of the moving cycle the vertical compression is the dominant effect which is 1.5 times higher than the horizontal stretch). Simulation results also approved these strain/compression ratios. The simulation has also shown that the horizontal compression data carry significant information on the applied force in pulling phase, although, horizontal and vertical strains have opposite directions.

5. Conclusions and further development

Test measurements and results of finite element simulations has shown significant nonlinear relationship between the cutting force and the output voltage of the signal conditioner when using pair of strain gauges in a compensated Wheatstone bridge. This nonlinearity is caused by the arrangement of the gauges, because vertical force causes also significant deformation in the horizontal gauge which is originally used for temperature compensation. Therefore, using of independent for each sensors precise current-source type driving method is necessary, although we should use external temperature compensation and enhanced electrical shielding. Because the geometry of the driving unit is varying point by point using complex mathematical models and relationships are necessary for accurate calculation of the momentary cutting force. For practical realization, an improved signal conditioning unit should be used, and also more accurate finite element simulation models should be built. The simplified block diagram of the signal conditioner mentioned above is shown on Figure 8.

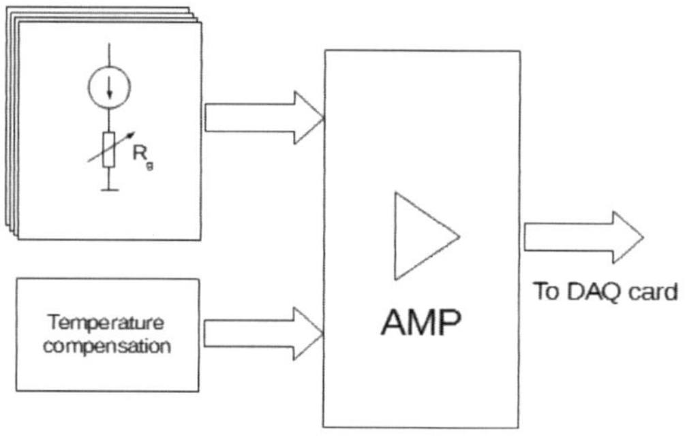

Figure 8. Block diagram of the improved signal conditioning circuit

Since the exact properties of materials applied in the driving unit are unknown, we need to calibrate the complete measuring system including the jigsaw to get the correct relationships between measured parameters and the cutting force. One of the further goals is to extend the measuring system with a piezoelectric accelerometer to measure the vertical vibration. It will help to compare the characterized cutting force and the tool's vertical vibration providing good basics for development of a fairly new method for jigsaws lifetime tests.

Acknowledgments

Authors wish to acknowledge the professional and financial support of NI Hungary Ltd. and also they wish to express their sincere thanks to general manager of the company Dr. László Ábrahám for his consistent help and support to the electrical engineering field at the University of Debrecen.

References

[1] Norton H N 1969 Handbook of Transducers for Electronic Measuring Systems *pp. 42ff., Prentice-Hall, Inc., Englewood Cliffs, N.J.*

[2] Hannah R L, and S E Reed 1992 The Strain Gage Users' Handbook *Elsevier Applied Science (London and New York)*

[3] Váradiné Sz A 2010 Linear Magnetic Break of Special Test Requirements with Dynamic Performance *Journal of Electrical and Electronic Engineering 3:(2) pp. 237-240*

[4] Earl J W 1976 Strain-Gage Instrumentation *Shock and Wibration Handbook 2nd Edition Chapter 17*

[5] L Kazup A Váradiné Szarka 2008 Control of Dynamic Linear Magnetic Break *ENELKO International Conference on Energetics and Electrotechnics, Romania, pp.21-25.* ISSN 1842-4546

Motor transport related harmful PM2.5 and PM10: from on-road measurements to the modelling of air pollution by neural network approach on street and urban level

O Lozhkina[1,3], V Lozhkin[1,3], N Nevmerzhitsky[1], D Tarkhov[2,3] and A Vasilyev[2,3]
[1] St. Petersburg University of State Fire Service of EMERCOM of Russia, 196105, Moskovsky, 149, St. Petersburg, Russia
[2] Peter the Great St. Petersburg Polytechnic University, 195251, Polytechnicheskaya, 29, St. Petersburg, Russia
[3] Project № 14-01-00733-A supported by the grant of The Russian Foundation for Basic Research

E-mail: olojkina@yandex.ru

Abstract. The level of PM10 and PM2.5 concentrations in the air on seven roads in St. Petersburg, Russia, were investigated using gravimetry and nephelometry measurement techniques in 2013-2015. The effects of meteorological conditions (temperature, relative humidity, wind direction, and speed) and the intensity of traffic flows on the results of the measurements were also evaluated. On the base of the measurements, there was developed a neural network modelling approach that allowed to quantify exhaust / non-exhaust PM10 and PM 2.5 emissions and carry out numerical investigations of air pollution by transport related PM2.5 and PM10 on street and urban level in St. Petersburg.

1. Introduction

PM10 are the particulate matter with a diameter of 10 μm (or less) and PM2.5 are the particulate matter with a diameter of 2.5 μm (or less) suspended in the atmosphere. PM10 are also known as "inhalable coarse particles" able to penetrate into the human lungs. PM2.5 ("fine particles") can pass not only into the respiratory system but also in the blood. Numerous scientific studies [1, 2] have linked particle pollution exposure to a variety of health problems, including premature death in people with heart or lung disease, nonfatal heart attacks, aggravated asthma, etc.

Road traffic is one of the most important sources of PM10 and PM2.5 emissions. Road traffic emissions include not only PM released from motor exhausts, but also "non-exhaust" particles derived from the brake wear, tire wear, road surface wear, and from the resuspension due to the turbulence generated by vehicle wheels [3-5].

The usage and the comparison of direct gravimetry and indirect nephelometry methods for the estimation of harmful PM10 and PM2.5 in the air near roads and highways allow to improve the quality and the accuracy of the experimental measurements and to develop mathematical models, allowing to carry out numerical investigations and monitoring of air pollution, on their base [6, 7]. A neural network approach is a highly attractive technique to build multi-level models of air pollution in megacities [8-11].

2. Measurement techniques and modelling methodology

2.1. PM10 and PM2.5 on-road concentration measurements

The results of proper measurements of PM concentrations as well as the data of measurements from mobile and stationary monitoring stations, kindly provided by the Committee on Environment Management, Environmental Protection and Ecological Safety of the Government of St. Petersburg (hear and after the Committee), were analyzed in the present investigations.

PM10 and PM2.5 on-road concentrations were determined using two main measurement techniques – gravimetry and nephelometry. Gravimetry is a direct analytical approach used for the routine estimation of PM mass concentrations. Its main advantage is that the results of measurements are not influenced by the chemical and physical properties of particles, but it is a highly laborious method. Moreover, it generally requires long averaging times (> 12h) to collect enough amount of the analyte on the filters. So, it can't show short-term temporal variations in PM concentrations. Nephelometry techniques based on the light scattering, on the contrary, allow real-time estimation of PM concentrations. However, light scattering is sensitive to particle size distribution, shape, and composition (refractive index and absorption) [12-14].

Nephelometry measurements. Proper simultaneous measurements of PM10 and PM2.5 concentrations were carried out in 2015 on four busy roads of St. Petersburg (Moskovsky pr., Mitrofan'evsky road, Koltushsky road and St. Petersburg Ring Road) using two laser DUSTTRAK 8530 spectrometers (TSI Incorporated / USA). The validation of DUSTTRAK 8530 measurements against gravimetric measurements has shown that DUSTRAK 8530 overestimated the daily mean ambient PM10 by a factor of 1.3-1.7 and PM2.5 by a factor of 1.7-2.1, like in previous investigations [12-14]. The measurements of PM2.5 and PM10 concentrations by the mobile monitoring laboratory were performed in 2012-2013 on the other busy roads (Ligovsky pr., Slavy pr. and Piskarevsky pr.) by means of an Air Pollution Monitor APM-2 nephelometer (Comde-Derenda GmbH / Germany).

The data of gravimetric measurements of PM10 (from the monitoring station № 3 in Karbysheva str.) and PM2.5 (from the station № 18 in Olga Forsh str.) for the March 2015 were provided by the Committee. A PNS16/18MT-3.1/6.1 (Comde-Derenda GmbH / Germany), equipped with a magazine with 15 filter cassettes, is used for the automatic continuous sampling of PM10 and PM2.5 on glass fiber filters, 47 mm in diameter. The filters are weighed before and after the sampling at ambient temperature 20 ± 1°C and relative humidity 50±3 %. We didn't validate APM-2 nephelometric measurements against gravimetric measurements in this study, but according to the information of the Committee, the gravimetric and real-time nephelometric measurements are reasonably well correlated [15].

2.2. Investigation of traffic flows on St. Petersburg roads

We examined traffic situation on seven roads in St. Petersburg with intensive traffic at the morning and evening rush hours. Such parameters as the average speed of traffic flows, location geometry (average height of buildings and width of the street), meteorological parameters (temperature, relative humidity, wind direction, and velocity) were also considered.

2.3. Neural network approach

On the base of the measurements, there was developed a neural network model with parameters (weights) tuned via optimization methods [9, 10] (the RProp method and the combination of cloud and RProp methods were in use). The neural network model of the complex system can gather pieces of heterogeneous information – differential equations, conservation laws, equations of state, symmetry conditions, etc. The information exchange via neural network parameters between different levels of hierarchy makes computing less resource consuming. A neural network approach is an attractive technique to build multi-level models of air pollution in mega-cities. There was developed a neural network model for the estimation of air pollution in St. Petersburg by PM10 and PM2.5 on the city- and street-level. The modelling technique allowed also to estimate the impact of different PM sources related to road transport: exhaust PM, non-exhaust PM (including brake wear, tire wear, road surface).

3. Results and discussion

The results of on-road measurements of PM concentrations are presented in Table 1.

Table 1. Average PM concentrations, μg/m^3, on roads of St. Petersburg.

Road	Average PM concentrations, μg/m^3, at different weather conditions					
	Rainy days, slow/moderate wind		Sunny/drizzly days, slow/moderate wind		Sunny dry days in spring, moderate wind	
	PM10	PM2.5	PM10	PM2.5	PM10	PM2.5
St. Petersburg Ring Road	27.3±2.2	21.8±1.5	63.4±8.1	30.0±6.3	230.8±31.4	116.7±15.8
Moskovsky pr.	8.0±3.6	4.4±0.6	34.2 ±3.4	17.3±2.5	57.6±4.1	19.1±3.9
Ligovsky pr.	7.8±1.3	3.8±1.4	20.1 ±6.1	11.4±4.3	88.2±7.2	48.5±6.0
Pr. Slavy	9.9±3.1	5.0±1.9	28.7±5.6	14.3±4.9	72.4±7.1	46.4±8.1
Piskarevsky pr.	10.9±1.1	4.2±1.3	30.5±5.8	12.3±3.4	124.5±7.2	64.2±8.3
Koltushsky road	18.8±5.2	14.1±2.9	53.4±7.3	21.7±3.4	187.5±9.1	54.8±7.3
Mitrofan'evsky road	17.0±2.5	8.9±2.4	49.2±8.1	26.3±3.1	255.6±13.9	50.9±7.5

The results of on-road measurements have shown that weather conditions in St. Petersburg are mostly favorable for the precipitation of PM pollutants and usually the concentrations of PM10 and PM 2.5 are below national 24h Limit Values (60 μg/m^3 for PM10 and 35 μg/m^3 for PM2.5). Average PM10 and PM2.5 concentrations are about 7-20 μg/m^3 and 3-15 μg/m^3, respectively, at rainy/drizzly days on the roads with moderate traffic volume (2500-4500 vehicle per hour) free from heavy duty vehicles (Moskovsky and Ligovsky prospects) and on the roads with the share of HDV ranging within 1-10 % (Prospect Slavy, Piskarevsky prospect, Koltushsky road, Mitrofan'evsky road). PM10 and PM2.5 concentrations on the St. Petersburg Ring Road (SPbRR) with traffic volume 7500-12000 vehicle per hour and the share of HDV ranging within 10-25% mostly suit national annual Limit Values (35 μg/m^3 for PM10 and 25 μg/m^3 for PM2.5) but exceed the European annual Limit Values (25 μg/m^3 for PM10 and 12 μg/m^3 for PM2.5).

The problem of excess PM10 and PM2.5 air pollution become significant in early spring when the roads are covered with huge dust-deposits accumulated through the winter. The investigations carried out on the SPbRR at different weather conditions in March, April and May 2015, namely, (1) Dry sunny day after 8 dry days in March (high visible level of dusting), (2) Dry sunny day after night drizzle in March (high visible level of dusting), (3) Dry sunny day after some rainy days and road washing in April, (4) Rainy day after road washing in April, (5) Dry sunny day after some rainy days and road washing in May, showed that the measured PM10 concentrations exceeded 24 hours Limit Value by 2.8-3.9 times and PM2.5 – by 2.4-3.3 times in March 2015. After some rainy days and washing, the PM concentrations have fallen below LV (Table 2). At the same time, the share of PM2.5 has increased from 49.6 % in March to 77.7 % in May, indirectly indicating the trend of the increasing share of exhaust PM.

Table 2. Average PM at different weather conditions on the St. Petersburg Ring Road

Average PM, μg/m^3	Conditions of PM monitoring / date				
	1/19.03.15	2/20.03.15	3/06.04.15	4/14.04.15	5/24.05.15
PM10	230.8±31.4	167.3±34.2	63.4±8.1	38.3±9.5	27.3±2.2
PM2.5	116.7±15.8	84.6±21.7	30.0±6.3	25.5±7.6	21.8±1.5
PM2.5/PM10	0.496	0.503	0.525	0.657	0.777

Journal of Physics: Conference Series **772** (2016) 012031 doi:10.1088/1742-6596/772/1/012031

The results of PM10 and PM2.5 gravimetric measurements from the monitoring stations №3 and № 18, respectively, for the period from 12 to 19 March 2015 confirmed that the average measured PM10 and PM2.5 concentrations exceeded National 24-hour Limit Values, by a factor of 1.2 and 1.6, respectively.
On the basis of the measurements, there was developed a city- and a street-level neural network model for the estimation of air pollution in St. Petersburg by PM10.
The street model was developed on the base of the Gaussian dispersion [16, 17]. Taking into account the principle of superposition and replacing the integral by a cubature formula, we receive the formula:

$$q_n(t,x_0,y_0,z_0;u_x,u_y)=\sum_{i=o}^{n}K_i\frac{Q\exp\left(-\frac{(x_i-x_0-u_xt)^2}{2(\sigma_x)^2t}-\frac{(y_i-y_0-u_yt)^2}{2(\sigma_y)^2t}-\frac{(z_i-z_0)^2}{2(\sigma_z)^2t}\right)}{\left(\sqrt{2\pi t}\right)^3\sigma_x\sigma_y\sigma_z},(1)$$

where q is the concentration of pollutant (g/m^3); Q is the emission of pollutant (g m^{-1}s^{-1}); t is time (s), x_0, y_0, z_0 are the coordinates of the emission source (m); K_i are numerical coefficients; x_i, y_i, z_i are the integration points (m); u_x, u_y are the components of wind speed and direction (m/s); σ_X, σ_Y, σ_Z are the Gaussian horizontal and vertical dispersion parameters. It is evident that this model corresponds to the artificial neural network with Gaussians as radial basis functions.
The calculations on city-level allow to identify the districts of St. Petersburg mostly affected by PM10 air pollution (taking into account the dependence of air pollution on wind direction and speed) and receive air pollution diagrams on city-level (Fig. 1).

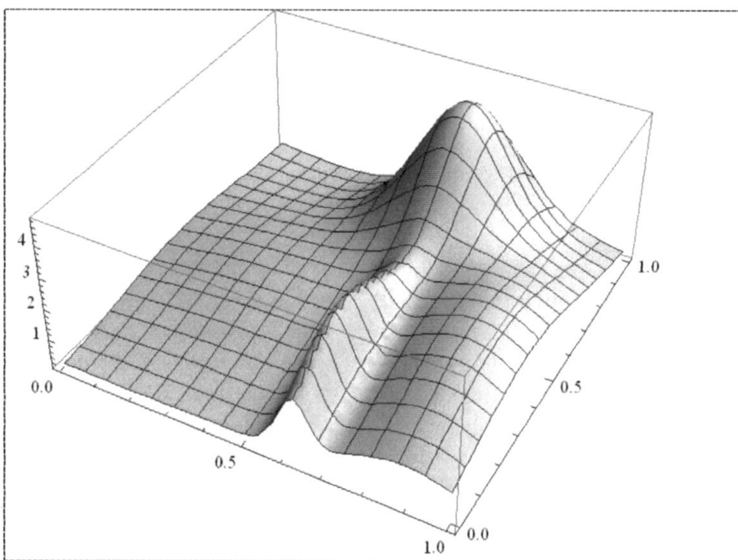

Figure 1. Computed PM10 concentration field for St. Petersburg (in terms of Limit Value Units)

Figure 1 demonstrates that moderate south-western winds contribute to the accumulation of PM10 in the central and, mostly, in the eastern parts of the city, and can lead to increased concentrations of PM10 exceeding the National Limit Value by 1.1-3.5 times at adverse weather conditions.
The on-street model was developed on the base of the Russian normative document for the calculation of the dispersion of harmful substances in the atmospheric air - OND-86 [18, 19].

OND-86 is based on the analytical approximation of the results of joint numerical integration of the equation of atmospheric diffusion and the system of equations of hydro thermodynamics for the atmospheric boundary layer [18, 19].

Using this approach, also known as K-theory, together with some approximations and assumptions [6, 18, 19], there was established that the concentration of a pollutant emitted from an unregulated line source, such as a traffic flow, is as follow [6]:

$$C_{м} = \frac{AMFm\eta}{H^{7/3}}, \qquad (2)$$

where C_M is the concentration of pollutant (g/m^3); A is the coefficient considering the temperature stratification of the atmosphere; M is the joint emission of pollutant from traffic flow (g/s), F is the dimensionless coefficient considering the velocity of gravitational sedimentation of particulate matter; $m' = 0{,}9$; η is the dimensionless coefficient considering the influence of the terrain relief; H is the source height (m).

The calculations on street-level predict that the adjacent area of the St. Petersburg Ring Road may be exposed to the enhanced levels of PM_{10} exceeding National Limit Value by 1.1-3.0 times at intensive traffic and unfavorable weather conditions (Fig. 2).

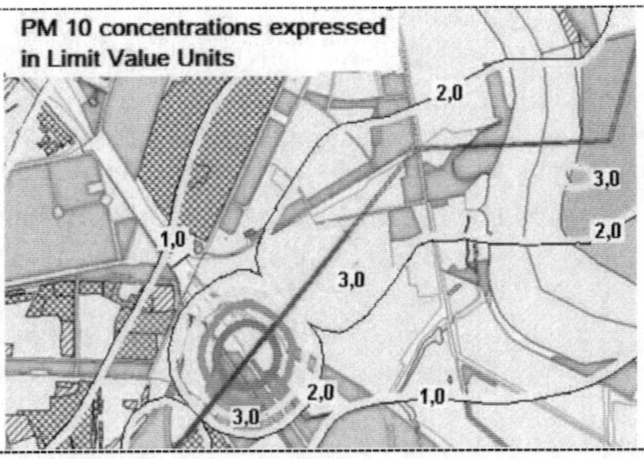

Figure 2. Visualization of the air pollution by exhaust and non-exhaust PM10 at adverse weather conditions on the busiest segment of the St. Petersburg Ring Road (near the prospect of Obukhovskoy Oborony).

Conclusions

The results of the measurements have shown that the roadside concentrations of PM10 and PM2.5 in St. Petersburg are mostly within the National and the European Limit Values. Short-term exceedances may occur in early spring when road dust resuspension contributes considerably to particle concentrations in the air. Vehicle-related PM air pollution in St. Petersburg is, generally, lower than in the European capitals and big cities because of the lower share of diesel passenger vehicle fleet (5.4 % in St. Petersburg against 70.4 % in Paris). Direct-reading nephelometric instruments are suitable for the real-time monitoring of daily variation of the roadside PM10 and PM2.5 concentrations, however, this method is sensitive to particle shape and composition. Recalibration for the specific aerosol, considering also seasonal variations, is necessary to produce valid measures of airborne particulate matter.

The validation of the results of the numerical investigations on street-level showed the underestimation of PM concentrations in comparison with the measurements. The underestimation of the modelled PM concentrations by 5-15 % may result from the neglect of dust resuspension on roads. Hence, there is a need to develop a model forecasting road resuspension emissions in order to contribute to the optimization of environment quality management in the Russian Federation.

References

[1] Pope III C A 2000 Review: Epidemiological basis for particulate air pollution health standards *Aerosol Science and Technol.* **32 (1)** 4-14

[2] Pope III C A, Dockery D V 2006 Health effects of fine particulate air pollution: lines that connect *J. Air & Waste Manage. Assoc.* **56** 709-742

[3] Amato F, Pandolfia M, Morenoa T, et al 2011 Sources and variability of inhalable road dust particles in three European cities *Atmos. Environ.* **45** 6777-6787

[4] Biswas S, Ntziachristos L, Moore K, Sioutas C 2007 Particle volatility in the vicinity of a freeway with heavy-duty diesel traffic *Atmos. Environ.* **41** 3479-3493

[5] Lozhkina O, Nevmerzhitsky N, Lozhkin V 2016 Evaluation of air pollution by PM10 and PM2.5 on St. Petersburg ring road: mobile measurements and source apportionment modelling *Proc. 10th Int. Conf. on Air Quality: Science and Application (Milano 14-18 March 2016)* Ed S Finardi, A Farrow et al (Hertfordshire: University of Hertfordshire) p 176

[6] Ložkin V, Ložkina O, Ušakov A 2013 Using K-theory in geographic information investigations of critical-level pollution of atmosphere in the vicinity of motor roads *World Applied Science Journal* **23 (13)** 96-100

[7] Lozhkina O V, Lozhkin V N 2015 Estimation of road transport related air pollution in Saint Petersburg using European and Russian calculation models *Transport. Res. Part D* **36** 178-189

[8] Tarkhov D, Vasilyev A 2005 New neural network technique to the numerical solution of mathematical physics problems. II: Complicated and nonstandard problems *Optical Memory and Neural Networks (Information Optics)* **14** 97-122

[9] Vasilyev A, Tarkhov D 2009 *Neural Network Modeling. Principles. Algorithms. Applications* (SPbSPU Publishing House, Saint-Petersburg) p 528 (in Russian)

[10] Vasilyev A, Tarkhov D 2014 Mathematical Models of Complex Systems on the Basis of Artificial Neural Networks *Nonlinear Phenomena in Complex Systems* **17** 327-35

[11] Haykin S 2008 *Neural Networks and Learning Machines* (Prentice Hall) p 936

[12] Kim J Y, Magari S R, Herrick R F et al 2004 Comparison of fine particle measurements from a direct-reading instrument and a gravimetric sampling method *J. Occup. Environ. Hyg.* **1 (11)** 707-715

[13] Zhu Y, Smith T J, Davis M E et al 2011 Comparing gravimetric and real-time sampling of $PM_{2.5}$ concentrations inside truck cabins *J. Occup. Environ. Hyg.* **8 (11)** 662–672

[14] Language B, Piketh S J, Burger R P Comparison of respirable particulate (PM4) measurements from direct-reading photometric instruments and a gravimetric sampling method *Proc. Conf. Test and Measurement Conference 2015: Quality of Life – What is the measure of it? (Cape Town 12-14 October 2015)*

[15] Official website of the Committee on Environment Management, Environmental Protection and Ecological Safety of the Government of St. Petersburg www.infoeco.ru

[16] Benson P 1992 A review of the development and application of the CALINE 3 and 4 models *Atmos. Environ.* **26 B (3)** 379-390

[17] Berkowicz R 2000 OSPM – a parameterized street pollution model *Environmental Monitoring and Assessment* **65 (2)** 323-331

[18] Genikhovich E L, Gracheva I G, Onikul R I, Filatova E N 2002. Air pollution modeling at urban scale - Russian experience and problems. *Water, Air & Soil Pollution: Focus* **2 (5-6)** 501-512

[19] Genikhovich E L, Sciermeier F A 1995. Comparison of the United States and Russian complex terrain diffusion models developed for regulatory applications. *Atmos. Environ.* **29 (17)** 2375-2385.

Acknowledgments

We thank Ivan Serebritsky and Alexander Zaporozets from the Committee on Environment Management, Environmental Protection and Ecological Safety of the Government of St. Petersburg for the provision of the data from the mobile and the stationary measuring stations.

Journal of Physics: Conference Series **772** (2016) 012032 doi:10.1088/1742-6596/772/1/012032

Neural network model of rupture conditions for elastic material sample based on measurements at static loading under different strain rates

I Bolgov, T Kaverzneva, S Kolesova, T Lazovskaya, O Stolyarov and D Tarkhov

Peter the Great St-Petersburg Polytechnic University, 29 Politechnicheskaya Str, 195251 Saint-Petersburg, Russia

E-mail: `tatianala@list.ru`

Abstract. The article deals with the problem of predicting of the temporal elongation law of the sample under dynamic loading. The determination of tensile behavior of samples under uniaxial loading is performed by a standard tensile method. The neural network approach is applied to construct an approximate elongation-force dependence using measurement data and posterior model of the dependence of rupture conditions on the neural network parameters. The considered approach can be used in the building industry.

1. Introduction

The statistic of accidents indicated that despite the safety requirements, which are connected with the implementation of risky operations, which exists in the world community, more or less serious traumatizing of the workers still happens. A number of these requirements concern the technical objects, which must satisfy, for example, the requirements of solidity and stability. Also, there are requirements, which regulate the behavior of the people in the execution of the works in different situations: in the normal operating regime and in an emergency situation. It is a question of so-called human factor, which influence we are trying to decrease as far as we can. Erroneous human actions can lead to serious consequences, which are difficult to predict.

There is a set of security tools which is used in relation to a particular workplace with the aim to reduce the risk of personal injury. In the case of work at height, we can name the following security tools:harnesses, strapping for positioning and retaining a person falling, ropes. Rescue stretchers and scarves are widely used in the organization of rescue operations on the cable cars. All materials from which these security measures were made must withstand certain load regulations throughout all the time of use. They should have the declared characteristics of strength and elastic, keeping them all the time of operation. The material from which is made the security system with the passage of time under the influence of external conditions aging and loses its mechanical properties.

Elasticity of various materials depend on many factors to take which into account is not always possible. Especially if we consider the materials with a complex structure. Elastic properties modeling allows us, for example, to predict the loss of material strength (rupture of a sample), deformation conditions, which, in turn, can lead to breakdowns and emergencies. This gives us the opportunity of promptly acting on changes in the elastic characteristics of considered objects and creating such constructs to which high requirements for safe operation are applicable.

High extensible materials are continually gaining importance owing to their application in various fields including civil engineering, sport, marine, and safety. These materials include a variety of ropes, slings, cords [1-4]. The application of such material in as high-deformable elements in various structures provides new possibilities in various applications. The properties of these materials depend on their structures and constituent materials and may be significantly varied. They may possess high extensibility under low load or have resistance to high strength, high energy absorption depending on the application. A detailed study of their properties is required for optimal use of these materials in a certain application.

2. Problem statement

In this work, we approximate the initial part of the dependence of the elongation of a rubber thread on a tensile force by the neural network. The aim is to study the dependence of rupture conditions on the neural network parameters. The neural network is used in two stages. The first stage is to approximate the experimental dependence of the elongation on the load. In the second stage, we determine the dependence of the sample rupture conditions on neural network parameters (weights) obtained in the first stage.

From our point of view, building a model, reflecting the circumstances of disruption of sample, which is made from an elastic material, can help us to estimate the probability of loss of mechanical properties.

Especially difficult to predict conditions of destruction, which usually occurs suddenly and are determined by the rapid growth of small damages the integrity of the object - cracks, irregularities, cavities, etc. Permanent diagnosis of such injuries in the course of operation of the facility is often difficult. It is desirable to diagnose the state of the object.

However, theoretical models do not allow to establish a similar relationship, as general considerations leave aside the peculiarities of the individual elements and the actual state of the samples.

3. Measures

The tensile test was carried out on an Instron-5965 universal testing machine (Figure 1) with an automatic data acquisition system. The determination of tensile behavior of samples under uniaxial loading is performed by a standard tensile method. The thin elastomeric yarns were selected as samples. Considered in the work methods can be applied to other samples of an elastic material.

The sample is rigidly fixed at one end in a stationary clamp and at another end in a movable clamp. It should be noted that tensile strains were calculated from crosshead displacement. When testing we set different strain rates. Plots of deformation (mm) against force (N) were obtained with a 50 and 80-mm sample gauge length and strain rates of 50, 250, 500 (Figure 2) and 1000 mm/min.

Elongation of the sample is proportional to the initial length of the sample. These graphs were able to establish that the break point and the elongation-force curve are weakly dependent on the strain rates. In addition, the tensile load is not dependent on the length of the samples.

Figure 2 shows the three phases of the elongation process. The dependence between the elongation and the load does not match the Hooke's law and is essentially nonlinear. In the first section, we observe the maximum rigidity of the sample at small deformations, gradually turning into the second stage, characterized by less deformation rigidity. In the third section, with increasing deformation, the dependency is almost linear with small perturbations before the destruction characterized by an obvious deformation rigidity increase. The rupture happens without the occurrence of a significant non-linear section. Such behavior of the sample

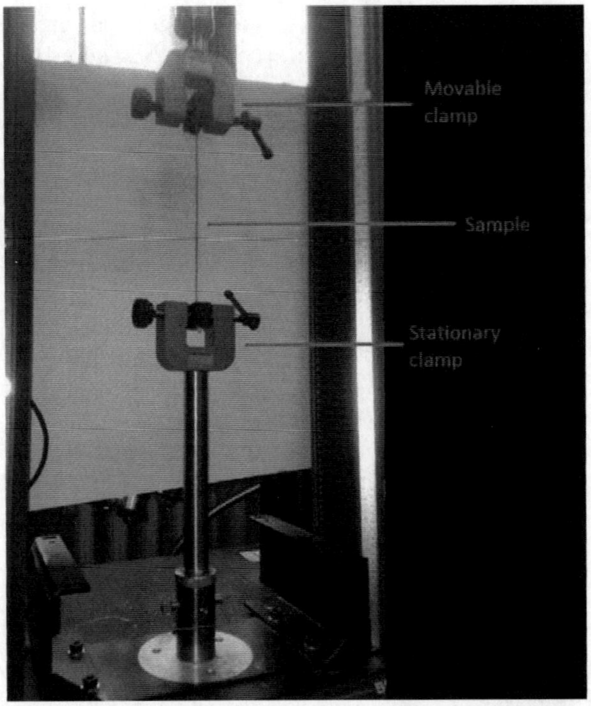

Figure 1. Instron-5965 universal testing machine. Position Measurement Accuracy: ± 0.01 mm or 0.05 percent of displacement (whichever is greater), Load Measurement Accuracy: ± 0.5 percent of reading down.

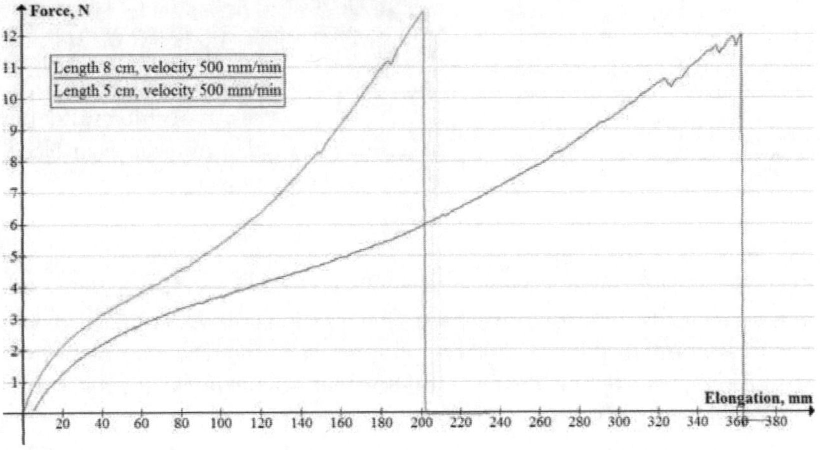

Figure 2. The dependencies of the elongation (mm) against force (N), sample length 50 and 80-mm, strain rate of 500 mm/min.

corresponds to the results of [9]. The lack of non-linear section before rupture it significantly complicates predicting.

4. Neural network approach

4.1. Network 1

Elongation-force dependence is looking as a neural network approximation with two neurons in the form

$$u(x, \mathbf{c}, \mathbf{a}, \mathbf{xc}) = c_1 \operatorname{th}(a_1(x - xc_1)) + c_2 \operatorname{th}(a_2(x - xc_2)), \tag{1}$$

where neural network weights, linearly incoming parameters c_i and nonlinear input parameters a_i, xc_i, are determined in the process of gradual learning network based on the minimization of the error functional given in a discrete form

$$J = \sum_{j=1}^{M} (u(x_j, \mathbf{c}, \mathbf{a}, \mathbf{xc}) - F_j)^2. \quad (2)$$

The error functional optimization is conducted by RProp algorithm or by combination of Particle Swarm and RProp algorithms [6].

Figure 3 shows an example of a plot of deviation of neural network approximation from the measurements. Figure 4 shows a typical situation: before the rupture, the neural network approximation error increases significantly, which may serve as another indicator of closeness to the point of rupture. For all experiments, neural network approximation relative error does not exceed 3 percent.

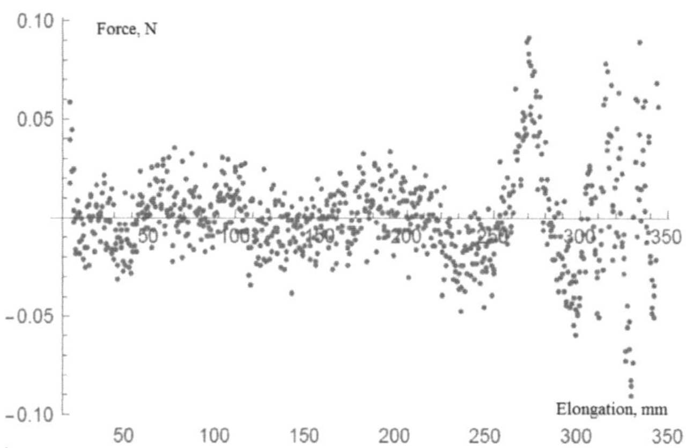

Figure 3. The error of the neural network approximation of the experimental dependence of the elastomeric yarn elongation on the tensile.

4.2. Network 2

For each of the eight approximations, we have the tensile force at rupture of the sample and the neural network parameters. Using these values, we build the neural network models of the dependence of the rupture tensile force on the neural network approximation parameter $c = \max(c_1, c_2)$ in the form $f(x, \mathbf{c}, \mathbf{a}, \mathbf{xc}) = d\,\mathrm{th}(g(c - h))$. To determine the weights of the neural network used are the same as before functional errors and such as nonlinear optimization algorithms. Figure 4 shows one of the obtained dependencies.

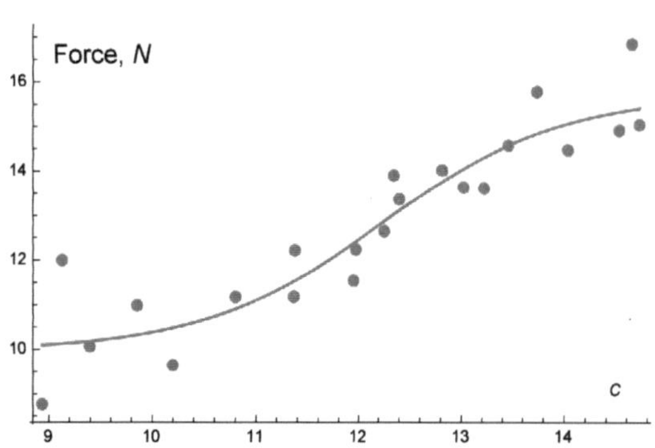

Figure 4. The dependence of the tensile force at which the elastomeric yarn rupture on the parameter $c = \max(c_1, c_2)$ of the neural approximation of elongation-force dependence.

Taking into account the complex nature of rupture processes and a strong influence on it of a particular sample defects, the result can be considered satisfactory. If we know the behavior of the sample in the initial part of the load, it is possible to predict the limit force and elongation at which rupture occurs. The obtained relationships can be used in predicting of the temporal elongation law of the sample under dynamic loading using the techniques discussed in the works [5-8].

5. Prediction

To predict the rupture conditions, we can use measurements taken not on the whole interval but on a small fraction of the points preceding the rupture. Further, we present the result corresponding the sample measuring points from 100 to 50 before the rupture.

In the first stage the dependence of the elongation on the tension force is approximated by a neural network, as previously. The simplicity of dependence allows to use the neural network with only one neuron $F(x) = c_0 \,\mathrm{th}(a_0(x - xc_0))$. In the second step, the dependence of the tensile force on the neural network coefficients obtained in the first stage for each sample is approximated by new neural network. Figure 5 shows an example of such dependence for the parameter xc_0.

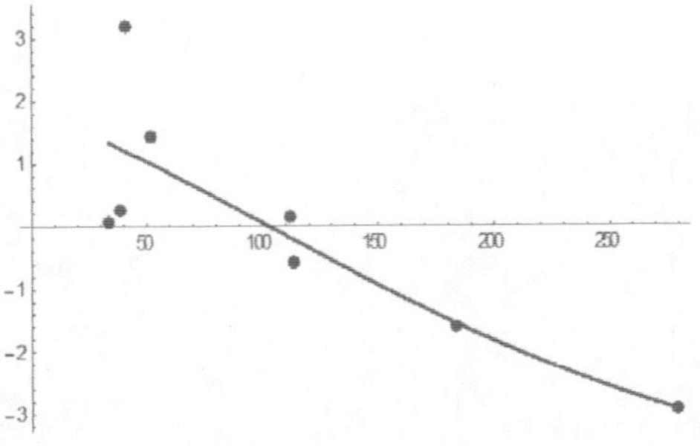

Figure 5. The the tensile force dependence on the parameter xc_0 of the neural network approximated the elongation-force dependence corresponding the sample measuring points from 100 to 50 before the rupture.

These dependencies allow the dynamic prediction of the elastomeric yarn rupture, relearning the specific sample neural network used the newly entered data in the process of strain.

6. Conclusions

The considered approach can be applied in the construction industry in justifying the choice of lifeline, which is used during construction work including industrial alpinism, in the calculation of risk during the evacuation of people by "jumping on the tent" [10]. Work at height is a dangerous type of work, which is associated with a large number of accidents caused by the fall of man, despite the regulated security measures [11]. That is why it is important to provide the strength of safety and amortization elements (slings, ropes, cables), which will depend on the elastic properties of the materials making up the harnesses elements. It should also be borne in mind that the strength characteristics can change significantly during the process of wear [12].

Acknowledgments

The work is supported by the Russian Foundation for Basic Research (Grant no. 14-01-00660).

References

[1] Hearle JWS 2016 *One-dimensional textiles. Handbook of Technical Textiles* (Elsevier) 345-360
[2] McKenna HA, Hearle JWS, OHear N 2004 *Handbook of Fibre Rope Technology. Handbook of Fibre Rope Technology* (Elsevier) 1-34
[3] Weller SD, Johanning L, Davies P, Banfield SJ 2015. *Synthetic mooring ropes for marine re-newable energy applications. Renew Energy.* Nov; **83:126878**

[4] McLaren AJ 2006 *Design and performance of ropes for climbing and sailing. Proc. of the Institution of Mechanical Engineers, Part L: Journal of Materials: Design and Applications* **220 (1)** 1-12
[5] Vasilyev A, Tarkhov D 2014 Mathematical Models of Complex Systems on the Basis of Artificial Neural Networks *Nonlinear Phenomena in Complex Systems* **17** 327335
[6] Riedmiller M and Braun H 1993 *A direct adaptive method for faster backpropagation learning: The Rprop algorithm. Proceedings of the IEEE International Conference on Neural Networks* (IEEE Press) 586-591
[7] Tarkhov D, Vasilyev A 2005 New neural network technique to the numerical solution of mathematical physics problems. I: Simple problems *Optical Memory and Neural Networks (Information Optics)* **14** 59-72
[8] Tarkhov D, Vasilyev A 2005 New neural network technique to the numerical solution of mathematical physics problems. II: Complicated and nonstandard problems *Optical Memory and Neural Networks (Information Optics)* **14** 97-122
[9] Kainov N, Tarkhov D, Shemyakina T 2014 Application of neural network modeling to identification and prediction problems in ecology data analysis for metallurgy and welding industry *Nonlinear Phenomena in Complex Systems* **17 (1)** 57-63
[10] Kaverzneva T, Savchenkova L 2012 Selection of materials for reliable evacuation by the method "jumping on the tent" *LI Science Week SPb GPU Materials of the international scientific-practical conference with international participation.: P.II* (Publishing house Politehn. University)
[11] Kaverzneva T, Mazurenko K 2015 Security control when working on hight *Science forum with international participation "Science Week SPbSPU": materials of scientific-practical conference. Institute of Military-Technical Education and Safety SPbSPU* (Publishing house Politehn. University) 187-190
[12] Kaverzneva T, Smirnova O 2013 The Impact of wear of construction equipment and hand tools on the working conditions *Safety in technosphere* **3 (42)** 14-18

Journal of Physics: Conference Series **772** (2016) 012033 doi:10.1088/1742-6596/772/1/012033

Development of portable measuring system for testing of electrical vehicle's heat energy recovery system

K Sarvajcz[1], A Váradiné Szarka[2]

[1,2]University of Debrecen, Department of Electrical and Electronic Engineering, Debrecen, Hungary

[1]sarvajcz@science.unideb.hu, [2]angela.varadi@science.unideb.hu

Abstract. Nowadays the consumer society applies a huge amount of energy in many fields including transportation sector. Internal combustion vehicles contribute substantially to the air pollution. An alternative solution for reducing energy consumption is replacing the internal combustion vehicles by electrical or hybrid vehicles. Today one of the biggest disadvantages of the electrical vehicles is the finite capacity of batteries. The research topic presented in this paper is the „Energy Harvesting", and development of energy recovery system for electrical vehicles which largely contributes in increasing the driving range. At the current phase of the research efficiency analysis of the heat energy recovery devices are investigated in real driving circumstances. Computer based mobile and wireless measurement system for the analysis was developed, tested and installed in a real vehicle. Driving tests were performed and analysed in different circumstances.

1. Introduction

In this research we examine methods of Energy Harvesting, which contributes to reduction of energy consumption's losses with special attention to the investigation and development of energy recovery systems in electric vehicles. In this paper authors introduce detailed operation of thermoelectric devices and provide detailed information on development of a new calibration and test system for TEGs. This newly developed testing system is able to analyse thermoelectric processes applying different semiconductor elements and changing temperature difference between the cold and hot points in range of 0 and 400 °C. The paper introduces workpoint measurement of different TEG types defining its maximum extractable electrical output power and also it includes detailed examination of measurement system developed for energy recovery, which can be applied in hybrid and electrical vehicles. The measuring system is able to detect two temperatures, current and voltage of TEG, GPS data and dynamic data of vehicle in the same time. Measured data are sent via wireless communication to cloud based storage.

2. Thermoelectric generator

The largest part of electrical losses is dissipated in form of heat while electrical and electronic producers spend huge resources to develop ideal cooling systems for electrical equipment. As thermoelectric generators use temperature difference, besides of the energy harvesting the cooling effect is an additional advantage of the application.

TEGs have three main parts: N-type semiconductor, P-type semiconductor, and also a copper conductor providing contact surface between semiconductors. The most commonly used semiconductor is Bismuth Telluride (Bi2Te3). The element is installed between two insulating surfaces made of ceramic material (Al2O3) in most of cases. [1] Theory of operation of TEGs is based on thermoelectric phenomena and processes described by Seebeck, Peltier and Thomson effects. [2], [3], [4]

3. Calibration system for thermoelectric generators

Validation of TEG modules of different types and sizes require a calibration system of high reliability and repeatability therefore the research group decided to develop a new system for our special needs.
Temperature of the heated side is generated by a 450 W Infrared Top Ceramic Heater. It can generate 750 °C temperature, therefore housing with special heat insulation should be used. The house is made of Bakelite plate with width of 15mm. The Bakelite is one of the best heat insulating material with 0.23 [W/mK°] heat conductivity coefficient. Inside of the Bakelite housing a secondary insulation is applied which made of a special non-combustible asbestos alloy. This material is used for inner insulation of combustion chamber in household furnaces. Between the heater piece and TEG an aluminium inner spreader is included in order to provide homogeneous temperature on the surface of the TEG. As there are many different TEG sizes are available from different producers, modular aluminium spacers are designed for general applicability of TEGs. These spacers are used for flexible assembling and insertion of K type thermocouples measuring temperature of hot point. The spacer also ensures the homogeneous temperature on the full surface of TEG's.
For cooling of the cold point a flow water cooler is used which is connected to household water tube. Also an aluminium spacer measuring temperature of cold point is inserted between the cooler and TEG module. Aluminium spacers are designed for all standard sized TEGs in order to ensure homogeneous temperature distribution on the full surface.

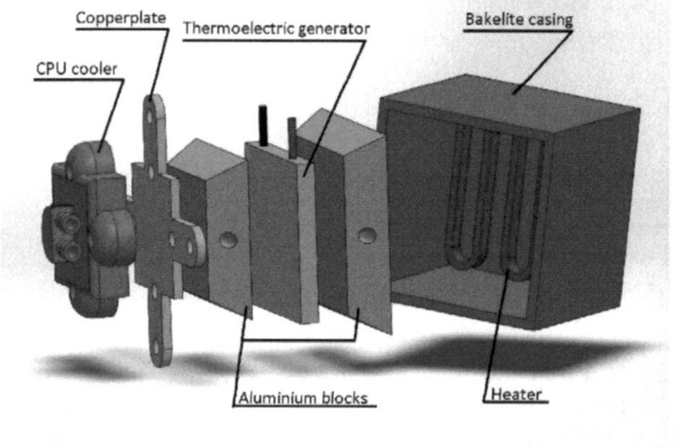

Figure 1. Measuring and calibrating equipment

Control of tests are developed using National Instruments hardware and software. The system was also presented at IMEKO's World Congress [5]. Different TEG modules tests were performed including workingpoint tests of two different types of TEG module provided by the same manufacturer. In these measurements eight different temperature differences were applied and the load resistance values were set by 0.1 Ω steps in the 0-5 Ω range. The aim of our examination was determination of maximum power of TEG modules at different temperature differences. These tests have proved that the maximum extractable electric power is 2.8 W in case of the 200 °C temperature difference and 0.002809 m^2 surface which is shown on Figure 2 and Figure 3.

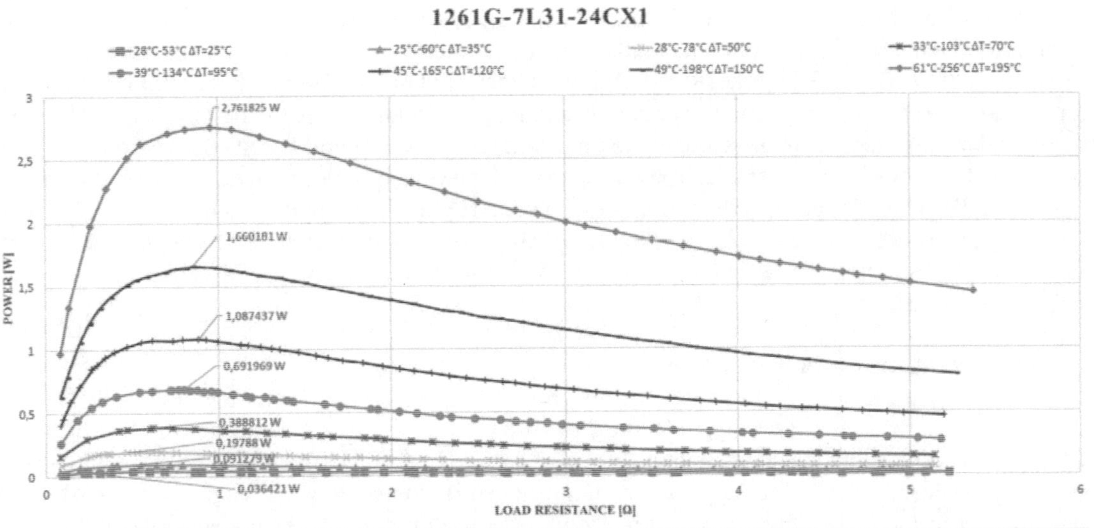

Figure 2. 1261G-7L31-24CX1 TEG module operating point measurements

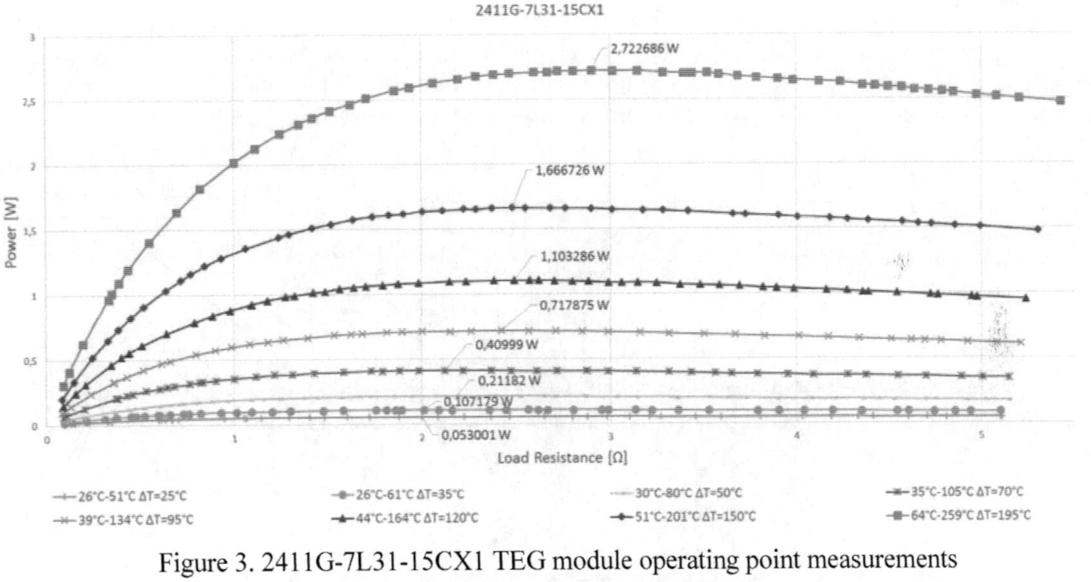

Figure 3. 2411G-7L31-15CX1 TEG module operating point measurements

4. Measurement system for heat energy recovery system tests

The main points of the system development included reliability and accuracy of the measurement, transfer, storing and processing of data but also the mobility and wireless communication possibility were in the centre of our concept. Another important condition was using the vehicle's DC power network for powering the instrumentation. The measuring system is based on compact RIO system from National Instruments. The cRIO module can be operated by 9-30 VDC power-supply voltage, which is also provided by the cigarette lighter charger of the vehicle. Four different modules are applied. The first is a voltage-measuring module, which has ±10Vfs input range and 24 bits resolution. The module measures the TEG's output voltage. The second is a current-measuring module, which has 5 Afs input and also 24 bit resolution. This module measures the TEG's output current value. The third module is a thermocouple module, which receives 2 K type thermocouples' signals. One thermocouple measures the engine-produced hot side temperature, the other thermocouple measures the cold side temperature,

Journal of Physics: Conference Series **772** (2016) 012033 doi:10.1088/1742-6596/772/1/012033

which is originated from the wind. The last module is a CAN bus module, which is applied to process dynamics parameters of the engine via OBD diagnostic interface. GPS module is also linked to the measuring system via USB interface for storing the GPS coordinates. The measured data are uploaded to a cloud-based storage using mobile internet. The first three modules measure the TEG's instantaneous power, and some other instantaneous operational parameters of the vehicle received also from the CAN module via OBD diagnostic interface. These are the instantaneous power, revolution, temperature of the engine and the vehicle's speed. GPS coordinates are added to the instantaneous data in order to get road characteristics and conditions in later processing. Two different driving tests are compared in each diagnostic phases. Data are stored in database and processed using a special self-developed data analysing software.

5. Testing of the measurement system

After building up the measurement system real tests were performed. Unfortunately until now the research team hasn't got possibility for testing in an electrical vehicle. The subject of the test was a combustion engine vehicle instead of an electrical vehicle. The most difficult task was simulating and producing as closest as possible temperature circumstances in the combustion engine vehicle's engine compartment to the electrical vehicle's drive train system. Different types of electrical motors were tested in order to define characteristic curves of variable load and temperature on the basis of which surface of the combustion engine was defined. This surface has almost the same temperature behavior as the electrical engine. The temperature of this surface does not exceed 100 °C and the surface is not cooled sharply by the wind. The surface size is 60x60 mm and it has appropriate thermal conductivity and sufficient strength material. The combustion engine's heat shield of exhaust manifold satisfies these parameters. According to our measurements, the temperature of the heat shield should not exceed 100 °C with no cooling, furthermore it has an adequate surface for inserting the TEG on.

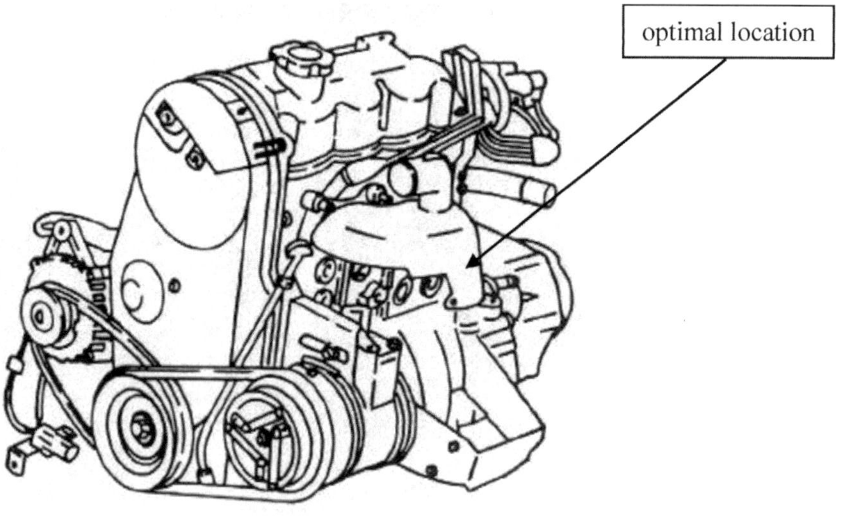

Figure 4. Select surface

After allocating appropriate surface testing periods and routes were defined. Tests were performed in the winter period, in order to ensure cool enough ambient temperature on the cold side of the TEG. The first part of the testing route included urban conditions while the second part did the highway conditions. Each measurements were performed on the same path. Saving GPS coordinates helped to fit the routes on the map shown on figure 5.

Figure 5. Test routes

One of the most critical tasks was definition of the correct fixture method of TEG. Several solutions were elaborated to ensure proper holding of the hot and cold sides. There were three possibilities for fixing the hot side. The first one is when the TEG module was fixed on the heat shield in direct way; the second is when an aluminum piece was put between the TEG module and the heat shield, and the third case is when a brass piece was put between the TEG and the heat shield.
For the cold side also several alternatives were elaborated. Similarly to the hot side in the first version no additional piece is attached to the cold side of the TEG, so the wind cools directly the surface of the cool side. After that, different types of heatsinks were applied. The heatsinks are different in their shapes and sizes. Finally, a thin aluminum and brass cuboid was set on the cool side of the TEG, which was blown by the wind. According to our measurements at high velocities the strong overcooling was experienced, therefore an air deflector box was prepared preventing direct cooling of the heat shield. The thermometers to both sides were attached. Figure 6. shows installation of the instrumentation.

Figure 6. Test installation

6. Results

Result of the tests (Figure 7.) shows the generated power in function of temperature differences. Results of four measurements performed by different concepts are presented on the diagram. In the first concept aluminium piece on the hot and on the cold side was applied without wind deflector. In the second concept brass piece on the hot and on the cold side together with the wind deflector is applied. The third is the hot and the cold side without any metal pieces and without wind deflector. In the forth one aluminium on the hot side and brass on the cold side without wind deflector is used. The third concept has much more dynamic changes, because the cold and the hot side did not contain thermally conductive pieces. In the case of fourth measurement heat conductive paste was not applied, the contact of the surfaces was not appropriate. The measured electrical power values belong to a determined temperature difference. The deviation of electrical power can be explained that the temperature differences can be produced by difference of several temperatures. Figure 7. shows an example for this. On Figure 7. in case of Concept 4 the cold side's temperature can range between 20 °C and 45 °C, and the hot side's temperature can range between 27 °C and 70 °C. According to the operating characteristic of TEG in an other temperature range other electrical power will be measured. According to the measurement's results the generated power can not reach the results of laboratory testing. Applying only one TEG without any energy investment from a 52x52 mm^2 surface 15-16 mW energy was produced. In the further research phase, we would like to reach a better surface contact, and examine the results of several TEG modules connected in series and parallel.

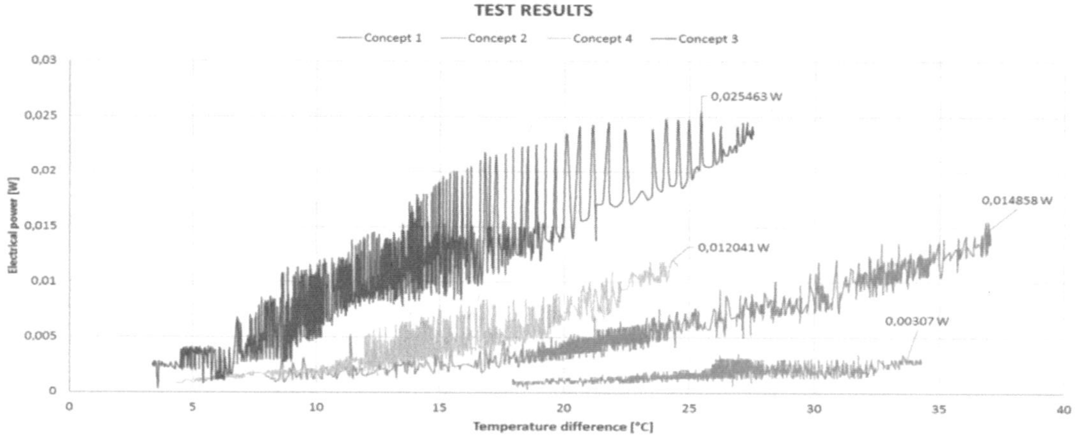

Figure 7. Test result

7. Summary

Results of the research project show positive effect of using TEGs as tool of energy harvesting in electrical vehicles. Simulation and calibration system is an essential instrument of the research providing reference characteristics and basics for comparison and evaluation of real tests in vehicles. In the phase of the project presented in this paper a mobile and wireless measurement system was developed and tested. The system includes sensors and transducers for TEG's measurement and uses CAN module for receiving the car's performance parameters. Efficiency comparisons underline differences between ideal and practical application, therefore in the next phase of the research increasing efficiency of practical application will be considered.

Acknowledgments

Authors wish to acknowledge the professional and financial support of NI Hungary Ltd. and also they wish to express their sincere thanks to general manager of the company Dr. László Ábrahám for his consistent help and support to the electrical engineering field at the University of Debrecen.

8. References

[1] Chena L., Gonga J., Suna F. and Wub C., 2002.: Effect of heat transfer on the performance of thermoelectric generators, *International Journal of Thermal Sciences Volume 41*, Pages 95–99

[2] Matthews J. E. 2011.: Thermoelectric and heat flow phenomena in mesoscopic systems, *Doctor of Philosophy DISSERTATION*

[3] Yan D. 2011.: Modeling and Application of a Thermoelectric Generator, *Masters of Applied Science Thesis*, University of Toronto

[4] Nieswand S. 2011.: A Peltier cooling system for SiPM temperature stabilization, *Masters of Applied Science Thesis*, University of Toronto

[5] Sarvajcz K. and Váradiné Szarka A. 2015.: simulation and calibration test of thermoelectric generators, *XXI IMEKO World Congress Measurement in Research and Industry*, Prague, Czech Republic, Pages 1003–08.

The reversed problem of migration streams modeling

D A Tarkhov[1], I K Shanshin[1] and D O Shakhanov[1]

[1]Peter The Great Saint-Petersburg Polytechnic University, Saint-Petersburg, Russian Federation, Politechnicheskaya st., 29.

Contact person: I K Shanshin, ivan.fizik92@yandex.ru

Abstract. This article discusses methods of solving the reversed problems of migration streams modeling: the analytical solution, the formula for the two-dimensional linear regression and neural networks. The cases with different amount of input data and number of neurons: 2, 10, 100, 1000 input points for linear regression method, and 5, 15, 25 neurons for network are considered. Also we have made numerical calculations for a period of time, which were increased by 10 times. The article proves the application of neural networks for constructing the models under consideration, especially for a non-linear model.

1. Introduction

Designing the complex development program of some region, for example Russia's Arctic zone, requires consideration of a wide range of problems – economic, environmental, and social ones [1], [2], [3]. Large mining projects should be not only economically viable, but also it is necessary to take into account their impact on the environment, population, etc. Rapid changes of demographic patterns intensify the imbalance in the population of the regions which are not inhabited densely enough. For a reasonable forecast of relevant processes, it is necessary to develop an adequate medium-term model. Problems in development of such a model are caused by an insufficient amount of demographic data. For solving the problem, application of neural network is proposed.

Models of migration streams, represented as non-linear dynamic system,

$$\begin{cases} \dfrac{dx}{dt} = sh(kx + k_1 y) - x \cdot ch(kx + k_1 y) \\ \dfrac{dy}{dt} = sh(k_2 x + ky) - y \cdot ch(k_2 x + ky) \end{cases} \quad (1)$$

and linearized system at a point of zero equilibrium position:

$$\begin{cases} \dfrac{dx}{dt} = (k-1)x + k_1 y \\ \dfrac{dy}{dt} = k_2 x + (k-1) y \end{cases} \quad (2)$$

Journal of Physics: Conference Series **772** (2016) 012034 doi:10.1088/1742-6596/772/1/012034

are outlined in [4]. A case of two populations interaction in two regions is considered. Here k_1 shows intensity of the first population tendency to coexist in the same region with the second population, k_2 shows intensity of the second popultaion tendency to coexist in the same region with the first population, k shows intesity of population tendency to coexist with the people of their own population group. Variables x and y are determined as follows: $x = (n_1^{\mu} - n_2^{\mu})/N, y = (n_1^{\nu} - n_2^{\nu})/N$ where n_1^{μ} and n_2^{μ} - the number of people in the first population in the first and second regions respectivly, n_1^{ν} and n_2^{ν} are the same for the second population. N – is the number of people in each population group., i.e. $N = n_1^{\mu} + n_2^{\mu} = n_1^{\nu} + n_2^{\nu}$.

To compare the methods described below by constructing the migration dynamics according to equations (1)-(2) it is necessary to determine the coefficients of the given model equations in accordance with the available data.

Here we analyze three methods of solving the given problem:

1) The first method assumes analytical solution to equation (2) with subsequent identification of coefficients by the best correspondence of the result to the data. This method has a number of shortcomings: firstly, it is impossible to apply the method to equation (1) and, secondly, in the case of models of a higher order, it is necessary to get a solution anew.

2) The second method is a discretization of equation (2) and subsequent identification of coefficients by formulas of two-dimensional linear regression. Using this method allows us to solve the problem in the case of models of higher order, but it is still difficult to apply it to equation (1).

3) The third method implies application of a neural network. It can be trained on the basis of numerical solution to the equation system (1 or 2) for quite a big set of parameters. Thus, the calculation of the required coefficients will be reduced to substitution of the data in the neural network. This method has many advantages including its universality, which makes the solution also appropriate for a non-linear system.

2. Results of calculations

2.1. Linear regression

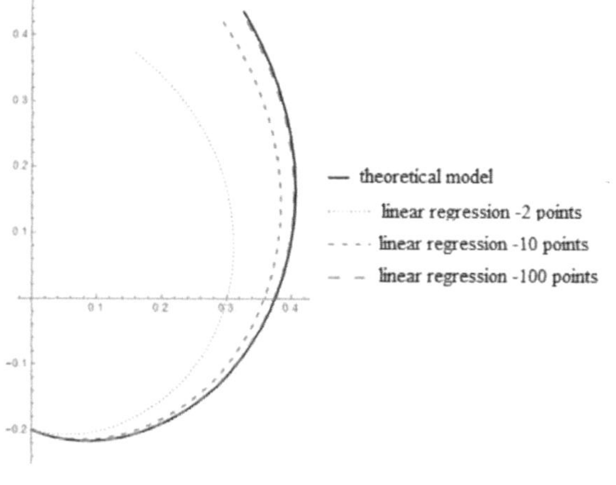

Figure 1. Linear regression, 2-10-100 points.

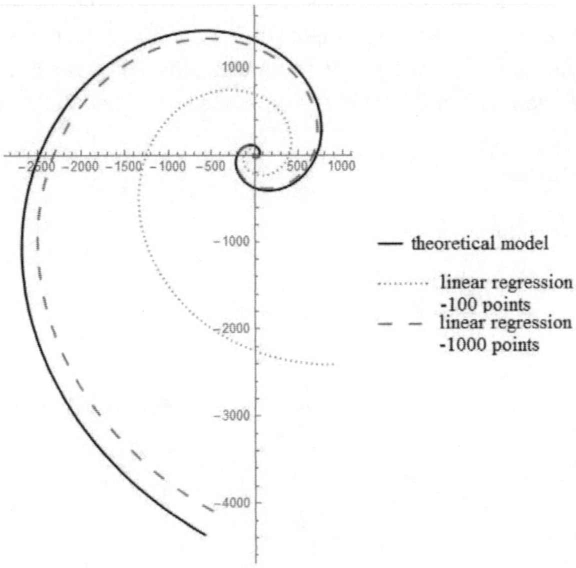

Figure 2. Linear regression 100-1000 points.

When the number of points is 2, we can see that the main trend of migration is the same as in the model, but we have big divergence between graphs of theoretical and calculated models (figure 1). Having increased the number of points to 10, we can see that error reduction is significant, but still it is not big enough. By increasing the number of the input points to 100 we achieve excellent accuracy: graphs of theoretical and calculated models almost coincide. However, tenfold augmentation of the forecast period, according to which the model is constructed, leads to significant growth of error figure 2. In order to reach good approximation, we need to increase the number of points to 1000. Nevertheless, in practice collecting such an amount of input data is rarely possible.

2.2. Neural networks

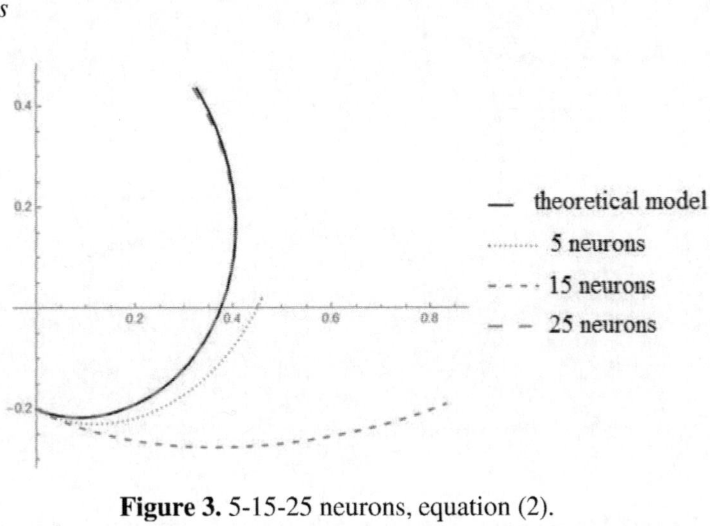

Figure 3. 5-15-25 neurons, equation (2).

First, we take a number of neurons equal to five for equation (2). A model which was plotted with this neural network by three points of an interval (0, 0, 5 and 1 moments of time) significantly differs from a theoretical model in the migration rate, but basic tendency is preserved (figure 3). Working with equation (2), we increase the number of neurons to twenty-five. The resulting model perfectly approximates the theoretical one; this result is comparable to the linear regression, which was plotted by 100 points. In case of equation (1), we get the same results. We can see that at the origin of coordinates these models are identical. So, let's carry out the same calculations, having multiplied the forecast period by 10.

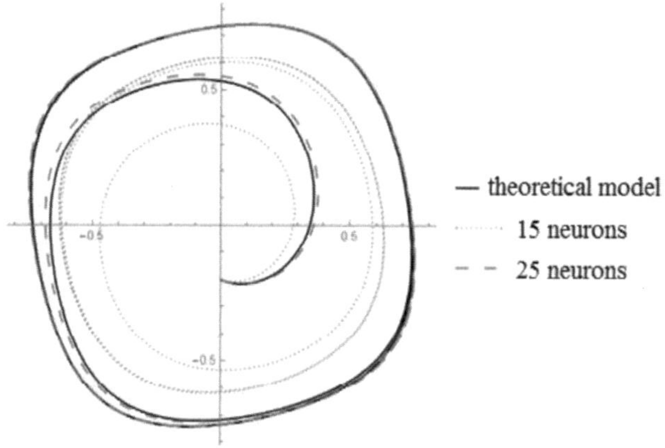

Figure 4. 15-25 neurons, equation (1), forecast period multiplied by 10.

Considering the work of a neural network that consists of fifteen neurons, in the case of equation (1), we increase the forecast period tenfold. Without changing the coefficients, we can see that the accuracy of the forecast decreases, but the main tendency of migration is preserved (figure 4). Using neural networks that consist of twenty-five elements, in case of equation (1), we get a truly amazing approximation in all parts of the function, even if we increase the forecast period tenfold.

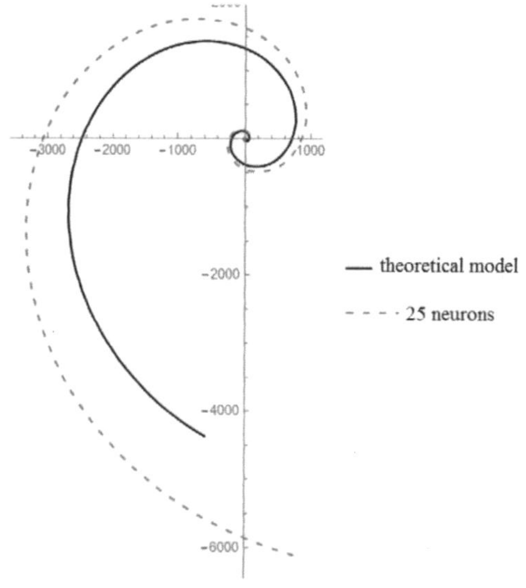

Figure 5. 25 neurons, equation (2), forecast period multiplied by 10.

Journal of Physics: Conference Series **772** (2016) 012034 doi:10.1088/1742-6596/772/1/012034

We analyze equation (2), using 25 neurons (figure 5). We can see that the result is better than a linear regression plotted by one hundred points, but worse than the one plotted by a thousand points, yet the divergence between the obtained result and the theoretical model is still quite big.

3. Conclusions

If the number of neurons in a neural network is greater than or equal to 25, it is possible to restore the solutions to equations (1) and (2) by data on three points in contrast to a linear regression, which needs a large amount of input data to reach the same accuracy. Neural networks can work with both linear and non-linear models; the accuracy of approximation depends more not on the volume of input data, but on the number of neurons in a neural network. In addition, neural networks have many other appreciable advantages over other methods of forecasting, such as its learning capability and universality. Many practical problems were solved or are solved by applying neural networks. This study has confirmed the universality and superiority of this method.

Our approach may be used for identification of dynamic systems' models (e.g. socio-economic, mechanical, physical, etc.) based on measurements or observations, when a model type is known and it is necessary to determine its parameters. System dimension, equations order, its linearity or non-linearity etc. do not have a great impact on the determination procedure. For example, this technique is used in [5], [6], [7].

Acknowledgments

The paper is based on the research carried out with the financial support of the grant of the Russian Science Foundation (Project No. 14-38-00009, the program-targeted management of the Russian Arctic zone development). Peter the Great St. Petersburg Polytechnic University.

References

[1] Rudenko D, Didenko N and Skripnuk D 2015 *Environmental security issues in the Russian Arctic*. 15th International Multidisciplinary scientific Geoconf. (SGEM 2015). Ecology, Economics, Education and legislation. Conference Proc., vol 3 (Bulgaria, 18-24 June 2015) pp 267-74

[2] Didenko N, Kunze K and Skripnuk D 2015 Russian Export Strategy and Social Sector: Consequences of Resource-Oriented Exports on Population of Russia *J. Mediter. of Soc. Sci.* **6(5)** 473-81

[3] Didenko N and Skripnuk D 2014 The impact of energy resources on social development in Russia J. *WIT Transact. on Eco. and the Environ.* **190** 151-9

[4] Weidlich W 2000 *Sociodynamics - A Systematic Approach to Mathematical Modelling in the Social Sciences* (Amsterdam: Harwood Academic) pp 107-168

[5] Vasilyev A and Tarkhov D 2014 Mathematical Models of Complex Systems on the Basis of Articial Neural Networks *J. Nonlin. Phen. in Compl. Syst.* **17** 327-35

[6] Kainov N, Tarkhov D and Shemyakina T 2014 Application of neural network modeling to identification and prediction in ecology data analysis for metallurgy and welding industry *J. J. Nonlin. Phen. in Compl. Syst.* **17** 57-63

[7] Lazovskaya T and Tarkhov D 2015 Fresh approaches to the construction of parameterized neural network solutions of a stiff differential equation *St. Peter. Polytech. Univ. J. Phys. and Math.* **1** 192-8

Neural network modeling of air pollution in tunnels according to indirect measurements

T Kaverzneva, T Lazovskaya, D Tarkhov and A Vasilyev

Peter the Great St-Petersburg Polytechnic University, 29 Politechnicheskaya Str, 195251 Saint-Petersburg, Russia

E-mail: a.n.vasilyev@gmail.com

Abstract.
The article deals with the problem of providing the necessary parameters of air of the working area in dead-end tunnels in the case of ventilation systems powered off. An ill-posed initial-boundary problem for the diffusion equation is used as a mathematical model for a description and analysis of mass transfer processes in the tunnel. The neural network approach is applied to construct an approximate solution (regularization) of the identification problem in the case of the approximate measurement data and the set of interval parameters of the modeled system. Two types of model measurements included binary data are considered. The direct problem solution and the inverse problem regularization for the offered neural network approach are constructed uniformly.

1. Introduction

Air pollution of the human environment reduces the duration of his life, leading to a variety of serious diseases. In his professional activity, a person is exposed to a greater number of harmful factors that constantly requires the development of additional security measures, which can not be carried out without research and theoretical work.

During the construction of underground structures, tunnels for different purposes (for example, underground roads and railways) one of the most important tasks to ensure the safety of the operations is to control the parameters of the air environment, which may be contaminated sources of different origin. Normalization of air mainly achieved using rational ventilation system. However, during the construction and operation may be necessary to conduct a number of studies, for example, repair, in the absence of ventilation or disable it.

2. Problem statement

The paper goes on to study the problems with a nonstandard statement [1-6]. Here we construct a mathematical model for the dissemination of harmful substances from the source, which is located in the depths of the tunnel. This model allows for the concentration of harmful substances at the outlet to determine the distribution of dangerous concentration zones for such substances through the tunnel.

The problem of identification arises when environmental forecasting the state of air quality in the tunnel; a diffusion equation in a moving medium (and some pieces of heterogeneous information about a concentration of the pollutant) is the basis of a mathematical model:

Let the pollutant concentration $\Phi = \Phi(x,t)$ in the area $I = [0; L] \subset R$ at the time $t \in (0; T)$ satisfies the following initial-boundary-value problem

$$\begin{cases} \frac{\partial \Phi}{\partial t} + u\frac{\partial \Phi}{\partial x} - \sigma\frac{\partial^2 \Phi}{\partial^2 x} + \tau\Phi = P, \\ \Phi(x,0) = \Phi_0(x), \\ \Phi(0,t) = \Phi_1(t), \end{cases} \tag{1}$$

here the function $P(x,t) = \sum_{s=1}^{k} p_s(t)\delta_s(x;x_s)$ characterizes the power of pollution sources, $p_s(t)$ – the power of the s-th contamination source distributed in a neighborhood of x_s with a density $\delta_s(x;x_s)$, u – the speed, σ – the coefficient of turbulent diffusion, $\tau > 0$ – the parameter that determines the intensity of the absorption of the pollutant due to its entrainment, deposition, chemical reactions, etc.

It is expected that there is no initial contamination throughout the tunnel $\Phi_0(x) = 0$, $0 \le x \le L$. No contamination is at the entrance to the tunnel $\Phi_1(t) = 0$, $0 \le t \le T$.

By the replacement of variables $\Phi(x,t) = \exp(ax+bt)U(x,t)$, $P(x,t) = \exp(ax+bt)Q(x,t)$, where $a = u/2\sigma$, $b = -\tau - u^2/4\sigma$, the differential equation can be transformed into the heat equation

$$\frac{\partial U}{\partial t} - \sigma\frac{\partial^2 U}{\partial x^2} = Q \tag{2}$$

with appropriate initial and boundary conditions on the function U

$$U(x,0) = U_0(x) = 0, \;\; U(0,t) = U_1(t) = 0. \tag{3}$$

Note that the assignment of boundary conditions of the form $(\frac{\partial}{\partial x}\Phi - \frac{u}{2\sigma}\Phi)(L,t) = 0$ at the end of the tunnel results in the function U to the homogeneous Neumann condition $(\frac{\partial}{\partial x}U)(L,t) = 0$.

Next, we will consider the transformed task for the function U. The coordinates x, t are normalized so that $L = 1$, $T = 1$. The problems with the proposed statement (as power sources and part of the boundary conditions at the end of the tunnel are not listed) are ill-posed - they are in need of additional data.

Suppose that, instead of the missing data conditions, we know some experimental observations

$$U(\tilde{x}_j, \tilde{t}_j) = \varphi_j, \; j = 1, \ldots, m, \tag{4}$$

obtained with a set of sensors, $\{(\tilde{x}_j, \tilde{t}_j)\}_{j=1}^{m} \in [0;1] \times [0;1]$. The challenge is to find an approximate solution to the problem – the function U and the restoration of the source – the function Q as well. Another option of setting the identification problem is the restoration of the initial conditions.

3. Neural network approach

The approach proposed [1-4] allow us to bring together pieces of heterogeneous information about the system in the neural network model, use the regularizing properties of neural networks for solving inverse and ill-posed problems.

An approximate solution of the identification problem can be found in the form of a system output of two artificial neural networks of a given architecture

$$U(x,t;\sigma) = \sum_{i=1}^{N_1} c_i g_i(x,t;\sigma,\mathbf{a}_i), \tag{5}$$

$$Q(x,t) = \sum_{i=1}^{N_2} d_i h_i(x,t;\mathbf{b}_i), \tag{6}$$

which weights – linearly incoming parameters c_i, d_i and nonlinear input parameters $\mathbf{a}_i$, $\mathbf{b}_i$ – be determined in the process of gradual learning network based on the minimization of the error functional given in a discrete form

$$J = \sum_{j=1}^{M} \left(\frac{\partial U}{\partial t} - \sigma \frac{\partial^2 U}{\partial x^2} - Q\right)^2 (x_j, t_j) + \lambda_0 \sum_{j=1}^{M_0} U^2(x_j, 0) + \lambda_1 \sum_{j=1}^{M_1} U^2(0, t_j) + \lambda \sum_{j=1}^{m} (U(\tilde{x}_j, \tilde{t}_j) - \varphi_j)^2. \quad (7)$$

Here $\{(x_j, t_j)\}_{j=1}^{M}$ – periodically regenerated sampling points in $[0;1] \times [0;1]$, $\{(x_j, 0)\}_{j=1}^{M_0}$, $\{(0, t_j)\}_{j=1}^{M_1}$, – sampling points on the border areas; $\lambda_k > 0$ – positive penalty coefficients.

In this problem of modeling air pollution in tunnels, the assessment of the concentration in the main part of the tunnel is more important than identifying the source, so we can find the neural network approximation to the function U minimizing the error functional J without the source Q in the first term. In this case, the first sum is taken over the sampling points of the part of the variable region such that it is known that $Q = 0$.

4. Measurement data modeling

We found some analytical solutions of the problem (for the given $Q(x,t)$ and the boundary conditions added at the end of the tunnel); these solutions were used when setting the "experimental measurements" φ_j. We choose, for example, nonhomogeneous Neumann's condition (linear on t increase in derivative) as an additional one $\partial_x U(1,t) = t$, then the approximate model solution is given by [7]

$$\tilde{U}(x,t) \approx tx - \frac{x}{2} + \frac{x^3}{6} + \sum_{n=0}^{100} \frac{\sin \frac{\pi(2n+1)x}{2}}{(2n+1)^4} (-1)^n \exp \frac{-\pi^2(2n+1)^2 t}{4}. \quad (8)$$

Using $\tilde{U}$ we generate the data of observations $\tilde{U}(x_*, \tilde{t}_j) = \varphi_j$, $j = 1, \ldots, m$, taken near the exit of the tunnel $x_* = 0.1$. Also, this function is used to check the result of solving the problem.

For modeling are offered two cases. If the special device measures data and captures the accurate model values we use the values of the function $\tilde{U}(x,t)$ at equidistant points $\tilde{t}_i$ of the time interval at fixed x_*

$$G_1(t_i) = \tilde{U}(x_*, \tilde{t}_i), \quad (9)$$

In another case, the binary data is created. We consider a concentration above the critical level as the availability and lower as the absence of harmful emissions in the air. So this data has a certain error. The data generation held by the next rule

$$G_2(t_i) = \begin{cases} \beta, \ \tilde{U}(x_*, t_i) + \varepsilon(2\xi - 1) > \alpha; \\ 0, \ \tilde{U}(x_*, t_i) + \varepsilon(2\xi - 1) \le \alpha, \end{cases} \quad (10)$$

where ε is a measurement error and can take the big values, ξ is a standard uniformly distributed random value, α is a sensitivity threshold, $t_i \in [0,1]$, $\beta = 2\alpha$.

In the second case we considered a condition corresponded to the impermeability of the right end of the tunnel. The allocation of constant intensity is on the right end of the tunnel at the part of $d =$ length.

Then the approximate model solution is given by [7]

$$\hat{U}(x,t) = \frac{16}{\pi^3} \sum_{n=0}^{+\infty} \frac{\sin \frac{\pi(2n+1)x}{2} \sin \frac{\pi(2n+1)d}{2}}{(2n+1)^3} \cdot (-1)^n (1 - \exp \frac{-\pi^2(2n+1)^2 t}{4}), \quad (11)$$

where $d = 0.1$.

5. Calculations

We construct a neural network model for our task on all available data. This is the equation (2), the boundary conditions (3), and measurement data (9)–(10). The selection of the model parameters (the weights of the neural network) is based on the error functional minimizing. The error functional optimization is conducted by RProp and Particle Swarm algorithms [9]. As basic elements of the neural network g_i, different types and number of functions were applied: Gaussians, sigmoids with activation function $\mathrm{th}(\cdot)$, etc.

5.1. Exact Data

Here we present results obtained by the neural network solution in the case of 10 neuro elements and two types of functions for the accurate "measurements" (9). The first type – sigmoid functions

$$u(x,t,x_c,t_c,a,b,c,d) = c(\mathrm{th}(a(x-x_c)+b(t-t_c))+d),$$

where x_c, t_c, a, b, c, d are the parameters (weights) of the neural network. As the result, we have the perceptron network with one hidden layer.

The basis functions of the second type are the solutions of the equation (2)

$$u(x,t,x_c,t_c,a,c) = c(\frac{1}{\sqrt{t-t_c}}\exp\frac{-0.25(x-x_c)^2}{t-t_c}+a). \tag{12}$$

The given error functional J (7) takes the form

$$J_0 = \lambda_0\sum_{j=1}^{M_0}U^2(x_j,0)+\lambda_1\sum_{j=1}^{M_1}U^2(0,t_j)+\lambda\sum_{j=1}^{m}(U(x_*,\tilde{t}_j)-G_1(\tilde{t}_j))^2. \tag{13}$$

The counterplot of the neural network solution and the real concentration function is presented in Figure 1 (the perceptron network) and Figure 2 (the special network) at the level of the harmful substances concentration value 0.05. In this model, this level acts as an example of critical exceeding of which is dangerous to humans.

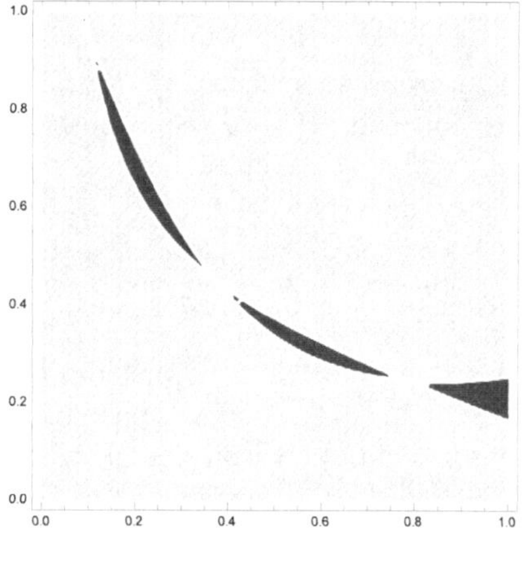

Figure 1. The counterplot of the neural network solution and the model concentration function $\tilde{U}$ of level 0.05. The perceptron network with $n = 10$ neuro elements.

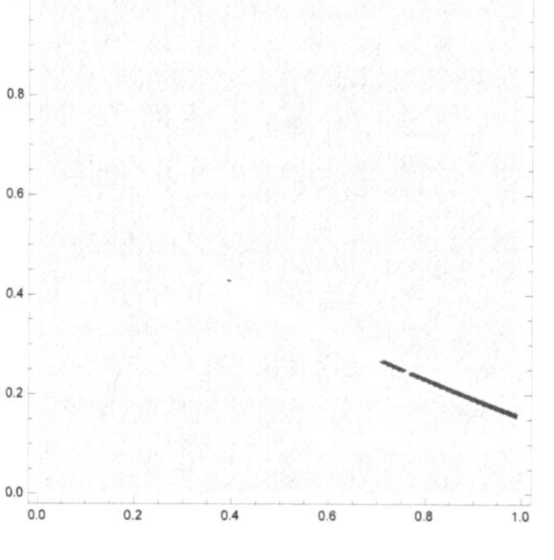

Figure 2. The counterplot of the neural network solution and the model concentration function $\tilde{U}$ of level 0.05. The neural network with $n = 10$ basis neuro elements. The basis functions used are fundamental solutions (12) of a diffusion equation

Journal of Physics: Conference Series **772** (2016) 012035 doi:10.1088/1742-6596/772/1/012035

The plots represent the elevated concentration prevalence through the tunnel from the source ($x = 1$) towards the entrance ($x = 0$) over time ($0 \leq t \leq 1$). As expected, the special features provide a more accurate solution (mismatched area marked by dark color).

Note, that with an increase in the number of neurons up to 20 the universal perceptron network provides a good approximation of a function $\tilde{U}(x, t)$ in an area near the exit of the tunnel. But there is an accumulation of error at the far end of the tunnel.

In the case of the model concentration function $\hat{U}$ (11), we have obtained a good neural network solution using fundamental solutions (12) of a diffusion equation only.

5.2. *Binary Data*

Neural network solution of considered problem is constructed for the data generated by the formula (10) at the parameters values $\varepsilon = \alpha = 0.03$ and $\beta = 2\alpha$. Note that in evaluating the results it must be taken into account the character of the data and the selected error.

A good result was achieved by the neural network with $n = 5$ special basis functions (12). Figure 3 illustrates the approximation of the concentration function $\tilde{U}(x, t)$ in the measuring point ($x_* = 0.1$).

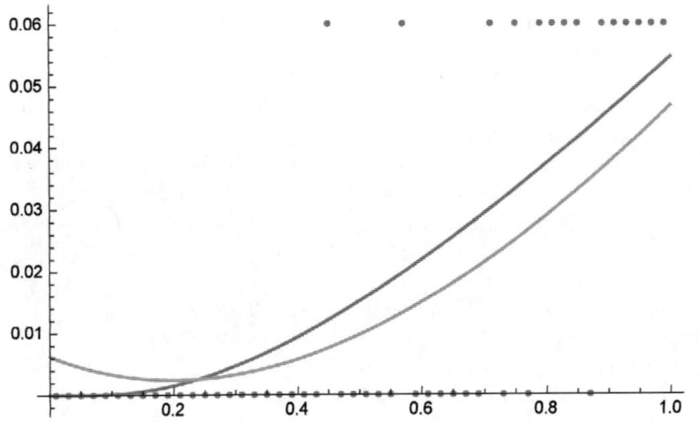

Figure 3. The plot of the neural network solution, the concentration function $\tilde{U}$ at $x_* = 0.1$ and the point data. The special network with $n = 10$ neuro elements.

The approximation of the concentration function $\hat{U}(x, t)$ (11) has a similar appearance.

6. Conclusions

The approximate neural network solutions of the considered problems are constructed. Computing experiment showed only a slight discrepancy between the boundaries of areas in excess of the maximum permissible concentration of harmful substances, obtained by the neural network solution in the case of 10 neuro elements and modeling solution.

As basic elements of the neural network, different types and number of functions were applied, but it is preferable to use the functions that in advance satisfy the diffusion equation, then the first term can be eliminated from the error functional. This approach is less versatile but allows us to get a more accurate solution with a smaller number of basic functions. The insignificant modification of methods described [2, 4] allows considering the case of interval specified parameters.

The statement of problems discussed in our work includes the differential equations. This fact causes the lack of overfitting. Usually, overfitting occurs when it is training a neural network on the finite set of input data in a situation when the capacity of this set does not exceed the number of weights. As a rule, the problem is solved by the division of a data set for training and test sets and the selection of the network structure to minimize the error on the test set. Differential equations are valid on some infinite set of points (in some domain). Therefore, the set of points from which training and test sets are selected is infinite. It makes possible the regeneration of the training and test sets and dimensional changes to eliminate overfitting.

In our problem, the number of times at which the measurements are made, finite and the points, from which we can select a test set, form a square $[0; 1] \times [0; 1]$, that is, an infinite set.

We are going to consider the possibility of creation forecast of harmful substances spread in a tunnel from any source and definition of zones with dangerous concentrations for protection of the workers who are in a tunnel from the influence of these pollutants.

Acknowledgments

The work is supported by the Russian Foundation for Basic Research (Grant no. 14-01-00660).

References

[1] Kainov N, Tarkhov D and Shemyakina T 2014 Application of Neural Network Modeling to Identification and Prediction Problems in Ecology Data Analysis for Metallurgy and Welding Industry *Nonlinear Phenomena in Complex Systems* **17** 57-63

[2] Lazovskaya T and Tarkhov D 2015 Fresh approaches to the construction of parameterized neural network solutions of a stiff differential equation *St. Petersburg Polytechnical University Journal: Physics and Mathematics* http://dx.doi.org/10.1016/j.spjpm.2015.07.005

[3] Vasilyev A, Tarkhov D and Guschin G 2004 Neural Networks Method in Pressure Gauge Modeling *Proc. of the 10th IMEKO TC7 Int. Symp. on Advances in Measurement Science (Saint-Petersburg, Russia vol 2)* 275-279

[4] Vasilyev A and Tarkhov D 2014 Mathematical Models of Complex Systems on the Basis of Artificial Neural Networks *Nonlinear Phenomena in Complex Systems* **17** 327 35

[5] Tarkhov D and Vasilyev A 2005 New neural network technique to the numerical solution of mathematical physics problems. I: Simple problems *Optical Memory and Neural Networks (Information Optics)* **14** 59-72

[6] Tarkhov D and Vasilyev A 2005 New neural network technique to the numerical solution of mathematical physics problems. II: Complicated and nonstandard problems *Optical Memory and Neural Networks (Information Optics)* **14** 97-122

[7] Polyanin A and Nazaikinskii V 2016 *Handbook of Linear Partial Differential Equations for Engineers and Scientists* (Chapman and Hall/CRC) p 1643

[8] Haykin S 2008 *Neural Networks and Learning Machines* (Prentice Hall) p 936

[9] Riedmiller M and Braun H 1993 *A direct adaptive method for faster backpropagation learning: The Rprop algorithm. Proceedings of the IEEE International Conference on Neural Networks* (IEEE Press) 586-591

Measurement model as a means for studying the process of emotion origination

R Taymanov[1], **Iu Baksheeva**[2], **K Sapozhnikova**[1] **and A Chunovkina**[1]

[1] D.I. Mendeleyev Institute for Metrology, St. Petersburg, Russia
[2] St.Petersburg State University of Aerospace Instrumentation, Russia

E-mail: k.v.s@vniim.ru

Abstract. In the last edition of the International Vocabulary of Metrology the concept "measurement" was spread outside the field of physical quantities. This fact makes it relevant to analyze the experience of developing the models of multidimensional quantity measurements. The model of measurements of expected emotions caused by musical and other acoustic impacts, is considered. The model relies upon a hypothesis of a nonlinear conversion of acoustic signals to a neurophysiological reaction giving rise to emotion. Methods for checking this hypothesis as well as experimental results are given.

1. Introduction

In the last edition of the International Vocabulary of Metrology [1] the concept "measurement" was spread outside the field of physical quantities. This fact can be explained by changing society priorities. In the 21-st century the need for a quantitative estimate of properties characterizing health, knowledge, and abilities of human, as well as that of economical and social processes, has significantly increased [2, 3]. In many cases these properties are multidimensional (multiparametric). For specialists in the field of metrology the above mentioned fact results in the necessity to cooperate with specialists who got used to other approaches to the quantitative estimates of experimental data, as well as to increase attention to the problems of measurement model development.

While developing a model for measuring a non-physical quantity characterizing human or society, it is desirable to apply a hypothesis demonstrating a mechanism of this quantity formation. Particularly, this statement can concern any human physiological reaction, including emotion. Such an approach enables the most important parameters determining a measurand value to be singled out. At the first stage of development, the model can be substantiated only by theoretical ideas, calculations and assumptions. The subsequent experimental justification of the model paves the way for measurement of properties immeasurable before, thereby contributing to better understanding of the scientific concepts concerning the investigation subject. A model considered below is an example demonstrating the efficiency of such an approach.

2. Model of emotion measurements

The model described in [4] and developed later on in [5-8] is a model of a hypothetical mechanism linking a musical or some other acoustical impact with an expected emotion of a listener. It was developed on the basis of an evolutional analysis of bioacoustic signals, a number of hypotheses and calculations. It relies upon a well-known fact that within the process of evolution of living organisms in water, they felt a need to perceive some signals which alerted the others to approaching natural disasters, enemies, "food" or a male (female). The earliest sensor system provided the sense of touch (the perception of water pressure changes within the infrasound frequency range). When amphibia and reptiles left water to occupy dry land, where the density of a medium became significantly lower, the sensitivity to the oscillations signaled about the vitally important situations reduced. A new possibility was formed on the basis of a sensitivity growth within a higher frequency range, nonlinear conversion of the perceived oscillations and selection of intermodulation oscillations with infra-low frequency, which simulate a reflex reaction of an organism, i.e., "proto-emotion".

The emotion in this model is defined according to a neurophysiological response being identified with the help of memory and as a consequence, giving rise to a transfer to a state an emotional colour of which can be identified by feeling or behavior of a listener.

The measurement model consists of three steps. The first one reflects a "mechanism" of forming the simplest (earliest) basic emotions that are common for people and animals. This step is connected with the reflex brain activities. The second and third steps give an idea of the way how complicated emotions are formed. They consider universal human and culture-dependent emotional images concerning cognitive activity, history of people, upbringing, education, belonging to a definite socio-culture group, etc. The first step of the measurement model and 3 steps model are shown in figure 1 and figure 2, accordingly.

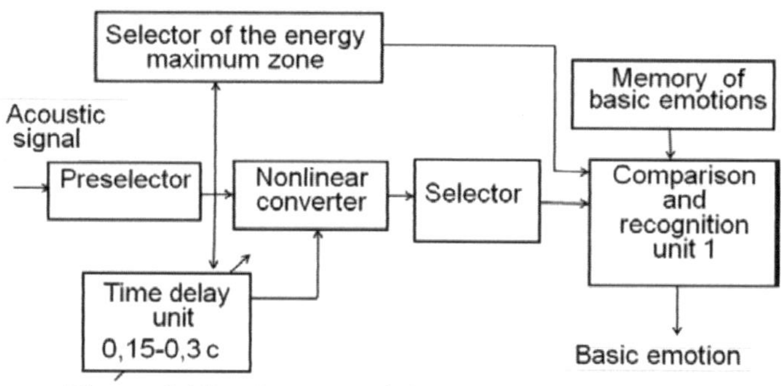

Figure 1. The first step of the measurement model.

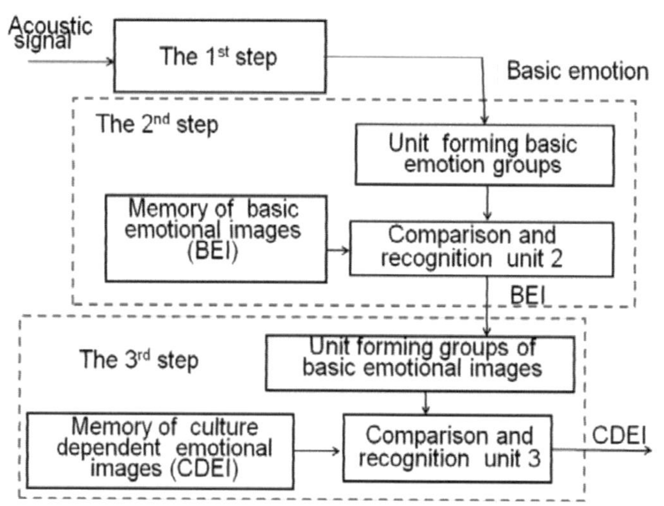

Figure 2. 3 steps measurement model.

The question arises with regard to the localization of the non-linear converter. The mechanism of the nonlinear conversion of acoustic signals in ear is well-known [9]. However, one of the main assumptions that underlies the model is as follows. The basic emotion is activated as a result of the joint nonlinear conversion of acoustic signals that are perceived at the moment and were perceived earlier. This procedure and the subsequent selection of intermodulation products within the range of brain biorhythms, can be performed only in the brain. (Assumptions on the non-linear conversion in a part of the brain directly connected with an organ of hearing, in recent years have been also pointed out by specialists in the field of neurophysiology [10].

3. Methods of checking up the main blocks of the first step of the measurement model

Experiments were aimed at proving the existence of characteristic neurophysiological reactions to sound stimuli in accordance with the model proposed. Stimuli were synthesized with the help of computer from sinusoidal oscillations in order to exclude the influence of subjective features of a performer and instrument timbre parameters. As the stimuli, tonic triads and sounds containing modulation at frequencies within the range of brain biorhythms (oscillations with single-sideband and frequency modulation) were used. It is known that musical intervals and sets of them have emotional colour [5]. Preliminary calculations [4, 5] have shown that after nonlinear conversion of sounds of major tonic triads or sounds with the corresponding modulation, components of a comparatively high level within the infra-low frequency range are formed. According to the above-mentioned hypothesis, they lead to the basic emotion generation. In experiments, the stimuli that cause such reactions played the role of the "deviant" ones. Minor tonic triads do not cause the components of such a level within the considered frequency range. Therefore, they were used as "standard" stimuli. The experiments included presentation of standard and deviant stimuli to participants and synchronous registration of electroencephalograms (EEGs).

During the studies, test procedures were improved by optimizing a duration and presentation order of the stimuli as well as data processing method.

The first experimental procedure can be called classical. Tonic triads of 4 tonalities were applied as the stimuli, measurement series being different in the tonalities. Each series included 20 equal triads. EEGs of the participants were recorded in pre test eyes closed and eyes open conditions in silence, then in eyes closed (to minimize external influence) during the stimulus presentations. The analysis of the EEGs was performed using a Short-time Fourier Transform with a rectangular time window. After that, spectrograms were built. The analysis of the spectrograms obtained revealed that the repetition of the same stimuli used in this procedure causes the "loss of interest" in them. As a consequence, a signal/ noise ratio degrades when the number of stimuli increases, although it was impossible to expect this ratio to be very high even under ideal conditions. The thing is that on one hand, the emotional colour of an individual triad is weak and cannot result in neurophysiological reactions of a significant level within the infra-low frequency range. On the other hand, EEG is influenced by many various random factors of physiological and cognitive character, such as respiration, muscle activity, tiredness, attraction of attention to irrelevant thoughts or side impacts, etc. It is rather hard to diminish their influence during the experiments. To decrease the "loss of interest", a quasi-random sequence of stimuli was formed, in which different stimuli were presented in an order unknown for participants.

The second procedure used the presentation of stimuli as a continuous record. It included major tonic triads (B-dur in the small octave, C-dur in the first octave, D-dur in the second octave) and oscillations with the corresponding modulation. According to the calculations, these sound combinations generate oscillations with the frequencies 5 Hz, 8.45 Hz, and 12.65 Hz. The deviant stimuli were presented in a quasi-random order. Minor tonic triads (b-mol in the small octave, c-mol and d–mol in the first octave) played the role of standard stimuli and were inserted between the deviant ones. According the third procedure, the stimuli were also presented in a quasi-random order, but after each stimulus a participant performed simple arithmetic calculations to remove an emotional mark left by the previous stimulus.

4. Analysis of results

The non-stationary character of EEGs caused the necessity to modify the data processing in the spectral field and to refuse from usage of the standard Fourier-analysis, since it does not permit a high resolution capability with regard to time and frequency to be simultaneously achieved. It "averages" a great number of nonrandom influence factors making the results worse. An analysis of special features of the measurements gave grounds to carry out the data processing within the time domain, and to evaluate the relative variation of EEG signal power within band ±0.5 Hz around the calculated frequency value. The process of analysis was realized in several stages:

a) Revealing outliers in the EEG (noise, pickups) and removing these outliers from records.

b) Calculation of values of the ratio of signal powers U_s и U_d for each standard and deviant stimuli, correspondingly. U_s and U_d were calculated for two time windows characterized by duration T_{win}: the 1st window was taken during the stimulus presentation (with time delay T_{st} from the stimulus beginning in order to determine the moment of starting participant's perception of a stimulus), while the 2nd one was taken just before the 1st one.

c) Revealing and removing outliers in the obtained samples of values U_s and U_d.

d) Evaluation of samples medians: $med(U_d)$ for the deviant stimuli and $med(U_s)$ for the standard stimuli.

The procedure reflected in point b), taking into account the possible variation of the moment of individual reaction, allows this moment of time to be determined the best way. It corresponds to the maximal ratio of the power values (from the values obtained at different T_{st} and T_{win}).

The problem concerning the choice of a method for removing outliers became extremely important. In deviant samples the number of values is rather small. At the same time, the laws of distribution of the quantities U_s and U_d are unknown. In the calculations, medians of samples were applied [11, 12] since the median is robust with regard to the outliers under conditions of an unknown distribution law.

The results of data processing on 12 tested participants are illustrated in figure 3. The median of $med(U_d)/med(U_s)$ calculated for all the participants (the time window position optimal for each participant was chosen) is 2.56 for the frequency of 5.0 Hz and 1.97 for the frequency of 12.65 Hz.

Application of the Bayesian decision theory allowed additional confirmations of the above-mentioned hypothesis that the nonlinear conversion is a part of the emotion formation process (see table 1). Calculations were carried out for a set of values T_{win} within the range from 0.1 s to 0.5 s and T_{st} in the range from 0 s to 0.4 s. For each pair, the frequency probabilities of correct "detection" of standard (pH_{ss}) and deviant (pH_{dd}) stimuli were calculated. As a result, there was obtained a pair of values T_{win} and T_{st}, for which pH_{ss} and pH_{dd} are maximal.

The probabilities pH_{ss} and pH_{dd} were calculated in the following way. It was considered that in spite of unknown laws of the U_s и U_d distribution, some assumptions can be applied:

1) For the standard stimuli $mode(U_s)=1$, i.e., under the influence of the standard stimuli at the frequencies of interest, there is no neurophysiological reaction, since in the standard stimuli themselves after the nonlinear conversion there are no infra-low-frequency components.

2) For the deviant stimuli $mode(U_d)=1.2$, i.e., the power of EEG at the frequency of interest, when the deviant stimulus is presented, exceeds the EEG power at this frequency before the presentation of the stimulus at the least by 20 %.

3) Dispersion of the U distributions for the standard and deviant stimuli are equal.

Under such conditions the threshold U_0 was calculated as

$$U_0 = \frac{mode(U_s) + mode(U_d)}{2} = 1.1 \tag{1}$$

For each standard stimulus at $U_s<U_0$ the decision was taken concerning the correct "detection" of the standard stimulus. For each deviant stimulus at $U_d \geq U_0$ the decision was taken concerning the correct "detection" of the deviant stimulus. In other words, under these conditions the reaction of the participant caused by the standard and deviant stimuli corresponded to the expected reaction within the framework of the hypothesis proposed. In realization of this procedure there were calculated values of n_{ss}, i.e., the number of the standard stimuli the response to which corresponded to the expected standard stimuli reaction, and values of n_{dd} , i.e., the number of the deviant stimuli the response to which corresponded to the expected response to the deviant stimuli.

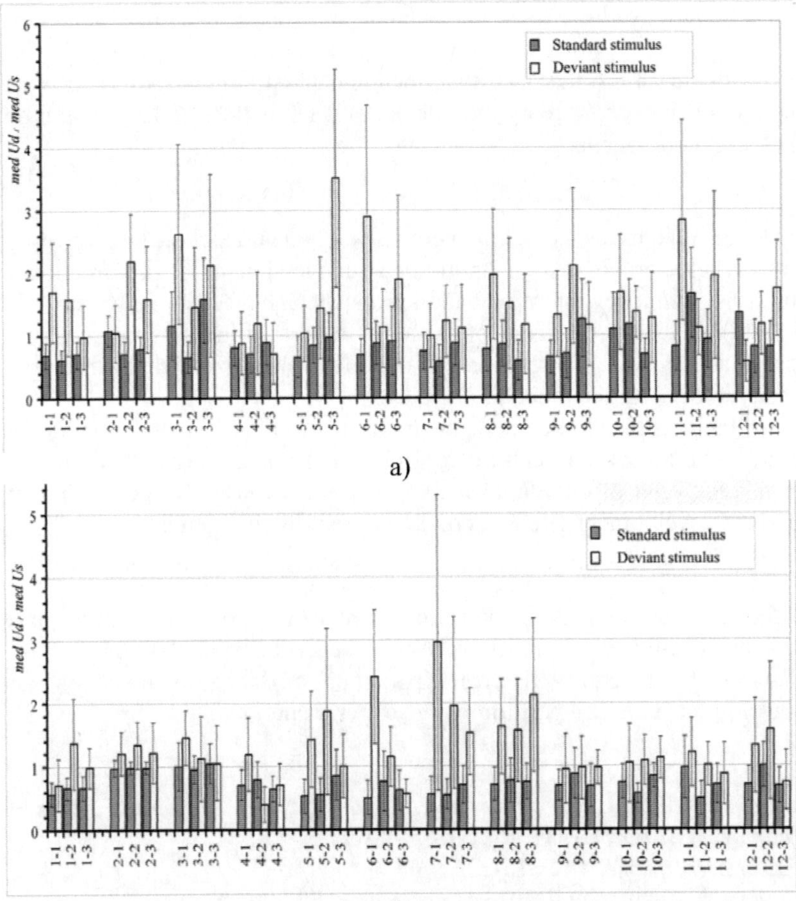

Figure 3. Medians of the EEG power ratios in the 1st and 2nd time windows (T_{win} = 0.3 s) for standard $med(U_s)$ and deviant $med(U_d)$ stimuli.

Along the *X*-axis the symbolic notations of participants (1 – 12) are indicated (participants 1 and 2 were tested according to the second procedure, participants 3-12 were tested according to the third one); the number given through a hyphen is a symbolic notation of the moment when the 1st time window starts relative to the stimulus beginning (0.1 s, 0.2 s, and 0.3 s).
a) 5 Hz;
b) 12.65 Hz

$$pH_{ss} = \frac{n_{ss}}{N_s},\ pH_{dd} = \frac{n_{dd}}{N_d}, \tag{2}$$

where N_s, N_d are the number of the standard and deviant stimuli.

For determining a pair of T_{win} and T_{st}, values at which both standard and deviant stimuli are "detected" in the best way, the total probability of the correct "detection" was calculated:

$$p = \frac{n_{ss} + n_{dd}}{N_s + N_d}. \tag{3}$$

Table 1. Calculation results.

	5 Hz					12.65 Hz				
No	T_{st} (s)	T_{win} (s)	p	pH_{ss}	pH_{dd}	T_{st} (s)	T_{win} (s)	p	pH_{ss}	pH_{dd}
3	0.1	0.2	0.73	0.71	0.76	0.1	0.2	0.64	0.64	0.65
4	0.1	0.2	0.67	0.66	0.69	0.1	0.4	0.66	0.69	0.57
5	0.4	0.2	0.69	0.65	0.78	0.2	0.1	0.71	0.69	0.79
6	0.2	0.4	0.72	0.69	0.80	0.1	0.2	0.69	0.69	0.71
7	0.2	0.4	0.75	0.76	0.75	0.2	0.2	0.70	0.68	0.72
8	0.2	0.4	0.63	0.64	0.63	0.2	0.1	0.68	0.68	0.69
9	0.1	0.3	0.69	0.71	0.67	0.4	0.2	0.64	0.65	0.63
10	0.1	0.4	0.66	0.64	0.71	0.0	0.4	0.60	0.54	0.73
11	0.3	0.1	0.70	0.68	0.74	0.4	0.2	0.75	0.81	0.61
12	0.4	0.4	0.71	0.73	0.67	0.1	0.3	0.62	0.67	0.53

Then, for each participant of the test they chose those values of T_{win} and T_{st}, at which the p value is maximal.

The results obtained are the evidence of the fact that the deviant stimuli steadily generate components with the infra-low frequencies predicted by the calculations, while the standard stimuli do not give rise to the generation of such components.

5. Conclusion

It has been shown that the impact of tonic triads causes the neurophysiological reaction forecasted by the measurement model. These results prove that the non-linear conversion of sound frequency oscillations and selection of infra-low frequency intermodulation products take place in the brain.

In combination with other data, this result permits a conclusion to be drawn that the measurement model proposed earlier can serve as a model of the neurophysiological mechanism of forming the basic emotions, which are common for people and animals.

In its turn, this conclusion enables scientists to begin the study of the 2nd and 3rd steps of the model that reflect the mechanism of universal human and culture-dependant emotional images. Then it will be possible to go over to the investigation of emotional feeling processes caused by perception of music, and later on to proceed to the development of corresponding measuring instruments.

Acknowledgements

The research is supported by the Russian Foundation for Humanities (RFH), project "Forming of a measurement model interrelating musical influence and emotions of listeners" (No. 15-04-00565).

The authors of this paper would like to note the active participation of Dr. Bazanova and Nikolenko in the developing the experimental procedures and performing the experiments.

References

[1] *International Vocabulary of Metrology – Basic and General Concepts and Associated Terms — 3rd edn — 2008 version with minor corrections* BIPM, JCGM 200(2012)

[2] Rossi G B, Berglund B 2011 *Measurement* **44** 815-22

[3] Sapozhnikova K, Chunovkina A, Taymanov R 2013 *Measurement* **50** 390-96

[4] Sapozhnikova K, Taymanov R 2003 *Proc. of the XVII IMEKO World Congress (Dubrovnik, Croatia, 22-27 June 2003)* 2049-53

[5] Sapozhnikova K, Taymanov R 2010 *Measurement Science Review* **10 No 3** 78—88

[6] Taymanov R, Sapozhnikova K 2014 *EPJ Web of Conferences* **77** 00016(2014)

[7] Baksheeva Y, Bazanova O, Kostromina S, Nikolenko E, Sapozhnikova K, Taymanov R 2015 *Ninth Triennial Conference of the European Society for the Cognitive Sciences of Music (Manchester, UK, 17-22 August 2015)* pp 202-203

[8] Baksheeva Iu, Danilova I, Sapozhnikova K, Taymanov 2016 *Proc. XVIII International Conference "Neuroinformatics-2016" (Moscow, 25-29 April 2016)* pp 222-231

[9] Smith C U M 2008 *Biology of sensory systems: 2nd Edition* (John Wiley & Sons Ltd)

[10] Lerud K D, Almonte F V, Kim J C, Large E W 2014 *Hearing Research* **308** 41-49

[11] Huber P J 1981 *Robust Statistics* (Wiley)

[12] Maronna R A, Martin D R, Yohai V J *Robust Statistics: Theory and Methods* (http://bookre.org/reader?file=493731&pg=23)

Some conceptual problems in the CCU 2016 draft of the New SI

Franco Pavese

formerly Consiglio Nazionale delle Ricerche, Istituto di Metrologia, Torino, Italy

Abstract. After a short introduction to the present SI (Système International des Unités), this poster presentation illustrates a digest of the basics of the proposed New SI and of its difference with respect to the present SI. This involves a review of some unresolved problems in the CCU 2016 draft, concerning: the role of the constants of physics in it and their role in the conceptual construction of this international standard; the implications for science of the New SI implementation. Consequences and new duties for the users are illustrated, involving a possible hierarchy between countries that would be installed by the new definition.

1. Introduction

The SI is a system evolving in time, as shown in [1]. The original definition of the SI can be found on the website of the BIPM, under "Metre Convention" [2] and under "SI Brochure" [3]:

"The Convention was signed in Paris in 1875 by representatives of seventeen nations. … the Metre Convention established a permanent organizational structure for member governments to act in common accord on all matters relating to units of measurement. The Convention, … remains the basis of international agreement on units of measurement. The BIPM now has fifty-seven Member States, including all the major industrialized countries." [2].

"This SI consists of a set of base units, prefixes and derived units, as described in these pages. The SI base units are a choice of seven well-defined units which by convention are regarded as dimensionally independent: the metre, the kilogram, the second, the ampere, the kelvin, the mole, and the candela. Derived units are formed by combining the base units according to the algebraic relations linking the corresponding quantities." [3].

"Les unités choisies doivent être accessibles à tous, supposées constantes dans le temps et l'espace, et faciles à réaliser avec une exactitude élevée" (from the official French text) [3].

2. Mandatory features of a modern system of units, and the present SI

A sound system of units of modern conception must fulfil specific requirements, some explicitly indicated in the above citations from [2, 3]. The units should be:

1. constants in time and space
2. organised in a rational system
3. easy to realise with high precision
4. accessible to everybody.

The present SI is built according to the above principles. The definition of the system consists in the definition of seven base units, in which a numerical indication of the magnitude of each unit is specified. In the specific jargon, it is of the type called "Explicit Unit Definition" (EUD).
Each of the base unit is individually defined. The present definitions are of different types:
(a) Make use of an artefact: e.g., for mass "The kilogram is the unit of mass; it is equal to the mass of

the international prototype of the kilogram";
(b) Make use of a physical state or condition: e.g., for temperature "The kelvin, unit of thermodynamic temperature, is the fraction 1/273.16 of the thermodynamic temperature of the triple point of water". Similar approach for time, amount of substance and luminous intensity;
(c) Define a realisation method of the unit, e.g. for length: "The metre is the length of the path travelled by light in vacuum during a time interval of 1/299 792 45 8 of a second". Similar approach for the electrical current.
(Actually, also the type (a) and (b) base units define indirectly a so-called "realisation method", to be implemented with a procedure—not to be confused with the "mise en pratique" that is a (set of) approximations of the definition).

At present, four base units violate the indication of the SI Brochure that "les grandeurs de base sont, par convention, considérées comme indépendantes", because, using for convenience also some derived units, the definition of: the unit of length involves the second; the unit of amount of substance involves the kilogram; the unit of electric current involves the newton and the metre; the unit of luminous intensity involves the hertz, the watt and the steradian.

The paper schematically illustrates the four above requirements, in particular in respect to the conceptual structure proposed in [4].

3. A summary of the New SI structure

The proposed change of definition, normally named "New SI", is present planned to be promulged in autumn 2018. The use of some 'constants' has been proposed—with stipulated numerical values.

3.1 Definition of the New SI

The following "Definition of the SI" is reported in the last available draft [5]:

"The International System of Units, the SI, is the system of units in which:
- the unperturbed ground state hyperfine splitting frequency of the caesium 133 atom $\Delta\nu_{Cs}$ is 9 192 631 770 Hz,
- the speed of light in vacuum c is 299 792 458 m/s,
- the Planck constant h is $6.626\ 070\ 040 \times 10^{-34}$ J s,
- the elementary charge e is $1.602\ 176\ 620\ 8 \times 10^{-19}$ C, (1)
- the Boltzmann constant k is $1.380\ 648\ 52 \times 10^{-23}$ J/K,
- the Avogadro constant N_A is $6.022\ 140\ 857 \times 10^{23}$ mol^{-1},
- the luminous efficacy K_{cd} of monochromatic radiation of frequency 540×10^{12} hertz is 683 lm/W.

The numerical values of the seven defining constants have no uncertainty".

The constants (no more labelled "fundamental") are of five different types:
– for time a specific electronic transition $\Delta\nu_{Cs}$ of a specific substance, ^{133}Cs—the same used at present;
– for luminous intensity, a technical constant K_{cd}, of physiological origin;
– for electrical current, the elementary charge e, via the fine-structure constant $\alpha = e^2/(2c\varepsilon_0 h)$;
– for temperature and amount of substance, the Boltzmann constant k_B and the Avogadro constant N_A, respectively, considered two 'conversion factors';
– for length and mass, the speed of light in vacuum c_0—the same as presently—and the Planck constant h, respectively, the only two real "fundamental" constants.

This can be called the 'group-definition'—in jargon, it is of the type called "Explicit Constant Definition" (ECD) or "Global Constant Definition" (GCD), since these constants should be considered and treated as a group.
It is indicated as the sufficient new SI definition. Table 1 in [5] indicates that these constants define the following units: Hz = s^{-1}, m s^{-1}, J s = kg m^2 s^{-1}, C = A s, J K^{-1}, mol^{-1}, lm W^{-1} = cd sr W^{-1}.

3.2 Definitions of the New SI units: Base units

Additionally to the group definition of the constants, in [5] there is also stated that the "description in terms of base and derived units is maintained in the present definition of the SI, but has been reformulated as a consequence of adoption of the defining constants". A reason may be found in the other statement: "Preserving continuity is an essential feature of any changes to the International System of Units". Then "the definitions of the traditional base units of the SI ... *follow* from the definition of the seven defining constants" (emphasis added). Therefore, also the definition for each single base unit is provided. At present, the above statement is implemented in [5] for each of the seven base units as follows:

- "The second, symbol s, is the SI unit of time. It is defined by taking the fixed numerical value of the caesium frequency $\Delta\nu_{Cs}$, the unperturbed ground-state hyperfine splitting frequency of the caesium 133 atom, to be 9 192 631 770 when expressed in the unit Hz, which is equal to s^{-1} for periodic phenomena."
- "The mole, symbol mol, is the SI unit of amount of substance of a specified elementary entity, which may be an atom, molecule, ion, electron, any other particle or a specified group of such particles. It is defined by taking the fixed numerical value of the Avogadro constant N_A to be 6.022 140 857 $\times 10^{23}$ when expressed in the unit mol^{-1}."
- "The metre, symbol m, is the SI unit of length. It is defined by taking the fixed numerical value of the speed of light in vacuum c to be 299 792 458 when expressed in the unit m s^{-1}, where the second is defined in terms of the caesium frequency $\Delta\nu_{Cs}$."
- "The kilogram, symbol kg, is the SI unit of mass. It is defined by taking the fixed numerical value of the Planck constant h to be 6.626 070 040 $\times 10^{-34}$ when expressed in the unit J s, which is equal to kg m^2 s^{-1}, where the metre and the second are defined in terms of *c* and $\Delta\nu_{Cs}$."
- "The ampere, symbol A, is the SI unit of electric current. It is defined by taking the fixed numerical value of the elementary charge e to be 1.602 176 620 8 $\times 10^{-19}$ when expressed in the unit C, which is equal to A s, where the second is defined in terms of $\Delta\nu_{Cs}$."
- "The kelvin, symbol K, is the SI unit of thermodynamic temperature. It is defined by taking the fixed numerical value of the Boltzmann constant *k* to be 1.380 648 52 $\times 10^{-23}$ when expressed in the unit J K^{-1}, which is equal to kg m^2 s^{-2} K^{-1}, where the kilogram, metre and second are defined in terms of *h*, *c* and $\Delta\nu_{Cs}$."
- "The candela, symbol cd, is the SI unit of luminous intensity in a given direction. It is defined by taking the fixed numerical value of the luminous efficacy of monochromatic radiation of frequency 540 $\times 10^{12}$ Hz, K_{cd}, to be 683 when expressed in the unit lm W^{-1}, which is equal to cd sr W^{-1}, or kg^{-1} m^{-2} s^3 cd sr, where the kilogram, metre and second are defined in terms of *h*, *c* and $\Delta\nu_{Cs}$."

The above definitions imply that the *present* base units—second, metre, kilogram, ampere, kelvin, mole, and candela—remain in the New SI, however with a totally differently-based definition and function: they are *not anymore the definition* of the SI.
They are maintained in order to fulfil the "continuity condition" and are indicated to "follow" from the SI definition—the 'group' one of the constants. However, some units remain not defined even in the above definitions: Hz, J, lm, W, sr: they are defined as derived only later in the conceptual roadmap.

4. Basic features of the New SI

The proposal of the New SI, to anchor the magnitude of the base units to constants included in basic laws of physics [8], entails some basic new features that are the pillars of the new definition and can be considered an interesting advancement with respect to the present SI.

Internal consistency of the base-unit system. Only a degree of consistency can be checked, because four of the seven constants (c_0, h, e, k_B) are multi-dimensional, so an arbitrary choice of the unit magnitudes in not allowed. However, for stipulation, the present values of the seven constants were obtained by taking advantage of the CODATA studies, which fix the (minimum) degree of consistency that is acceptable for the SI from now on. This is an important asset of the proposal.

Continuity (constancy) of the unit magnitude. The continuity requirement (also called constancy in time of the unit magnitudes) is stressed in the proposal as a basic need. Because of this requirement, the set of basic units is taken un-altered from the present SI, but only as to the base quantities and unit denomination, while the definition of the base units is now founded on the same constants of the group definition. In addition, since the stipulated values of the constants come from the present-SI, the magnitudes of those constants are assumed to remain un-altered.
Actually, only the use of three of the seven constants is new: h, k_B and e. In fact, $\Delta\nu_{Cs}$, c_0, N_A and K_{cd}

are already included in the present-SI definitions, though in an Explicit Unit Definition (EUD) type.

Lack of definitional methods. In the New SI there is not anymore a definitional method specified for each base unit. This is maybe the most important practical change in the features of the SI of interest of the users. It means that "the realizations are separated conceptually from the definitions" [5].
In addition, to be at the top step of the hierarchical ladder, an institute should now demonstrate that its standards comply with the new condition set by New SI: that, by using them, they are able to obtain the stipulated value of each relevant constant—in the past they had to demonstrate that they own consistent realisations of the relevant "definitional methods".

Lack of reference to any specific substance: uniqueness. The preference of using of constants in strict sense—though of different types—was born from the fact that they are not a property of any specific substance. Only for the time, whose definition is formulated differently from the present one but is basically the same, the atomic transition is one of ^{133}Cs. The advantage is that a unique property of the constants: their lack of definitional non-uniqueness. Consequently, the "value of a constant" is the only relevant property here, which is intrinsically invariant in time and space. Its numerical value, instead, is contingently assigned as the 'best' one available at the time of the SI change in definition.

5. Main criticalities in the CCU 2016 Draft of the New SI

5.1 Concerning unit definitions

Only the main issue will be illustrated, concerning the conceptual frame of the new definition: it is based on an analysis that can be found in [6].

The SI-2018, differently from the present one, is structured *into two conceptual 'frames'*, which will be called here A (constants) and B (base units), which can be found in the text of the CCU 2016 Draft [5] in the following order:

❖ *General principles*

A1) Some present advantages considered achieved in the present SI "led to the decision to define all units with the help of defining constants";
B2) The "*description* in terms of base and derived units is *maintained* in the present definition of the SI, but has been *reformulated* as a consequence of adoption of the defining constants" (emphasis added);

❖ *"Definition of the SI"*

A3) The 'group-definition'. For each constant, it is of the type: "the [constant] [symbol] is [numerical value] [present-SI units]", where the indicated units, single or algebraic combination, can be formed by either present base units or derived units. For example, "the Planck constant h is 6.626 070 040 $\times 10^{-34}$ J s";
A4) Then, Table 1 in [5] shows the caption "The seven defining constants of the SI, and the seven corresponding units they define" (emphasis added), where the name, symbol, numerical value and "Unit" are reported, for example for h is J s = kg m^2 s^{-1}, where the initial units are then spelled out in terms of only the present base units, except for the luminous efficacy K_{cd}, which is lm W^{-1} = cd sr W^{-1}, where a derived unit and a special name unit are indicated;

❖ *Definitions of the base units*

A5-B5) "Preserving continuity is an essential feature of any changes to the International System of Units, which has always been assured in all changes to the definitions. The numerical values of the defining constants have been chosen to be consistent with the earlier definitions insofar as advances in science and knowledge allow";
B6) "The definitions of the traditional base units of the SI, as listed in Table 2 [of [5]], follow from the definition of the seven defining constants" (emphases added);
B7) The new definition of each of the seven present base units is provided. They are reported here above. For each base unit, it is of the type: "[The base unit] is defined by taking the fixed numerical value of [a constant, one and different for each base unit], to be [numerical value] when expressed in

Journal of Physics: Conference Series **772** (2016) 012037 doi:10.1088/1742-6596/772/1/012037

the [present-SI units], where the [other present-SI base units indicated] is [or are] defined in terms of the [(other) relevant constant indicated in the group definition]". For example, "the kilogram … is defined by taking the fixed numerical value of the Planck constant h to be 6.626 070 040 $\times 10^{-34}$ when expressed in the unit J s, which is equal to kg m^2 s^{-1}, where the metre and the second are defined in terms of c and $\otimes/_{\mathrm{Cs}}$";

❖ *Definitions of the derived units*

B8) Table 4 [5] "The [derived] 22 SI units with special names and symbols": "Derived units are defined as products of powers of the base units", namely the joule (J), hertz (Hz), watt (W), coulomb (C), lumen (lm) and steradian (sr).

The splitting of the New SI structure into *two frames* entails several difficulties in its correct implementation, many still unresolved.

Group definition (frame A). There is an obvious mismatch in the conceptual sequence. In fact, at step A3 concerning the definition of the SI, the new definitions of the base and derived units were not yet provided: it is incorrect in a definition to refer to a prior knowledge of the reader about the existence of certain previous units—statement B2 in itself is not sufficient to alleviate the problem.
It is not possible to disentangle the kind of definition A3 and resolve the mismatch, because the reason is the fact that the numerical value of the constant is indicated, entraining the need to specify the units in which it is expressed. The reason for the choice cannot be found within frame A, but is due to frame B according to step B5.
Actually, the choice of the constants does not depend on frame B, except for the fact they must be seven according to the proposers as the present base units—according to principle B5— and involving all the seven present ones, but *without one-to-one correspondence* with respect to the base units. However, considering that four of the seven are multidimensional, in principle, one could select instead less than seven constants, provided that they include all the dimensions of the base units.
From a conceptual viewpoint, the group definition does not need—and cannot—anticipate any numerical value. It should be considered as the "SI-fundamental" definition [4], and it would be sufficient to say:
"The SI-fundamental is the coherent system of units in which C_1 is the fundamental-unit of frequency, C_2 is the fundamental-unit of speed and C_3 is the fundamental-unit of action, …" [4], where the C_i are the chosen constants. **"In this way, according to this definition, for example the Planck constant has a value 1 fundamental unit of action**". [4]
Actually, the seven constants do not define the (previous) units indicated at step A4, but, as indicated in [7], *frequency, speed, action, electrical charge, heat capacity, reciprocal amount of substance and luminous intensity*, which *are the true new "base" units of the SI-fundamental system*.
However, due to the principle B5, the base units cannot be changed. That principle is *not* arising from the underlying physical principles governing the choice of the SI-fundamental, but from a *practice* that is considered highly desirable in metrology (and in measurement science): the practice issues are what can be labelled "SI-conventional" [4], here frame B, which is the *only* frame of the present-SI.

Then the numerical values come, originating from the present base units.

Base units (frame B). The need to maintain the use of base units is not justified in B1, but in B5, where it is also clear that the magnitude of each base unit should remain unchanged through the change of definitions. This is the (implicit in the 2016 Draft [5]) reason for the use of numerical values for the constants that are simply the *'best' available* at the moment of the change, as said in the second sentence in B5.
However, in B6 there is a logical inversion of this justification, where it is said that their definitions, as implemented in B7, "… *follow* from the definition of the seven defining constants" (emphasis added). In fact, in B7 the reported numerical value is obviously the one before the change of the SI definition, retained in the group definition: again a consequence of the inversion of the conceptual order illustrated in the previous subsection.

Journal of Physics: Conference Series **772** (2016) 012037 doi:10.1088/1742-6596/772/1/012037

This inconsistency also reflects in B7, the text of the new base unit definitions, when they refer to the group definition for the definition of the other base units appearing in the definition of the relevant one: e.g., in the definition of the kilogram, which is based on the Planck constant only, unit "kg m^2 s^{-1}, where the metre and the second *are defined in terms* of c and $\otimes\nu_{\mathrm{Cs}}$" (emphasis added). Here a circular reasoning occurs, because the group-definition is obliged to refer to the present-SI units while reporting the numerical values, but the latter refers now to the group-definition.
Thus, the use of a separate "SI-conventional" indicated above [4], logically coming after frame A, the SI-fundamental, would also resolve the problem in the base unit definitions [4]:
"The SI-conventional is the coherent system of units in which
(A) the unit of time (duration) is the second, symbol s, the unit of length is the metre, symbol m, the unit of mass is the kilogram, symbol kg, ...;
(B) the frequency $\otimes\nu_{\mathrm{Cs}}$ is k_1 s^{-1}, the speed c is k_2 m s^{-1}, the action h is k_3 kg m^2 s^{-2}, ...;
(C) the numerical values are k_1 = [num. value]×10^{-10}, k_2 = [num. value]×10^{-9}, k_3 = [num. value]×10^{-34}, ...",
where the numerical values are stipulated according to the 'best' values at the moment when the new definition comes into effect. Notice that also the base units should be seen *as a group*.

In practice, *the numerical values spelled out in the CCU group-definition are the conversion factors from the "SI-fundamental" to the "SI-conventional", the two distinct frames.*

What is intended for units in the SI:2018. The Introduction in [5] states: "The definition of the SI units is *established* in terms of a set of seven defining constants. From the *units of these defining constants* the complete system of units can be *derived*. These *seven defining constants* are the most fundamental feature of the definition of the entire system of units" (emphases added). The expression "the units of these defining constants" is ambiguous: does it refer to the present SI units (e.g., present-SI "velocity" for c_0), or to the present set of base units (e.g., present-SI "length" and "time" for c_0), or to a New SI unit for c_0?
In addition, the third sentence is a statement about the prevalence of the definition of the group of constants (called the group-definition hereinabove) with respect to the 'base units'. In fact, the period then states: "The specific constants have been identified as the best choice *reflecting* the previous definition of the SI based on seven base units and the progress in science" (emphasis added), putting the base units as the alternative way "reflecting" the previous SI definition.
According to this approach, Section 1.1 "Motivation for the use of defining constants to define the SI" [5] states: "*the realizations are separated conceptually from the definitions*" (emphasis added), a basic statement implicitly meaning that the definition does not indicate any 'definitional method' [5] like the previous SI, where instead in practice "the definition and the realization" were "equivalent".
The lack of 'definitional methods', indicated as the effect of using constants in the definition, looks one of the few—important—practical reasons for the latter choice. However, the implementation problems are actually moved, from the realization by definitional methods to the relationship between definition and realizations, and to their link to the previous SI.

6. Final remarks

With respect to the CCU 2013 Draft, the 2016 draft is improved because: the constants are not labelled as "fundamental" anymore; the term "mise en pratique" is replaced by the term "realization"—though considered equivalent; the "continuity condition" is introduced explicitly. However, other conceptual issues still need further consideration. In particular one can notice that many units are defined by using the *reciprocal* of a constant. This is not without consequences, because, as the reciprocal of a rational number is normally a real number, so the reciprocal of a stipulated number is also a real number, not a stipulated one unless a decision is taken to truncate the latter. This issue may reflect an inversion of the reasoning with respect to the correct conceptual flow.

References

[1] http://www.bipm.org/en/measurement-units/history-si/
[2] http://www.bipm.org/en/worldwide-metrology/metre-convention/
[3] http://www.bipm.org/en/measurement-units/, http://www.bipm.org/en/publications/si-brochure/
[4] Mari L. and Pavese F., A conceptual roadmap for setting up a system of units in the "New SI" context, TC1-TC7-TC13 Joint Symposium 2016, to be published on J. of Physics Conf. Series
[5] BIPM, CCU, 9th Brochure of the SI, draft of 11 December 2015, published on January 4, 2016 (2016 Draft), http://www.bipm.org/en/measurement-units/new-si
[6] Pavese F., A critical review of the CCU 2016 Draft of the Brochure for the "New SI" definition using (fundamental) constants, arXiv.org, physics.gen.ph, arXiv: 1601.00857v.3, 24 February 2016, pp. 22
[7] Newell D.B., A more fundamental International System of Units, Physics Today 67(7), 35 (2014); doi:10.1063/PT.3.2448. View online: http://dx.doi.org/10.1063/PT.3.2448
[8] Bordé Ch.J., Reforming the international system of units: On our way to redefine the base units solely from fundamental constants and beyond, arXiv: 1602.01752, 29 June 2016, pp. 29

Comparing the effects of combined numerical and visuo-spatial psychoeducational trainings conducted by curricular teachers and external trainers. Preliminary evidence across kindergarteners.

M Agus[1], M L Mascia[1], M C Fastame[1], V Napoleone[1], A M Porru[1], F Siddu[1], D Lucangeli[2], M P Penna[1]

[1]Department of Pedagogy, Psychology, Philosophy - Faculty of Humanistic Studies, University of Cagliari, Via Is Mirrionis, 1, 09123 Cagliari, Italy
[2]Department of Developmental Psychology and Socialization, University of Padova, Via Venezia, 8, 35131 Padova, Italy
E-mail: mirian.agus@unica.it

Abstract. The aim of this study was to verify the efficacy of two pencil-and-paper trainings empowering numerical and visuo-spatial abilities in Italian five-year-old kindergarteners. Specifically, the trainings were respectively carried out by the curricular teacher or by an external trainer. The former received a specific training in order to use the psychoeducational programmes with her pupils, whereas the latter received a specific education about the role of numerical and visuo-spatial abilities for school achievement and she was also trained to use psychoeducational trainings in kindergarten schools. At pre-test and post-test nonverbal functions and numeracy knowledge were assessed through a battery of standardized tests. The results show that both the numerical psychoeducational programme and the visuo-spatial one are useful tools to enhance mathematical achievements in kindergarteners. However, when the trainings were proposed by the external trainer, the efficacy of the psychoeducational programmes was more significant. These outcomes seem to be related both to the expertise and the novelty effect of the external trainer on the classroom.

1. Introduction

The measurement is a complex activity in which the researcher try to acquire specific and detailed information on the world, in order to describe the nature of the relationships among concepts and proprieties [1]. In the educational and psychological fields, the application of measurement presents a series of difficulties, related to different dimensions, such as the nature of the attributes, the theoretical and abstracted features of the cognitive processes underlying human behaviour in life span [2]. Indeed, the measurement of knowledge, skills and abilities (KSAs) [2] in human cognition is related to a social construction of the inquired concepts, which can be assessed indirectly [3]. Therefore, the evaluation of the KSAs is related to the measurement of "soft" systems [1,4], that is, a complex set of domains in human cognition directly involved in different aspects of behaviour [5]. Thus, from a theoretical perspective, in order to support the measurement, it is useful to assess each construct using many indicators, considering the multi-faceted dimensions of each behaviour and applying appropriate and validated instruments [6]. It follows that the measurement needs to be characterised by *objectivity* and *intersubjectivity* [7], in order to support the application of a functional concept of the Measurement

System [8]. Embracing this approach, objectivity is crucial in order to meaningfully discriminate conceptual entities that eventually share some properties through the application of universal procedures measuring a particular variable [9]. Moreover, intersubjectivity implies that within each measurement system, all the measurements about some psychosocial aspect "are comparable with each other" [10, p. 3773]. Overall, the abovementioned aspects tied to the measurement system require an agreement among the scientists in relation to the psychological concepts measured [2]. Therefore, the application of validated instruments of measurement allow us to acquire a specific knowledge about the world through the experience (i.e., Empiricist approach to measurement [2,11]), the demonstration that useful theories are linked with practice (i.e., Pragmatic approach to measurement [2,12]) and finally thought the intuition that scientific research is devoted "to gain knowledge about a natural world" (i.e., Realism to the measurement [e.g. 2, p.8]).

Applying the abovementioned theoretical assumptions in the psychoeducational field, the current study points out how complex the measurement of several aspects of numerical knowledge in pre-schoolers can be. That is, embracing a *representational viewpoint* [7,14], the measurement of numeracy learning in kindergarteners can be thought like an homomorphism between the "empirical relational structures" [7] and a "symbolic relational structure" [7, p.71].

There is evidence that mathematics learning is made up of a number of components, such as basic knowledge of numbers, memory for arithmetical facts, understanding of mathematical concepts, and the ability to follow procedures [15]. Concerning this, several studies show that math performance has high stability and the development of numeracy skills among children is predicted by counting ability [16]. A further body of psychological research has identified the cognitive processes (e.g., counting ability, visual attention, metacognitive knowledge, and listening comprehension) underpinning the development of numerical knowledge in early childhood [17]. Specifically, there is evidence of a strong relationship between mathematical achievements and the development of visuo-spatial working memory abilities in life span [18,19], that is, since early childhood children learn to represent and manipulate visual (e.g., shapes) and spatial (i.e., locations) information in order to perform geometry and math (e.g., comparing quantities) tasks [20]. Overall, this suggests that the conduction of specific trainings empowering non-verbal mnestic processes since early childhood [21] can play a fundamental role in order to promote the empowerment of numeracy skills. Indeed, a body of studies carried out in educational settings underlines the importance to train visuo-spatial working memory processes through targeted interventions and specific teaching strategies, in order to support the development of numerical abilities both in typically and atypically developing children [22]. In this regard, a long research tradition concerning the measurement of mathematics learning in pre-schoolers [23,24] as a function of the measurement of psychoeducational trainings efficacy suggests that the effect size of the trainings enhancing numerical skills since the early childhood highlights the practical significance of the outcomes in terms of improvements of specific aspects of mathematics achievements (e.g., mental calculation, counting sequences, quantity comparisons) [25]. Concerning this, there is evidence that the effectiveness of the psychoeducational intervention in primary school depends also upon the expertise and competences showed by the teacher [26], such as her/his knowledge about the curriculum contents, the strategies selected to promote numeracy learning and therefore the way in which educational goals are presented to the pupils [27].

Some experimental evidence suggests that since the age of three the adoption of the investigative approach seems to facilitate math achievement. Besides, a further crucial factor impacting school achievements is the level of novelty of the stimuli to be learned, because the originality of the material can drive and then enhance motivation, attention and attainment of the students both when the curricular activities and specific psychoeducational programs are conducted in the classroom [28–30].

However, to our knowledge, no studies have been conducted in order to investigate whether a specific expertise of the trainer about the relationship between visuo-spatial abilities and the development of numeracy can influence mathematical learning in preschoolers.

2. Aims and method

This work was aimed at exploring: 1) the effect of a combined visuo-spatial and numerical psychoeducational training on the development of math skills in pre-schoolers; 2) whether the expertise of the trainer about the contribute of visuo-spatial skills in promoting numeracy development can impact the empowerment of math skills in kindergarten children.

Therefore, for the aims of the study, the curricular teacher received a specific training about the administration of the mathematical psychoeducational program, but she was not specifically trained about the role of visuo-spatial functions for the development of numeracy skills. In contrast, an external trainer received a specific education on the relationship between mathematics and non-verbal abilities for math achievements in pre-schoolers as well as she was trained to use a psychoeducational programme empowering numeracy in preschool children.

Thus, it was hypothesised that: 1) children trained with a combined numerical and visuo-spatial programme would have outperformed children carrying out only the curricular activities (i.e., control group); 2) after a number of training sessions (i.e., post-test), children trained by the external trainer would have shown better math skills than children trained by their curricular teacher [27,31,32].

Before the training (i.e., at pre-test) and after that (i.e., at post-test), pre-schoolers were collectively administrated a battery of tests assessing their numerical and visuo-spatial abilities.

2.1. Participants, tools and procedure

After the parents provided written consent, seventy-two 5-year-old children were recruited in two kindergarten schools located in Sardinia (Italy). The participants (56.9% male) were divided into two experimental groups, which were trained by a combined mathematical and visuo-spatial programme by the curricular teacher (G1, n=27) or by an external trainer (G2, n=23). Moreover, there was a further group (G3, n=22) that carried out only the curricular activities. Both the trainers and children took voluntarily part in the study (i.e., non-probabilistic sampling).

At pre and post-test, the participants individually completed a battery of standardised tests. The Battery for Numerical Intelligence (BIN) [33] measuring different facets of mathematical abilities in children, related to cognitive and metacognitive aspects. Specifically, this tool assesses four dimensions of numerical skills: the lexical numeric knowledge (i.e., evaluating the skill to read and write Arabic numbers); the semantic numeric knowledge (i.e., appraising the ability to associate numerical sizes, dots and Arabic digits); the pre-syntactical numeric knowledge (i.e., considering the competency to link numbers to their number representation); counting skill (i.e., assessing the ability to recite the number–words sequence forward and backward). Each subscale revealed worthy psychometric characteristics, showing high test-retest reliability (lexical: r=.89; semantic: r=.69; pre-syntactic: r=.79; counting: r =.74).The evaluation of these multiple scales, related to different aspects of numerical abilities, supports a deeper measurement of the potential differences among the groups. The Coloured Progressive Matrices (CPM) [34] evaluated the fluid intelligence and the visuo-spatial abilities, which are strongly related to the development of numerical abilities. The child is asked to point out across a series of alternatives, the item necessary to complete 36 visuo-spatial patterns, that is, he/she has to relies on his/her imagery abilities in order to solve each visuo-spatial problem. The battery of tests was selected because of the specific homomorphism between the presented problems and the numeracy and visuo-spatial skills. Each individual session lasted about 40 minutes.

After pre-test, both the experimental groups were collectively presented the pencil-and-paper numerical training "L'intelligenza numerica I" developed by Lucangeli et al. [35] and the visuo-spatial psycho-educational training "Conosco le forme" by Lucangeli et al. [36], for 10 weekly sessions lasting thirty minutes each. Both the psychoeducational programmes were developed in order to enrich counting, semantic, lexical, and syntactic numeracy skills and non-verbal functions, respectively. Moreover, the abovementioned training were developed to enrich the metacognitive processes underpinning the math and nonverbal tasks [27].

3. Results

The descriptive statistics for the scales of BIN and CPM [33,34] were observed. Then the linear correlation (Pearson's *r)* was computed between the dimensions investigated before and after the training. As expected, at pre-test and post-test positive linear correlations between visuo-spatial and mathematical abilities were detected. A Manova with mixed design on the scores of the BIN and CPM with the *'training group'* as the between factor, the *'scale'* and the *'time'* as within factors. The multivariate tests were significant for the *'time'* (Wilks' lambda $_{[1;54]}$ = .651, p=.0001), for the *'scales'* (Wilks' lambda $_{[4;\ 51]}$ = .487, p=.0001), for the interaction between *'time'* * *'training group'* (Wilks' lambda $_{[2;\ 54]}$ = .686, p=.0001), between *'scale'* * *'training group'* (Wilks' lambda $_{[8;\ 102]}$ = .700, p=.017) was applied. Then by the application of the Bonferroni Post Hoc Procedure, significant differences between the groups at the post-test were detected (Table 1 summarises these main results).

Table 1. Main results of the Bonferroni Post Hoc Procedure

group		group	Dependent variable
G1	>	G3	General mean at post-test
G1	>	G3	Counting scale at post-test
G2	>	G3	General mean at post-test
G2	>	G3	Lexical scale at post-test Semantic scale at post-test Counting scale at post-test Pre syntax scale at post test
G2	>	G1	CPM at post test
G2	>	G3	CPM at post test

Legend : G1 combined training administered by the curricular teacher; G2 combined training administered by the external trainer; G3 control group

4. Discussion

Applying the concepts of *objectivity* and *intersubjectivity* in the measurement [7], this study attempted to overcome the disagreement among scientists about the measurement methods, typical of the *soft systems* [5,14]. The selected instruments referred to accurate explanatory theories shared among scientists [14,33,37], on which there is the agreement in the educational field [19]. The combination of all these aspects support us in the challenge to apply a "theoretical control over the reproduction of the measured construct" [19, p.54].

Considering the above-mentioned aspects, we believe that the current investigation highlights at least two very relevant outcomes. First, the use of psychoeducational trainings developed by Lucangeli and coll. [35,36,38,39] are effective to enrich mathematical and non-verbal skills promoting the development of numeracy in pre-schoolers. Indeed, compared to the control groups, at post-test both the experimental groups showed better numeracy and visuo-spatial skills, as well as a greater metacognitive competence in carrying out math and non-verbal tasks. Therefore, overall these outcomes support the hypothesis concerning the role played by visuo-spatial processes for the development of numeracy in pre-schoolers [19,40]. Second, the expertise of the trainer seems to play a crucial role in favouring mathematic learning in kindergartners. That is, children trained by an external trainer showing a specific knowledge about the role played by visuo-spatial abilities in promoting math learning outperformed pre-schoolers trained by their curricular teacher [28,41]. Although current findings highlight the relevance of adequate educational materials and the competence of the trainer for the promotion of numeracy learning in kindergarten children, future research has to clarify the long-term effects of the interventions aimed at promoting the development of math skills in pre-schoolers, through measurements conducted at follow-ups.

References

[1] Mari L 2013 A quest for the definition of measurement *Measurement* **46** 2889–95
[2] Maul A, Irribarra D T and Wilson M 2015 On the philosophical foundations of psychological measurement *Measurement*
[3] Borsboom D 2005 *Measuring the mind: Conceptual issues in contemporary psychometrics* (Cambridge University Press)
[4] Finkelstein L, Morawski R Z and Mari L 2005 Logical and philosophical aspects of measurement *Measurement* **38** 257–8
[5] Finkelstein L 2005 Problems of measurement in soft systems *Measurement* **38** 267–74
[6] Sijtsma K 2011 Introduction to the measurement of psychological attributes *Measurement* **44** 1209–19
[7] Mari L 2000 Beyond the representational viewpoint: a new formalization of measurement *Measurement* **27** 71–84
[8] Wilson M 2013 Using the concept of a measurement system to characterize measurement models used in psychometrics *Measurement* **46** 3766–74
[9] Fisher W P J 2000 Objectivity in psychosocial measurement: what, why, how. *J. Outcome Meas.* **4** 527–63
[10] Wilson M 2013 Seeking a Balance Between the Statistical and Scientific Elements in Psychometrics *Psychometrika* **78** 211–36
[11] Stevens S S 1957 On the psychophysical law. *Psychol. Rev.* **64** 153
[12] Messick S 1996 Validity and washback in language testing *Lang. Test.* **13** 241–56
[13] Michell J 2005 The logic of measurement: A realist overview *Measurement* **38** 285–94
[14] Fisher W and Wilson M 2015 Building a Productive Trading Zone in Educational Assessment Research and Practice *Pensam. Educ. Rev. Investig. Educ. Latinoam.* **52** 55–78
[15] Dowker A 1998 Individual differences in normal arithmetical development.
[16] Aunola K, Leskinen E, Lerkkanen M-K and Nurmi J-E 2004 Developmental Dynamics of Math Performance From Preschool to Grade 2. *J. Educ. Psychol.* **96** 699–713
[17] Mammarella I C, Lucangeli D and Cornoldi C 2010 Spatial working memory and arithmetic deficits in children with nonverbal learning difficulties. *J. Learn. Disabil.* **43** 455–68
[18] Passolunghi M C and Costa H M 2014 Working memory and early numeracy training in preschool children *Child Neuropsychol.* 1–18
[19] Pendrill L R and Fisher W P 2015 Counting and quantification: Comparing psychometric and metrological perspectives on visual perceptions of number *Measurement* **71** 46–55
[20] Clements D H 2004 Geometric and spatial thinking in early childhood education *Engaging young children in mathematics: Standards for early childhood mathematics education* ed D H Clements and J Sarama (Lawrence Erlbaum Associates) pp 267–97
[21] Kroesbergen E H, van 't Noordende J E and Kolkman M E 2014 Training working memory in kindergarten children: Effects on working memory and early numeracy *Child Neuropsychol.* **20** 23–37

[22] Crocker N, Riley E P and Mattson S N 2015 Visual-spatial abilities relate to mathematics achievement in children with heavy prenatal alcohol exposure *Neuropsychology* **29** 108–16
[23] Salminen J, Koponen T, Räsänen P and Aro M 2015 Preventive Support for Kindergarteners Most At-Risk for Mathematics Difficulties: Computer-Assisted Intervention *Math. Think. Learn.* **17** 273–95
[24] Räsänen P, Salminen J, Wilson A J, Aunio P and Dehaene S 2009 Computer-assisted intervention for children with low numeracy skills *Cogn. Dev.* **24** 450–72
[25] Kirk R E 1996 Practical Significance: A Concept Whose Time Has Come *Educ. Psychol. Meas.* **56** 746–59
[26] Slavin R E 2008 Perspectives on Evidence-Based Research in Education—What Works? Issues in Synthesizing Educational Program Evaluations *Educ. Res.* **37** 5–14
[27] Ko J and Sammons P 2013 *Effective Teaching: A Review of Research and Evidence.* (ERIC)
[28] Schomaker J and Meeter M 2012 Novelty Enhances Visual Perception *PLoS One* **7** e50599
[29] Mascia M L, Agus M, Fastame M C and Addis A 2016 Enhancement in Mathematical Abilities: A System Approach *Towards a Post-Bertalanffy Systemics SE - 25* Contemporary Systems Thinking ed G Minati, M R Abram and E Pessa (Springer International Publishing) pp 243–9
[30] Rudd L C, Lambert M C, Satterwhite M and Smith C H 2009 Professional Development + Coaching = Enhanced Teaching: Increasing Usage of Math Mediated Language in Preschool Classrooms *Early Child. Educ. J.* **37** 63–9
[31] Brown L and Coles A 2010 Mathematics teacher and mathematics teacher educator change—insight through theoretical perspectives *J. Math. Teach. Educ.* **13** 375–82
[32] Sakonidis C and Potari D 2014 Mathematics teacher educators'/researchers' collaboration with teachers as a context for professional learning *ZDM* **46** 293–304
[33] Molin A, Poli S and Lucangeli D 2007 *Batteria Intelligenza Numerica 4-6 [Battery for Numerical Intelligence]* (Trento: Erickson)
[34] Belacchi C, Scalisi T G, Cannoni E and Cornoldi C 2008 CPM-Coloured Progressive Matrices Standardizzazione Italiana [CPM-Coloured Progressive Matrices Italian Standardization] *Firenze Giunti OS Organ. Spec.*
[35] Lucangeli D, Poli S and Molin A 2003 *L'intelligenza numerica* (Trento: Erickson)
[36] Lucangeli D, Mammarella I C, Todeschini M, Miele G and Cornoldi C 2009 *Conosco le forme* (Firenze: Giunti)
[37] Raven J C 1958 *Guide to using the Coloured Progressive Matrices* (Oxford, England: H. K. Lewis & Co)
[38] Agus M, Fastame M C, Penna M P, Mascia M L, Cocco I and Ruggiu A 2014 EPA-0319 – Combined cognitive enrichment programs of numerical and visual-spatial abilities as a primary prevention method for learning difficulties in mathematics: an explorative study. *Eur. Psychiatry* **29** 1
[39] Agus M, Mascia M L, Fastame M C, Penna M P, Napoleone V, Porru A M and Pessa E 2015 The measurement of treatment effects in mathematical abilities: Comparing trainings in pencil-and-paper and computerised formats *XXI IMEKO World Congress* (IMEKO)
[40] Agus M, Mascia M L, Fastame M C, Melis V, Pilloni M C and Penna M P 2015 The measurement of enhancement in mathematical abilities as a result of joint cognitive trainings in numerical and visualspatial skills: A preliminary study *J. Phys. Conf. Ser.* **588** 012041
[41] Van Oers B and Duijkers D 2013 Teaching in a play-based curriculum: Theory, practice and evidence of developmental education for young children *J. Curric. Stud.* **45** 511–34

Overview of existing algorithms for emotion classification. Uncertainties in evaluations of accuracies.

H Avetisyan, O Bruna, J Holub
Department of Measurement, Faculty of Electrical Engineering, Czech Technical University in Prague, Technicka 2, Prague 6, 166 27, Czech Republic

E-mail: avetihak@fel.cvut.cz

Abstract. A numerous techniques and algorithms are dedicated to extract emotions from input data. In our investigation it was stated that emotion-detection approaches can be classified into 3 following types: Keyword based / lexical-based, learning based, and hybrid. The most commonly used techniques, such as keyword-spotting method, Support Vector Machines, Naïve Bayes Classifier, Hidden Markov Model and hybrid algorithms, have impressive results in this sphere and can reach more than 90% determining accuracy.

1. Introduction
Sentiment analysis (SA) is a method of extracting, analyzing, and determining emotions and sentiments of either text or speech information. It is widely used in e-shop or reviews of services to know customers' opinion about specific products or services taking into account specific criteria. It also can be used in social networks to analyze users' attitude in different situations (sadness, happiness, irony, etc.). Existing many approaches and algorithms trying to deal with SA. All of them have their advantages and disadvantages and different accuracies.
In this paper, information about some of commonly used current algorithms with their related works with obtained results has been collected. It was also paid attention to that fact that most of techniques were not evaluated the same way so problems during comparisons of accuracies of individual techniques may occur.

2. Used algorithms and their assessment

2.1. Keyword based approach

2.1.1. Description
One of the most used technique is keyword based approach. It detects emotions in the text at the basic word level and it is highly reliable mostly for analyzing emotion-bearing words and for simple sentences with clearly expressed emotions.

2.2. Learning based approaches

2.2.1. Descriptions

Learning based approaches are trained from a training set. Then the classifier determines either emotion directly for a word or a more complex structure of classifiers. Several types of learning based algorithms exist.

2.2.2. Support Vector Machines

2.2.2.1. Description

One of learning based approaches is the Support Vector Machines (SVM) binary classification technique. This algorithm uses training examples which contain information about their category. SVM creates a model which linearly classifies input data in already existing categories [1].

2.2.2.2. Related works and results

Table 1. Related works using SVM algorithm and obtained results

Paper name	Year	Algorithm	F-Score, %	Accuracy, %
[2]	2002	SVM with features based on unigrams	N/A	82.9
[3]	2007	SVM with General Inquirer, WordNet Affect, and other features	N/A	73.89
[4]	2009	SVM with Information Gain feature extraction	92.86 (POS) 88.28 (NEG)	91.15
[5]	2011	SVM with a linear kernel	N/A	80.29
[6]	2015	Speaker-dependent SVM with thresholding fusion	N/A	75.67

2.2.3. Naïve Bayes Classifier

2.2.3.1. Description

Another learning based approach is the Naïve Bayes Classifier (NBC). Different types of NBC exist depending on representation of input text. However, multinomial Naïve Bayes model has better results and is more accurate than other event models [7]. In this model, the frequencies of occurrences of specific words have been counted and demonstrated as a vector [8].

2.2.3.2. Related works and results

Table 2. Works related to Naïve Bayes Classifier and obtained results

Paper name	Year	Algorithm	F-Score, %	Accuracy, %
[2]	2002	NBC with features based on unigrams	N/A	78.7
[9]	2012	NBC and Naïve Search	N/A	~85
[10]	2013	NBC with Facebook Query Language query	72	N/A
[11]	2014	ERR-based NBC	84	N/A

[12]	2014	Multinomial NBC with features	91.2	N/A

2.2.4. Hidden Markov Model

2.2.4.1. Description

The Hidden Markov Model (HMM) is also simple and frequently used learning based algorithm for emotion detection. Basically, HMM is a technique which is able to distribute classes over sequence of observations.

2.2.4.2. Related works and results

Table 3. Works related to Hidden Markov Model and obtained results

Paper name	Year	Algorithm	F-Score, %	Accuracy, %
[13]	2003	Continuous HMM	N/A	77.8
[14]	2012	A High-order HMM with Viterbi algorithm	35.3	N/A
[15]	2013	3 states HMM	N/A	82.95

2.3. Hybrid approaches

2.3.1. Description

This type of approaches is combination of already mentioned approach types. It is able to detect emotions based on detected keywords, learned patterns, and other additional information from various dictionaries and thesauri [16].

2.3.2. Related works and results

Table 4. Works related to hybrid approaches and obtained results

Paper name	Year	Algorithm	F-Score, %	Accuracy, %
[17]	2004	Hybrid SVM (PMI/Osgood and Lemmas), 100 folds	N/A	89
[18]	2013	Keyword-spotting method and rule-based method	76.97	N/A
[19]	2013	Multinomial NBC with greedy search	N/A	85
[20]	2014	NBC and SVM using Information Gain and Chi-Square methods	N/A	71
[21]	2015	SVM and CRF with applied rules	N/A	91

3. Assessment of algorithms

3.1. Definition of the problem

It is important to mention that the researchers use different methodologies to assess the algorithms but use the same terminology, which may cause confusion for a reader or other researchers. In Tables 1-4, columns F-score and Accuracy are mentioned. The most common way of evaluating result of the AI is called F-score. Some authors calculated them in a different way than others and named obtained results as "accuracy". This can cause a problem while comparing obtained results of different algorithms and approaches. For example, authors mostly assessed their algorithms using F-score [4], [12], [14], etc.

In some cases authors calculated the percentage of prediction of individual emotions and presented their results [6], sometimes with calculation of average result [13]. Note that first mentioned article calculates 6 types of emotions and the second mentioned article calculates 7 types. This also makes comparison of two algorithms impossible.

Other authors presented average results of their approaches without mentioning any specific information about steps or formulas by which they got their results [17].

These are few examples of fact that the term "Accuracy" is highly figurative and can't be used as a standard term for comparing various algorithms.

3.1.1. Assessment using F-score

The first type is F-score which is calculated using 2 parameters called precision and recall. Precision and recall are measured using 4 parameters:

TP – the number of true positive samples
TN – the number of true negative samples
FP – the number of false positive samples
FN – the number of false negative samples

Precision (equation (3.1)) is the number of true positive samples divided by the number of predicted positive samples.

$$Precision = \frac{TP}{TP + FP} \tag{3.1}$$

Recall (equation (3.2)) is the number of true positive samples divided by the actual positive samples.

$$Recall = \frac{TP}{TP + FN} \tag{3.2}$$

The harmonic mean of precision and recall is called F-score and determined by equation (3.3). The F-score unit is percent.

$$F = \frac{2 \cdot Precision \cdot Recall}{Precision + Recall} \tag{3.3}$$

3.1.2. Assessment using term "Accuracy"

The second type of result is called Accuracy. This parameter presents the level of accuracy of specific algorithms. In this paper, simple average of all emotions accuracies were counted using the equation (3.4):

$$Average = \frac{1}{n} \cdot \sum_{i=1}^{n} Emotion_i \tag{3.4}$$

where n – the number of calculated emotions
$Emotion_i$ – the value of each individual emotion accuracy
The unit of measurement of Accuracy is percent.

3.2. Idea for solution of a problem
To resolve this incompatibility there is an idea to make standardized testing system for emotion classification algorithms and obtain results for them on the same conditions and criteria, and using the same linguistic resources [22]. This way is possible to prevent false accuracies and bring all the results to the same unit of measurement.
At this time there are no specific recommendations about approaches for algorithms testing standardization or make above mentioned results comparable. Nevertheless, this article should help other researchers to take into account above mentioned problem in their future works.

4. Conclusions
In this paper it was observed existing algorithms for emotion classification. Most commonly used and most efficient approaches have been taken into consideration and tables with information about them have been made. However, it has been noticed that between results of different approaches incomparableness can occur. To prevent this problem, it was recommended to make standardized testing system for emotion classification algorithms to be sure that those techniques are obtaining same type of results.

5. References
[1] H. Binali, C. Wu, and V. Potdar, "Computational Approaches for Emotion Detection in Text," *4th IEEE Int. Conf. Digit. Ecosyst. Technol. - Conf. Proc. IEEE-DEST 2010, DEST 2010*, vol. 37, no. 5, pp. 498–527, 2010.
[2] B. Pang, L. Lee, and S. Vaithyanathan, "Thumbs up?: sentiment classification using machine learning techniques," *Proc. Conf. Empir. Methods Nat. Lang. Process.*, pp. 79–86, 2002.
[3] S. Aman and S. Szpakowicz, "Identifying Expressions of Emotion in Text," in *Text, Speech and Dialogue*, vol. 4629, Berlin, Heidelberg: Springer Berlin Heidelberg, 2007, pp. 196–205.
[4] W. Zheng and Q. Ye, "Sentiment classification of Chinese traveler reviews by support vector machine algorithm," *3rd Int. Symp. Intell. Inf. Technol. Appl. IITA 2009*, vol. 3, pp. 335–338, 2009.
[5] R. Burget, J. Karásek, and Z. Smékal, "Recognition of emotions in Czech newspaper headlines," *Radioengineering*, vol. 20, no. 1, pp. 39–47, 2011.
[6] S. Gupta, A. Mehra, and Vinay, "Speech emotion recognition using SVM with thresholding fusion," in *2015 2nd International Conference on Signal Processing and Integrated Networks (SPIN)*, 2015, pp. 570–574.
[7] A. McCallum and K. Nigam, "A Comparison of Event Models for Naive Bayes Text Classification," *AAAI/ICML-98 Work. Learn. Text Categ.*, pp. 41–48, 1998.
[8] K.-M. Schneider, "On Word Frequency Information and Negative Evidence in Naive Bayes Text Classification," *Adv. Nat. Lang. Process.*, pp. 474–485, 2004.
[9] M. M. Itani, R. N. Zantout, L. Hamandi, and I. Elkabani, "Classifying sentiment in arabic social networks: Naïve search versus Naïve bayes," *2012 2nd Int. Conf. Adv. Comput. Tools Eng. Appl.*, pp. 192–197, 2012.
[10] C. Troussas, M. Virvou, K. J. Espinosa, K. Llaguno, and J. Caro, "Sentiment analysis of Facebook statuses using Naive Bayes Classifier for language learning," in *IISA 2013 - 4th International Conference on Information, Intelligence, Systems and Applications*, 2013, pp. 198–205.
[11] S. Shaheen, W. El-Hajj, H. Hajj, and S. Elbassuoni, "Emotion Recognition from Text Based on Automatically Generated Rules," in *2014 IEEE International Conference on Data Mining*

Workshop, 2014, pp. 383–392.
[12] S. Yoshida, J. Kitazono, S. Ozawa, T. Sugawara, T. Haga, and S. Nakamura, "Sentiment analysis for various SNS media using Naïve Bayes classifier and its application to flaming detection," in *2014 IEEE Symposium on Computational Intelligence in Big Data (CIBD)*, 2014, pp. 1–6.
[13] B. Schuller, G. Rigoll, and M. Lang, "Hidden Markov model-based speech emotion recognition," in *2003 IEEE International Conference on Acoustics, Speech, and Signal Processing, 2003. Proceedings. (ICASSP '03).*, 2003, vol. 2, pp. II–1–4.
[14] D. T. Ho and T. H. Cao, "A High-Order Hidden Markov Model for Emotion Detection from Textual Data," in *Knowledge Management and Acquisition for Intelligent Systems*, D. Richards and B. H. Kang, Eds. Berlin, Heidelberg: Springer Berlin Heidelberg, 2012, pp. 94–105.
[15] S. Rustamov, E. Mustafayev, and M. a. Clements, "Sentiment analysis using Neuro-Fuzzy and Hidden Markov models of text," *2013 Proc. IEEE Southeastcon*, pp. 1–6, 2013.
[16] E. C.-C. Kao, C.-C. Liu, T.-H. Yang, C.-T. Hsieh, and V.-W. Soo, "Towards Text-based Emotion Detection A Survey and Possible Improvements," in *2009 International Conference on Information Management and Engineering*, 2009, pp. 70–74.
[17] T. Mullen and N. Collier, "Sentiment Analysis using Support Vector Machines with Diverse Information Sources," *Proc. 2004 Conf. Empir. Methods Nat. Lang. Process. (EMNLP 2004)*, pp. 412–418, 2004.
[18] U. Krcadinac, P. Pasquier, J. Jovanovic, and V. Devedzic, "Synesketch: An Open Source Library for Sentence-Based Emotion Recognition," *IEEE Trans. Affect. Comput.*, vol. 4, no. 3, pp. 312–325, 2013.
[19] N. Chirawichitchai, "Sentiment classification by a hybrid method of greedy search and multinomial naïve bayes algorithm," in *2013 Eleventh International Conference on ICT and Knowledge Engineering*, 2013, pp. 1–4.
[20] X. Sun and C. Li, "Hybrid model based sentiment classification of Chinese micro-blog," in *2014 International Conference on Audio, Language and Image Processing*, 2014, pp. 358–361.
[21] D. S. Nair, J. P. Jayan, Rajeev R.R, and E. Sherly, "Sentiment Analysis of Malayalam film review using machine learning techniques," in *2015 International Conference on Advances in Computing, Communications and Informatics (ICACCI)*, 2015, pp. 2381–2384.
[22] C. Strapparava and A. Valitutti, "WordNet-Affect: an affective extension of WordNet," *Proc. 4th Int. Conf. Lang. Resour. Eval.*, pp. 1083–1086, 2004.

6. Acknowledgements

This work was supported by the Grant Agency of the Czech Technical University in Prague, grant No. SGS16/095/OHK3/1T/13

Journal of Physics: Conference Series **772** (2016) 012040 doi:10.1088/1742-6596/772/1/012040

Cultural capital as a measurand

R Taymanov, K Sapozhnikova

D.I. Mendeleyev Institute for Metrology, 19 Moskovsky pr., St. Petersburg, Russia

E-mail: k.v.s@vniim.ru

Abstract. The necessity for developing metrology due to extension of its application sphere is noted. The efficiency of the metrological approach to measurement of multidimensional quantities in the field of humanities is shown using the development of cultural capital interpreted by L. Harrison. The cultural capital is defined as a measure of the society structure complexity and adaptive capacity.

1. Introduction

Many years ago the need for interaction between society members has resulted in originating the concept of "measurement" and developing metrology. At the end of the 20th century only physical quantities were subjected to measurement, while metrological traceability being provided solely by the comparison with measurement standards or reference materials. At the beginning of the 21st century chemical, biological and other nonphysical quantities, the most of which are of multidimensional (multiparametric) character, were recognized to be measurable and the concept of "reference measurement procedure" was introduced [1].

Metrology is shifting to a higher level, which will enable metrology to be used as an efficient cognition instrument in a wider range of knowledge, including humanities [2-4]. Metrology experts will have to deal with new fields of knowledge studying, in particular, person and society including the processes of their development. At the same time, it will be necessary for experts in the humanitarian sciences to develop understanding of metrology requirements providing recognition of the estimates obtained by experts all over the world.

The most important stage of the multidimensional quantity measurement is the development of a measurement model using an equation, algorithm, or block scheme linking the measurand with its key parameters measured directly.

An analysis of the society development dynamics is of interest as an example of the metrological approach to investigations in the humanitarian field.

Actually, to perform such an analysis is possible to apply human capital ("the quality of the work force") associated with G. Becker, social capital ("the tendency of a society to encourage association of its members") addressed by G. Loury, J. Coleman, F. Fukuyama, and R. Putnam, or financial/resource/property capital studied by A. Smith, K. Marx, T. Piketty, and others.

In particular, in [5] based on statistics data characterizing the dynamics of financial capital development T. Piketty proves the relation between emerging political and economical crises, on one hand, and the concepts of the modern model of market economy, on the other hand. A rate of return on capital is significantly higher for long periods of time than the rate of growth of income and output. This fact leads to high concentration of wealth. If modern economic tendencies continue, by 2050 a thousandth of world population will own (40-60) % of global wealth. T. Piketty considers that the best thing to do is to impose a global tax on capital, although at present "a truly global tax on capital is no doubt a utopian ideal". In our opinion, just in a future, it will be hardly ever probable to impose such a tax without military confrontation.

Investigations of the financial capital dynamics give an opportunity to look at one aspect of future social prospects and reveal some problems due to the capital of such a type. It is also of interest to consider other ways of thinking. The concept of social capital appeared approximately 100 years ago and at present it is characterized by various definitions.

L. Harrison's book [6] has been issued recently in the USA. In this book the author uses a concept of "cultural capital" (CC) that is a multidimensional quantity being formed by "a set of values, beliefs, and attitudes". According to his opinion, the CC drives evolution of society toward the goals of the UN Universal Declaration of Human Rights: "democratic governance, including rule of law; social justice, including education, health care, and opportunity for all; and, elimination of poverty".

In contrast with the social capital, the CC concept has been formulated recently, it is of a complex character and opens an opportunity to take into account a greater number of parameters affecting socio-economic processes. Book [6] aroused much controversy in society. It contains a lot of research information, but metrology experts have many questions with regard to the details of this concept. All in all, these issues inspired the authors of this paper to rely upon this book while considering the features of the metrological approach to studying the society development dynamics.

2. Cultural capital from metrology standpoint

L. Harrison evaluates the CC by 25 parameters that represent values, beliefs, and attitudes dominating in a given society. L. Harrison's parameters are as follows (inverted commas show exactly his own words):

- "worldview" that reflects the perception of "religion", "destiny", "time orientation" (planning the future), "wealth", and "knowledge";
- "values and virtues" that include the attitude to "ethical code", "the lesser virtues" (everyday virtues), and "education";
- "economic behavior" that is defined by the perception of "work/achievement", "frugality", "entrepreneurship", risk, "competition", "innovation", and "advancement";
- "social behavior" that includes the attitude to "rule of law/corruption", "radius of identification and trust", "family" (role of a family in the society), "association" (cooperation), the concept of "individual/the group" (the very role of a person in society), "fertility" as well as features of "authority", "role of elites", "church-state relations", "gender relationships".

A relationship of the CC with the parameters mentioned may be considered as a measurement model.

The conclusions drawn by L. Harrison are of great impression:

- The author proves that growth of the CC value leads to the economic and social development.
- Cultures that contribute to society development or impede it are brought to light.
- Among the main features distinguishing these cultures, religion is definitely mentioned.
- The author stresses that the Protestants, Confucians, Jews, Mormons, etc. are the most successful groups.

However, these conclusions are of a phenomenological character. L. Harrison does not explain the criterion of the selection of parameters. In many cases, the parameters chosen are interrelated. For example, a positive value of "frugality" corresponding to the high level of the CC ("the mother of investment and prosperity") overlaps the parameter of "time orientation" ("future focus…, deferred gratification); "wealth" (wealth is a "product of human creativity") is linked with "work/achievement" ("work leads to wealth"), and so on.

The influence of various parameters on the CC is different, which is noted in [6], but this different significance is not taken into account. For a number of scales high and low limits are not antonyms. Of the 25 parameters only 11 received "strong confirmation" and 3 received "moderately strong confirmation". There is "no significant support" for 2 of them and no data are available for others. The essence of the CC has not been disclosed.

The metrology standpoint requires that a multidimensional measurand is to be defined and a measurand model is to be developed. To provide a model that would represent the "mechanism" of the measurand formation is desirable since it is likely to be the case when only a justified parameter set is taken into consideration. While choosing a parameter set, to apply the analogy between the dynamics of evolution of society as well as evolution of biological and technical systems, can be effective.

A concept of the interaction of a developing system containing a number of units, with its environment, which relies upon the "principle of the maximum of average mutual information" [7], makes it possible to approach the development of the measurement model of a multidimensional measurand. This principle enables the system to adapt adequately to unpredictable changes of the environment, when new threats emerge, and survive. It follows from this concept that 3 fundamental tendencies are inherent in functioning of such a system: growing diversity of the system behavioral reactions, maximization of the behavioural accuracy, and economy of the resource.

However, to achieve the growth of the diversity of the behavioural reactions seems to be possible only by the system complication. This comes from observed complication of multicellular organisms, communities of living creatures, human societies and even technical systems for thousands of years.

In nature, an "evolutionary change is not a continuous thing". Evolution "has indeed shown at least one vector: toward increasing complexity" [8]. As applied to technique, an illustration is a transition to the "Internet of Things" era that is characterized by a fast growth of production and service spectrum due to a significant growth of the number of links between both the animate and inanimate.

The Harrison's conclusion that the CC growth stimulates the evolution progress [6] as well as the results of research [7] enable the authors to define the CC as a quantity that characterizes the complexity and adaptive capacity of society.

Within the frames of this paper, as a measure of complexity, the number of reactions contributed to the system functioning, i.e. adequate ones, that the system itself can form as a response to threats having been arisen, may be proposed.

In its turn, these definitions give an opportunity to analyse the society development dynamics using an approximate evaluation of the CC on the basis of the number and efficiency of connections between the units of society. The units, basing on the totality of connections that have already existed and future connections that will arise in case of a new threat to the society, can form an adequate behavioural reaction neutralizing the threat. In our opinion, one should judge the efficiency of the connections taking into account the expected time delay of the adequate reaction.

Coming back to the CC measurement model being developed as a model of a "mechanism" of the CC formation, it should be noted that while determining its parameters a situation can arise which is typical for measurement of a number of other multidimensional quantities related to the humanitarian field: selection of parameters may be connected with the assumed method of their evaluation.

In particular, intention to be guided by usage of well-known statistical data may result in the fact that originally selected parameters, in their turn, will have to be considered as multidimensional quantities and then this leads to determining their relations with the others, the values of which are linked with the CC indirectly. As a consequence, the obtained parameters of the measurement model are of different significance.

In our opinion, the groundlessness of some part of the 25 parameter set given in [6] is possible to be explained, to a significant extent, by this feature.

However, for a number of other tasks, the list of the original parameters may be useful.

The authors of the present paper propose a version of the CC measurement model as a model of the "mechanism" of the CC formation given below. In this version the CC is determined taking into account [6] by the following parameters that significantly influence the processes of origination and formation of the connections between the society units (the high and low limits of parameters are given in brackets through a slash):

- to characterize the worldview:

  - life credo (active/ passive),
  - need for knowledge (getting knowledge: continuously, in a broad area/ pragmatically, in a narrow area);
- to characterize the ethical values:
  - responsibility (personal/ conferred upon somebody else or collective responsibility),
  - attitude to the rights of other people (respect/ depreciation),
  - thinking style (independent judgements/ adherence to authorities);
- to characterize standards of behavior:
  - propensity to risk (moderate/low),
  - attitude to entrepreneurship (investment in creativity/ rent-seeking),
  - attitude to innovations (interest/ rejection),
  - attitude to laws (reasonable law abiding/ corruption,
  - association (trust, cooperation / mistrust to others),
  - features of authority (responsibility towards society; dispersed authority/ out-of-control, authoritarianism).

In contrast with [6], religion is not in the list of the parameters proposed since not religion but standards established in a number of confessions restrict the participation of many believers in the processes of origination and formation of the connections between the society units.

One should evaluate the parameters mentioned by relative units, i.e., numbers on special scales, and take them into account using the coefficients of significance. The increase in the values of these parameters like in those chosen by L. Harrison is an evidence that the society is approaching to the goals of the UN Universal Declaration of Human Rights and that its ability to withstand threats and adapt to new conditions is improving.

3. System structure, its complexity, and influence of inertance

Growing threats are a stimulus to increase the complexity of a developing system. The main reason for the threat origination is the system itself. The more the number of system units is, the more likely a defect will prevent the system from development.

The simplest multichannel measuring system including no more than a few dozens of channels with sensors as well as a data processing block, as a rule, has a hierarchical structure and usually performs better, having more sensors. An operator's observation together with traditional metrological maintenance may provide the system efficiency. In such a case, the growth of the sensor number increases the complexity of the system keeping a connection type. However, if one means a system including hundreds or thousands of sensors, which is typical for the "Internet of Things", then the traditional methods of the metrological maintenance are inappropriate. Meanwhile, an unnoticed metrological defect in one or two sensors or their connections can cause serious human and material losses. Such a threat leads to complicating measuring systems, supplying connections between channels (between the units included in various channels that play the role of the branches of hierarchical structures), i.e. introducing elements of decentralization, and organizing a metrological self-check with these connections as a basis [9]. The role of the metrological self check is to detect and correct defects, which enables enhancing the efficiency of the measuring system.

Similar tendencies are typical in society. The development of any society is accompanied by increasing the need for available resources. Attempts to obtain them by expansion lead to migration of population, wars, epidemics, technogeneous catastrophes, etc. Long before, the increase in the CC value, in general, was caused by external expansion resulting in the increase in the number of society units and forming the simplest hierarchical structures in the form of states, which facilitated taking new resources and providing self-defense. At first, such structures were built relying upon a common language, but the expansion resulted in forming inhomogeneous society.

Appearance of religions contributed to strengthening the hierarchical structures, including those with people spoken different languages. Confessions specific for various states were formed.

The development of nationalism also contributed to strengthening the hierarchical structures. In nonhomogeneous society with large population, establishment of a national state with a dominant religion resulted in originating the elements of decentralization in the society structure. Local communities were formed basing on different religious and ethical doctrines with the focus on getting the resource for life by using the connections between the units related to various branches of the hierarchical structure. In particular, such doctrines are characteristic for the Protestants, Jews, Confucians, Mormons, Russian old believers, etc.

The development of technologies, enlargement of states, and their aggregation have led to diminishing the benefits from getting new resources in comparison with the resource losses caused by expansion processes. Since the 19th century, the role of the resource connected with intellectual activities has been increasing quickly and in the 21st century it became dominating. As a result, the tendency finally spread out leading to the complicating of society by origination and activation of the connections between the units at the low and middle levels of various branches of the hierarchical system, that is the system decentralized and a net started forming. However, such a process requires significant time.

In techniques, the activation of the connections between system units if such activation is provided in automatic mode can occur for a few seconds or faster. If not, the activation can be performed with the help of a man for a significantly longer time interval.

In structures of living creatures, in particular, in the nervous system, such activation can take several months being a basis for curing, e.g., after stroke. The time needed for activation depends on individual features of a living creature.

In society the activation of connections between units and corresponding opportunity to form an adequate reaction to the threat arisen with the minimum consumption of resources at the highest speed, depends on worldview, ethical values, and standards of behaviour [6]. The parameters characterizing the worldview, values, and behaviour are of inertial nature. They can change gradually as new generations grow up, even in case of abrupt changes of the society governance system.

For example, the society with a strict multi-step hierarchical structure unlike the democratic society, has significantly lesser number of effective links among the units and is more inertial in development of management decisions. Moreover, such society possesses a less adaptation ability to changing conditions of existence.

However, a formal transition to a partially decentralized system with the increased number of links which although being mentioned in laws, actually will not result in a quick growing of the CC. Memory about strict standards of behaviour in recent times will support the tendency of many people to isolate their interests within the frames of their families and friends.

There are many examples of such a kind in history, particularly in the 20th century.

It should be noted that ~~a~~ significant variation of the parameters determining worldview, ethical values, and standards of behaviour even in case of favourable conditions may require many years.

The inertance of formation of new connections between society units should be taken into account analyzing the dynamics of the society development in a future.

The threats the society has to answer in the nearest time are well-known. Among them there are the growth of economic inequality, mass migration, growth of terrorism, increase in the power of energetic facilities accompanied by the risk of damage. However, this list is not exhaustive.

For example, at the beginning of 2016 S. Hawking in one of his interviews said that the main future threats should be associated with the artificial intelligence, robots-soldiers, nuclear weapons, and genetically modified viruses. In the opinion of R. Kurzweil, during the 21st century a new breakthrough in science and technique will result in reaching such a scientific and technical level that would have been reached in 20000 years, if a present rate of development remained unchanged. Hardly ever, new threats are possible to foresee.

The avalanche character of technology development and globalization of links between people have resulted in the situation that for the first time in history, the threat growth pace has become higher than the pace of adaptation to them, meanwhile the range of the threats has become universal.

The idea concerning the dynamics of the society development has to attract the attention of metrology specialists to researches in the humanitarian field. In our view, first of all, new programs of upbringing and education, beginning from kindergartens up to universities, have to be developed and substantiated by scientific investigations with the emphasis placed on formation of independent points of view, interest in new knowledge, and ability for quick perception of changes in priority directions in science, production, and services.

4. Conclusion

In the 21th century one may notice the transition of the society interests to the field of knowledge related to person and society including the processes of their development. While performing investigations in the humanitarian field, the metrological approach became actual.

Using the analysis of the society development dynamics as an example and involving the concept of "cultural capital" interpreted by L. Harrison, the authors of this paper show the features of the metrological approach and appropriateness of applying the analogy between biological, social, and technical evolution.

The definition of the cultural capital is grounded as a quantity that characterizes the complexity and adaptive capacity of society. The dependency of the quantity multidimensional parameter set being measured directly and their coefficients of significance on an intended method of their measuring, is noted.

A criterion revealing the parameters determining a value of the cultural capital and a version of the parameter list is given as applied to the measurement model reflecting the "mechanism" of the cultural capital formation, is proposed. The role of activation inertance with regard to the connections between society units is emphasized for the cultural capital measurements.

Evolution of society is accompanied by the increase in the number of situations that may result in tragic consequences for the mankind. An increased danger is due to the fact that the rate of the growth of cultural capital becomes lower than that of the threats to the society development.

References

[1] *International Vocabulary of Metrology – Basic and General Concepts and Associated Terms — 3rd edn. — 2008 version with minor corrections* BIPM, JCGM 200(2012)
[2] Rossi G B, Berglund B 2011 *Measurement* **44** 815-22
[3]Sapozhnikova K, Chunovkina A, Taymanov R 2013 *Measurement* **50** 390-96
[4] Sapozhnikova K and Taymanov R 2015 *Proc. XXI IMEKO World Congress* (Prague, Czech Republic August 30 - September 4, 2015) 6 p.
[5] Piketty T 2014 *Capital in the Twenty-First Century* (Cambridge: Harvard University Press)
[6] Harrison L 2013 *Jews, Confucians, and Protestants. Cultural Capital and the End of Multiculturalism* (Lanham, MD: Rowman & Littlefield publishers, Inc.)
[7] Golitsyn G A and Petrov V M 2005 *Social and Cultural dynamics: Long-term Trends (Information Approach)*, (Moskva: KomKniga).
[8] Ward P 2001 *Future Evoluton (*New York: Nimes Books, Henry Holt and Company, LLC)
[9] Taymanov R, Sapozhnikova K, Danilova I, Druzhinin I, 2015 *Proc. XXI IMEKO World Congress* (Prague, Czech Republic August 30 - September 4, 2015) 5 p.

Journal of Physics: Conference Series **772** (2016) 012041 doi:10.1088/1742-6596/772/1/012041

Neural network modeling of conditions of destruction of wood plank based on measurements

V Filkin, T Kaverzneva, T Lazovskaya, E Lukinskiy, A Petrov, O Stolyarov and D Tarkhov

Peter the Great St-Petersburg Polytechnic University, 29 Politechnicheskaya Str, 195251 Saint-Petersburg, Russia

E-mail: tatianala@list.ru

Abstract. The paper deals with the possibility of predicting the ultimate load breaking timber sample based on the loading force dependence on the deflection before destruction. Prediction of mechanical properties of wood is handicapped by complex anisotropic structures. The anisotropic nature of the material and, in a great measure, the random nature of wood grain local features defining moment of destruction lead to a significant dependence of the required load on the individual characteristics of a particular bar. The ultimate load is sought as a function of the coefficients of the neural network approximation of the loading force dependence on the deflection. For this purpose, a number of experiments on timber sample loading until the destruction is conducted. Modeling of the conditions of material destruction may provide the required safety control in building industry.

1. Introduction

Analysis of accidents in the industrial sector of human activity makes it possible to classify the three main types of causes of industrial injuries.

- Organizational;
- Technical – bad order of scaffolding, jigs and tools, as well as structural defects of machines, lifting tools, building structures and equipment;
- Psychophysiological and other.

Prevention of occupational injuries has always been, and remains a complex problem that requires special attention of specialists in different fields. The accidents related to human fall on height works, have especially severe consequences. So the security development for such works is especially important.

Falling of workers from a height may happen under different conditions: finding working near open doorways, at the edge of the roof and the absence at the same time (or bad order) the necessary barriers; the use of occasional guard rails and materials for closing of hatches and openings; exploitation of untested or defective scaffolding and cradles; movement of workers from one work site to another along girders, beams and other structures; elements of frail construction and others.

Traditionally, wood is used as the material of the structural elements of guard rails, scaffolding, beams, steps. In addition, the use of wooden structures in the coverage of

Journal of Physics: Conference Series **772** (2016) 012041 doi:10.1088/1742-6596/772/1/012041

modern buildings of public and industrial purposes (sports, exhibition and entertainment halls, markets, warehouses) and apartment buildings (mansard floor), based on technologies combining maximal use of structural prefabricated elements with the convenience their transportation and installation in situ [5].

2. Problem statement

The study of the strength properties of building wood materials [1-4] is handicapped by their complex anisotropic structures. Tensile strength of a wooden beam depends strongly on the type of wood, its manufacturing technology, the specific features of planks (such as knots), etc. At the same time, it is necessary to predict the destruction of such constructions in view of their wide use in the building industry and in the case of historic construction timber assessment [5]. In this regard, interest is a prediction of an ultimate force by its behavior of a sample under load nondestructive.

In the work [6], an effort was made to model the deformation process and the destruction of the wooden sample using the commercial FE software ABAQUS. For this purpose, authors used a general approach combining fracture mechanics approaches within a continuum framework, classical flow theory of plasticity in combination with the Hill criterion, multi-surface plasticity models, and others [6], to construct one single 3D material model taking into account the anisotropy of wood. The results of calculations are compared with experimental data. Finally, the predicted value of critical displacement corresponding to the moment of destruction and experimental value differ in several times. This result suggests that it is necessary to use a much more complicated model for more realistic simulation.

In the paper [7], bending properties of beech wood are predicted using the stress-wave method. The authors developed the linear regression models based on coefficients calculated using load deformation curves, but they did not approve this choice.

The approaches described above have a number of restrictions on the experimental conditions, especially the moisture, and equipment used. We offer the neural network approach for modeling the destruction of conditions of the destruction of wood plank taking into account the different experimentation conditions in situ in the form of additional neural network parameters.

3. Measures

We have performed experiments (three-point bending) for loading the wood planks to the initial stage of destruction inclusive. The softwood beams were selected as the samples. The techniques used in the work can be applied to other species of wood. The sample is placed on a lower fixed support. The load is transmitted via the movable loading rod. The beam deflection is defined as the movement of the load transmitting rod. Figure 1 shows the scheme of loading.

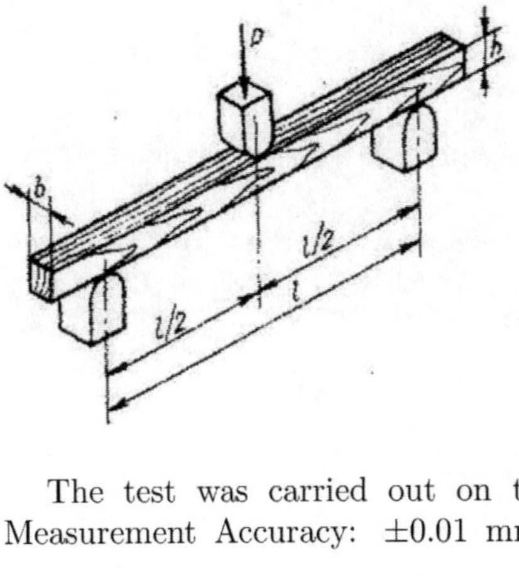

Figure 1. The wood plank loading scheme.

The test was carried out on the double column testing system Instron-5965, Position Measurement Accuracy: ±0.01 mm or 0.05 percent of displacement (whichever is greater),

Load Measurement Accuracy: ± 0.5 percent of reading down. Experiments were conducted with samples of sections of 20×40 , 15×20 and 15×30 mm and loading rates of 10, 50 and 100 mm/min.

Figure 2. 1 – an experimental sample (softwood timber), 2 – lower fixed supports, 3 – the moving loading rod

4. Neural network approach

4.1. network 1

Neural networks applying showed a good effectiveness in data hidden dependencies determinations [6,8]. Deflection-force dependence is looking as a neural network approximation with one neuron in the form

$$F(x, c, a, xc) = c\text{th}(a(x - xc)). \tag{1}$$

The network weights a, c and xc are determined during the minimization of the error functional

$$J = \sum_{i=1}^{m} (F(x_i) - F_i)^2, \tag{2}$$

where x_i is a measured deflection and F_i – corresponding loading force. The error functional optimization is conducted by RProp algorithm or by combination of Particle Swarm and RProp algorithms [9].

Figure 3 shows an example of a plot of neural network approximation and the measurements. We can build a much more precise approximation using a network with two neurons, but it does not allow posterior predicting the load accurately, which leads to the destruction of the wood plank.

4.2. network 2

The neural network approximation coefficients c, a and xc were calculated for the results of all $n = 27$ experiments. So, we have the data array $(F_i^{max}, D_i^{max}, c_i, a_i, xc_i)$, $i = 1, \ldots, 27$,

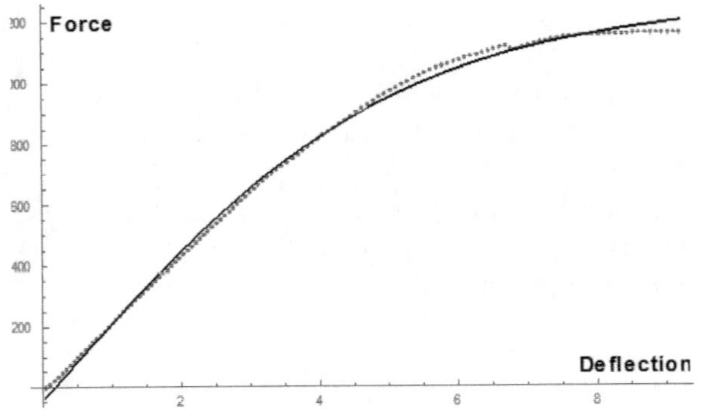

Figure 3. The dependencies of the deflection (mm) against force (N), sample section 15×20 mm, loading rate 50 mm/min. Experimental data – dots, neural network approximation – solid line.

where F_i^{max} is the ultimate force limit power and D_i^{max} is the corresponding deflection in the experiment number i.

Neural network approximations of F^{max} and D^{max}-dependencies were constructed using the neural network of type (1). The explicit connection appears only in the case of force. Figure 3 illustrates the ultimate force limit power dependence on the parameter c.

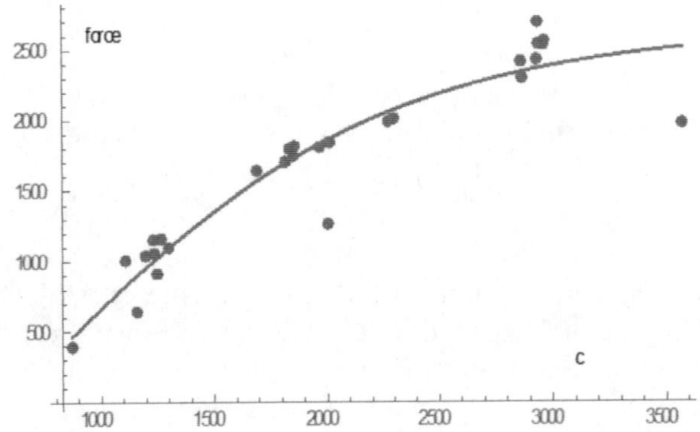

Figure 4. The dependence of the ultimate force at which the wood planks destruct on the parameter c of the neural approximation of deflection-force dependence.

Note that while the sample standard deviation of the ultimate force is 643.3, the deviation of force measurements from neural network model corresponded to the parameter c is equal to 197.1, to the parameter a – 518.4, to the parameter xc – 412.9.

In addition, we have obtained the common neural network model of the dependence of the ultimate force limit power from the three parameters together in the form

$$F(c, a, xc) = 2638.9\text{th}[0.000617c - 0.36] + 5504.7\text{th}[0.39a - 0.08] - 57.2\text{th}[12.9xc + 23.6], \quad (3)$$

where $c \in (500; 3500)$, $a \in (0; 0.3)$, $xc \in (-2.5; 0)$.

The sample deviation of ultimate force from common neural network model is 167.1, so it can be possible to use neural network model instead of mean values of strength and modulus of elasticity from the tests of small tensile samples [5].

5. Prediction

The disadvantage of previous dependencies is that they are based on the whole sample until destruction. In the context of application it is desirable to predict the rupture before it occurred. One approach is to build a dynamic neural network approximation. The corresponding algorithm assumes the dynamic addition of terms in the error functional. Another approach is the approximation of the moving sample.

Let us carry out a new experiment. The neural network model of the load force dependence on deflection is constructed by the first measurements before the destruction of the sample. Thus, we have the parameters c_0 a_0 and xc_0 for a given wood plank.

Then we can determine the critical value of the load force for the sample using the dependence of the ultimate force at which the wood planks destruct on the parameters (the previous section). Figure 5 shows an example of such dependence for the parameter c_0.

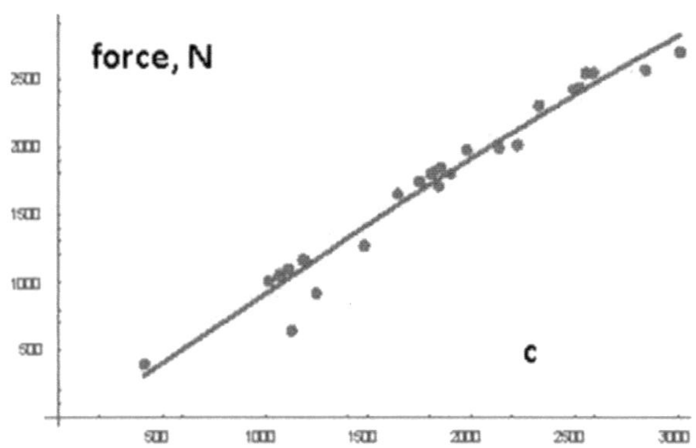

Figure 5. The dependence of the ultimate force at which the wood planks destruct on the parameter c_0 of the neural approximation of deflection-force dependence based on the first fraction of measurements before destruction.

Such dependencies allow the dynamic prediction of timber sample destruction, relearning the specific sample neural network used the newly entered data in the process of loading.

6. Conclusions

Results of this study can find application in the building industry in justifying the choice of bricklayers scaffold constructions, used in conducting different kinds of building and repair work at height or height discontinuity. Work at heights is a dangerous type of work, which is associated with a large number of accidents caused by man fall in spite of the regulated safety measures. Therefore, it is important to ensure the structural solidity of timber floors; the material used in construction must hold certain loads both dynamic and static nature throughout the allotted time or be in time replaced. Modeling of the conditions of material destruction will provide the required safety control. To model behavior of such constructions under dynamic loads can be applied the methods and results of this work, together with the methods described in [8, 9].

Acknowledgments

The work is supported by the Russian Foundation for Basic Research (Grant no. 14-01-00660).

References

[1] Hoadley RB 1980 *Understanding Wood: A Craftmans Guide to Wood Technology* (Newtown, Connecticut: The Taunton Press) 109-135

[2] Yoshihara H 2007 Simple estimation of critical stress intensity factors of wood by tests with double cantilever beam and three-point end-notched flexure *Holzforschung* **61** 182189,

[3] Guinea GV, Planas J, Elices M 1992 Measurement of the fracture energy using three-point bend tests: Part 1Influence of experimental procedures *Materials and Structure* **25(4)** 212-218

[4] Guinea GV, Planas J, Elices M 1992 Measurement of the fracture energy using three-point bend tests: Part 2 Influence of bulk energy dissipation *Materials and Structure* **25(5)** 305-312

[5] Kloiber M, Drdacky M, Machado J S, Piazza M, Yamaguchi N 2015 Prediction of mechanical properties by means of semi-destructive methods: A review *Construction and Building Materials* **101** 12151234

[6] Sandhaas C, van de Kuilen J W G 2013 Material model for wood *HERON* **58 (2/3)** 171-191

Journal of Physics: Conference Series **772** (2016) 012041 doi:10.1088/1742-6596/772/1/012041

[7] Guntekin E, Ozkan S, Yilmaz T 2014 Predictionof bending properties for beech lumber using stress wave method *Maderas. Ciencia y tecnologia* **16(1)** 93-98

[8] Tarkhov D, Vasilyev A 2005 New neural network technique to the numerical solution of mathematical physics problems. I: Simple problems *Optical Memory and Neural Networks (Information Optics)* **14** 59-72

[9] Tarkhov D, Vasilyev A 2005 New neural network technique to the numerical solution of mathematical physics problems. II: Complicated and nonstandard problems *Optical Memory and Neural Networks (Information Optics)* **14** 97-122

[10] Riedmiller M and Braun H 1993 *A direct adaptive method for faster backpropagation learning: The Rprop algorithm. Proceedings of the IEEE International Conference on Neural Networks* (IEEE Press) 586-591

[11] Haykin S 1999 Neural Networks. A Comprehensive Foundation (Prentice Hall.) p 823

Probabilistic reasoning under time pressure: an assessment in Italian, Spanish and English psychology undergraduates

M Agus[1], P K Hitchcott[2], M P Penna[1], M Peró-Cebollero[3], J Guàrdia-Olmos[3]
[1]Department of Pedagogy, Psychology, Philosophy, Faculty of Humanistic Studies, University of Cagliari, Via Is Mirrionis, 1, 09123 Cagliari, Italy
[2]Department of Psychology, Southampton Solent University, East Park Terrace, Southampton, SO14 0YN, UK
[3]Department of Methodology of the Behavioural Sciences, Faculty of Psychology, University of Barcelona, Pg. Vall d'Hebron, 171, 08035 Barcelona, Spain
E-mail: mirian.agus@unica.it

Abstract. Many studies have investigated the features of probabilistic reasoning developed in relation to different formats of problem presentation, showing that it is affected by various individual and contextual factors. Incomplete understanding of the identity and role of these factors may explain the inconsistent evidence concerning the effect of problem presentation format. Thus, superior performance has sometimes been observed for graphically, rather than verbally, presented problems. The present study was undertaken to address this issue. Psychology undergraduates without any statistical expertise ($N = 173$ in Italy; $N = 118$ in Spain; $N = 55$ in England) were administered statistical problems in two formats (verbal-numerical and graphical-pictorial) under a condition of time pressure. Students also completed additional measures indexing several potentially relevant individual dimensions (statistical ability, statistical anxiety, attitudes towards statistics and confidence). Interestingly, a facilitatory effect of graphical presentation was observed in the Italian and Spanish samples but not in the English one. Significantly, the individual dimensions predicting statistical performance also differed between the samples, highlighting a different role of confidence. Hence, these findings confirm previous observations concerning problem presentation format while simultaneously highlighting the importance of individual dimensions.

1. Introduction

Difficulties encountered by psychology undergraduates studying statistics have been documented [1,2]. The study of probabilistic reasoning provides an empirical basis for the development of effective interventions for improving statistical performance [3,4]. There is, however, ongoing debate around the factors that enhance vs impede the probabilistic reasoning is in the literature [5,6]. Some studies have highlighted that specific features of the problems (e.g., the use of graphical-pictorial illustrations) can enhance performance [5,7,8], while other studies have shown that mild time pressure also increases task accuracy [9,10]. Some progress has been made through the recognition that such factors may be either contextual (i.e., format of problem presentation, time pressure) or individual (i.e., abilities, anxiety, attitudes, confidence) [7,9] in nature. Previous evidence suggests a better understanding of the role of individual factors may resolve the effect of problem presentation format [9,10]. The present study was undertaken for this purpose.

2. Aim and method

This investigation assessed the accuracy of probabilistic reasoning under a time pressure condition. In similar previous studies we have assessed student performance with and without time pressure and observed that the presence of time pressure does not increase anxiety, but instead enhances performance in Italian undergraduates [9–11]. We matched the performances of the same student through the use of similar problems administered in two formats (verbal-numerical – N – and graphical- pictorial – G), attempting to detect the potential effect of graphical facilitation. Three samples (Italian, Spanish and English) were tested, in part, to establish the generality of the findings. All students were psychology undergraduates without prior statistical training. A key stimulus for the present study was the observation that the metacognitive dimension of confidence affected probabilistic reasoning performance [9]. Therefore, in order to confirm and extend this observation all students also completed measures assessing various individual characteristics (visuo-spatial and numerical abilities, attitudes towards statistics, statistical anxiety and confidence in the correctness of response). We hypothesized that (1) an effect of graphical facilitation would be observed in all three samples; and (2) that confidence would affect performance regardless of problem presentation format.

2.1. Participants, instrument and procedure

We studied 346 undergraduates, enrolled in the first year of degree courses in Psychology, in three countries: Italy (University of "Sapienza", N = 173; female 60.1%), Spain (University of Barcelona, N = 118; female 72%), England (Solent University, N = 55; female 89.1%).
All participants were recruited via opportunity sampling. The mean age of the participants was in the Spanish sample 20.9 ± 6.4 years old (mean ± standard deviation), in the English sample m= 20.4 ± 5.4 years old, in the Italian sample m= 19.6 ± 1.6 years old. Participants were tested in a single session in large groups; all questionnaires were completed in pencil-and-paper format. All measures were translated into participants' native language; all responses were recorded in the participants' native language. We administered a specific instrument to inquire the probabilistic reasoning in two formats (N and G) [11,12], devised in the previous phases of our investigation. For each problem, participants were required to report their level of confidence in the accuracy of response. To evaluate participants' numerical and visuo-spatial abilities, we administered two scales of the Primary Mental Abilities Questionnaire (PMA) [13]. The Survey of Attitudes Towards Statistics (SATS-28) [14] and the Statistical Anxiety Scale (SAS) [15], were used to measure the participant attitudes and anxiety related to statistics; both these questionnaires demonstrated the cross-country validity between Italian and Spanish samples [16,17]. The descriptive statistics of the samples are showed in the Table 1.
To evaluate the role of the study variables in predicting performance (in the N and G formats, respectively), as in our previous work [9], two Multivariate Linear Regressions were applied, independently for each format and for each sample. Subsequently, a mixed design Ancova was carried out (separately for each sample), using as repeated measures the performances in two formats (N and G) [18].

3. Results

We calculated the descriptive statistics and the linear Pearson's r correlations between the variables. Consequently, to explore the effect of the variables in probabilistic reasoning, two Multivariate Linear Regressions (by using the Forward Method) were applied, by means of the number of correct responses (firstly in the N format as the criterion, and then, subsequently, the number of correct responses in the G format). The predictors were students' abilities, attitudes, anxiety and confidence (Table 2). The most significant predictor of performance in the Italian and Spanish samples was the confidence about the correctness of one's own responses in both formats, differently from the English students, which show a significant effect of attitudes towards statistics and of visuo-spatial ability.

Table 1 – Descriptive statistics in each sample

	Spanish sample		English sample		Italian sample^	
N	118		55		173	
Percentage of women	72		89.1		60.1	
Age Mean	20.920		20.420		19.690	
Age Standard deviation	6.451		5.415		1.638	
Lowest age	17		17		18	
Highest age	54		53		36	
University affiliation	Barcelona		Southampton		Rome	
	Spanish sample		**British sample**		**Italian sample ^**	
	M	**SD**	**M**	**SD**	**M**	**SD**
PMA VISUO-SPATIAL SCALE	24.805	11.410	20.314	13.123	23.595	11.695
PMA NUMERICAL SCALE	15.871	5.740	12.907	5.864	18.500	6.173
SATS AFFECT	21.347	4.866	19.723	4.689	19.432	5.336
SATS COMPETENCE	26.898	5.055	25.723	5.261	25.845	5.894
SATS VALUE	46.550	8.718	43.425	8.334	44.214	9.167
SATS DIFFICULTY	25.474	5.447	26.425	5.999	25.880	6.164
SAS EXAMINATION	32.855	6.244	31.028	8.392	28.542	9.714
SAS INTERPRETATION	19.711	6.461	21.117	6.830	15.505	6.572
SAS HELP	18.771	8.086	17.882	7.642	16.226	8.074
FN CORRECT RESPONSES	1.593	1.248	1.355	1.131	2.277	1.472
FG CORRECT RESPONSES	2.196	1.240	1.409	1.106	2.572	1.435
FN CONFIDENCE	3.216	.944	2.899	.751	3.248	.962
FG CONFIDENCE	3.542	1.000	3.004	.835	3.530	.893

^ Data presented in Agus et al. [9]

Then we applied a Mixed Design Ancova, wherein the repeated measures were the performances in two formats (N and G), and the covariates were identified among the significant predictors recognized in the former regressions.

In the Spanish sample significant effects of Format [F= 8.204, df= 1;111, p=.005, η^2= .069, Observed Power=.810] and of covariates were detected [Format*Sats_Affect, F= 5.560, df= 1;111, p=.020, η^2= .048, Observed Power=.647; Format*Confidence_G, F= 4.104, df= 1;111, p=.045, η^2= .036, Observed Power=.519]. Indeed, for Spanish undergraduates we observed a higher number of correct responses in G format (m=2.217, sd=1.240) than in the N format (m=1.617, sd=1.246), highlighting an effect of graphical facilitation (Table 3).

Conversely, in the English sample no format effect and no effect of covariates were identified; it is interesting to observe that in this sample the number of correct responses is insignificantly higher in the N format (m=1.545, sd=1.148), than in the G format (m=1.393, sd=1.116) (Table 3).

A significant difference in the means of the correct responses in the N and G formats was found in the Italian students, as yet observed in our previous research [9]. They showed a significant effect of Format [F=8.736, df= 1;154, p=.004, η^2= .054, Observed Power=.836] and of Format*Confidence_N [F=9.513, df= 1;154, p=.002, η^2= .058, Observed Power=.865]. Indeed, they registered a better performance in the G format (m =2.791, sd =1.252) than in the N format (m=2.474, sd=1.408) (Table 3).

In summary, under a time pressure condition graphical facilitation was observed in Italian and Spanish samples, but not in the English sample. However, the computed effect sizes call for a degree of caution in interpreting certain findings (Figure 1).

Journal of Physics: Conference Series **772** (2016) 012042 doi:10.1088/1742-6596/772/1/012042

Table 2 – Multivariate Linear Regressions (Forward Method). Criterion variable: number of correct responses in N format and in G format

	R_2	SE	Variables	β	t	p
Criterion variable: number of correct responses in N format						
Spanish sample	.178	1.142	MEAN_CONFIDENCE_N	.258	2.812	.006**
			SATS_AFFECT	.252	2.744	.007**
English sample	.481	.949	PMA_VISUO-SPATIAL	.522	3.073	.007**
			SATS_COMPETENCE	.440	2.592	.018*
Italian sample ^	.315	1.243	PMA_NUMERICAL	.184	2.222	.028*
			MEAN_CONFIDENCE_N	.481	5.795	.0001**
Criterion variable: number of correct responses in G format						
Spanish sample	.129	1.157	MEAN_CONFIDENCE_G	.359	4.089	.0001**
English sample	.394	.846	SATS_DIFFICULTY	.627	3.780	.001**
Italian sample ^	.218	1.119	MEAN_CONFIDENCE_G	.467	5.464	.0001**

Note: * $p<.05$; ** $p<.001$ ^ Data presented in Agus et al. [9]

Table 3 – Results of mixed design Ancova

Sample	Source	Wilks' Lambda	df	F	p	η^2	N FORMAT M (SD)	G FORMAT M (SD)
Spanish sample	FORMAT	.931**	1	8.204	.005**	.069		
	FORMAT * SATS AFFECT	.952*	1	5.560	.020*	.048		
	FORMAT* CONFIDENCE N	.972	1	3.153	.079	.028	1.617 (1.246)	2.217 (1.240)
	FORMAT*CONFIDENCE G	.964*	1	4.104	.045*	.036		
	ERRORS		111					
English sample	FORMAT	1.000	1	.000	.998	.000		
	FORMAT * VISUO-SPATIAL PMA	.999	1	.017	.897	.001		
	FORMAT * SATS COMPETENCE	1.000	1	.009	.924	.000	1.545 (1.148)	1.393 (1.116)
	FORMAT * SATS DIFFICULTY	1.000	1	.001	.982	.000		
	ERRORS		29					
Italian sample ^	FORMAT	.946**	1	8.736	.004**	.054		
	FORMAT * NUMERICAL PMA	.979	1	3.241	.074	.021		
	FORMAT* CONFIDENCE N	.942**	1	9.513	.002**	.058	2.474 (1.408)	2.791 (1.252)
	FORMAT*CONFIDENCE G	.981	1	2.987	.086	.019		
	ERRORS		154					

Note: M = mean; SD= standard deviation; **p < .01 ^Data presented in Agus et al. [9]

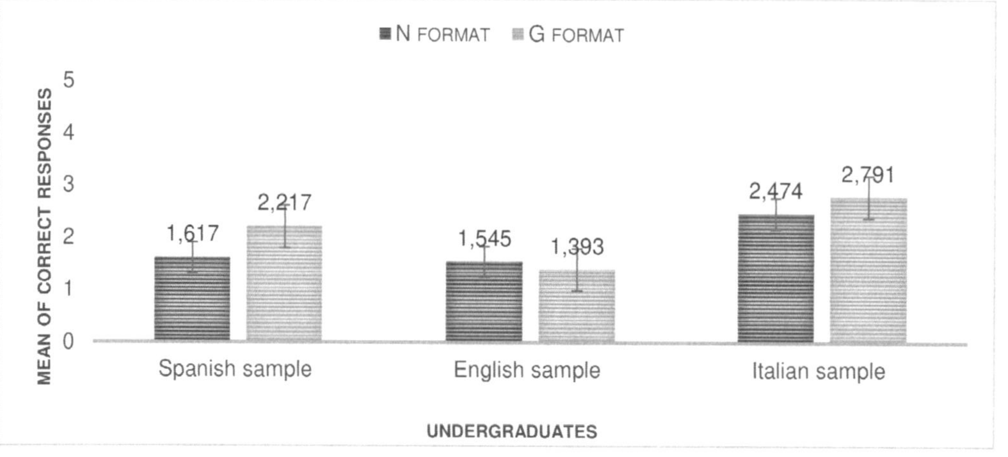

Figure 1. Means of correct responses in N and G formats in the samples.

4. Discussion

In broad terms the present data confirm and extend previous findings that have pointed to similarities between Italian and Spanish students [9]. Thus, in these two samples performance was affected by the confidence (related to the metacognitive dimension) in verbal-numerical and graphical formats and graphical facilitation effect was observed [9]. In contrast English undergraduates showed different characteristics in that their performances were affected by attitudes and abilities and not by the confidence [9]; moreover, they did not show graphical facilitation. These dissimilar outcomes might be related to the presence of effective differences among these populations, highlighted in the scores obtained in our assessments in the PMA and also in the PISA assessments [19] (i.e., the mean score of performance in mathematics is similar in Italy and Spain, but different in UK). In addition, differences in the pre-university teaching of probabilistic reasoning in these countries, including the manner in which probabilistic data are presented in text books (*e.g.,* the use of tree diagrams were used) could be a factor [20–24]

Despite the interesting pattern of findings, the results need to be approached with some degree of caution because of the low number of English students that participated in our research. Nevertheless, these preliminary findings suggest that the investigation of the probabilistic reasoning is a complex matter, having specific and different characteristics compared to the mathematical performance. Previous works highlighted that probabilistic reasoning is influenced by a wide array of interacting factors concerning the individual and the context. The inconsistent findings about the effect of graphical facilitation may be because the importance of some individual and contextual factors has been overlooked. In this investigation, the role of individual factors appears crucial; then, it seems reductive to talk about the effect of graphical facilitation in probabilistic reasoning related only to the presentation of the graphical devices, but it might be interesting to deepen the interactions among some individual factors (i.e., abilities, metacognition, anxiety and attitudes) [9,11]. Such evidences looks essential because our participants did not have any statistical education. Then, as consequence, in order to improve the probabilistic reasoning performance, it might be useful to apply the evaluation of individual specificities to support an adequate presentation of this type of problems.

References

[1] Guàrdia-Olmos J, Freixa-Blanxart M, Peró-Cebollero M, Turbany J, Cosculluela A, Barrios M and Rifà X 2006 Factors related to the academic performance of students in the statistics course in psychology *Qual. Quant.* **40** 661–74

[2] Chiesi F and Primi C 2010 Cognitive and non-cognitive factors related to students' statistics achievement *Stat. Educ. Res. J.* **9** 6–26

[3] Sirota M, Kostovičová L and Vallée-Tourangeau F 2015 Now you Bayes, now you don't: effects of set-problem and frequency-format mental representations on statistical reasoning *Psychon. Bull. Rev.* 1–9
[4] Sirota M, Kostovičová L and Juanchich M 2014 The effect of iconicity of visual displays on statistical reasoning: evidence in favor of the null hypothesis. *Psychon. Bull. Rev.* **21** 961–8
[5] Brase G L and Hill W T 2015 Good fences make for good neighbors but bad science: a review of what improves Bayesian reasoning and why *Front. Psychol.* **6** 340
[6] Knauff M and Johnson-Laird P N 2002 Visual imagery can impede reasoning *Mem. Cognit.* **30** 363–71
[7] Mandel D R and Navarrete G 2015 Introduction to Improving Bayesian Reasoning: What Works and Why? *Front. Psychol.* **6**
[8] Garcia-Retamero R and Hoffrage U 2013 Visual representation of statistical information improves diagnostic inferences in doctors and their patients. *Soc. Sci. Med.* **83** 27–33
[9] Agus M, Peró-Cebollero M, Penna M P, Guàrdia Olmos J and Guàrdia-Olmos J 2015 Comparing Psychology Undergraduates' Performance in Probabilistic Reasoning under Verbal-Numerical and Graphical-Pictorial Problem Presentation Format: What is the Role of Individual and Contextual Dimensions? *Eurasia J. Math. Sci. Technol. Educ.* **11** 735–50
[10] Agus M, Penna M P, Peró-Cebollero M, Guàrdia-Olmos J and Pessa E 2016 Investigating the probabilistic reasoning in verbal–numerical and graphical–pictorial formats in relation to cognitive and non-cognitive dimensions: The proposal of a model *Pers. Individ. Dif.* **94** 44–53
[11] Agus M, Penna M P, Peró-Cebollero M and Guàrdia-Olmos J 2016 Assessing Probabilistic Reasoning in Verbal-Numerical and Graphical-Pictorial Formats: An Evaluation of the Psychometric Properties of an Instrument. *Eurasia J. Math. Sci. Technol. Educ.* **12** 2013–38
[12] Agus M, Peró-Cebollero M, Penna M P and Guàrdia-Olmos J 2015 Towards the development of problems comparing verbal-numerical and graphical formats in statistical reasoning *Qual. Quant.* **49** 691–709
[13] Thurstone L L 1948 *Primary Mental Abilities* (Chicago, IL: University Of Chicago Press)
[14] Dauphinee T L, Schau C and Stevens J J 1997 Survey of attitudes toward statistics: Factor structure and factorial invariance for women and men *Struct. Equ. Model. A Multidiscip. J.* **4** 129–41
[15] Vigil-Colet A, Lorenzo-Seva U and Condon L 2008 Development and validation of the statistical anxiety scale. *Psicothema* **20** 174–80
[16] Chiesi F, Primi C and Carmona J 2011 Measuring Statistics Anxiety. Cross-Country Validity of the Statistical Anxiety Scale (SAS) *J. Psychoeduc. Assess.* **29** 559–69
[17] Chiesi F and Primi C 2009 Assessing statistics attitudes among college students: Psychometric properties of the Italian version of the Survey of Attitudes toward Statistics (SATS) *Learn. Individ. Differ.* **19** 309–13
[18] Hair J F, Tatham R L, Anderson R E and Black W 2006 *Multivariate data analysis* vol 6 (Pearson Prentice Hall Upper Saddle River, NJ)
[19] Binder K, Krauss S and Bruckmaier G 2015 Effects of visualizing statistical information – An empirical study on tree diagrams and 2 x 2 tables *Front. Psychol.* **6**
[20] Phillips J L 1999 *How to think about statistics* (New York: W. H. Freeman)
[21] Tout D and Gal I 2015 Perspectives on numeracy: reflections from international assessments *ZDM* **47** 691–706
[22] Gal I 2005 Towards "probability literacy" for all citizens: Building blocks and instructional dilemmas *Exploring Probability in School* ed G A Jones (Springer) pp 39–63
[23] Agasisti T and Cordero-Ferrera J M 2013 Educational disparities across regions: A multilevel analysis for Italy and Spain *J. Policy Model.* **35** 1079–102
[24] Agasisti T and Pérez-Esparrells C 2010 Comparing efficiency in a cross-country perspective: the case of Italian and Spanish state universities *High. Educ.* **59** 85–103

The group expert evaluation of the metrological assurance of electric power measurements

O M Velychko[1], S R Karpenko[2]
[1]Director of the Institute, [2]Head of laboratory
[1,2]State Enterprise "Ukrmetrteststandard", 4, Metrologichna Str., 03143, Kyiv, Ukraine

E-mail: velychko@ukrcsm.kiev.ua

Abstract. The results of group expert evaluation of the metrological assurance of electric power measurements are considered. Evaluated competence of experts on metrology was taken into account during group expert evaluation. Special software and universal software are used for processing received expert data.

1. Introduction

The aim of group expert evaluation in metrology is to assess the quality of a metrological work and specialists on metrology. This aim is implemented by the method of group expert evaluation, the essence of which is to set out a quality level on the basis of the common criteria for quality evaluation and expert questionnaires (prepared for the particular measurements) is determined by the total level of quality.

Metrological assurance (MA) of electric power measurements should be considered in two approaches: traditional approach (verification and calibration of equipment with the definition of the metrological characteristics) [1] and not traditional approach (group expert evaluation of state of MA). The group expert evaluation (GEE) is widely used in various fields. They are intended to resolve problematic issues concerning certain activities to find solutions (or ways of solving them). For this is expedient to consider the opinions of qualified experts with special skills or knowledge in particular field [2-3]. Considering the practical competence of each expert is involved for GEE taking into account their objective professional data allows increasing the reliability and accuracy of such GEEs.

MA is the establishment and application of metrological rules and regulations also the development, production and application of technical means needed to achieve the necessary unity and accuracy of certain measurements [4-7]. Authentic knowledge of the real status of MA of certain physical quantity measurements is very important. GEE involving experts on metrology (i.e., highly qualified metrology specialists can be one of the useful means solution of noted issue.

2. National standard base of electrical power measurements

The head of the national standard base of electrical power measurements (EPM) is National Standard of the unit of electric power and power factor (DETU 08-08-02). Secondary Standard of the unit of electric power at industrial frequency range (VETU 08-08-01-08) and Secondary Standard of the unit of electric power at extended frequency range (VETU 08-08-02-10) are created, keeping and operated in State Enterprise "Ukrmetrteststandard".

Transfer of the unit of electric power is going by the state verification scheme in accordance with national standard DSTU 4116. Each year about 250 working standards are verified and calibrated by

Journal of Physics: Conference Series **772** (2016) 012043 doi:10.1088/1742-6596/772/1/012043

using National Standard and about 320 working standards by using Secondary Standards. Verified working standards of the unit of electric power are widely used for verification of the most common measuring instruments (MI) in the energy sector – meters of active and reactive energy.

The evaluation of state of MA of EPM on national level is of extreme importance. An important issue for calibration MI of electric power is provision metrological traceability (MT) to National Standard. State Enterprise "Ukrmetrteststandard" has internationally recognized calibration and measurement capabilities (CMC) for calibration MI of electric power [8]. This CMC on measurement of electric power were received by positive results of international comparisons of National Standards for projects of European and Asian-European regional metrology organizations (projects EURAMET.EM-K5.1 and COOMET.EM-S2) [9].

3. Main results of group expert evaluation

GEE of MA conducted by the methodology which described in [10]. For GEE involved a group of 26 experts in field of metrology whose competence were previously estimated Table 1. Statistical data a group of 26 experts in field of metrology is shown in Figure 1. 16 experts (61.5 %) are from national state-owned enterprises of technical regulation, and 10 experts (38.5 %) are from the enterprises of national energy sector from 26 involved experts. Special software (Expertise CE 1.0) [10] and universal software (Microsoft Excel 2010) were used to process the obtained data estimates.

The view of these software windows are shown in Figures 2–3 with evaluated results.

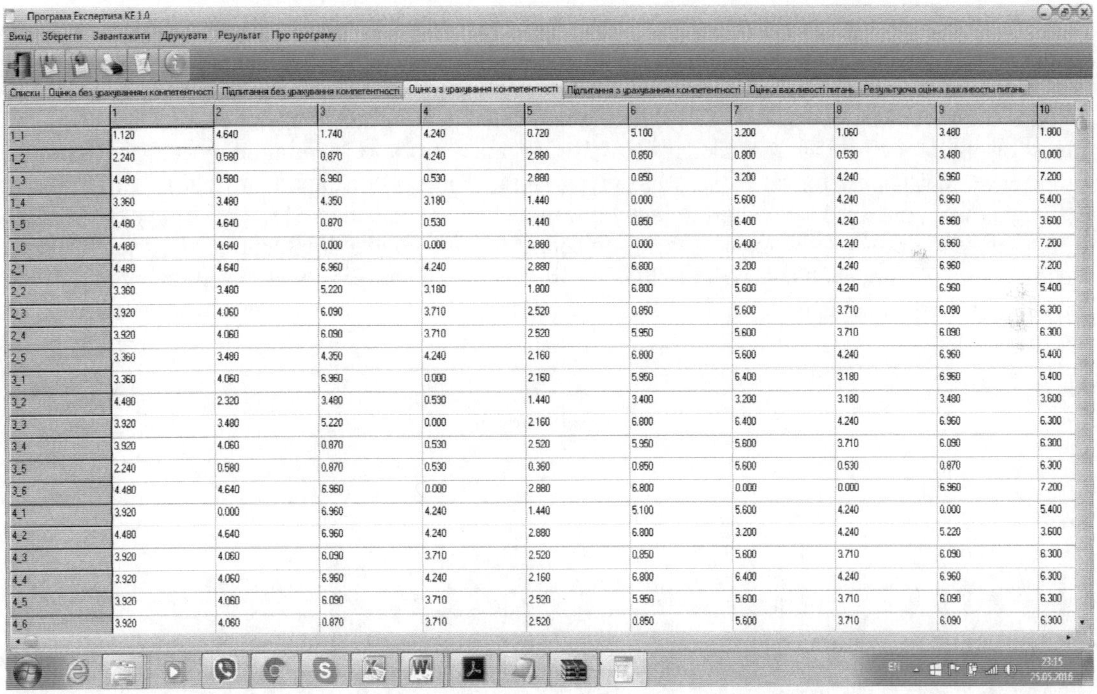

	1	2	3	4	5	6	7	8	9	10
1_1	1.120	4.640	1.740	4.240	0.720	5.100	3.200	1.060	3.480	1.800
1_2	2.240	0.580	0.870	4.240	2.880	0.850	0.800	0.530	3.480	0.000
1_3	4.480	0.580	6.960	0.530	2.880	0.850	3.200	4.240	6.960	7.200
1_4	3.360	3.480	4.350	3.180	1.440	0.000	5.600	4.240	6.960	5.400
1_5	4.480	4.640	0.870	0.530	1.440	0.850	6.400	4.240	6.960	3.600
1_6	4.480	4.640	0.000	0.000	2.880	0.000	6.400	4.240	6.960	7.200
2_1	4.480	4.640	6.960	4.240	2.880	6.800	3.200	4.240	6.960	7.200
2_2	3.360	3.480	5.220	3.180	1.800	6.800	5.600	4.240	6.960	5.400
2_3	3.920	4.060	6.090	3.710	2.520	0.850	5.600	3.710	6.090	6.300
2_4	3.920	4.060	6.090	3.710	2.520	5.950	5.600	3.710	6.090	6.300
2_5	3.360	3.480	4.350	4.240	2.160	6.800	5.600	4.240	6.960	5.400
3_1	3.360	4.060	6.960	0.000	2.160	5.950	6.400	3.180	6.960	5.400
3_2	4.480	2.320	3.480	0.530	1.440	3.400	3.200	3.180	3.480	3.600
3_3	3.920	3.480	5.220	0.000	2.160	6.800	6.400	4.240	6.960	6.300
3_4	3.920	4.060	0.870	0.530	2.520	5.950	5.600	3.710	6.090	6.300
3_5	2.240	0.580	0.870	0.530	0.360	0.850	5.600	0.530	0.870	6.300
3_6	4.480	4.640	6.960	0.000	2.880	6.800	0.000	0.000	6.960	7.200
4_1	3.920	0.000	6.960	4.240	1.440	5.100	5.600	4.240	0.000	5.400
4_2	4.480	4.640	6.960	4.240	2.880	6.800	3.200	4.240	5.220	3.600
4_3	3.920	4.060	6.090	3.710	2.520	0.850	5.600	3.710	6.090	6.300
4_4	3.920	4.060	6.960	4.240	2.160	6.800	6.400	4.240	6.960	6.300
4_5	3.920	4.060	6.090	3.710	2.520	5.950	5.600	3.710	6.090	6.300
4_6	3.920	4.060	0.870	3.710	2.520	0.850	5.600	3.710	6.090	6.300

Figure 1. Statistical data a group of 26 experts in field of metrology (horizontal - expert number; vertical – sub-question number)

Journal of Physics: Conference Series **772** (2016) 012043 doi:10.1088/1742-6596/772/1/012043

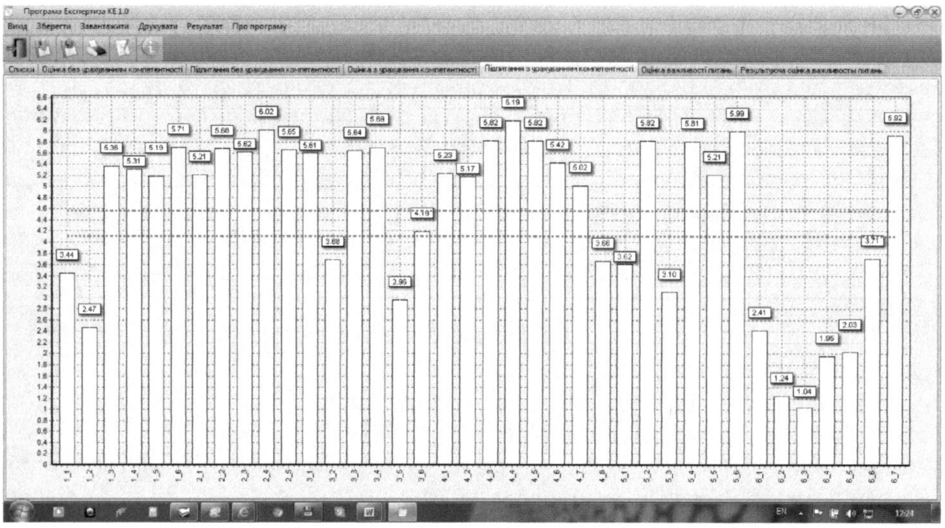

Figure 2. Evaluated average grades by using special software (Expertise CE 1.0) taking into account the competence of experts

Evaluation was done for 6 problematic questions (X1–X6) which contain 38 sub-questions, taking into account the established grade evaluations. Reference values of expert evaluations are shown as dashed lines on Figures 2–3. Reference value for evaluated average grade is 4.70. It was calculated by using universal software (Microsoft Excel 2010) taking into account the competence of experts.

Values of used competence coefficients k_k are in range from 0 – minimum to 1 –maximum. Competence coefficients for all experts are given in the Table 1 for expert in field of EPM. These coefficients were obtained by using the methodology which described in [11, 12]. Also experts were asked to make their own assessment of their competence during conducting mentioned questionnaire.

19 experts from 26 involved experts in field of metrology (73 %) have overestimated their competence according to the results as compared with the specific objective estimates and including all 10 least competent experts.

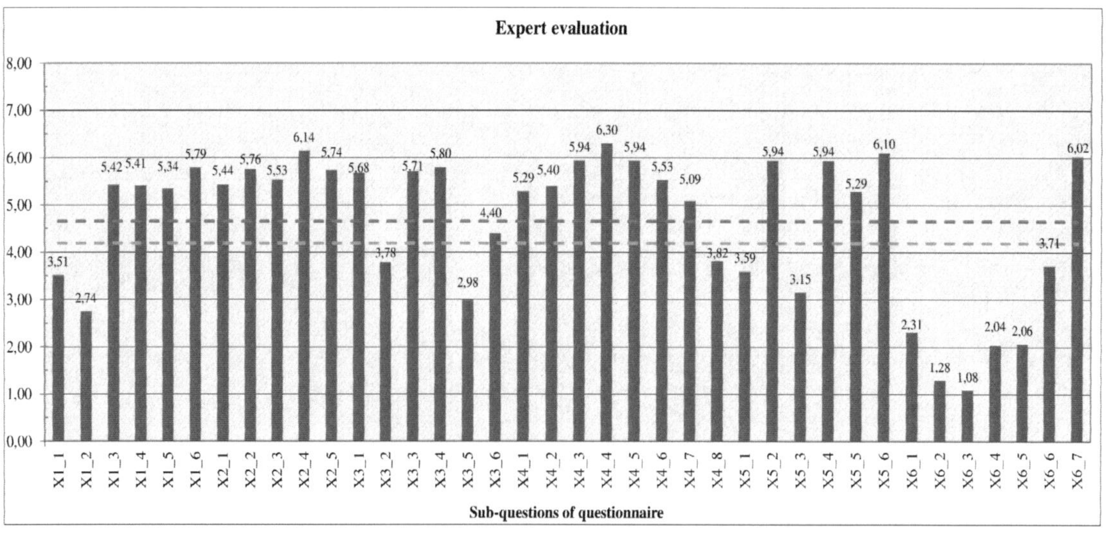

Figure 3. Evaluated average grades by using universal software (Microsoft Excel 2010) taking into account the competence of experts

Table 1. Competence coefficients for all experts

Expert	01	02	03	04	05	06	07	08	09	10	11	12	13	14
k_k	0.58	0.60	0.88	0.56	0.40	0.86	0.81	0.57	0.88	0.91	0.81	0.91	0.88	0.84
Expert	15	16	17	18	19	20	21	22	23	24	25	26	–	–
k_k	0.86	0.65	0.88	0.81	0.86	0.60	0.91	0.84	0.67	1.00	0.70	0.70	–	–

Overall analysis of the results (Figures 2–3) showed that the least important issues for consideration: X2 (average grade with taking into account the competence of experts is 5.72); X4 (5.41); X5 (5.00); X3 (4.73); X1 (4.70). But the most important issue is X6 (2.64).
The resulting evaluation of the importance of questions are shown in the Table 2.

Table 2. The resulting evaluation of the importance of questions

Question number	Average value without taking into account the competence of experts	Reference value without taking into account the competence of experts	Result without taking into account the competence of experts	Average value with taking into account the competence of experts	Reference value without taking into account the competence of experts	Result without taking into account the competence of experts
X2	7.523	6.088	1.435	5.636	4.569	1.067
X4	7.067	6.088	0.979	5.292	4.569	0.723
X5	6.590	6.088	0.502	4.926	4.569	0.357
X3	6.109	6.088	0.021	4.628	4.569	0.059
X1	6.083	6.088	-0.005	4.579	4.569	0.010
X6	3.500	6.088	-2.588	2.615	4.569	-1.954

By the results of analysis of the resulting evaluation of the importance of questions was built histogram for degrees deviation of evaluated average scores from the reference value with and without taking into account the competence of experts by using universal software (Microsoft Excel 2010) (Figure 4).
By the results of analysis were also evaluated average grades of expert evaluation for questions (X1–X6) without and with taking into account the competence of experts by using special software (Expertise CE 1.0) (Figure 5) [11].
The results obtained show small variation of average grades of expert evaluation for questions (X1–X6) that testifies to its quite good consistency.
Also analysis of the results (Figures 2–3) showed that in all cases 13 sub-questions (34 %) were priority for further detailed analysis in order to take the necessary decisions. But 25 sub-questions (66 %) did not have priority or not matter at all for their further analysis.
Considering competence coefficient of experts did not influence the end result evaluation given the rather homogeneous of expert's evaluation on problematic questions that were discussed.

4. Priority problematic questions for improvement of the metrological assurance of electrical power measurements

Only the problematic question of MA (X6) is attributed for further more detailed researching by the results of GEE on problematic questions of MA of EPM.

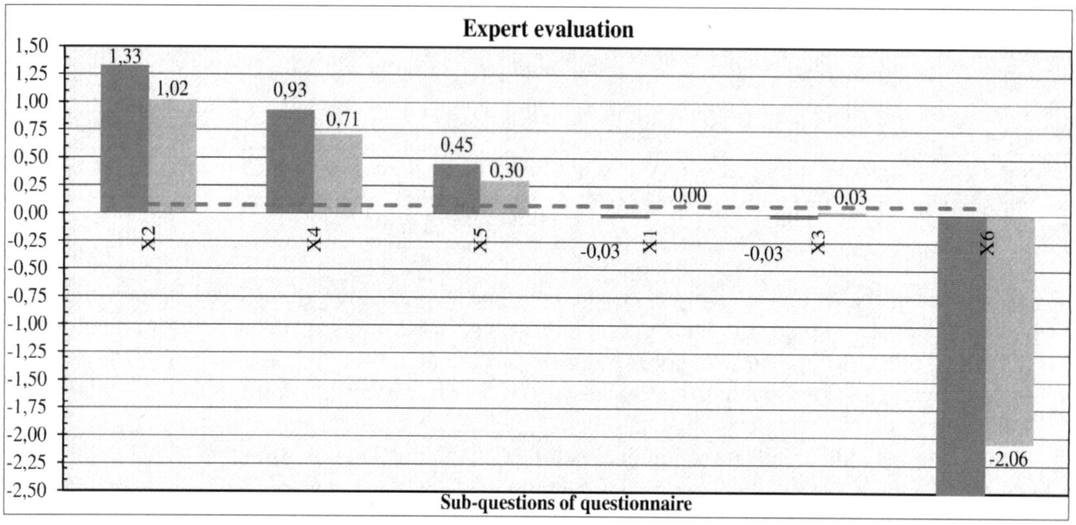

Figure 4. Histogram of degrees deviation of evaluated average grades from the reference value with and without taking into account the competence of experts

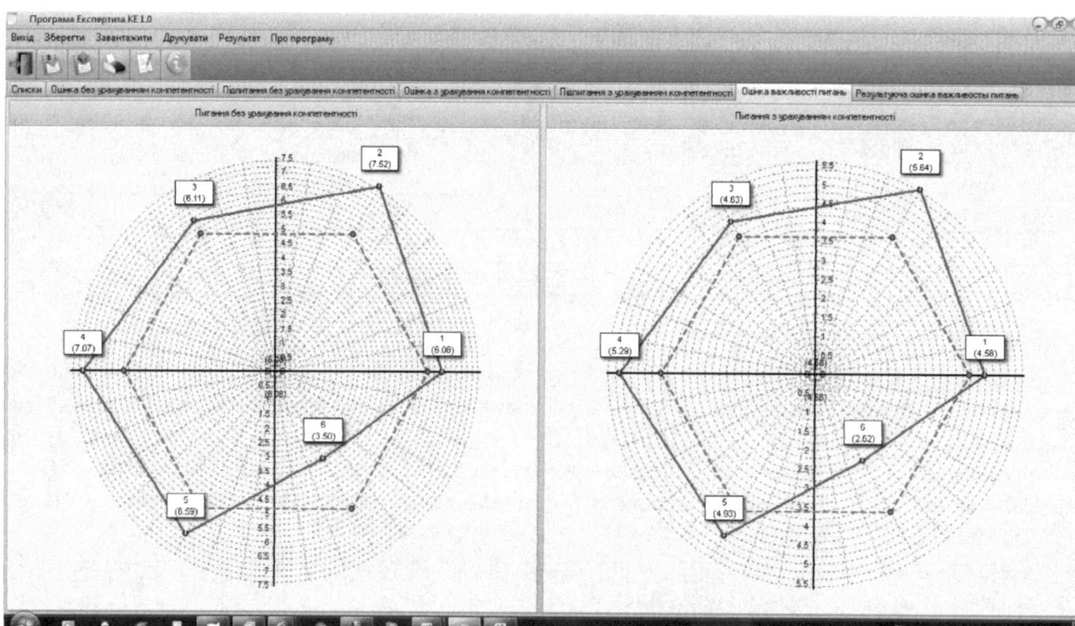

(a) (b)

Figure 5. The chart for average grades of expert evaluations
(a – without taking into account the competence of experts;
b – with taking into account the competence of experts)

The following sub-questions are attributed for further more detailed researching by the results of GEE (in order of importance):
- correlation between the number of verified and calibrated MI by the enterprise (X6_3);
- calibration of working standards (X6_2);
- the use of calibration methods of MI (X6_4);

- status of evaluation uncertainty during calibration of MI (X6_5);
- verification of working standards (X6_1);
- number of experts who conduct or participate in testing (X1_2);
- availability methods that require development or review (X3_5);
- use of verification protocol forms (X5_3), etc.

The other problematic questions of MA are referred to the ones that have no of primary importance.

5. Conclusion

The group expert evaluation with involved experts in field of metrology can be a useful tool to establish the real status of metrological assurance for specific measurements. Special software (for example, Expertise CE 1.0) and universal software (for example, Microsoft Excel 2010) can be used for mathematical processing of obtained expert data.

Positive status of metrological assurance for electrical power measurements by the results of the group expert evaluation can be stated generally. However, it should be noted there are some problematic questions regarding: calibration of working standards, methods of calibration and evaluation of uncertainty during calibration of MI.

6. References

[1] Velychko O, Karpenko S, Gachok V. 2013. Evaluation of the uncertainty of calibration of the three phase high precision comparator of the electrical power. *19th Symp. IMEKO TC 4 Symp. and 17th IWADC Workshop Advances in Instrumentation and Sensors Interoperability* (Barcelona, Spain, July 18–19, 2013). 41–46.

[2] Velychko O, Gordiyenko T. 2015. Evaluation of competence of the experts in field of metrology and instrumentations. *XXI IMEKO World Congress "Measurement in research and industry"* (Prague, Czech Republic, August 30 – September 4, 2015). 5.

[3] Grabovetskii B. E. 2010. Methods of expert estimations: theory, methodology, directions of the use. Vinnytsa, VNTU. 171.

[4] Kalinina I. O., Gozhii O. P., Musenko G. O. Taking into account of competence of experts in the methods of multi-criterion analysis in the tasks of rational choice. *Scientific works of Bleak Sea StateUniversity. Computer technology*. 191. 116–123.

[5] Chernyshova T. Y. 2009. Hierarchical model of estimation and selection of experts. Reports of TUSMR. *Managements, computing engineering and informatics.* 1 (19). Part 1. 168–173.

[6] Kolpakova T. A. 2011. Determination of competence of experts at the acceptance of group decisions. *Radioelectronics,informatics, management.* 1. 40–43.

[7] Drake P. R. 1998. Using the Analytic Hierarchy Process in Engineering Education. *International Journal of Engineering Education*. 14 (3). 191–196.

[8] Velichko O N. 2010. Calibration and measurement capabilities of metrological institutes: features of preparation, examination, and publication. *Measurement Techniques*. 53. 721–726.

[9] Velychko O, Karpenko S, Gachok V. 2013. The Results of Supplementary Comparison of 50/60 Hz Power COOMET.EM-S2. *10th International Congress on Electrical Metrology. Congress Digest* (Buenos Aires, Argentina, September 25–27, 2013). 4 p.

[10] Velychko O M, Gordiyenko T B, Kolomiets L V. 2015. Methodologies of expert's competence evaluation and group expert evaluation. *Metallurgical and Mining Industry*. 2. 262–271.

[11] Velychko O M, Gordiyenko T B, Kolomiets L V. 2015. Hardware-software complexes for evaluation of competence level of experts. *Metallurgical and Mining Industry*. **7**. 396–401.

[12] Velychko O M, Karpenko S R, Gordiyenko T B, Kolomiets L V. 2015. Special software for experts evaluation. *Metallurgical and Mining Industry*. 7. 401–407.

Journal of Physics: Conference Series **772** (2016) 012044 doi:10.1088/1742-6596/772/1/012044

Is there anything nontrivial about caring in nursing that is rigorously measureable?

J Sumner[1], **W P Fisher Jr**[2]

[1]Louisiana State University School of Nursing, New Orleans, Louisiana, USA
E- mail: jsumne@lsuhsc.edu

[2]University of California, Berkeley, California, USA
E- mail: wfisher@berkeley.edu

Abstract. Is there anything about the well-developed body of theory around caring in nursing that leads towards improved measurement and management? Nursing is primarily practical in character; remarkably, the theory of caring in nursing remains to this day disconnected from practice in fundamental ways. Habermas' developmental theory of moral consciousness, rooted in the work of Kohlberg, supports a theory of caring in nursing as a form of communicative action. The genesis of this connection between caring in nursing and communicative action is deeply rooted in human vulnerability and the need for 'considerateness.' Vulnerability is always present in any human interaction and none more so than in nursing. The theory components were the personal selves of both nurse and patient, the professional self of the nurse, the illness self of the patient and interaction. It is within the latter that moral maturity emerges across preconventional, conventional and post conventional developmental levels. Metrologically, concept analysis of the theory had to be focused by the demands of scale construction on writing testable items. Pilot studies were conducted in three countries; seven subscales were theorized and identified in the data. Cronbach's Alpha was over .85 for all scales, with acceptable model fit. The subscales measure the theoretical constructs in ways utilizable in nursing practice.

1. Introduction

Nursing care is predicated on understanding the lived or subjective experience of the patient and with this knowledge nurses apply appropriate knowledge and skills to alleviate suffering. It is this combination of compassion and practical know-how that is regarded as caring. Caring in nursing is the crux of clinical practice, with both nurse and patient as the key participants in a complicated and complex communicative relationship. Despite its central role in nursing, extensive efforts at measuring the processes and outcomes of caring in nursing [1-2] have to date not generally resulted in practical tools measuring in a well-defined unit, though there have been some successes in other areas of nursing measurement [3].

Recent work frames caring in nursing as a maturational process [4-7], providing a basis for improved measurement that implements advances in developmental psychology [8]. Available data from a survey study of 193 nurses in the US, UK, and New Zealand have the potential for providing initial and provisional cross-sectional insights into nurses' experiences in caring. With record shortfalls of available nurses looming, recruitment and retention of skilled RNs should be a high priority for public and private policymakers, health system and community leaders, and managers of healthcare organizations who want to improve the delivery of healthcare services by reducing unnecessary costs (waste) and increasing efficiency, while maintaining or improving healthcare quality. We manage what we measure, and what counts gets counted; in nursing, fulfilling the demands of caring management requires systematic measurement of the highest quality.

2. Caring in nursing

As a concept, caring in nursing remains 'murky' [9]. Paley [10] observed that descriptive studies continue, but also sensed that nothing really new has emerged to move the ontology and epistemology of caring in nursing forward. In recent literature, the foci of caring in nursing has ranged across theoretical constructs [11], ethics [12], caring in nursing education [13], and other areas. There are accordingly many instruments that test caring in nursing [2]. As Watson [2] herself indicates, there has been a tendency not to investigate philosophic issues and questions before developing an instrument but "this collection of caring instruments is a means to bridge opposing viewpoints, dualisms, and conflicting paradigms" (p. 5). The need for a firm philosophical, ontologic-epistemic basis is imperative for construct definition prior to instrument development and testing. Nevertheless each piece contributes to the tapestry of caring in nursing knowledge.

The ideal is not approached without maturation, however. Three stages or levels of development are especially relevant. At Kohlberg's [14] Pre-Conventional level of moral maturity and Benner's [15] Novice and Advanced Beginner stages, the nurse who has just begun practice has to synthesize the various forms and areas of knowledge he or she has acquired. Initially, these will be limited to theory and task skills; experiential knowledge can only come with time in practice. He or she has to think through the rules or processes of each step of the skills, as these have not yet become intuitive or instinctive. The practice is 'by the book,' with concrete thinking. This nurse tends to be inward thinking, wondering how he or she is going to cope with all the demands being made. Patients are somewhat faceless, they are 'illness objects,' depersonalized, and the pre-conventional nurse is not able to see beyond this. The nurse is protective of his or her human vulnerability, which may not be acknowledged, as there is no understanding that the need for considerateness [16] applies to him or herself as well as the patient.

At the middle level of developmental moral maturity the patient is emerging as a total human being for the nurse, who is beginning to identify needs beyond those of the illness and the treatment requirements. Kohlberg [14] refers to this level as Conventional (Level B: Stages 3 & 4) and Benner [15] as Competent & Proficient. The nurse is beginning to have confidence in his or her task performance and is managing the routines of the unit more easily. No longer does the nurse need to think through each step of every procedure and consult a detailed time line for the routine of the day. He or she remains, however, conscious of the need for routine and the relative inflexibility of the organizational policies and regulations, which demands obedience from the nurse. The duties and obligations to the organization are paramount, apparently at the expense of the patient, i.e., medication administration is at set times and the nurse will be questioned if this is not followed. The nurse is unlikely to be unable to reflect on the logic of this or why physician rounds in acute-care hospitals are so early that the patient has to be awakened, when he or she needs the rest.

At the third level, Kohlberg's Post-Conventional (Level C: Stages 5 & 6) and Benner's Expert stage, the nurse is now experienced and has synthesized the three areas of knowledge: theory, practice and experience. The nurse has learned how to utilize his or her emotions optimally in the nurse-patient interactions. Rather than having to consciously think through every task and procedure, the nurse is able to think about how to use emotion, which is a different skill. The nurse identifies the patient as a complete human being with both illness self and personal self needs, and he or she knows how to meet both. He or she has developed a high level of competence with associated confidence and comfort in self. 'Experiential intuition' [15] has developed, so the nurse can anticipate illness problems and initiate preventive action. The nurse is capable of 'standing outside' the interaction and can reflect on what is happening within it, and can consciously steer it in directions where both he or she and the patient are comfortable with the course of treatment. This is not done in a coercive fashion, but rather occurs in ways considerate of the needs and obligations of both. He or she is capable of abstract objective thinking. At this stage the nurse has developed the ability to reflect on what is just and fair within the interaction.

3. Measurement models and methods

Measurement requires the calibration of instruments that are objective in the sense of representing units of comparison that remain invariantly sized and ordered across samples, to within a specified range of uncertainty. What this means in practical terms is that, so far as possible, inferences about autonomous constructs should be framed and expressed in terms that remain constant across the particulars of given data sets. When multiple tests demonstrate the strength of a variable, and its ability to stand alone on its own apart from the questions asked and persons responding, as has been demonstrated for many psychological constructs, including developmental ones, over many years [8], then a theory becomes a persistently useful tool in predicting the future behavior of a class of phenomena. Commonly employed statistical presentations of summed or averaged ratings presume measurement has taken place when no experimental tests have substantiated the existence of a meaningful, additive unit mapped onto a number line.

Theories and models for each measured construct were developed in relation to a Rasch perspective on the three requirements of axiomatic conjoint measurement theory [17-18]. For the caring in nursing constructs, (1) data from each of the seven scales must approximate unique monotonic functions, (2) the mapped differences between nurses on each scale must be random, and (3) random perturbations must characterize the observations. The additive form of the dichotomous model is:

$$\ln[Pni / (Pni-1)] = Bn - Di$$

which effectively says that the log-odds of observed success (lack of symptoms or clinical signs) for person n on item i is equal to the difference between the estimate B of person n's ability and the difficulty estimate D of item i [Rasch, 1980].

A bank of 185 items was developed, 121 of which were related to either the nurse (72) or the patient (49). The remaining 64 items focused on the nurse-patient interaction. Ten demographic items were also included in order to capture not only personal data but also how long a nurse had been in practice, in what specialty and with what educational preparation.

Given the complexity of the theory, it was important to ensure that the items incorporated specific features and that these features were coded appropriately before any data were gathered. For instance, for the nurse's personal self, theoretical features of interest in the items included both positive and negative (a) outwardly expressed emotional characteristics (such as altruism), and (b) inwardly acknowledged emotional characteristics (such as a need to help others), combined with (c) a sense of sharing oneself, to be acknowledged as a human being as well as being a nurse. Each of these features is included, for example, in an item like: "I am passionate about nursing as a profession."

Cognitive processes were included in items involving knowing how to make a patient comfortable, knowing pathophysiology and treatment, and knowing how to practice skills. Some items were coded to reflect both emotion and cognition: "All of who I am wants to get involved in helping the patients." Items involving the nurse's professional self are intended to address the knowledge of the health/illness problems that brought patients to the care of nurses, e.g. the illness or disease diagnosis and its treatment: "When tending to patients I get satisfaction in using my medical diagnosis and treatment knowledge." Practice skills required from the highest technical procedures to the simplest comfort measures are recognized: for example "I know how to make a patient comfortable."

For the nurse items, it is hypothesized that the presentation of the professional self and cognitive factors will come before the presentation of the personal self and emotional factors (i.e., the professional self and cognitive items will garner more agreeable ratings and so calibrate lower on the scale than the personal self and emotion items), and that the integration of professional and personal selves will follow (with these items calibrating highest of all).

For the patient, items had to include the personal self, e.g. "a patient's sense of self is important to him or her," and the illness self, e.g. "the patient knows he or she has illness needs which are not human needs." The illness self items included physical, emotional, social and spiritual needs, e.g. "having emotional needs met is important to the patient" or "having spiritual needs met is important to the patient." From the emotional perspective, items must reflect the patients' needs for control, dignity, trust, and ordinary things notable simply as symbols of an everyday reality formerly taken for granted.

Table
Caring in Nursing Subscale Summary Statistics

Construct	# Items Total/Avg[4]	Measurement Reliability Alpha/Rasch[5]/Separation[6]	Avg MnSq InFit[7](SD)/Outfit[8](SD)
Caring Identity	44/35	.97/.91-.95/3.2-4.5	1.05(.5)/1.02(.5)
Patient Focus	38/29	.95/.89-.92/2.9-3.4	1.01(.5)/1.01(.5)
Professionalism	11/8	.87/.56-.67/1.1-1.4	1.03(.6)/1.01(.7)
Partnership	29/23	.92/.73-.90/1.7-3.1	1.01(.4)/1.04(.4)
Communicative Mutuality	10/7	.93/.60-.68/1.2-1.5	1.01(.6)/1.00(.6)
Responsibilities of Care	15/12	.86/.55-.61/1.1-1.3	0.98(.4)/0.97(.5)
Nursing Practice	28/21	.93/.84-.88/2.3-2.7	0.99(.5)/1.01(.5)

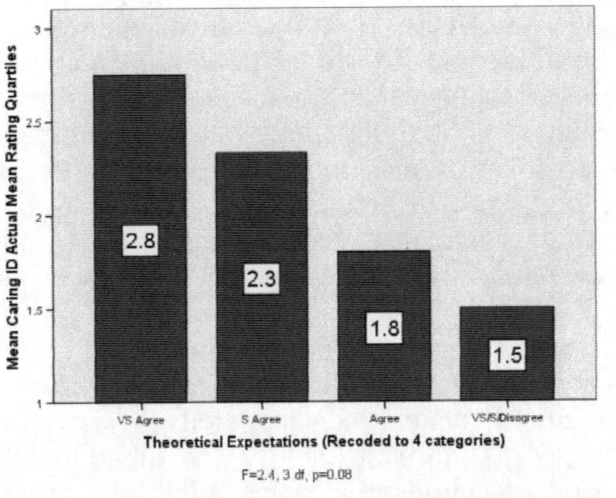

Figure 1. Theoretical vs Empirical Mean Rating Quartiles: Caring Identity Scale

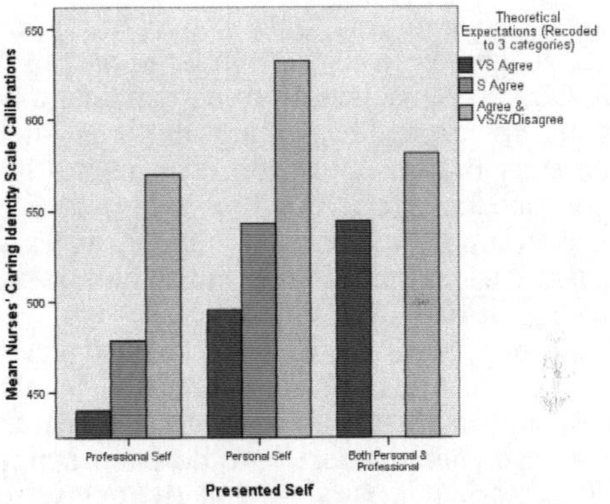

Figure 2. Theoretical vs. Empirical Nurses' Caring Identity Calibrations by Self Categories

4 The average number of items responded to differs from the number of items available because two of the three survey forms used included only about two-thirds of the total number of items.

5 Rasch reliabilities vary in a range depending on whether or not extreme measures are included, and depending on whether the root mean square error is inflated by outfit, or not. Rasch reliability coefficients are estimated from individual error terms, and so are often more conservative than Cronbach's alpha, which typically underestimates error at the extremes. Alpha is furthermore adversely affected by missing data; about 20-30% of the data in the present study are missing, as not all respondents were administered all items.

6 The separation statistic, G, indicates how many ranges in the scale with centers four errors apart (a 99% confidence interval) are defined.

7 Information-weighted model fit. A mean square statistic is a chi-square divided by its degrees of freedom. Both InFit and OutFit have expected values of 1.00. The standard deviations are typically more informative than the averages as to overall data quality.

8 Outlier-sensitive model fit.

4. Results

The Table shows summary statistics for the seven Caring in Nursing constructs: Caring Identity, Patient Focus, Professionalism, Partnership, Communicative Mutuality, Responsibilities of Care, and Nursing Practice. For each subscale, the Disagree-Agree rating scale was optimized from the original six categories to four. All of the Disagree categories combine to form a single category capturing about 7% of the responses.

Reliability, separation, and model fit, as indicated by the information-weighted and outlier-sensitive mean square statistics, were satisfactory on all scales (see Table). Cronbach's alpha was over .85 for all scales. Results obtained for the Nurses' Caring Identity subscale will serve to illustrate the general quality of progress with the data so far. A lack of resources has prevented development of a complete picture of the functioning of all seven scales, and of the shared multidimensional framework.

Figure 1 shows that the mean rating quartiles correspond with the theoretically expected ratings, with the latter recoded from the original six to four categories in the same way the data were. The observed pattern (observed mean ratings of 2.8, 2.3, 1.8, and 1.5 for Very Strongly Agree, Strongly Agree, Agree, and any Disagree response, respectively) conforms with the theory (F=2.4, 3 d.f., p=.08).

Figure 2 shows that items theoretically expected to elicit responses in specific categories calibrate on average in the expected order.

5. Conclusions

Cultivating institutional resources in support of the primary mission in nursing, caring, requires methods for identifying, gauging, reporting, and managing the effects of caring on the nurses who provide it. These effects stem in large part from systematic tendencies in the health care system, and among nurses themselves, to foreground cognitive and professional aspects of their identities as nurses, while minimizing and neglecting emotional and personal aspects. Though the nurse-patient relationship is inherently bi-directional, too often nurses do not receive the emotional and personal supports they need to function effectively. Providing nurses and nurse managers with practical information on how moral conceptions of caring in nursing differ across years in service and specialty areas could be an important means of reducing nurse burnout and turnover rates, retaining experienced nurses, and improving the overall quality of care.

Preliminary results from the survey suggest consistent associations between levels of development in moral conceptions of caring practice. The data invite analysis using multidimensional item response models, and deployment via a caring in nursing management system administered in an online environment enabling real time decision supports relevant to (a) more efficiently and meaningfully managing caring in nursing, and (b) to maximizing nurses' capacities for long, satisfying careers. Conclusions as to the developmental nature of caring will, of course, require new research tracking nurses' caring experiences over time, as is the case for any developmental or growth phenomenon [19]. In addition, it will be important in future research to frame maturation in caring in nursing in terms of theory providing more rigorous and decisive explanatory power (such as hierarchical complexity [8]) than can be obtained from the perspective employed in the development of the present caring in nursing survey.

References

[1] Watson J 1985 *Nursing: The philosophy and science of caring* (Boulder, Colorado: Colorado Associated University Press)
[2] Watson J 2009 *Assessing and measuring caring in nursing and health science, 2nd Ed.* (New York: Springer-Verlag)
[3] Holm K & Kavanagh J 1985 *Research in Nursing and Health* **8** 13-18
[4] Sumner J 2006 *Internat J Human Caring* **10** 8-16
[5] Sumner J 2007 *Internat J of Human Caring* **11** 30-37
[6] Sumner J 2008a *Nursing Admin Qtrly* **32** 92-101
[7] Sumner J 2008b *The moral construct of caring in nursing as communicative action* (Saarbrücken, Germany: VDM Verlag)
[8] Dawson T L 2002 *Internat J Behav Dev* **26** 154-166
[9] Morse J M 1995 *Advanced Nursing Sci* **17** 31-46

Journal of Physics: Conference Series **772** (2016) 012044 doi:10.1088/1742-6596/772/1/012044

[10] Paley J 2001 *J Advanced Nurs* **36** 188-198
[11] Finch L P 2009 *Internat J Human Caring* **12** 25-32
[12] Cortiss J D & Kendrick K 2003 *Nursing Ethics* **10** 77-88
[13] Blum C A & Gordon S C 2009 *Internat J Human Caring* **13** 29-35
[14] Kohlberg L 1981 Essays on moral development In J Habermas 1995 *Moral Consciousness and Communicative Action* pp. 116-188 (Cambridge, MA: The MIT Press)
[15] Benner P 1984 *From novice to expert* (Menlo Park, CA: Addison-Wesley)
[16] Habermas J 1995 *The Theory of Communicative Action and Moral Consciousness* (Cambridge, Mass: The MIT Press)
[17] Massof R W & McDonnell P J 2012 *Investig Ophthal & Visual Sci* **53** 1905-1916
[18] Rasch G 1980 *Probabilistic models for some intelligence and attainment tests* (Chicago: University of Chicago Press)
[19] Hilton T L & Patrick C 1970 J Ed Meas **7** 15-24

Modeling Visual Symptoms and Visual Skills to Measure Functional Binocular Vision

M K Powers[1], W P Fisher Jr[2], R W Massof[3]

[1]Gemstone Foundation, Rodeo, CA USA, [2]University of California, Berkeley, CA USA, [3]Johns Hopkins University, Baltimore, MD USA

maureenpowers@gemstonefoundation.org, wfisher@berkeley.edu, bmassof@jhmi.edu

Abstract. Obtaining a clear image of the world depends on good eye coordination ("binocular vision"). Yet no standard exists by which to determine a threshold for good vs poor binocular vision, as exists for the eye chart and visual acuity. We asked whether data on the signs and symptoms related to binocular vision are sufficiently consistent with children's self-reported visual symptoms to substantiate a construct model of Functional Binocular Vision (FBV), and then whether that model can be used to aggregate clinical and survey observations into a meaningful diagnostic measure. Data on visual symptoms from 1,100 children attending school in Los Angeles were obtained using the Convergence Insufficiency Symptom Survey (CISS); and for more than 300 students in that sample, 35 additional measures were taken, including acuity, cover test near and far, near point of convergence, near point of accommodation, accommodative facility, vergence ranges, tracking ability, and oral reading fluency. A preliminary analysis of data from the 15-item, 5-category CISS and 15 clinical variables from 103 grade school students who reported convergence problems (CISS scores of 16 or higher) suggests that the clinical and survey observations will be optimally combined in a multidimensional model.

Introduction

Using two eyes together to provide merged and unconfused images to the brain is a neurological challenge. Though we rarely think about it, the visual system must be trained to achieve ocular coordination, just as the body must be trained to swing a bat or ride a bicycle. It may seem surprising, therefore, that no gold standard akin to the eye chart for assessing monocular visual acuity exists for a first-level assessment of functional binocular vision.

The clinical diagnosis of "convergence insufficiency" has classically been used to connote binocular dysfunction that is not as extreme as strabismus yet indicates an inability to use the two eyes together in an optimally coordinated fashion. Convergence ability is assessed in a variety of ways, including near point tests or fusion ranges. Although each vergence measure has a continuum of possible values, the criteria for positioning a diagnosis on the continuum can be wide-ranging. While all agree that a tropia (an eye that does not converge under stimulation) indicates strabismus [1-2], and tables exist to assess

[1] To whom any correspondence should be addressed.

near point with regard to age [3], and most might agree that accommodative or tracking problems contribute to non-strabismic binocular discomfort and function, there is no universal agreement on the basic criteria for overall functional binocular vision.

A body of previous research investigated these issues without the benefit of probabilistic construct models [4-5]. Results demonstrated:

- A significant relationship between symptoms of visual skill problems as measured by the Convergence Insufficiency Symptom Survey (CISS) [6] and three separate measures of reading fluency. One measure was administered by the school, and thus was completely free of experimenter bias.
- Reading speed slowed significantly when subjects with normal convergence skills read through prisms. Magnitude of reduction was directly related to the degree of vergence instability. Stress on the vergence system, as in conditions such as FBV, may result in slower reading rates for children in school.
- Significant correlations between symptom scores from the FBV symptom survey and actual signs measured during school visits on the same subjects, which suggests that the CISS is a valid indicator of visual skill problems in children of school age.

Those familiar with Rasch models may assume that they are used only to evaluate questionnaire or educational test data. But, by adopting innovative theoretical developments [7-8] and previous applied research methods [9-13], the aim here is to evaluate FBV as a measureable state variable that can be estimated from a combination of clinical and survey observations. The availability of a large dataset with multiple variables and the CISS responses will support the formulation and testing of an underlying model that could ultimately support development of a standard FBV test.

Original study design

Students were recruited from four elementary schools in Los Angeles, grades 3 through 6 (N=1062). The CISS was administered to all students. Subjects were selected for inclusion in the study if the CISS score was less than 15, indicating convergence problems, reducing the sample size to 418. Clinical variables, including aspects of acuity and binocular vision, were organized into primary (11 items) and secondary (four items) dimensions of signs and symptoms. Clinical variables were measured for 312 cases. Some subjects participated in a binocular visual skill training program via internet at their school (N=51; six did not complete the intervention). All variables were measured at each of four time points: pre-intervention, and after each of three training sessions. The analysis presented here involves 103 students from one school; of these, 42 had measures at all four time points.

Frequencies of the original optometric measures were examined and ratings were assigned to express the inferences usually made about the impact of the variable on FBV. Data were fit to a multidimensional random coefficients multinomial logit model [14-15] to assess the visual and survey dimensions measured by the clinical and survey variables, and the relationship between them.

Preliminary results

Preliminary reanalyses of subsamples of this study's CISS data have provisionally established reasonable fit to Rasch rating scale models. Comparisons of item difficulty locations with similar items from other tests were used to investigate whether the underlying latent variables are similar: Pre-training CISS item location estimates based on 202 students with CI from three schools correlated 0.71 with CISS item logit values computed from Borsting, et al.'s [6] "Always" category response frequencies for 47 CI students. The correlation was 0.93 for item locations estimated from Borsting et al.'s 56 NBV student responses and our previous 635 NBV student responses. Borsting et al. [6] do not report item location reliabilities, so it is not possible to disattenuate these correlations; however, given the low sample sizes in the Borsting et al. study, calibration error is likely high, so disattenuation would increase the correlations. In addition, Rasch item location estimates reported by Conlon, et al. [16] for the Visual Discomfort Scale correlated 0.71 with nine CISS items having matching content.

Of 103 students from one school in grades 3-5, 257 measures were made across four time points. Students ages ranged from 8-12, with a mode = 9; 91% were between ages 8-10. No students had had eye surgery, 7 of the 42 with measures at all four time points had worn glasses/contacts at some time, 5 of 42 owned glasses at that time, and 3 of 42 wore glasses/contacts regularly in class and had them that day.

Measures changed in the expected directions of improved FBV but were not statistically significant due to the small sample size. Measures of the primary FBV signs correlated 0.23 with the secondary signs, and -0.13 with the CISS measures (indicating worsening conditions as the CISS scores dropped and the FBV measures increased). The secondary signs correlated -0.38 with the CISS measures.

Conclusion

The small sample size in this analysis, from only one school, precludes assessing a multidimensional model that parameterizes time points. A larger sample from all 4 schools will enable comparison of changes in FBV with reading outcomes. Analyses presented elsewhere show a poor correlation ($r = 0.17$ to 0.23; r^2 = approx. 0.05) between acuity and visual reading symptoms. Thus, monocular eye chart measurements are a poor predictor of reading problems in school, supporting the need for broader FBV assessments and standards.

Neither Powers nor the developers of the CISS hypothesized a unified FBV construct or proposed to develop a theoretical model of FBV. Powers' existing data will enable us to understand whether the high correlation between the CISS and FBV orthoptics implies that the integration of an FBV symptom survey (a revised CISS) with an orthoptics assessment would be likely to produce a common or separate self-report and clinical measure(s) of FBV, and also how the resulting measure or measures relate to reading outcomes.

The significance of the project lies in its attempt to provide a measurement tool to assess FBV in the classroom. The results will be important both to the field of Special Education, where there is great need to assess physical readiness for reading in order to avoid over-categorization as well as promote appropriate remediation, and also for classrooms at large, where many students' reading fluency may be poor due to slow or inaccurate visual skills during reading. Teachers are more likely than anyone to notice that a child seems uncomfortable while reading. This work could provide them with an assessment tool for their use. In most states (California included) students are screened for visual acuity and referred to an eye care professional outside of school if they fail. But the test only assesses distance acuity and does not allow screening for visual skills close to the face that require both eyes.

Hence it is important for those in education to have the ability to assess or screen for FBV so that they can take the appropriate action. Part of the present work will use existing data to answer such questions as: How much of reading behavior is determined by acuity? By binocular skills? By oculomotor function? Answers to questions like these can help shape health and education policy in the future as well as give teachers and parents the insights they need to enable them to provide the best guidance for their child's education. And the development of a tool that can be used to easily assess binocular vision ability would be a welcome addition to educators, who struggle daily with determining which children might need help with basic sensory processing issues before they are ready to learn.

References
[1] Duke-Elder S 1973 *System of Ophthalmology, Vol 6: Ocular Motility and Strabismus* Mosby, London
[2] Hoyt C S, and Taylor D 2013 *Pediatric Ophthalmology and Strabismus* 4th Edition Elsevier
[3] Griffin J R, and Grisham J D 2002 *Binocular Anomalies* New York: Butterworth-Heinemann, Fourth Edition
[4] Morita Y, Hoffman R and Powers M 2010 Visual skills and reading: Symptoms and fluency in elementary school students In Low Vision, Task Performance, Plasticity and Rehabilitation Session 3632/A241, Association for Research in Vision and Ophthalmology, Fort Lauderdale, Florida, Tuesday, May 4
[5] Powers M K, Miner G L, Morita Y and Tyler C W 2010 Vergence stress significantly affects reading rates In Reading in Aging and Disease 3068/A263 conducted at the Association for Research in Vision and Ophthalmology, Fort Lauderdale, Florida, Tuesday, May 4
[6] Borsting E J, Rouse M W, Mitchell G L, Scheiman M, Cotter S A, Cooper J, et al 2003 *Optom. and Vision Sci.* **80** 832-838
[7] Christensen K B 2007 In *Advances in Statistical Methods for the Health Sciences: Applications to Cancer and AIDS Studies, Genome Sequence Analysis, and Survival Analysis* pp 95-108 New York: Springer-Verlag
[8] Zwinderman A 1991 *Psychometrika* **56** 589-600
[9] Best W R 2008 In J N Cadwallader Ed, *Crohn's Disease: Etiology, pathogenesis and interventions* Chapter 5 New York: Nova Science Publishers, Inc
[10] Cipriani D, Fox C, Khuder S and Boudreau N 2005 *J. Appl. Meas.* **6** 180-201
[11] Fisher W P Jr, and Burton E 2010 *J. Appl. Meas.***11** 271-287
[12] Hughes L, Perkins K, Wright B D, and Westrick H 2003 *J. Alzheimer's Dis.* **5** 367-373
[13] Massof RW, and McDonnell PJ 2012 *Investigative Ophthal. & Visual Sci.* **53** 1905-1916
[14] Adams R J, Wilson M R, and Wang W-C 1997 *Appl. Psychol. Meas.* **21** 1-24
[15] Briggs D, and Wilson M 2003 *J. Appl. Meas.* **4** 87-100
[16] Conlon E G, Lovegrove W J, Chekaluk E, & Pattison P E 1999 *Visual Cognition* **66** 637-663

Acknowledgments

This study was funded by the National Eye Institute (R01 EY017414).

Changes in the clinical measurement of visual acuity

I L Bailey[1] and A J Jackson[2]

[1] School of Optometry, University of California, Berkeley, CA 94720-2020, USA
[2] Ophthalmology, Royal Victoria Hospital, Belfast BT12 6BA, UK

E-Mail: ibailey@berkeley.edu

Abstract. In 1862, Hermann Snellen introduced his letter chart for the clinical measurement of visual acuity. His chart presented letters,or optotypes, arranged in a progressively diminishing size sequence, and the visual acuity was determined by the smallest letters that could read at a specified distance. Numerous modifications of the design of the optotypes, the progression of size and the chart layout were suggested, and in 1976, Bailey and Lovie published a set of design principles that made the visual task the same at all size levels, so that size become the only significant variable. This required the same number of letters at each size level, fixed spacing ratios and a logarithmic progression of size. This facilitates more precise quantification of visual acuity by giving credit for every letter read correctly, and this gives clinicians tighter confidence limits for determining changes or differences in visual acuity. However, optotype choices, and associated spacing arrangements can have significant effects on visual acuity scores as can viewing conditions and testing protocols. Computer based visual acuity tests are becoming more commonplace, there will be more variety in test charts and procedures which will create some problems for making comparisons between tests.

1. Introduction

In 1862, Hermann Snellen introduced his famous "Snellen chart" for the clinical measurement of visual acuity. With it, he introduced the "Snellen fraction" as a measure of angular size. His chart presented a series of serifed letters that varied from large to small. His letters, which he called "optotypes" were designed so that the stroke widths of the letters were 1/5 of the letter height. An individual's visual acuity was determined by the smallest letters that could be read correctly when the chart was presented at a standard viewing distance. The angular size of the threshold size optotype was specified by the "Snellen fraction" whose denominator is the viewing distance, and the numerator is the distance at which the height of the letters subtends 5 minutes of arc. In the USA, 20 feet is the standard viewing distance, so a visual acuity of 20/40 indicates that the testing distance was 20 feet and the height of smallest letters read would subtend 5 arcmin at 40 feet. Most individuals with good normal vision can achieve a visual acuity of 20/20 or better. Thus, an individual with "good normal vision" can just recognize letters that subtend 5 minutes of arc, and the component strokes and most of spaces within these letters subtend 1 minute of arc. The angular size (in minutes of arc) of the detail within the smallest recognizable optotypes is said to be the "minimum angle of resolution" (MAR). The MAR is equal to the reciprocal of the Snellen fraction. Thus for a visual acuity of 20/20, the MAR = 1 arcmin; for 20/40, the MAR = 2 arcmin; for 20/200, the MAR = 10 arcmin; and so forth. Following Snellen's original chart came many alternatives that offered numerous suggested "improvements" and modifications of the design of the optotypes, the progression of size, the range of sizes and the layout. (reviewed by Bennett, 1965).

Most of these alternative charts had only one or two optypes at the largest size, and the number of letters became more numerous as the sizes became smaller. And generally, the ratio between one size and the next became smaller at the smaller sizes. Despite wide variations in the chart designs, they were all said to be "Snellen charts". The term "Snellen chart" has become a generic descriptor. There is no standardization for Snellen charts.

Journal of Physics: Conference Series **772** (2016) 012046 doi:10.1088/1742-6596/772/1/012046

One of the more important design changes came from Green (1905) who advocated that the progression of size should be logarithmic with a ratio of 0.1 log units (ratio = 1.2589). Sloan (1959) also favored the same logarithmic size progression and she developed a set of 10 non-serifed letters with specified configurations, all of which were based an a 5x5 grid, and these have become the gold standard for letters on the visual acuity charts of today.

The majority of visual acuity charts use Roman letters as their optoypes. But there are many alternative optotypes including numbers and pictures or symbols and some optotypes, notably the Landolt ring and the Tumbling E, are figures whose orientations have to be identified.

Snellen Chart 1862 Bailey-Lovie Chart 1976

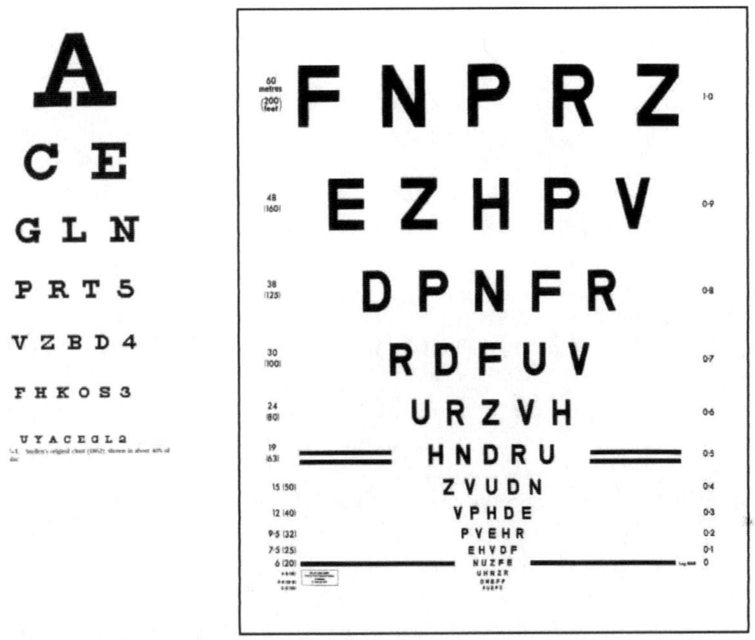

Figure 1. Letter charts for visual acuity measurement

The Bailey-Lovie chart design principles

In 1974, at the University of Melbourne, Jan Lovie-Kitchin and I embarked on a project to study vision in macular degeneration. We expected most of our subjects would have poor visual acuity, and we realized that the available visual acuity charts would not be suitable for our purposes. We expected that most our visual acuity business was going to be done at larger sizes where there were too few letters and the ratios between sizes was too large. It was necessary for us to design some new charts so that we could better test subjects with poor visual acuity.

As the optotypes, we chose to use the family of 10 letters recommended by the British Standard Institute in 1968. The letters were non-serifed and their designs were based on a 5x4 grid. First of all, we wanted as many letters as was practical at the largest size, and we decided to use 5 letters. We made the spacing between letters equal to the letter width. Then, we decided that all the subsequent rows should also have 5 letters. We used the logarithmic size progression as first advocated by Green. We realized that if the spacing between on size and the next were made proportional to the adjacent letter sizes, then the visual task would be the same at each size level. That is, as you go from one size level to the next, size becomes the only significant variable.

It had not been the intention, but we now had a chart design, where visual task was the same throughout. An important advantage is that, if the viewing distance is changed, the threshold size simply moves to different size level on the chart in proportion to the change in viewing distance.

LogMAR. A new unit for designating visual acuity.
On these new charts, we labeled the size the different optotypes by the common logarithms of the minimum angle of resolution (logMAR), assuming the letters were being viewing from the standard distance of 6 meters (or 20 feet). Thus, for 20/20 vision the MAR = 1 arcmin, so $\log_{10}$MAR = 0; at 20/200, the MAR = 10, so $\log_{10}$MAR = 1.0; at 20/40, the MAR = 2 so the logMAR = 0.30. The visual acuity range on these charts extended from 20/200 to 20/10; that is from logMAR = 1.0 to logMAR = -0.3.

When vision is so poor that the subject is unable to read the largest letters on the chart at the standard distance (6 m or 20ft), the chart could be moved closer, with an appropriate adjustment being to the logMAR score. For example, if the subject was unable to read the letters in the top row, the chart could be moved closer, and let us say to 1.5 meter (5 ft.). Then, a logMAR score correction of 0.6 needs be made to account for the 4-fold reduction in viewing distance (because $\log_{10}(4) = 0.6$). For the second largest letters on the chart at the standard 6-meter distance, the logMAR label is 0.9. If this is the smallest letter size that can be read at 1.5 meters, then the logMAR score for this subject is 0.9 + 0.6 = 1.5.

VAR . Visual Acuity Rating
An alternative method for designating visual acuity is the Visual Acuity Rating (VAR) which is an arbitrary scale for where VAR = 100 – 50xlogMAR. This is easier to use than the logMAR scale. On the VAR scale, useful benchmarks are VAR = 100 when VA = 20/20 (commonly considered as normal vision), VAR = 50 when VA = 20/200 (s commonly used as a border for classifying "legal blindness" and VAR = 0 when VA = 20/2000 (which is about the limit for measuring visual acuity with optotypes). For charts complying with the Bailey-Lovie design principles using a size progession in steps of 0.1 log units and with 5 letter at each size, then each extra letter read correctly increases the VAR score by 1 point. For visual acuities better than 20/20, the VAR is greater than 100. For each extra row carries a VAR value of 5 points. On the VAR scale, 15 points represents a visual acuity change by a factor of 2-fold (0.3 log units). For both the logMAR and VAR scales, the labels on the chart must specify the viewing distance for which the chart has been calibrated. Many charts today use 4 meters as the standard viewing distances, but viewing distances of 20 ft, and 6 meters are still widely used. If the charts are used at non-standard distances, an adjustment should be made to the logMAR or VAR score. The correction factors are $\log_{10}$(standard distance/new distance) added to the logMAR score and 50xlog(new distance/standard distance) added to the VAR score. Halving the viewing distance requires a 0.30 addition to the logMAR score and a 15 point subtraction from the VAR score.

Letter-by-letter scoring
This logMAR scale also facilitates giving visual acuity score that give credit for every letter read. Because there are 5 letters per row, and the size increment between rows is 0.10 then each letter can be assigned a value of 0.02. To give an example, consider a subject who correctly reads all the letters at the size where logMAR = 0.3 (this is equivalent to 20/40). Should this person attempt the next smallest row and get one letter correct, the visual acuity score would become 0.28. Had two extra letters been read in that next smaller row, the score would have become logMAR =0.26.

Giving credit for every letter read correctly provides more precise measurement of visual acuity, and this gives substantial improvement in the sensitivity to detect changes or differences. Unfortunately, it is still common practice for clinicians to assign visual acuity scores on a row-by-row basis. The

Journal of Physics: Conference Series **772** (2016) 012046 doi:10.1088/1742-6596/772/1/012046

assigned visual acuity score corresponds to the size the smallest letters at which a certain percentage (usually 50% or more) of letters are read correctly.
Comparing test and retest measurements of normally sighted subjects, when visual acuity scores are assigned on a row-by-row basis, it is found that there is perfect test-retest concordance about 60% of the time. For 20% of the time, the second score is 1-row better, and 20% of the time, one-row worse. The 95% confidence limits for identifying change become 2 rows (i.e., 2 size levels) on the chart. When scoring visual acuity of a letter-by-letter basis, as expected for a finer scale, the rate perfect concordance between test and retest is reduced and it is about 20%. The distribution of test/retest discrepancies has a standard deviation of about 0.04 log units, so the 95%confidence limits for change become 1.00 and this corresponds to 5 letters. If visual acuity scores are assigned on a row-by-row basis, the confidence limits for change are ±2 rows on the chart. But when the scores are assigned giving credit for every extra letter read, the confidence limits become about ±5 letters.

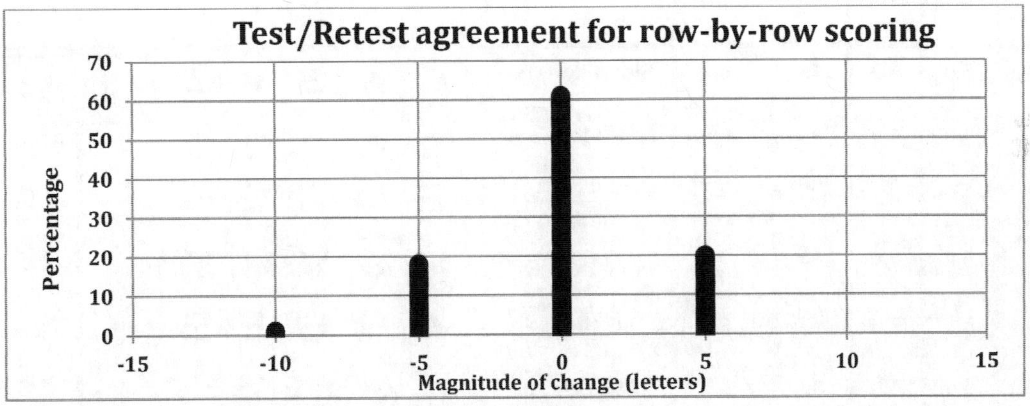

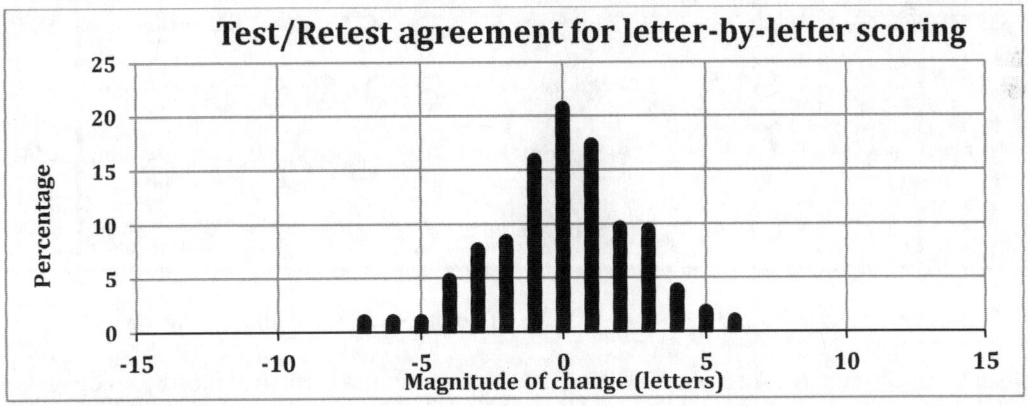

Figure 2. Histograms of test/retest discrepancies when assigning visual acuity scores using the row-by-row method and the letter-by-letter method.

The histograms of Figure 2 show the frequencies of test/retest agreements and discrepancies when scores were assigned by the row–by-row method and by the letter-by-letter method. Letter-by-letter scoring substantially enhances a clinician's ability to determine changes of difference in visual acuity.

Journal of Physics: Conference Series **772** (2016) 012046 doi:10.1088/1742-6596/772/1/012046

The clinical research community has adopted the Bailey-Lovie chart design principles that standardize the visual task, the logMAR units and scoring visual acuity by giving credit for every letter read. Most visual acuity charts used in clinical research have 5 optoypes at each size, and a size progression ratio of 0.1 log units (1.26x) . The spacings between optotypes and between rows is usually is kept proportional to the optotype size. However, there are many a different alternative optotypes being used. There are different families of letters, different families of numbers, some have symbols arranged in different orientations and some have sets of symbols or pictures. It is not widely appreciated that these alternative charts do not always give equivalent scores of visual acuity, even though the manufacture's labeling usually suggests otherwise. Figure 3 shows a row of optotypes from 10 different charts. The sizes of the optotypes are to scale in that the manufacturers' labels assign the same visual acuity value to each of these.

Optotypes *n*	Height *arcmin*	Width *arcmin*	Row (5) *arcmin*	Height/ Stroke	Dimensions of 10 sets of Optotypes	Letters different
10	5.0	5.0	45	5	Sloan Letters	Standard
4	5.0	5.0	45	5	Landolt Rings	-0.8
4	5.0	5.0	45	5	Tumbling E	-1.0
4	5.0	5.0	45	5	HOTV	+2.5
8	5.0	5.0	45	5	LVRC Numbers	-1.5
10	5.0	4.0	36	5	1968 British	-1.0
5	5.7	5.7	51	5	PV Numbers	+0.8
4	6.4	4.3	40	7.5	Lea Numbers	+0.6
4	5.0	5.0	51	5	Patti Pics	-1.6
4	6.0	6.0	56	7	Lea Symbols	-1.0

Figure 3. Illustration and a list of dimensions for rows of optotypes from 10 different charts

The Sloan letters a family is used in the ETDRS (Early Treatment of Diabetic Retinopathy Study) is generally taken as the standard optotype against which others should be compared. On a group of 20 normally sighted subjects, we measured visual acuity with 10 charts with these various optotypes. The right-hand column of Figure 3 indicates the magnitude of the average differences in the visual acuity scores for each optotype when compared with the Sloan letters. The differences are here expressed as number of letters – a one letter difference correspond to 0.02 log units. The difference between the easiest chart (HOTV) and the most difficult chart (Patti Pics) is substantial. (4.1 letters or 0.082 log units) These differences result from variations in legibility of the different sets of optotypes and also from the variations in sizes and spacing. These results come from normally sighted subjects, but it is well known that for many ocular disorders, spacing and congestion of the optotypes on the chart a can have more pronounced effects of their visual acuity scores.

Journal of Physics: Conference Series **772** (2016) 012046 doi:10.1088/1742-6596/772/1/012046

Need to identify tests and procedures.
Because visual acuity scores can be significantly affected by the choice of visual acuity chart and testing procedures, all reports and records of visual acuity results should include specific information about the test and testing protocols. The chart should be identified. Record should be made of the illumination levels and note should be made about optical correction being worn by the persons being tested. There should be rules about guessing, stopping, correcting and feedback. While there is no universally accepted protocol, it is common for clinicians to encourage but not oblige guessing. A recommended procedure is that if half or more letters on a given row are read correctly, the patient should be asked to try to read any letters in the next row, and explicit permission should be given to guess when not quite sure. Usually, the patient is not told whether responses are correct or not, and the clinician does not point in order to help the patient locate the row or a letter within a row.

Future changes in visual acuity testing
Most measurement of visual acuity in clinical research is using charts printed on trans-illuminated plastic panels or on cards with a width about 65 cm or more designed for a standard viewing distance of 4 meters. The size range extends from logMAR =1.0 to -0.3. (i.e., 20/200 to 20/10) In order for such a chart to be presented on a computer-controlled display, a screen with 4 or more megapixels is required. But both the research and clinical ophthalmic communities are beginning to use computer-controlled tests of visual acuity. Computer-controlled tests enable easy variations to the visual stimulus by allowing choices of different sets of optotypes, and changes of the layout from chart format, to single rows, to isolated letters or to single letters with nearby flanking lines. Stimulus variables such as luminance, contrast, color and presentation times become easily varied. With computerized control, detailed data can be recorded, instant scores and reliability values can be generated and different psychophysical testing strategies and procedures can be implemented. The subjects' responses could be generated by their own keyboard or button-press responses or by speech. And there can be more sophisticated and potentially useful performance information available from response times or detailed error analysis.
With computerization of visual acuity testing there should be much better control of the testing procedures but there will be much more diversity in the tests being used. It can be expected that, for a given test, better measurements will be obtained, but with an expanded diversity of tests it will become more difficult to compare the results from one test to another. Thus, it will become more important for clinicians and researchers to record and report the details of their visual acuity tests, and for collecting data for research or monitoring individual patients, the test procedures should remain consistent.

References
Bailey IL and Lovie JE 1976 New design principles for visual acuity letter charts. *Am J Optom Physiol Opt* **53** 740
Bailey IL 1988 Measurement of visual acuity: towards standardization. In *Vision Science Symposium: a Tribute to Gordon Heath, Indiana University 219*
Bennett,AG 1965 Ophthalmic test types. *Br J Pbysiol Opt* **22** 238
British Standard 1968 *Test charts for determining distance visual acuity: BS 4274-1968*, London, British Standards Institute.
Ferris FL, Kassoff A, Bresnick GH and Bailey IL 1982 New visual acuity charts for clinical research *Amer. J. Ophthal.* **94**, 92
Green J 1905 Notes on the clinical determination of the acuteness of vision including the construction and graduation of optotypes, *Trans Am Ophthalmol Soc* **10** 644
Sloan LL 1959 New test charts for the measurement of visual acuity at far and near distances. *Am J Ophthalmol* **48** 807

A strategy for measuring patient preferences to incorporate in benefit-risk assessment of new ophthalmic devices and procedures

R W Massof[1], C Bradley[2]
[1]Professor of Ophthalmology, Johns Hopkins University, Baltimore, MD, USA
[2] Research Associate, Wilmer Eye Institute, Johns Hopkins University, Baltimore, MD, USA

E-mail: bmassof@jhmi.edu

Abstract. The U.S. Food and Drug Administration recently released guidance documents explaining that measurement of patient preferences should be considered during the pre-market approval process to specify patients' tolerances for risk and perspectives on benefit when assessing the benefit-risk profile of new medical devices. For ophthalmological patients, the typical primary clinical outcome is a visual impairment measure. Especially for surgically-implanted devices, the benefit a specified improvement in vision measures must be translated to a patient-specific benefit of the improvement in ability to function in everyday life. We developed, and validated with simulations, a strategy for measuring an individual patient's ability to function and the overall benefit to that patient of specified improvements in functional ability. Our strategy employs Rasch analysis to measure changes in functional ability; multidimensional scaling to measure patient-specific benefits of changes in functional ability; and structural equation modeling to cross-walk patient preferences for functional ability changes to changes in visual impairment measures.

1. Introduction
Recently the U.S. Food and Drug Administration (FDA) issued guidelines for industry and FDA staff on the consideration of patient preferences when evaluating the benefit-risk profiles of medical devices during the pre-market approval process [1-2]. In ophthalmology, the potential benefits of surgically-implanted devices designed to improve vision (e.g., intraocular lens, retinal stimulator for prosthetic vision), typically are evaluated psychophysically with clinical visual impairment measures (e.g., visual acuity, visual fields). Measures of medical device benefits and risks to the patient necessarily entail potentially idiosyncratic personal preferences when judging the value of what is to be gained as a consequence of improved visual impairment measures. For example, visual acuity loss means different things to different people depending on which of the patient's activities are limited or precluded by the vision loss and the value to the patient of being able to perform those activities with a criterion level of ease. The challenge in addressing this problem has three parts: 1) identify which activities are important to the patient and are limited or precluded by the patient's visual impairment; 2) measure how much value the patient assigns to different levels of improvement of his or her ability to perform the identified activities; and 3) cross-walk the values assigned by the patient to improved ability to perform his or her self-selected activities to corresponding improvements in clinical visual impairment measures. Here we present our strategy for solving this three-part problem.

2. Background to Our Strategy
Relevant to part 1 of the problem, we have developed and routinely employ an adaptive rating scale instrument designed to measure a patient's functional ability to perform vision-dependent activities. This instrument, the Activity Inventory (AI), consists of an item bank with 510 items that describe everyday activities within a hierarchical framework [3-5]. At the top level of the hierarchy, the activities are organized according to the objective they serve (Daily Living, Social Interactions, or Recreation). Fifty of the 510 items are classified as "Goals", which describe what the person is intending to accomplish (e.g., prepare daily meals, manage personal finances, entertain guests). The other 460 items in the bank, which are nested under the Goals, are classified as "Tasks". Tasks describe specific cognitive and motor activities that must be performed to accomplish their parent Goal (e.g., cut food, read recipes, measure ingredients, read bills, write checks, sign name). First, the patient is presented a Goal item and asked to rate how important it is for them to be able to accomplish it without depending on another person (ranging from "not important" to "very important"). If the patient responds "not important" the interviewer moves on to the next Goal. If the Goal is of non-zero importance to the patient, the patient is then asked to rate how difficult the Goal is to achieve without depending on another person (ranging from "not difficult" to "impossible"). If the patient responds "not difficult", the interviewer moves on to the next Goal, otherwise the patient is asked to rate the difficulty of Tasks nested under that Goal. Difficulty ratings of AI items were obtained from 3200 low vision patients (disabling chronic visual impairments in both eyes) and item measures and response category thresholds were estimated from Rasch analysis for all 510 items in the AI item bank [6].

With respect to part 2 of the 3-part problem, we must estimate the utility (*i.e.*, a variable representing value, benefit, satisfaction, happiness, etc.) assigned by each patient to a hypothetical risk-free intervention that would make each of the identified important and difficult Goals in the AI not difficult to that patient. Psychometric tools that have been used in the past to estimate utilities include stated-preference methods such as partial profile discrete choice [7], pairwise comparison [8], best-worst scaling [9], and iterative bidding games [10]. The pairwise comparisons methods are inspired by and analysed within the framework of Thurstone's law of comparative judgment [11]; discrete choice methods are loosely grounded in conjoint measurement theory [12]; iterative bidding games are framed by game theory [13]; and best-worst scaling is justified by post hoc probabilistic measurement models

[14]. The underlying theory is important because the lack of consensus about the value of items requires that utilities be estimated from efficient repeated measures on each person, as opposed to the "crowd-sourcing" approach of Rasch analysis.

To solve the third part of the 3-part problem, we must be able to predict changes in the difficulty of Goals and Tasks in the AI from changes in visual impairment measures. From our previous work, we have learned that correlations between person measures estimated from difficulty ratings of AI Goals and different subsets of AI Tasks are the result of two independent vision factors – one factor represents visual processing underlying the recognition and identification of objects (the ventral stream or "what" visual pathway) and the other factor represents visual processing underlying perception of spatial relations and motion (the dorsal stream or "where" visual pathway) [6,11]. The first factor is affected by changes in visual acuity and the second factor is affected by visual field loss and scotomas (blind spots). By determining how much of each factor is demanded by each item in the AI, and specifying the patient's visual ability on each factor from visual impairment measures, we will be positioned to predict the change in difficulty of each Goal as a result of intervention that causes changes in the patient's visual impairment measures. A model giving us that capability will enable us to crosswalk utilities attached to the importance and difficulty ratings of activities back to clinical visual impairment measures.

3. Simulation of the Estimation of Utilities from Triadic Comparisons of Differences

The utility to an individual patient of an intervention that makes an activity easy to perform will depend on both the importance (I) and difficulty (D) of that activity to the patient, i.e., $U(I,D)$. The marginal utilities are then $mU(I)=\partial U(I,D)/\partial I$ and $mU(D)= \partial U(I,D)/ \partial D$. Relative utilities of individual activities range from 0 (when I=0 or D=0) to 1.0 (when $I=I_{max}$ and $D=D_{max}$). Activities in the AI having unique combinations of importance and difficulty ratings, i.e., (I_A,D_A), (I_B,D_B), (I_C,D_C), are presented to the patient three at a time and the patient is asked to rank-order the activities according to the amount (money, time, effort, pain, etc.) he/she is willing to invest in making the activity easy to perform. The patient is then asked if the amount he/she is willing to invest to improve the middle-ranked activity, (I_B,D_B), is closer to the amount for the top-ranked activity, (I_A,D_A), or to the amount for the bottom-ranked activity, (I_C,D_C). If the response is (I_C,D_C), then rank scores are assigned to the differences in utilities between each pair of activities, i.e., 3 for (I_A,D_A)- (I_C,D_C), 2 for (I_A,D_A)-(I_B,D_B), and 1 for (I_B,D_B)-(I_C,D_C) and entered into the appropriate cells of a dissimilarity matrix. All possible combinations of activities with unique (I,D) ratings are presented 3 at a time are presented in this way and difference ratings in each cell of the dissimilarity matrix are averaged. Multidimensional scaling (MDS) is used to estimate the position of each activity on a number line (or in a multidimensional space), which represents a continuous utility scale. These data collection steps were simulated with a computer program that had marginal utilities assigned to each importance rating and to each difficulty rating (ranging from 0 to 1). Total utility of each activity was defined as the product of the marginal utilities. The simulation proved that MDS applied to the dissimilarity matrix generated by ranking differences between pairs of utilities in triadic comparison trials can estimate the assigned marginal and total utilities.

4. A Structural Equation Model to Crosswalk Visual Impairments to Visual Ability

The primary outcomes from the ophthalmologist's perspective are visual impairment measures such as visual acuity, contrast sensitivity, and scotoma maps. For the ophthalmologist to interpret these measures in terms that are meaningful to the patient, they must be mapped onto visual ability. To cross walk visual impairments to visual ability, we need to develop a model that explains how vision factors 1 and 2 are related to visual impairments.

To evaluate how well the two vision factors can predict visual ability, we employed confirmatory factor analysis (*i.e.*, a structural equation model). The model we constructed has seven latent variables: the two independent vision variables identified with the exploratory factor analysis and five health state variables estimated from a wide variety of measures and indicators. The model for the two vision variables included visual acuity, contrast sensitivity, a variety of ocular and visual system disease diagnoses, and visual symptoms. The weights on the paths from the latent vision variables to the functional ability measures were set to the values estimated from the exploratory factor analysis, so that the health state variables were restricted to modeling the unexplained variance.

Vision factor 1 is equal to log binocular visual acuity and explains 73% of the observed variance in visual acuity. Vision factor 2 depends on the spatial distribution of binocular blind spots in the visual field relative to the area of retina in the dominant eye used for fixation (i.e., preferred retinal locus or PRL). The precise nature of this dependence is currently an active area of research. The aim of this work is to use clinical measures of visual impairments (binocular visual acuity and binocular scotoma maps) to predict the values of vision factor 1 and vision factor 2. Psychometric measures of psychological state, cognitive state, physical functioning ability, and diagnostic indicators (*i.e.*, clinical signs from tests and examinations and symptoms from a structured health history survey) are used to estimate the 5 latent health state variables (effect modifiers). Using the estimated latent vision and health state variables, the structural equation model is then employed to predict patient difficulty ratings of AI Goals and Tasks. The distribution across patients of the ratio of mean square residuals (observed uncertainty variance) to the expected uncertainty variance (*i.e.*, information weighted) is used to assess the validity of the structural equation model.

5. Discussion

A structural equation model is used to estimate the two latent visual ability variables ("what" and "where" components of visual processing in the cortex) from clinical visual impairment measures, as well as estimating latent visual ability effect modifiers from other health state data. These estimates are then used to predict patient difficulty ratings of activities in the AI. Given a particular patient's *(I,D)* ratings of activities in the AI, and population-based marginal utiliites on *I* and *D* ratings and on their combinations, the model can be used to predict the utility to the patient of improvements in clinical measures of visual impairments.

6. References

[1] FDA's Guidance for Industry and Food and Drug Administration Staff; Factors to Consider When Making Benefit-Risk Determinations in Medical Device Premarket Approval and De Novo Classifications, issued on March 28, 2012.

[2] FDA's Draft Guidance for Industry, Food and Drug Administration Staff, and Other Stakeholders; Patient Preference Information – Submission, Review in PMAs, HDE Applications, and De Novo Requests, and Inclusion in Device Labeling, issued on May 18, 2015.

[3] Massof R W, Hsu C T, Baker F H, Barnett G D, Park W L, Deremeik J T, Rainey C, Epstein C 2005 Visual disability variables. I. The importance and difficulty of activity goals for a sample of low vision patients *Arch Phys Med Rehabil* **86** 946-53

[4] Massof R W, Hsu C T, Baker F H, Barnett G D, Park W L, Deremeik J T, Rainey C, Epstein C 2005 Visual disability variables. II. The difficulty of tasks for a sample of low vision patients *Arch Phys Med Rehabil* **86** 954-67

[5] Massof R W, Ahmadian L, Grover L L, Deremeik J T, Goldstein J E, Rainey C, Epstein C, Barnett G D 2007 The Activity Inventory: An adaptive visual function questionnaire *Optom Vision Sci* **84** 763-74

[6] Goldstein J E, Chun M W, Fletcher D C, Deremeik J T, Massof R W 2014 Low Vision Research Network Study Group. Visual ability of patients seeking outpatient low vision services in the United States *JAMA Ophthal* **132** 1169-77

[7] Kessels R, Bradley J, Goos P 2011 Bayesian optimal designs for discrete choice experiments with partial profiles *J Choice Modelling* **4** 52-74

[8] Bradley R A, Terry M E 1952 Rank analysis of incomplete block designs, I. The method of paired comparisons *Biometrika* **39** 324-45
[9] Flynn T N, Louviere J J, Peters T J, Coast J 2007 Best–worst scaling: What it can do for health care research and how to do it *J Health Econ* **26** 171–89
[10] Brookshire D S, Thomas D C 1981 The advantages of contingent valuation methods for benefit-cost analysis *Public Choice* **36** 235-52
[11] Thurstone L L 1927 A law of comparative judgment *Psychol Rev* **34** 273-86
[12] Luce R D, Tukey J W 1964 Simultaneous conjoint measurement : a new scale type of simultaneous measurement *J Math Psychol* **1** 1-27
[13] Milgrom P, Weber R 1982 A theory of auctions and competitive bidding *Econometrica* **50** 1089-122
[14] Marley A A J, Flynn T N, Louviere J J 2008 Probabilistic models of set-dependent and attribute-level best-worst choice *J Math Psychol* **52** 281-96

7. Acknowledgments

This study is funded by Janssen Research and Development, Raritan, NJ (USA)

A Note on the Usefulness of the Behavioural Rasch Selection Model for Causal Inference in the Social Sciences

Matthew P. Rabbitt[1]

Economic Research Service, U.S. Department of Agriculture, 355 E Street, SW, Washington, DC, 20024-3221, USA

E-mail: matthew.rabbitt@ers.usda.gov

Abstract. Social scientists are often interested in examining causal relationships where the outcome of interest is represented by an intangible concept, such as an individual's well-being or ability. Estimating causal relationships in this scenario is particularly challenging because the social scientist must rely on measurement models to measure individual's properties or attributes and then address issues related to survey data, such as omitted variables. In this paper, the usefulness of the recently proposed behavioural Rasch selection model is explored using a series of Monte Carlo experiments. The behavioural Rasch selection model is particularly useful for these types of applications because it is capable of estimating the causal effect of a binary treatment effect on an outcome that is represented by an intangible concept using cross-sectional data. Other methodology typically relies of summary measures from measurement models that require additional assumptions, some of which make these approaches less efficient. Recommendations for application of the behavioural Rasch selection model are made based on results from the Monte Carlo experiments.

1. Introduction

The social sciences are often interested in measuring properties or attributes of individuals with no clearly defined physical referent (intangible concepts), such as attitudes, values, beliefs, well-being, health, or ability. For the measurement of these concepts to be valid and reliable, it must be supported by both social science and measurement theory. Social science theory allows the researcher to clearly define the concept that is to be measured and develop an instrument while measurement theory provides the foundation for establishing and validating the instrument's corresponding scale. Many social scientists rely on item response theory (IRT) models, such as the Rasch model [3] to develop and validate their scales.

Researchers in the social and physical sciences undertake the measurement of concepts with and without physical referents to confirm, reject, and refine research hypotheses. Findings from the testing of hypotheses advance theory and inform policymakers. Properly testing a hypothesis requires the careful design of an experiment, whether it is in the laboratory or using a social survey. Physical scientists often have the advantage of using randomized experiments with carefully constructed treatment and control groups. While randomized experiments are arguably the best way to test a hypothesis, they are often costly, time consuming, and for social scientists, one must consider the ethical

[1] The views expressed in this paper are those of the author and do not necessarily reflect those of the Economic Research Service or the U.S. Department of Agriculture.

issues associated with human subjects. Alternatively, researchers may rely on natural experiments or statistical methodology to identify causal relationships when randomized experiments are not feasible.

Estimating causal relationships is particularly challenging when using survey data, as key characteristics essential to social science models may be omitted. When this occurs bias is introduced into the analysis, potentially contributing to counterintuitive or misleading findings. This occurs because the treatment and control groups are not randomized, ensuring the sample is not representative of the population intended for analysis. Self-selection bias ensues because individuals select into treatment and control groups based on their observable and unobservable characteristics.

This paper examines the usefulness of a recently proposed model, the behavioural Rasch selection model [2], by conducting a Monte Carlo study of the model's properties. The Monte Carlo study undertaken here is the first to examine the behavioural Rasch selection model and will be used to provide guidance for researchers interested in implementing the model in their own analyses. The behavioural Rasch selection model is particularly useful for social science researchers because it is capable of estimating the causal effect of a binary treatment effect on an outcome that is represented by an intangible concept using cross-sectional data. The behavioural Rasch selection model maintains all of the Rasch model assumptions while incorporating a multivariate behavioural component and addressing selection on contemporaneous unobservable characteristics. Identification of the model is achieved using exclusion restrictions in the form of instrumental variables.

2. Methodology

IRT models, such as the Rasch model, are particularly useful for measuring an individual's properties or attributes when no clear physical referent exists by formulating them as latent traits. For the purposes of this paper, the Rasch model is framed to measure an individual's ability; however, it could just as easily be altered to measure an individual's well-being, health, food security, or attitudes and beliefs. The Rasch measurement model assumes individual i's underlying unobservable (latent) index of ability is θ_i, with the property that higher values of the index correspond to greater ability. Assuming there are N individuals administered J binary questions (e.g., yes or no.) that capture different levels of ability, then the probability that the individual affirms the jth question is

$$P(Y_{ij}=1|\theta_i,\delta_j)=\frac{\exp(\theta_i-\delta_j)}{1+\exp(\theta_i-\delta_j)}, i=1,...,N; j=1,...,J \quad (1)$$

where exp(·) is the exponential function and δ_j is a threshold or "calibration" parameter. The model further assumes that the individual's responses to each question are independent, conditional on θ_i, and that the factor loadings (discrimination parameters) are constrained to be equal across all items and normalized to one.

An attractive property of the Rasch model for statistical inference is that if all individuals' responses to the questions follow this model and if the individuals answer all J items, then their latent ability can be ranked and compared using simple counts of the number of affirmed items. Formally, the count of affirmed items, commonly referred to as the raw score, is a sufficient statistic for θ_i.

Summary measures have proven very useful for social scientists interested in examining the relationships between an intangible concept and other variables. Rasch scores may be used to estimate continuous regression models, such as Ordinarily Least Squares (OLS). However, for these models to be identified, all observations with extreme values (all affirmative or negative responses) to an instrument's questions (the majority of respondents for many instruments and surveys) may have to be dropped from the analysis sample prior to estimation depending on how the Rasch model is estimated (e.g., joint or conditional maximum likelihood.). Alternatively, the raw score may be used to estimate count data regression models for samples of individuals with no missing responses to an instrument's questions. Missing responses many also be imputed, but the accuracy of these imputations must be assessed.

The Rasch model can also be extended to incorporate a multivariate behavioural component for cases in which social scientists are interested in examining the relationship between the latent trait and other variables. Specifically, one may assume that individual i's latent ability, θ_i, depends on an observed treatment (T_i), control characteristics (X_i), and an unobserved variable (e_i) such that

$$\theta_i = \beta_T T_i + \beta_X{}' X_i + e_i. \quad (2)$$

It is common practice to assume that $e_i \sim N(0, \sigma^2)$ for identification purposes. Combining specifications (1) and (2) yields the behavioural Rasch model[2] [5].

Using the behavioural Rasch model has several advantages over summary measures for this type of inference. Methodologically, the behavioural Rasch model is similar to a regression model where the dependent variable is the Rasch score. The main difference is that the behavioural Rasch model estimates specifications (1) and (2) jointly, while regression models involving the Rasch score estimate these parameters in two steps. Regression models involving Rasch scores may also draw different samples, as individuals with extreme scores may be excluded, depending on the methodology used to produce the Rasch scores. The behavioural Rasch model, on the other hand, can incorporate individuals with extreme scores. Another advantage of the behavioural Rasch model is its ability to impute missing values by virtue of the Rasch measurement model's properties.

Estimating the causal effect of a treatment on an intangible outcome is further complicated by the endogeneity of the decision to select into the treatment group (to be treated). Behavioural Rasch models are able to control for selection on an individual's observable characteristics, but do not account for selection on an individual's unobservable characteristics. More formally, the individual's decision to select into the treatment group is

$$T_i = \begin{cases} 1 \text{ if } \alpha_X{}' X_i + \alpha_Z{}' Z_i + u_i > 0 \\ 0 \text{ otherwise} \end{cases} \quad (3)$$

where T_i and X_i are defined above, Z_i is a set of instrumental variables, and u_i is an unobserved variable. For identification purposes, one may assume $u_i \sim N(0, 1)$[3]; however, it need not be. The resulting model is consistent with a probit model for the decision to select into the treatment group.

The endogeneity of the treatment can be formalized, as in [2] and [4], by assuming the error-component in specification (2) can be decomposed into u_i and e_i^*, such that $e_i = \lambda u_i + e_i^*$. Thus, the individual's latent ability index can be respecified as

$$\theta_i^* = \beta_T T_i + \beta_X{}' X_i + \lambda u_i + e_i^* \quad (4)$$

where λ is an unknown factor loading (selection) parameter to be estimated, and e_i^* represents the new unobserved variable after controlling for observed and unobserved variables (heterogeneity). It is assumed that $e_i^* \sim N(0, \eta^2)$. Correlation between the treatment variable (T_i) and the latent ability is generated through the factor loading parameter, λ. If λ is estimated to be nonzero, u_i influences the individual's selection into the treatment group and the likelihood of affirming a question, rendering the baseline behavioural Rasch model inconsistent and biased.

An estimator for the behavioural Rasch selection model, which combines specifications (1)-(3), was proposed by [2] and tested on survey data. In this paper, [2] derived the likelihood function for the behavioural Rasch selection model, which is as follows:

[2] The behavioral Rasch model is equivalent to the person-level explanatory Rasch model described in [5].

[3] Alternatively, one may assume that u_i is time-invariant and estimate a model that combines specifications (1)-(4) using panel data methods. For an example of this approach, see [1].

$$L(\alpha,\beta,\delta,\eta^2,\lambda)=\prod_{i=1}^{N}\left\{T_i\int_{-\alpha_X X_i-\alpha_Z Z_i}^{\infty}\int_{-\infty}^{\infty}\prod_{j=1}^{J}\left[\frac{\exp(q_{ij}(\theta_i^*-\delta_j))}{1+\exp(q_{ij}(\theta_i^*-\delta_j))}\right]\frac{1}{\eta}\varphi\left(\frac{e_i^*}{\eta}\right)de_i^*\varphi(u_i)du_i\right.$$
$$\left.+(1-T_i)\int_{-\infty}^{-\alpha_X X_i-\alpha_Z Z_i}\int_{-\infty}^{\infty}\prod_{j=1}^{J}\left[\frac{\exp(q_{ij}(\theta_i^*-\delta_j))}{1+\exp(q_{ij}(\theta_i^*-\delta_j))}\right]\frac{1}{\eta}\varphi\left(\frac{e_i^*}{\eta}\right)de_i^*\varphi(u_i)du_i\right\} \quad (5)$$

where the ‘i’ subscript denotes the ith sample household (i = 1, …, N). The proposed model allows for all of the parameters to be jointly estimated using full information maximum likelihood (FIML). It is useful to note that the factor loading parameter, λ, is a sufficient statistic for exogeneity of the treatment decision.

In the current paper, I estimate the likelihood function specified in (5) using the Stata code provided by [2]. The Stata program approximates the two-dimensional integral using Gauss-Steen and Gauss-Hermite quadrature techniques for the first and second integrals, respectively. All models programs were executed using State 13 2-Core Multiprocessor statistical software.

3. Monte Carlo Study Design

The data generating process was designed to examine the performance of the behavioural Rasch selection model with samples of varying size (in terms of number of individuals). The instrument length was held constant across the samples to assess how many individuals would need to be surveyed in real-world data for reliable inference. Four separate experiments were performed using 500, 1,000, 5,000, and 10,000 observations and an instrument that consisted of 5 questions. For each sample size, 1,000 repetitions were completed and 8 quadrature points were used for each level of integration to approximate the integrals.

Data for each experiment were generated according the following assumptions.

1. $X_i \sim U(0,1)$
2. $Z_i \sim U(0,1)$
3. $T_i^* = \alpha_X X_i + \alpha_Z Z_i + u_i > 0 : u_i \sim N(0,1)$
4. $Y_{ij} = \dfrac{\exp(\beta_T T_i + \beta_X X_i + \lambda u_i + e_i^* - \delta_j)}{1+\exp(\beta_T T_i + \beta_X X_i + \lambda u_i + e_i^* - \delta_j)} : e_i^* \sim N(0,\eta^2)$
5. $[\alpha_X, \alpha_Z, \alpha_{Constant}] = [1,1,-1]$
6. $[\beta_{Treatment}, \beta_X, \beta_{Constant}, \delta_1, \delta_2, \delta_3, \delta_4, \delta_4, \eta^2, \lambda] = [1,1,-1,-0.6,-0.3,0.0,0.3,0.6,\exp(1),1]$

4. Results

Estimates for the Monte Carlo experiments were obtained by estimating the behavioural Rasch selection model using simulated data generated by the procedure outlined above. Summary statistics for the experiments are reported in Table 1, and include each parameter’s average, standard deviation, and mean square error. The top, middle, and bottom panels contain summary information for the selection, outcome, and error-component equations, respectively. The columns report the true value and summary information, separately, for each of the experiments by sample size.

Estimates from the selection equation presented in the top panel of Table 1, which was estimated jointly with the other equations, performed very well, regardless of sample size. Average parameter values were close to their true values, even in smaller sample sizes, suggesting modest bias. For example, the percent bias in the smallest sample (N = 500) ranged from .03 to 1 percent, while bias in the largest sample (N = 10,000) ranged from 0.1 to 0.2 percent. As the sample size increases, the standard deviation of the selection equation parameters declines in magnitude.

Journal of Physics: Conference Series **772** (2016) 012048 doi:10.1088/1742-6596/772/1/012048

Table 1.Monte Carlo Summary Statistics for the Behavioural Rasch Selection Model: Average, Standard Deviation, and Mean Square Error of the Coefficients.

	True Value	Sample Size			
		500	1,000	5,000	10,000
		Selection (Treatment) Equation			
α_X	1.000	1.010	0.999	1.003	1.001
		(0.206)	(0.137)	(0.067)	(0.045)
		[0.043]	[0.019]	[0.005]	[0.002]
α_Z	1.000	0.997	1.009	1.000	1.002
		(0.204)	(0.145)	(0.064)	(0.044)
		[0.041]	[0.021]	[0.004]	[0.002]
$\alpha_{Constant}$	-1.000	-1.006	-1.007	-1.002	-1.002
		(0.162)	(0.110)	(0.050)	(0.035)
		[0.026]	[0.012]	[0.003]	[0.001]
		Outcome Equation			
$\beta_{Treatment}$	1.000	0.979	1.031	0.988	0.999
		(0.933)	(0.615)	(0.268)	(0.191)
		[0.871]	[0.379]	[0.072]	[0.037]
β_X	1.000	1.031	0.996	1.010	1.003
		(0.541)	(0.370)	(0.154)	(0.113)
		[0.293]	[0.137]	[0.024]	[0.013]
$\beta_{Constant}$	-1.000	-1.006	-1.015	-0.997	-1.002
		(0.364)	(0.241)	(0.109)	(0.077)
		[0.132]	[0.059]	[0.012]	[0.006]
δ_1	-0.600	-0.608	-0.606	-0.598	-0.601
		(0.175)	(0.124)	(0.056)	(0.038)
		[0.031]	[0.015]	[0.003]	[0.001]
δ_2	-0.300	-0.304	-0.303	-0.299	-0.299
		(0.174)	(0.121)	(0.055)	(0.037)
		[0.030]	[0.015]	[0.003]	[0.001]
δ_4	0.300	0.305	0.296	0.302	0.299
		(0.170)	(0.117)	(0.056)	(0.039)
		[0.029]	[0.014]	[0.003]	[0.002]
δ_5	0.600	0.605	0.602	0.602	0.600
		(0.169)	(0.120)	(0.055)	(0.039)
		[0.029]	[0.014]	[0.003]	[0.001]
		Error Components			
$\ln(\eta^2)$	1.000	0.923	0.977	0.993	0.995
		(0.269)	(0.154)	(0.064)	(0.044)
		[0.078]	[0.024]	[0.004]	[0.002]
λ	1.000	1.027	0.990	1.008	1.001
		(0.606)	(0.402)	(0.172)	(0.125)
		[0.368]	[0.162]	[0.029]	[0.015]

Note: Models estimated using simulated data with varying number of individuals and a constant instrument length of 5 items. Standard deviations are in parenthesis and mean square errors are in brackets.

The primary coefficient of interest, an estimate of the treatment's effect, is shown in the middle panel of Table 1. The behavioural Rasch selection model is able to obtain reliable estimates of this effect, even in smaller samples, albeit biased downwards (with the exception of the N = 1,000 experiment where it is biased upwards). The model appears to perform best with samples that are greater than or equal to 5,000 observations. For the experiments with the two largest sample sizes, the percent bias ranges from -1.2 percent to -0.1 percent. Significant gains in terms of mean square error are also observed moving from a sample size of 1,000 to 5,000 or more observations.

The bottom panel of Table 1 contains estimates of the model's error components. Of particular interest are the parameter estimates for the factor loading, λ, which describes the degree of selection in one's sample. Estimates of the selection parameter are underestimated in the experiment with the smallest sample size and overestimated in the other experiments. The percentage bias for this parameter was relatively small across all experiments, ranging from 2.7 percent to 0.1 percent in the experiments with 500 and 10,000 observations, respectively. Once again, significant gains in terms of mean square error are observed for this parameter moving from a sample size of 1,000 to 5,000 or more observations. Estimates of the variance for the latent trait are reasonable in samples with 1,000 or more observations.

5. Conclusion

Social scientists are often interested in examining causal relationships for the purposes of informing theory or policy. Examination of these types of relationships become even more challenging when the outcome of interest is represented by an intangible concept. When intangible concepts are involved in research, the social scientist must appeal to measurement methodology, such as item response theory, to develop and validate a scale. After a scale is constructed, the social scientist may use various approaches to model the relationships of interest. In the current paper, I examine the usefulness of applying the recently proposed behavioural Rasch selection model to situations in which the researcher wishes to estimate the causal effect of a binary treatment effect on an outcome that is represented by an intangible concept using cross-sectional data.

Results from the Monte Carlo experiments suggest that the behavioural Rasch selection model is consistent and has modest bias in samples of size 500 or more. While the Monte Carlo experiments did not assess the effect of instrument length, the choice of an instrument with only 5 questions provides confidence that performance will only improve as the length of the instrument increases. While the behavioural Rasch selection model performed reasonably in all of the experiments, significant gains, in terms of mean square error, were observed for samples with 5,000 or more observations. Thus, it is recommended that the model be applied to samples with at least 5,000 observations. For those with smaller samples, careful attention should be paid to the statistical power they will have in their sample. In these cases it may be useful to conduct a power analysis.

Acknowledgements

The author would like to thank David C. Ribar at the Melbourne Institute of Applied Economic and Social Research for helpful comments.

References

[1] Moffitt, R. A., & Ribar, D. C. (2016). Rasch Analyses of Very Low Food Security Among Households and Children in the Three City Study. *Southern Economic Journal*, 82(4), 1123-1146.

[2] Rabbitt, M. P. (2013). Measuring the Effect of Supplemental Nutrition Assistance Program Participation on Food Insecurity Using a Behavioral Rasch Selection Model.Working Paper No. 13-20. University of North Carolina at Greensboro, Department of Economics.

[3] Rasch, G. (1960). Studies in Mathematical Psychology: I. Probabilistic Models for Some Intelligence and Attainment Tests.

[4] Terza, J. V. (2009). Parametric Nonlinear Regression with Endogenous Switching. *Econometric Reviews*, 28(6), 555-580.

[5] Wilson, M., & De Boeck, P. (2004). Descriptive and Explanatory Item Response Models. *Explanatory Item Response Models: A Generalized Linear and Nonlinear Approach*, 43-74.

IMEKO2016 TC1-TC7-TC13 IOP Publishing
Journal of Physics: Conference Series **772** (2016) 012049 doi:10.1088/1742-6596/772/1/012049

Assessment as a tool to Understand Students' Conceptions of the Structure of Matter

Linda Morell and Mark Wilson

BEAR Center
Graduate School of Education
University of California, Berkeley
California, USA

E-mail: lindamorell@berkeley.edu

Abstract. Learning progressions built and based on empirical evidence provide a foundation for advancing the state of the art of teaching. This paper contributes to the advancement of the idea that assessment can provide empirical evidence to examine the underlying theory of cognition and provides a method for developing and verifying the theory. Using a construct within the science domain of Structure of Matter as a basis for discussion, issues of how to address theoretical ambiguities and practical dilemmas are explored.

1. Introduction

What is the nature of the learning progression in the content domain of Structure of Matter? To address this, we put forth a hypothesized progression, which is based on the literature of how students understand the structure of matter [1-7].

Six components or constructs were identified and articulated into the learning progression. There are four core constructs and two auxiliary ones. Macro Properties (MAC), Changes of State and Other Physical Changes (PHS), Particulate Explanations of Physical Change (EPC), and Particulate Explanations of Chemical Change (ECC) are the core constructs while Measurement and Data Handling (MDH), and Density and Mass & Volume (DMV) are auxiliary. Multiple links connect these six variables to form the learning progression. The focus of this paper is on one of the core constructs, ECC.

Student understanding of the Particulate Explanations of Chemical Changes (ECC) construct, was initially hypothesized to have six progressive levels (Levels 0-5). At Level 0 or the lowest level of understanding, students think that total mass may change in all types of substance change. At Level 1, students think that, the total volume of the substances involved may change yet the total mass remains the same in both chemical and physical changes. At Level 2, students understand that chemical changes can be distinguished from physical changes, i.e., from dissolving, or heating, or changes of state, by the appearance of new substances. These may have properties quite different from the properties of the original substances, and the changes are often irreversible. Also, at Level 2, students also understand that, in chemical changes, while mass is conserved, many other properties may change. At Level 3, students think that the smallest possible particles of a substance may either be atoms, or combinations of different kinds of atoms. The latter are called molecules. Also, different substances can be made of the same types of atoms, but are composed of different combinations of these atoms, i.e., different molecules. At Level 4, students understand that, (a) in a chemical change the molecules/atoms of the products will be different from those of the reactants, but the total numbers of atoms will remain the same, and (b) in physical changes, the atoms/molecules remain the same, their numbers and their total mass remains the same, but they may be arranged in different ways. Finally, at Level 5, students understand that, in a chemical change the atoms will be unchanged, but may combine in new ways to form different molecules; the molecules of any substance may just break apart to give two new substances.

2. Methods and Data Sources

We approached this investigation using a construct modeling approach as recommended in *Developing Assessments for the Next Generation Science Standards* [8]. This approach allows for an integrated means of developing assessments that support and build on the goals for curriculum and intentions of instruction. Specifically, researchers used the BEAR Assessment System [9,10] to provide guiding principles for the development and revision of the assessment materials.

Researchers developed assessments to measure students' understanding for each of the theorized levels of the ECC construct. Before administering the assessments on a large scale, all items went through a rigorous review, which included review panels by teams of teachers, scientists, and other content experts; cognitive interviews with middle school students; and small-scale administrations to ensure students were responding to items as intended.

The final test consisted of 56 items, which included 17 multiple-choice, 9 open-ended, 3 drawing, and 27 fill-in-blank items. Each item was hypothesized to correspond to the theoretical developmental levels described within the construct map. Figure 1 shows an example of a multiple-choice item. The correct choice is "A" and the targeted construct map level is Level 4 on the ECC construct map.

SAMPLE ITEM: In a chemical reaction, the number of reactant atoms:

A. Is equal to the number of product atoms.
B. Is greater that the number of product atoms.
C. Cannot be determined.
D. Is less than the number of product atoms

Figure 1: Sample ECC item.

Student responses to the 56 items were scored and calibrated using the partial credit model. After calibration, the items were reviewed in light of the construct map. A total of 1,086 middle grade students participated in this research.

3. Results and Discussion

For the analysis each component including the construct map, items, scoring guides, student responses, and the analysis were examined. We focus on the construct levels and how the empirical evidence in this paper informs the levels. As the ordering of the construct levels is an important part of the empirical testing of the hypothesized construct map, we examined the relationship between the theoretical progression and the empirical results. We found that the a priori level assignments of items did not align as anticipated with the hypothesized levels in the construct map. Hence, we explored the possibility of an alternative structure within the item sequences in the construct. We started with a re-examination of the results from the psychometric analyses and also included a

Journal of Physics: Conference Series **772** (2016) 012049 doi:10.1088/1742-6596/772/1/012049

qualitative examination of the items within the construct and considered whether they might be grouped in relation to similarities within the items of each group. This work was done through research meetings with content experts and teachers. Interpretation of the analysis led researchers to three progressive levels rather than six. And, the types of chemical changes were parsed into "strands" to capture their differences.

The ECC items can be categorized into three groups, which are labeled as strands, with each strand being a distinct component of the main ECC construct. These three strands are described as follows:

- ECC-A: Chemical and physical changes in the inter-atomic combinations and in the arrangements of atoms and molecules
- ECC-B: Changes in macroscopic properties, which accompany chemical and/or physical changes
- ECC-C: Representations of elements, compounds, and different phases, in terms of arrangements of atoms and molecules

In ECC-A, students can explain chemical and physical changes in terms of the inter-atomic combinations and the arrangements of atoms and molecules. At the first level, students can recognize and explain molecular and atomic representations of physical and chemical changes. In the second level, students understand that in a chemical change, the atoms or molecules change the way they combine and form new materials. Students also know that there are empty spaces or a vacuum between atoms and molecules. Lastly, in the third level, students can construct diagrams to give molecular and atomic representations of both physical and chemical changes. They know that in a chemical change, the atoms stay the same, but in different molecules. The numbers of atoms involved remain the same for both physical and chemical changes.

In ECC-B, students can distinguish between chemical and physical changes through the observations of macroscopic properties. For the first level, students can distinguish between these changes for obvious, familiar cases, such as when a chunk of wood burns or when an ice cube melts. In the second level, students can do this for less obvious cases, such as when a candle is burned or when sugar is dissolved into tea. Students also understand that mass is conserved throughout these changes. In the last stage, students can recognize that the properties differ after a chemical change. They can also distinguish chemical from physical changes in unfamiliar cases, such as when solid iodine turns into a gas.

In ECC-C, students are able to represent elements, compounds, and different phases, in terms of the arrangements of atoms and molecules. At the first level, students can recognize representations of monatomic molecules. They are aware that atoms and molecules are not usually visible and may also have different sizes. In the second level, students can recognize representations of diatomic molecules. They also know that atoms and molecules have weight. At the topmost level, students can recognize and distinguish between diagrams of different polyatomic elements and compounds.

Within each of these three strands, the items could be divided into three levels, with the type of demand similar within each group, but different, both qualitatively and in expected level of demand, between the groups.

Figure 2 shows the initial ECC construct map (on the left) and how it changed based on the empirical results (on the right side).

Journal of Physics: Conference Series **772** (2016) 012049 doi:10.1088/1742-6596/772/1/012049

Initial ECC construct map

Levels	Description
Level 5: Chemical Changes Produce different molecules	In a chemical change, the atoms will be unchanged, but may combine in new ways to form different molecules: the molecules of any substance may just break apart to give two new substances
Level 4: Chemical & Physical Change - Particulate Explanations of Differences	In a chemical change, the molecules/atoms of the products will be different from those of the reactants, but the total numbers of atoms will remain the same In physical changes, the atoms/molecules remain the same, their numbers and their total mass remains the same, but they may be arranged in different ways.
Level 3: Atoms and Molecules	The smallest possible particles of a substance may either be atoms, or combinations of different kinds of atoms: the latter are called molecules. Different substances can be made of the same types of atoms, but are composed of different combinations of these atoms – i.e. different molecules.
Level 2: Chemical & Physical Change - Observable Differences	Chemical changes can be distinguished from physical changes (i.e. from dissolving heating, changes of state) by the appearance of new substances. These may have properties quite different from the properties of the original substances, and the changes are often irreversible. In chemical changes, whilst mass is conserved, many other properties may change; in those involving solids and liquids, products may be solid or liquid or both.
Level 1: Conservation of Matter	In both chemical and physical changes, the total volume of the substances involved may change but the total mass remains the same
Level 0: Matter not conserved	In all types of substance change, the total mass may change.

Empirically-Informed ECC construct map

Strand A: Properties of Atoms & Molecules	**Strand B:** Macroscopic properties	**Strand C:** Molecular representations
Knowing that in a chemical change the atoms stay the same but in different molecules, whilst in physical change the atoms stay in the same molecules, but that in both changes the numbers of atoms involved stay the same.	Can recognize that properties all differ after a chemical change Can distinguish chemical from physical change in unfamiliar cases	Can recognize and distinguish between representations of one or two elements or compounds, or of mixtures between these.
Understands that in a chemical change the atoms or molecules change the way they combine and so form new materials	Can distinguish chemical from physical change: where it is not clear whether or not a new material has been produced. Knowing that mass is conserved in all cases.	Knows that atoms and molecules have weight, that in solids and in liquids there are empty spaces between atoms and molecules. Knows that different molecules can be made of the same atoms or of different atoms. Can recognize a representation of an element with diatomic molecules.
Understands that two different materials will be made of the different molecules, but that these can be made of the same atoms or of different atoms.	Can distinguish chemical from physical change: where it is clear whether or not a new material has been produce	Knows that atoms and molecules are not usually visible and may have different sizes. Recognizes a representation of an element with monatomic molecules.

Figure 2: Initial ECC construct map (left) with empirically informed ECC construct map (right).

As mentioned above, the outcome of this regrouping is shown as a Wright map in Figure 3. The Wright map shows the student ability distribution and item thresholds on the same logit scale. Here the three strands are shown separately and with respect to the same overall student score scale for ECC. The other feature shown here is that the three groups with each strand are represented by the colors – blue, green and yellow, representing the three levels from the lowest to the highest, respectively. Figure 3 shows evidence of a progression sequence with each strand, but also shows that the overlaps between the three produced the apparent lack of order when they are represented all together. The color-coded items also indicate a progression sequence, in a clear way for ECC-C but not for ECC-A. These possible inconsistencies will have to be explored by more a detailed examination of the ways in which the demands of individual questions are perceived by students.

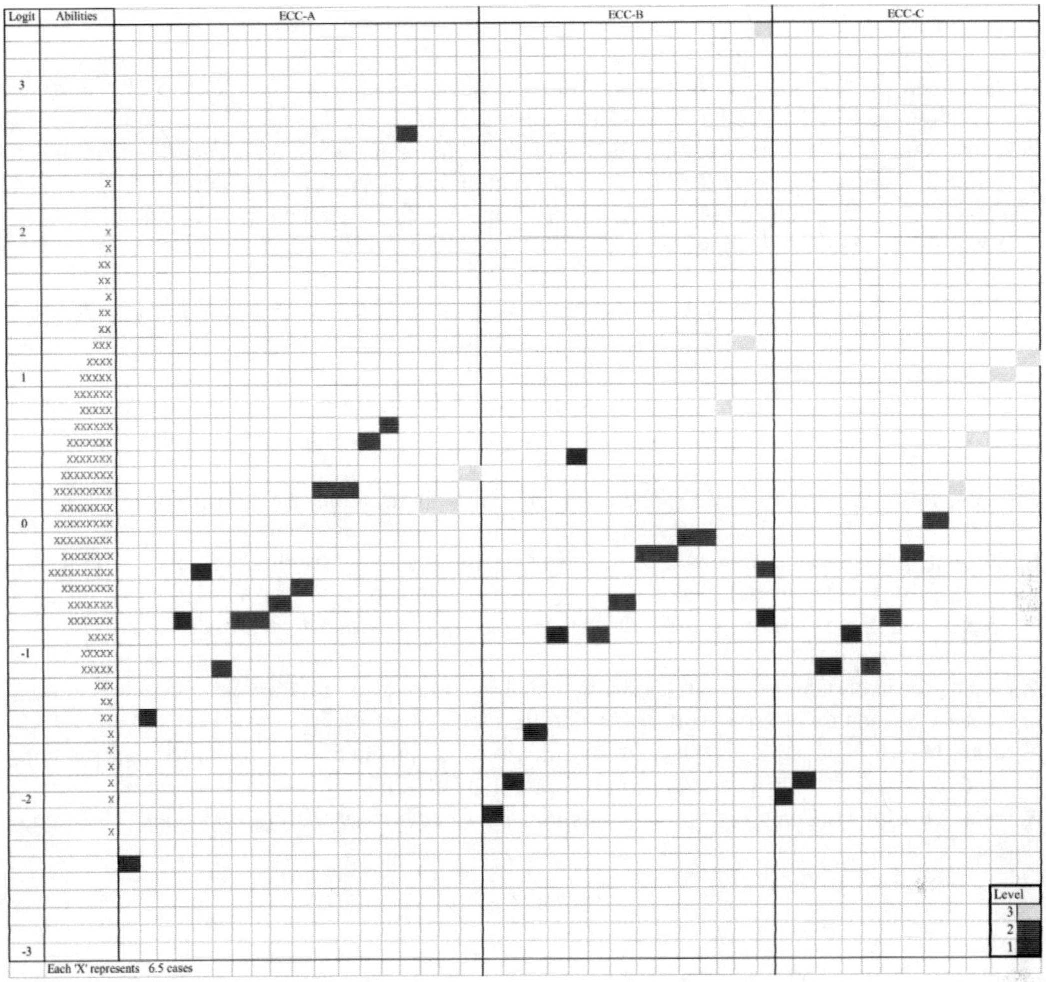

Figure 3: Wright Map of ECC construct separated into three strands (ECC-A, ECC-B and ECC-C).

4. Significance

This research contributes to the current literature on understanding how assessment can illuminate and inform theories of student understanding. This research illustrates how assessments can provide empirical evidence to examine the underlying theory of learning and reveals issues that arise from investigating learning progressions. The issue of how to handle the strand-structure within a larger context and, indeed, putting the pieces together coherently to advance our knowledge of how students come to understand complex science ideas is presented here.

Acknowledgements

The research reported here was supported by the Institute of Education Sciences, U.S. Department of Education, through Grant R305A100692 to University of California, Berkeley. The opinions expressed are those of the authors and do not represent views of the Institute or the U.S. Department of Education.

The authors thank Paul Black for his contribution to the development and revision of the assessment materials. The authors thank Shih-Ying Yao and Tina Chiu for their assessment development and analysis support for this paper.

References

[1] NGSS Lead States, (2013). *Next Generation Science Standards: For States, By States*. Washington, DC: The National Academies Press.

[2] National Research Council (NRC). (2012). *A framework for K-12 science education practices, crosscutting concepts, and core ideas*. Washington, D.C.: National Academies Press.

[3] Hadenfeldt, J. C., Neuman, K., Bernholt, S., Liu, X., & Parchman, I. (2016). Students' progression in understanding the matter concept. *Journal of Research in Science Teaching*, **53**(5), 667–708.

[4] Merritt, J., & Krajcik, J. (2013). Learning progression developed to support students building a particle model of matter. In G. Tsaparlis & H. Sevian (Eds.). *Concepts of matter in science education* (pp. 11–45). New York: Springer.

[5] Black, P., Wilson, M. & Yao, S (2011). Road Maps for Learning: A Guide to the Navigation of Learning Progressions, Measurement: *Interdisciplinary Research and Perspectives*, **9**(2-3), 71-123.

[6] Johnson, P. & Tymms, P. (2011). The emergence of a learning progression in middle school chemistry. *J Res Sci Teach*, **48**(849–877).

[7] Smith, C.L., Wiser, M., Anderson, C.W., & Krajcik, J. (2006). Implications of research on children's learning for standards and assessment: A proposed learning progression for matter and the atomic molecular theory. *Measurement: Interdisciplinary Research and Perspectives*, **14**, 1-98.

[8] National Research Council (NRC). (2014). Developing assessments for the Next Generation Science Standards. Washington, D.C.: National Academies Press.

[9] Wilson, M., & Sloane, K. (2000). From principles to practice: An embedded assessment system. *Applied Measurement in Education*, **13**(2), 181-208.

[10] Wilson, M. (2005). *Constructing measures: An item response modeling approach*. Mahwah, NJ: Erlbaum

Journal of Physics: Conference Series **772** (2016) 012050 doi:10.1088/1742-6596/772/1/012050

Axiomatic definition of quantity as a basis for teaching metrology in Mechanical Engineering

S V Muravyov[1] and B Ramamoorthy[2]

[1]National Research Tomsk Polytechnic University, Tomsk, 634050, Russia

[2]Vignan's University, Vadlamudi, Guntur, Andhra Pradesh, 522213, India

E-mail: muravyov@tpu.ru

Abstract. In spite of satisfactory availability of didactical material, there exist some problems in teaching metrology that are conditioned by rather outdated common view on the role and position of this subject in the overall discipline of Mechanical Engineering, as well as in other engineering branches. In fact, in the last ten years, many changes have occurred in metrology both as a science itself and as a practical activity on measurement uniformity support. Terminology, concerning essential elements of measurement process and measurement situation, has been revised to a large extent. In the paper, certain recommendations are given on a shift of the subject teaching paradigm and its contents.

1. Introduction

Metrology (and allied courses like quality control; standardization; and conformity testing), as an academic subject, holds a high position in any engineering educational program and, particularly, in Mechanical Engineering. Common comprehension of this course role in shaping necessary and sufficient set of competencies to provide successful professional activity of an engineer was not indigenous, though, at present, special importance of this role is undisputed [1–4].

Necessary professional competencies of modern-day mechanical engineers are rather versatile. They include abilities such as: participation in development of technical documentation in accordance with standards and engineering specifications; organization and management of metrological assurance of manufacturing process using generic inspection methods of process equipment and production quality; creating subsystems of quality management for technological processes in production floors; carrying out standardization work and preparatory operations for conformity tests of engineering tools, systems, processes, machinery and materials.

These competencies should rely on knowledge of generic metrological concepts; organizational, scientific and methodological foundations of metrological assurance; legal grounds of measurement uniformity and traceability. The knowledge should also be corroborated by such practical skills as to measure object parameters by means of typical methods and instruments; to estimate measurement uncertainties; to prepare equipment and documentation for certification and so on.

The listed knowledge and skills suppose student's comprehension of nature and very fundamentals of *quantity* notation and allied concepts. However, the issue has been given short shrift by of-the-shelf metrology textbooks, see, e.g. [3, 4]. They mainly are intended to provide understanding of the basics of mechanical measurements and various shop-floor measurement techniques.

Aim of the paper is to show how to introduce in rather artless way both important and contradictory [5] notation of *quantity* that can be used in teaching engineering metrology in order to explain to students what is difference between additive and non-additve, physical and non-physical quantities, etc. We profit here from *representational approach* to construction of measurement theory, since now, after many years of intensive interdisciplinary debates concerning its merits and demerits, see e.g. [6–12], our strong opinion is that this framework can be a good basis for justification of concepts related to measurement (surely, under creative way of thinking). Let us demonstrate this below as brief sketch.

2. Axiomatic introduction of quantity notation

For our consideration, we use definition of measurement in a wide sense [6]: *measurement* is the process of empirical objective associating numbers (or symbols) with properties manifestations of objects in order to describe them. Measurement is a mediator between object and subject (see figure 1), that is, object under measurement is not directly observable by a subject. *Object* under measurement is a body or system, or thing, or phenomenon, or process, etc. which is characterized by one or many properties to be measured. *Property* is a qualitative determinacy of the object; a property is measured, not an object.

2.1. Empirical and numerical relational systems

Object property can be *manifested* with different intensity. ***Example.*** Specific specimen of a green apple can be light green, green, dark green and so on; just as well a property of tables, length, for the individual table can be manifested as 1 meter or 2 meters.

One can consider manifestations a_1, a_2, ... of some property A as elements of appropriate set $A = \{a_1, a_2, \dots\}$. Then it can be noticed that measurement involves symbolization of binary relations on the set A. Thus, the concept of property is based on *empirical relation.*

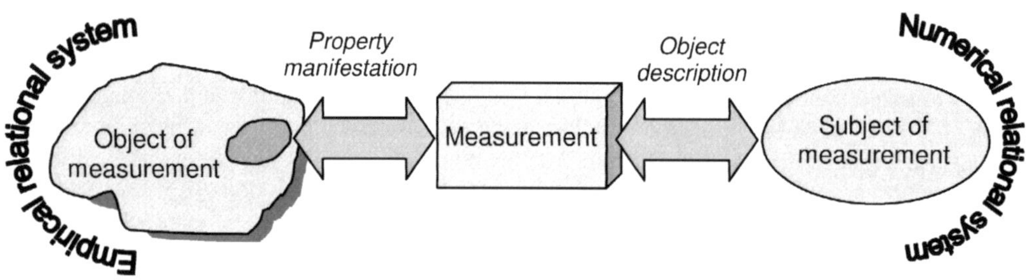

Figure 1. Measurement as mediator between object and subject.

Empirical relational system (ERS) is a set $A = \{a_1, a_2, ...\}$ of object property manifestations that possesses an *order relation* and provides a possibility of observable *comparison operation* over property manifestations.

Numerical relational system (NRS) is a set $B = \{b_1, b_2, ...\}$ of symbols (in particular case, numbers) that possesses an order relation; relations between the set elements are those between numbers including arithmetic operations.

2.2. Operations over property manifestations

As soon as the quantity concept is primary one it is introducing by axiomatic way. For axiomatic definition of quantity, two kinds of operations over property manifestations are necessary and sufficient:

- *comparison* operation and
- *additive* operation.

Mathematical operation (in NRS) must conform to physical operation (in ERS).

Comparison operation reveals relations "$\succ$", "$\prec$" and "$\sim$" in ERS; and relations ">", "<" and "=" in NRS. The operation implementation, as a rule, is *specific* for each of investigated properties. ***Example.*** (1) To compare two metallic rods by a *length* it is necessary: to juxtapose the rods; to locate the rods end-to-end on the one hand; to estimate location of the rods ends on the opposite hand; to shape appropriate judgment. (2) To compare *masses* of two items one should put them on the different pans of scales.

Additive operation intended to combine (or concatenate) property manifestations using operation of concatenation "÷" in ERS and addition "+" in NRS. Again, to compare manifestation of each particular property one should operate accounting its specific character. ***Example.*** (1) When combining *lengths* of two metallic rods it is necessary the end of one of the rods to fit tightly to the end of other rod and guarantee their collinearity. (2) To combine *masses* of two items one should put them on the same pan of scales.

2.3. Definition of quantity

Now we can introduce axioms of quantity. Let some property $A = \{a_1, a_2, \dots\}$ of some object has manifestations a_1, a_2, etc. Corresponding variable A is an *additive quantity* and existence of its numerical representation is feasible iff axioms of order and additivity are hold.

Axioms of order guarantee that an order associated with object by assigning numbers is namely that order revealed due to real observation or measurement:

Q1. *Transitivity* $a_i \succsim a_j$ and $a_j \succsim a_k \Rightarrow a_i \succsim a_k$.

Q2. *Antisymmetry* $a_i \succsim a_j$ and $a_j \succsim a_i \Rightarrow a_i \sim a_j$.

Q3. *Linearity* $a_i \succsim a_j$ or $a_j \succsim a_i$.

Axioms of additivity provide rules of combining manifestations of investigated property (axiom Q9 introduces a possibility to express property manifestations in terms of *unit* of measurement):

Q4. *Associativity* $(a_i \div a_j) \div a_k \sim a_i \div (a_j \div a_k)$.

Q5. *Commutativity* $a_i \div a_j \sim a_j \div a_i$.

Q6. *Monotonicity* $a_i \succsim a_j \Leftrightarrow a_i \div a_k \succsim a_j \div a_k$.

Q7. *Solvability* $a_i \succ a_j \Rightarrow$ there exists such a_k that $a_i \sim a_j \div a_k$.

Q8. *Positivity* $a_i \div a_j \succ a_j$.

Q9. *Archimedian postulate* there exists such $b \in \mathbb{N}$ that $ba_i \succsim a_j$,
where $1a_i \sim a_i$; $(b \div 1)a_i \sim ba_i \div a_i$; ba_i is b-fold concatenation of a_i with itself.

The axioms are adopted from [13] and, actually, are axioms proposed by O.L. Hölder [14, 15].

3. Classes of quantities

3.1. Additive quantities

If some variable satisfies axioms Q1–Q9, it is an *additive quantity*. ***Example.*** Weight is an additive quantity as weights can be ordered by their values and each weight manifestation is combined additively with other weight manifestations.

Additive variables are called also *quantitative* ones as their values (manifestations) always greater or less of each other in definite number of times. In terms of property, which can be qualitatively revealed and be estimated qualitatively or quantitatively, the quantitatively estimated properties are additive.

Among a variety of additive quantities, ***physical quantities*** play considerable part. Large volume of data in scientific papers, monographs, reports, and handbooks is represented by just physical quantities. Namely this class of quantities is a subject of conventional metrology.

International vocabulary of metrology (VIM3, [16]) gives the following definition of (physical or chemical) additive quantity:

quantity is property of a phenomenon, body, or substance, where the property has a magnitude that can be expressed as a number and a reference with a note that a reference can be a measurement unit, a measurement procedure, a reference material, or a combination of such.

Physical quantities are expressed in units. *Measurement equation* for physical quantities has the view:

$$a = b\,[a], \quad (1)$$

where a is a magnitude of the physical quantity, $[a]$ is a unit accepted for the given physical quantity, b is a numerical value of the physical quantity.

Magnitude is a quantitative determinacy being inherent to individual object. In accordance to VIM3, a unit of measurement is real scalar quantity, defined and adopted by convention, with which any other quantity of the same kind can be compared to express the ratio of the two quantities as a number; quantity value is number and unit (reference) together expressing magnitude of a quantity.

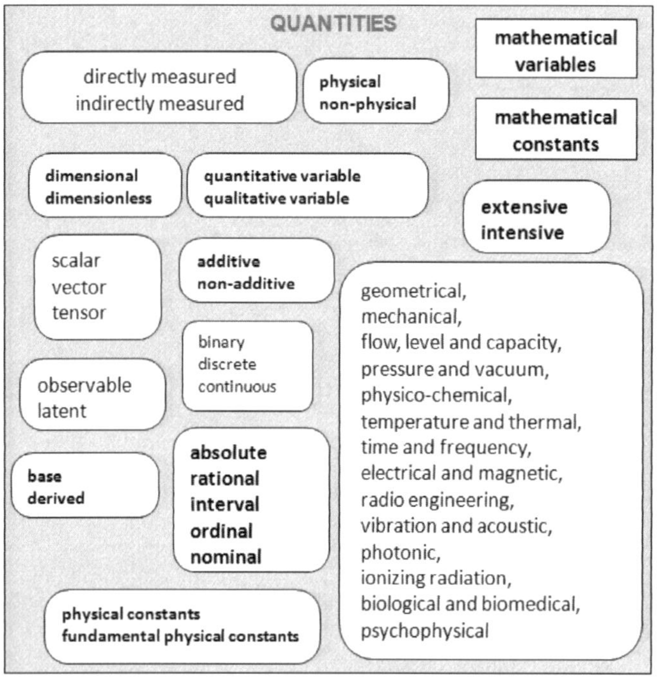

Figure 2. Map of basic classes of quantities.

Non-physical quantities. There usually supposed that physical quantities characterize objects of material world and are investigated in natural and engineering sciences and non-physical quantities are properties of immaterial objects and are studied in humanitarian sciences. However, what is really important whether axioms Q1–Q9 are valid for a given quantity. If an answer to this question is positive, the quantity can be treated as physical. For example, *cost* (as many other economical, certainly, 'non-physical' quantities) expressed in monetary units can be considered as a physical quantity because axioms Q1–Q9 are hold.

3.2. Non-additive quantities

There are quantities, for which only axioms of order Q1–Q3 are valid and additivity axioms are not satisfied. In fact, they ceased to be quantities, that is, entities having magnitude. Such "quantities" are called non-additive or qualitative. **Example.** (1) Ordinal variables: mineral hardness, earthquake intensity, Beaufort wind force. (2) Nominal variables: color, nationality, gender.

3.3. Extensive and intensive quantities

Extensive quantities are directly observable and accessible to compare; satisfy the *distributive rule*: if an object O characterized by value a of quantity A is divided into k parts O_1, O_2, …, O_k, which are represented by values a_1, a_2, …, a_k of the same quantity A then $a = a_1 + a_2 + \ldots + a_k$. ***Examples:*** length, mass, time, volume, force and others.

Thus, the property of interest does change when the object is divided or combined. An extensive property is one that is additive for independent, non-interacting sub-objects. The property is proportional to the amount of material in the object.

Intensive quantities are those the distributive rule is not applied to and expression for a is $a = a_1 = a_2 = \ldots = a_k$. An intensive property is a bulk property, meaning that it is a physical property of an object that does not depend on the object size or its amount of material. ***Examples:*** temperature, velocity,

Journal of Physics: Conference Series **772** (2016) 012050 doi:10.1088/1742-6596/772/1/012050

refractive index, density, hardness, magnetization and others. All specific quantities, i.e. calculated per unit of mass or volume, are intensive.

For intensive quantities axioms of additivity Q4–Q9 are not valid and they are not additive. However, intensive quantities can be measured indirectly, that is, are calculated using measured values of extensive quantities.

4. Measurement and its uncertainty

Measurement is an objective empirical operation f: $A \to B$, which maps the property manifestations onto numbers in such a way that the relations between numbers correspond to the relations between empirical elements (see section 2.1). It means that the mapping f is *homomorphic* [6].

Under homomorphism f the set A is broken up into non-intersected sets of preimages:

$$A = \bigcup_{b \in B} f^{-1}(b), \quad (2)$$

where $f^{-1}(b)$ consists of all elements of A, having the same image in $f(A)$. This fact allows to introduce the notation of *measurement uncertainty* into the representational treatment [17]. Figure 3 illustrates this for a case of length measurement using a ruler, where the empirical elements form the series of values of quantity and the numerical objects are the scale numberings. In figure 3 the function f is a correspondence of each ruler division to a definite numerical score. One can see that inverse mapping $a' = f^{-1}(b)$ never coincides with preimage a since there are *no empirical conditions able to guarantee* validity of the hypothesis $a' = a$.

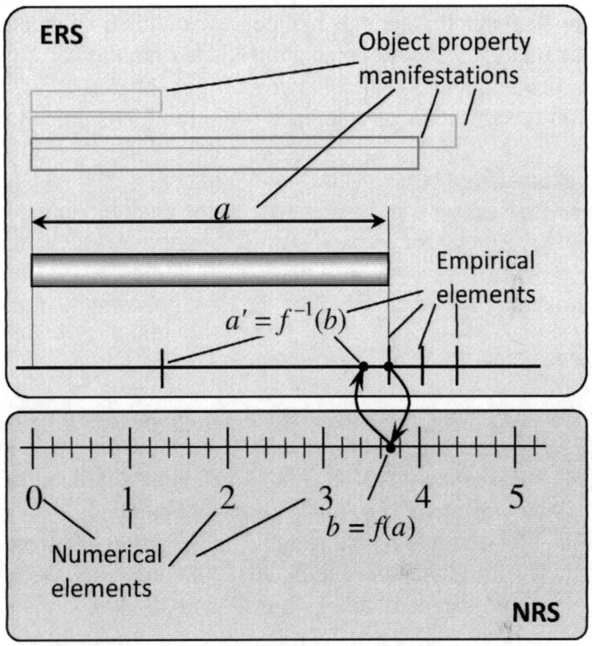

Figure 3. Rod length measurement as homomorphic mapping.

5. Possible metrology course outline

Based on the introduced notations course of metrology can have the outline as follows.

Initial mathematical concepts: Sets. Set of all subsets, set partitions and other combinatorial configurations. Cartesian product of sets. Binary relations. Main properties of binary relations. Equivalence relations. Order relations. Correspondences. Mappings. Mapping classes.

Quantities and scales: Quantity notation. Empirical and numerical relational systems. Object and its properties. Axiomatic definition of quantity. Classes of quantities. Measurement definition. Measurement problems (representation, uniqueness and adequacy). Scale notation. Main types of scale (absolute, ratio, interval, order and nominal scales). Comparative scale characterization. Invariance of feasible scale transformations.

Quantity calculus: Main equation of additive quantity; see equation (1). Number, unit, value, magnitude. Principle of representation of additive quantities. Quantity dimension. Dimension properties. Power form of dimensional formulae. Quantity structure. Dimensional analysis. System of units. International System of Units. Number of base units. Principles of establishing system of units. Measurement uniformity. Standards of units.

Error and uncertainty: True value of quantity. Systematic and random effects. Sources of uncertainty. Probabilistic model of uncertainty. Normal distribution. Uniform distribution. Estimation of distribution parameters (expectation and standard deviation). Consistency, unbiasedness and efficiency of estimators. Interval estimation. Standard normal distribution and Laplace function. Confidence interval and confidence bound.

Measuring instrument (MI): Generalized structure of MI. Classes of MI. Metrological characterization of MI. Instrumental error and its components. Accuracy classes. Calibration. Hierarchy of standards. Traceability. Calibration proceures.

Uncertainty analysis: Correction of measurements. Recommendations of ISO Guide to Expression of Uncertainty in Measurement (GUM). Type A uncertainties. Type B uncertainties.

6. Conclusion

Axiomatic introduction of quantity concept provides a possibility, in consecutive, consistent and understandable for students manner, to lay down all material of the metrology course. Necessary notation of dimension should be also introduced in axiomatic way allowing to study properties of dimensional and dimensionless (dimension one) quantities. In this case, further switching to topics of quantity calculus and systems of quantities (units) will be well-justified.

Acknowledgments

The first author is to convey feelings of gratitude to Prof. V.G. Knorring (St. Petersburg Polytechnic University, Russia) for perennial fruitful supporting interaction in measurement science and metrology.

Part of this work was performed while the first author was at Vignan University, India. Vignan University's, particularly Prof. B. Ramamoorthy's, support is gratefully acknowledged.

References

[1] Regtien P P L 2005 *Measurement Science for Engineers* (London: Kogan Page Science)
[2] Laaneots R and Mathiesen O 2006 *An Introduction to Metrology* (Tallinn: TUT Press)
[3] Raghavendra N V and Krishnamurthy L 2013 *Engineering Metrology and Measurements* (Oxford: Oxford University Press)
[4] Rajendra R 2008 *Principles of Engineering Metrology* (Mumbai: Jaico Publishing House)
[5] Mari L and Giordani A 2012 Quantity and quantity value *Metrologia* **49** 756–64
[6] Finkelstein L and Leaning M S 1984 A review of the fundamental concepts of measurement *Measurement* **2**(1) 25–34
[7] Michell J 2001 Teaching and misteaching measurement in psychology *Aust. Psychol.* **36**(3) 211–7
[8] Muravyov S V and Savolainen V 1997 Representation theory treatment of measurement semantics for ratio, ordinal and nominal scales *Measurement* **22**(1–2) 37–46
[9] Mari L 2003 Epistemology of measurement *Measurement* **34**(1) 17–30
[10] Rossi G B 2014 *Measurement and Probability. A Probabilistic Theory of Measurement with Applications* (Dordrecht: Springer)
[11] Velleman P F and Wilkinson L 1993 Nominal, ordinal, interval, and ratio typologies are misleading *Am. Stat.* **47**(1) 65–72
[12] Sarle W S 1997 *Measurement theory: Frequently asked questions* (Cary: SAS Institute Inc.) ftp://ftp.sas.com/pub/neural/measurement.html, accessed 21.03.2016
[13] Barett P T 2003 Beyond psychometrics: measurement, non-quantitative structure, and applied numerics *J. Manage. Psychol.* **3**(18) 421–39
[14] Michell J and Ernst C 1996 The axioms of quantity and the theory of measurement: Part I, an English translation of Hölder (1901) *J. Math. Psychol.* **40** 235–52
[15] Michell J and Ernst C 1997 The axioms of quantity and the theory of measurement: Part II, an English translation of Hölder (1901) *J. Math. Psychol.* **41** 345–56
[16] *JCGM 200:2008 International vocabulary of metrology – Basic and general concepts and associated terms (VIM)*
[17] Muravyov S V 2012 Chaotic results of multidimensional ordinal measurements *Proc. 20th IMEKO World Cong. (Busan, Republic of Korea, 9–14 September 2012)* vol 3 pp. 2071–4

Journal of Physics: Conference Series **772** (2016) 012051 doi:10.1088/1742-6596/772/1/012051

Calibrating Charisma: The many-facet Rasch model for leader measurement and automated coaching

Matt Barney, Ph.D.

LeaderAmp, 1091 Crystal Springs Drive, Vacaville, CA 95688

matt@leaderamp.com

Abstract. No one is a leader unless others follow. Consequently, leadership is fundamentally a social judgment construct, and may be best measured via a Many Facet Rasch Model designed for this purpose. Uniquely, the MFRM allows for objective, accurate and precise estimation of leader attributes, along with identification of rater biases and other distortions of the available information. This presentation will outline a mobile computer-adaptive measurement system that measures and develops charisma, among others. Uniquely, the approach calibrates and mass-personalizes artificially intelligent, Rasch-calibrated electronic coaching that is neither too hard nor too easy but "just right" to help each unique leader develop improved charisma.

1.Influential Instruments

It is not an overstatement to suggest that leaders architect the fate of homo sapiens. By way of persuasion and teamwork, leaders help bring the future to the present. They found important organizations like IMEKO, build civilizations, and inspire the creation of spacecraft that may land a human on Mars in 2030 (e.g. SpaceX). They are often the glue that instrumentally hold our social bonds together, and give our teams direction and purpose. And, when they are evil, they can destroy our species and others by the millions.[1]

For too long, however, the attributes and behaviors of leadership have been managed unscientifically. Recent reviews of the vast, inter-disciplinary literature on charisma suggest that poor measurement is a key culprit accounting for the fact that in spite of decades of attempts to study leadership in a wide range of social science disciplines, leadership theory is stuck in a quagmire[13]. Charisma, as a subset of leadership shares the same fate, in part, this is because the original scholars - Aristotle and Weber failed to be specific enough about what charisma actually is to provide a blueprint for instrument construction[13]. Earlier attempts at defining charisma were muddied by vague and multivariate definitions of charisma that attempted to mix values, emotions, personality, behaviors, processes, attributions, and even outcomes into one unidimensional latent trait[13,14,15]. Unlike other types of psychological measurement, the fact that followers are the ultimate judges of whether they find leader behaviors inspiring is a special challenge for traditional applications of Rasch Measurement.

This is consistent with contemporary definitions of charisma inspired by economic theory that suggest leader verbal and non-verbal behaviors are what signal follower attributions and labels of being "charismatic"[13,16]. In particular, research with this cleaner conceptualization of charisma suggests that leaders behave charismatically when they communicate shared values in vivid ways, through storytelling, metaphors, and symbolism, including emotional displays[16]. This behavioral definition also has the benefit of being concrete for skill development, and emerging evidence from randomized controlled experiments has shown that charisma, defined behaviorally, can be taught, and has improved outcomes for followers[4,16]. Fortunately, in 1989, J. Michael Linacre reformulated Fischer's

1973 linear logistic test model into a Rasch Measurement approach engineered for just this type of situation.[2,3]

2.Many-Facet Rasch Model for Leaders

The Many-Facet Rasch Model (MFRM) is foundational for the objective measurement of leaders, because it extends the Rasch family of models to include facets for severity/leniency of judges and other facets that could affect a rating situation. Because it conforms to the same objective standards as the rest of the Rasch measurement methods, it allows the estimation and removal of elements that distort measurement in a myriad of other ways, not limited to severity/leniency bias. The MFRM has been shown to be robust to many forms of misfit to the Rasch model, so small perturbations distort the ultimate result in trivial, unimportant ways[12]. The MFRM has been used to measure leader charisma with computer-adaptive assessments that are short, precise and based on the latest science[4,5,6].

2.1. A Many Facet Rasch Model for leadership

A mobile computer-adaptive measurement, coaching, and journaling platform uses a four-facet MFRM:

$$\log\left(\frac{P_{niqj}}{P_{niqj-1}}\right) = B_n - D_i - R_q - F_j$$

where

P_{niqj} = probability of observing category j when rater q response to item i for leader n

$P_{niq(j-1)}$=probability of observing category j-1 when rater q response to item i for leader n

B_n= attribute of leader n

D_i= difficulty of item i

R_q= difficulty of rater q

F_j= Rasch-Andrich threshold, the point of equal probability on the latent variable between categories j-1 and j

In earlier research, this model was applied such that an overlapping sample of raters judged the same set of leaders charismatic behaviors[1,4], as was originally intended for the MFRM[2]. But this was done in a large multinational corporation where the chairman directed it to be applied this way, which is rare. More commonly, different raters judge the leadership of entirely different, non-overlapping groups of people, making the classical MFRM approach impractical for many leader measurement situations. When the judges and leaders do not overlap, their measurement estimates are in different frames of reference, and are therefore uninterpretable. This situation is common in organizations, given that raters may switch teams, resign, go on maternity leave, become busy, or otherwise be unable or unwilling to rate a given leader in a particular timeframe. To make the measurement system more robust to these likely situations, and allow for objective comparisons of quarterly developmental results, an anchored, computer-adaptive approach was adopted.

2.1.1. *Computer-Adaptive Measurement.* Computer-adaptive measurement uses pre-calibrated item banks to measure individuals. It starts out with a question, typically in the middle of the scale, and depending on the users prior ratings, adjusts the next questions based on the new estimate of their location on the line. The algorithm stops when one or more stopping rules are met, such as having a trivial amount of uncertainty (measurement error), or after a certain number of questions have been asked. The major benefits to computer-adaptive approaches are that they are up to 90% shorter, more precise, and are harder to misappropriate than traditional approaches where everyone is asked the same questions[7]. In situations where the most expensive, and time-constrained human resources are involved, computer-adaptive measurements are especially practical. In particular, the brevity of

Journal of Physics: Conference Series **772** (2016) 012051 doi:10.1088/1742-6596/772/1/012051

computer-adaptive measures is especially crucial to be accepted by the senior-most leaders of an organization. [6]

2.1.2 Anchored Facets. The current approach uses a software platform that combines the MFRM and Computer-Adaptive methods to make a smartphone-based system convenient, and useful for the likely situations where leaders are rated by different people at different times. Just as item difficulties and Rasch-Andrich thresholds are already known apriori to closely approximate the Rasch Model, the system also first uses a computer-adaptive measure to calibrate raters. All raters to rate a short vignette about a leader, and rate him/her to also estimate their severity/leniency bias. In this way, the system is able to concatenate all raw data from raters in order to estimate the leader attribute measure.

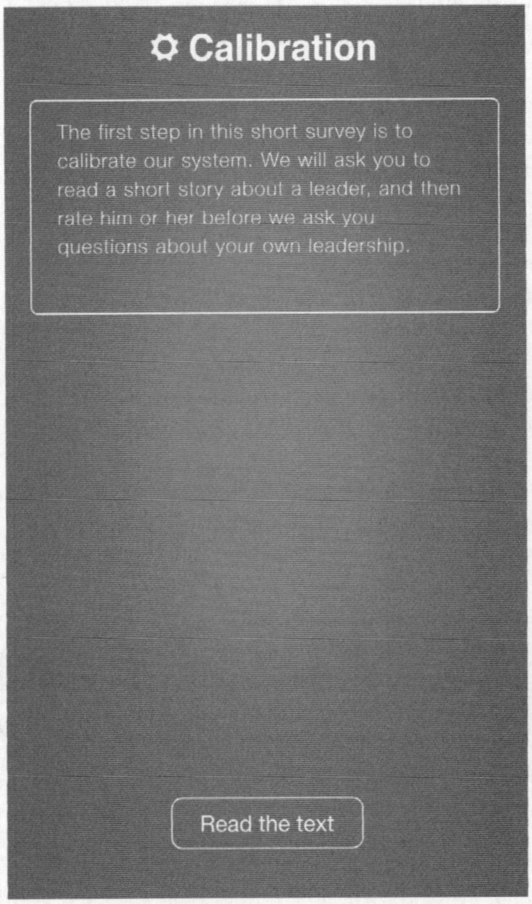

Figure 1. Rater Calibration via smartphone

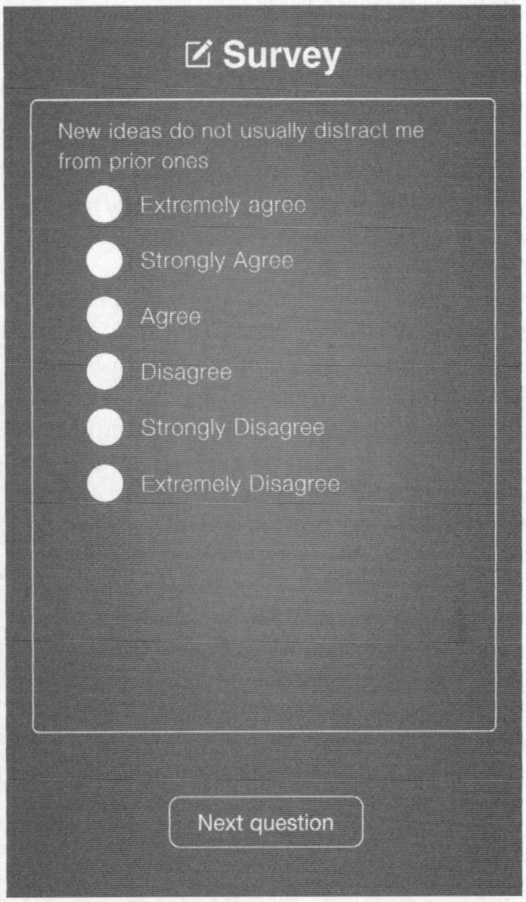

Figure 2: Mobile rating display

2.1.3 Calibrated Coaching. One traditional challenge in developing leaders is helping them develop on their jobs. Usually, people get a multi-page report that explains the result and recommendations for improvement. But as with any skill, it is easy to forget to practice communicating with metaphors, stories, and moral values so that a leaders charisma improves in the eyes of followers. Developmental Psychological literature suggests that it is especially important that individuals practice charismatic tasks that are neither too easy, nor too hard, but just right for their proficiency level.

To help leaders develop their charisma, based on their baseline, calibrated feedback content can be used to coach people through their smartphones to practice on the days and times that they desire. Each leader uses a smartphone graphical user interface, to schedule mornings they want personalized automated suggestions for practicing their charismatic behaviors. They also schedule a time later that evening when they reflect upon that deliberate practice, and journal about it in a mobile journal.

To make sure each person's artificially intelligent coaching is "just right" for their current level of proficiency, the MFRM is used to calibrate two and three-sentence "eCoaching" statements for each level of the latent trait, using a construct map[9], the latest domain-specific content, and science around persuasion[10] and delivering feedback[6]. For example, signifying the importance of a team's work is quite easy for leaders to do more frequently, if they're already seen as low in charisma, this is a good place for them to start to practice, more frequently. They're more likely to develop by mastering these easy tasks before they attempt more complex charismatic behaviors such as modulating the intonation of his/her voice to emphasize key points[6,11]. Only very highly charismatic leaders should practice behaviors that are exceptionally difficult, such as using alliteration and metaphors appropriately, with fervor. These are typically only done extremely well by famous leaders such as Martin Luther King Jr., JFK, and Steve Jobs who were considered especially charismatic by their followers because of their exceptional intuitive mastery of charismatic behaviors.

Table 1 shows a sample of eCoaching calibrations. Uniquely, this approach can be used with very senior and famous leaders, such as Elon Musk, who can give his verbatim eCoaching statements that can be calibrated with the MFRM, and mass-personalized to all SpaceX, Solar City and Tesla employees who work for him. In this way, the dual MFRM calibrations allow personalization of advice in each unique person's zone of proximal development ("Goldilocks Zone") from highly credible expert leaders whose time is scarce. While we have not yet tested the motivational utility of calibrating verbatim scripts from famous or highly credible leaders, it seems likely to be welcomed by prospective leaders, and worthy of future empirical tests.

Table 1: Example MFRM Calibrated Coaching Statements for Mass-Personalized eCoaching

logits	eCoaching Push Notifications	Mean-Square Infit	Mean-Square Outfit	Expected Point-Measure Correlation	Actual Point-Measure Correlation
-2.18	Today, underscore the importance of your team's work. What can you do today to put the value you all create in the top of your team's mind?	1.26	1.26	0.61	0.74
-0.66	What can you do to express your true emotions about how you feel about the future? Make an extra effort to talk about your emotions in an impassioned way today, as you work with your team.	0.76	0.74	0.73	0.74

Journal of Physics: Conference Series **772** (2016) 012051 doi:10.1088/1742-6596/772/1/012051

Table 1: Example MFRM Calibrated Coaching Statements for Mass-Personalized eCoaching

+0.02	Great leaders envision opportunities in otherwise hopeless situations. Work on seeing your biggest challenges as opportunities, and convince yourself about how you and your team can make them a reality. Be sure to communicate this optimism to your team	1.03	1.0	0.70	0.73
+0.65	Work on improving your storytelling today. Try writing out a story that communicates a key team lesson you want to underscore, and try it out with someone you trust. What suggestions do they have for making your story better?	1.14	1.21	0.75	0.73
+1.39	Metaphors electrify teams. They help people remember your leadership even when you're not around. What metaphors can you generate about how your team needs to work together to achieve your shared vision of the exciting future?	0.85	0.99	0.81	0.74

Summary

The leader assessment and development industry is worth over USD$20B in North America alone, but rarely uses engineering-worthy measurements on charisma or any other attribute. Rasch-measurement approaches to developing a leader's charisma to inspire his/her team is one subset of examples that can make a big difference to transform the workplace. This is acutely true when objective measurement is combined with technology that can scale to millions or even billions of people. The key is to ensure that the measurements and development are tailored to each unique persons' proficiency level, to mass-personalize development. Because mobile devices are ubiquitous, even in the developing world, mobile measurement and development systems present an improved path toward scaling evidence-based management development.

References

1. Barney M 2010 *Leadership @ Infosys* (Penguin India)
2. Linacre J M 1989 *Many-facet Rasch Measurement* (Chicago: MESA Press)
3. Fischer G H 1973 *Acta Psychologica* **37** 359-374
4. Day D V & Barney M 2012 ed W H Mobley, Y Wang, & M Li, *Advances in Global Leadership* (Vol. 7) (London, UK: Emerald)
5. Barney M 2013. *Leading Value Creation: Organizational Science, Bioinspiration and the Cue See Model* (New York: Palgrave)
6. Barney M 2015 *Amplifying Leaders: Using Quantified Self Technologies to Grow Leaders like Olympians* (CA: LeaderAmp Press)
7. Barney M 2012 Ed. J McHenry & M McCall. *Leaders Building Leaders: How Senior Leaders Develop Their Protégés*. (Soc. for I/O Psych, San Diego, CA)

Journal of Physics: Conference Series **772** (2016) 012051 doi:10.1088/1742-6596/772/1/012051

8. Barney M, & Fisher W 2016 *Ann Rev Org Psychol Org Behav* **3**
9. Wilson M 2005 *Constructing Measures: An Item Response Modeling Approach* (New York: Routledge)
10. Cialdini R 2008 *Influence: Science and Practice* (New York: Prentice Hall)
11. Antonakis J, Fenley M, Liechti S 2011 *Academy Mgmt L & E* **10** 374-396
12. Linacre, JM 2016. *Brief Explanation of the Theory behind Many-Facets Rasch Measurement.* Facet manual, http://www.winsteps.com/facetman/theory.htm
13. Antonakis J, Bastardoz N, Jacquart P, Shamir B 2016. *Charisma: an ill defined and ill-measured gift*. An. Rev. of OP & OB, 3, 293-319.
14. Bass BM 1985. *Leadership and Performance Beyond Expectations*. NY:Free Press
15. Bass BM, Avolio BJ 1995. *MLQ Multifactor Leadership Questionnaire*. Redwood City, CA: Mindgarden.
16. Antonakis J, Fenley M, Liechti S. 2011. *Can charisma be taught? Tests of two interventions*. Acad. Manag. Learn. Educ. 10:374–96

17.

Single-tone and Polyharmonic Eddy Current Metal Detection and Non-Destructive Testing Education Software

J Svatoš
Faculty of Electrical Engineering, Czech Technical University in Prague, Prague 6, Technicka 2, 166 27, Czech Republic

E-mail: svatoja1@fel.cvut.cz

Abstract. This paper describes the design of a measuring chain for polyharmonic metal detectors used for education in laboratory exercises at Czech Technical University in Prague, Faculty of Electrical Engineering, Department of Measurement. The Measuring chain is composed of DDS signal generator, Digitiser and PC with software programmed in Labview. Eddy current principles or more specifically eddy current metal detectors are an important part of non-destructive testing, instrumentations and measurement. A short introduction to the background and principles of eddy current metal detectors are presented. Next part of the article deals with a brief description of the most common methods, as well as, non-traditional polyharmonic methods for eddy current metal detection. The following part contains an implementation of the proposed algorithms in LabVIEW graphical programming language. Finally, the created program for education of eddy current metal detectors and results obtained on the metal detector ATMID are discussed.

1. Introduction

Eddy current principles can be used in a wide range of applications. They can be used in non-destructive testing for crack detection, material thickness or to measure conductive material properties (conductivity or relative permeability). The limitation of this method is depth of penetration. Since the depth of penetration depends on frequency, frequency is important parameter to consider.

They find also use as metal detectors in archeological and geological surveys, in security helps detect weapons (at airports and other public places) as well as in civil engineering to find steel bracing pipes in concrete walls, in agriculture to remove metal parts from cultivable areas, but are also used in pharmacy, food and chemical industries [1], [2] and [3]. There are many methods how to detect metal objects. Still, the most common detecting method is using eddy current principles. [4]

These currents arise thanks to varying primary magnetic fields transmitted as transmitting signal and create secondary magnetic fields. This secondary magnetic field is scanned as received signal and analyzed by the detector. There are two main groups of eddy current metal detector - Pulse Induction (PI) and Very Low Frequency (VLF) or continuous wave detectors. PI detectors evaluate signal in time domain, VLF detectors evaluate signal as a complex plane in frequency domain. Since VLF metal detectors are better in discrimination and recognition of metallic objects than PI detectors, the laboratory exercise is focused on these types of metal detectors. [5]

The information about metallic objects is mainly contained in the amplitude and the phase shift of the received signal [4]. It leads to limited discrimination ability due to the fact that some objects have a similar response. Possibility to excite the detector with different signals and with differenet frequencies

[6] or [7] together with advanced signal processing of received signal can improve the discrimination ability of the detector.

2. Theoretical Background

Eddy current methods used in metal detection or non-destructive testing usually use probe that consists of transmitting and receiving coils. Time-varying electric current I pass through transmitting coils and generates transmitting magnetic field. If the conductive object is presented, eddy currents generate at the surface of the object. This phenomenon is described by one of the Maxwell's equations - Ampere law (1).

$$\oint_l B_T dl = \mu_0 I \tag{1}$$

, where μ_0 is permeability of the vacuum and l is length of closed curve.

Eddy currents generate their own (opposite) magnetic field. The opposite magnetic field is detected usually by the receiver coil where is inducted voltage U by the second Maxwell's equation - Faraday's law of induction (2).

$$\oint_l Edl = -\frac{d\Phi_B}{dt} \tag{2}$$

, where E is electric field and Φ_B is magnetic flux.

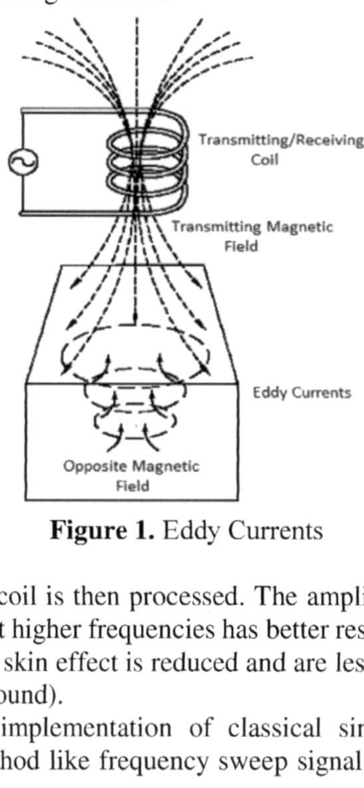

Figure 1. Eddy Currents

The induced voltage in receiving coil is then processed. The amplitude and the phase of the received signal is evaluated. It is known that higher frequencies has better resolution but on the other hand lower frequencies has better penetration, skin effect is reduced and are less affected by the surroundings (e.g. when the object is buried in the ground).

Presented approach lies in the implementation of classical single-tone methods as well as an unconventional polyharmonic method like frequency sweep signal, step sweep sine or sin(x)/x signal [6], [8] and [9].

Polyharmonic approach in non-destructive testing brings opportunity to evaluate tested object with multiple frequencies at once, e.g. to locate cracks in different depth.

Eddy current Metal detectors excited by polyharmonic signals offers to measure Response function of the detected object, which improve the discrimination ability of the detector. [10] The measured data can be also evaluated in both the time and the frequency domains.

The laboratory exercises are focused on both methods of the excitation signal and both methods of signal processing. Thanks to a combination of all these methods better understanding of eddy current principles and determination of conductive objects are done.

3. Implementation

All stated methods are implemented to one workplace. Testing measuring chain is composed of DDS signal generator, which can generate all proposed signals, digitiser for analogue data conversion, PC with LabVIEW software and different probes for testing (differential, reflection, encircling or searching head of ATMID metal detector produced by Schiebel Company). [11] Block diagram of a working place with ATMID search head is shown in figure 2.

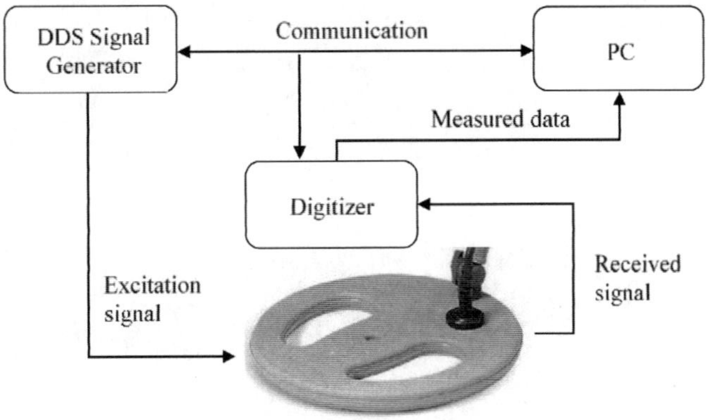

Figure 2. Blok diagram of testing workplace

As a source of an excitation signal, function generator AFG 3102 with a resolution of 14 bits is used. For data digitizing a four channel oscilloscope TDS 3044B (resolution of 12 bits in hi-res mode) is used. Communication with instruments via GPIB uses VISA commands. For a program implementation a graphical programming platform LabVIEW is used. LabVIEW integrates graphical, text-based and other programming approaches within a single environment. It offers large integration with a great number of hardware devices and provides many built-in libraries for advanced analysis and data visualization. Thanks to that LabVIEW is suitable as an excellent education tool.

The claim is to minimize the user interaction with the instruments (function generator, oscilloscope). To ensure proper program execution Flat Sequence Structure is used. It consists of several frames, where parts of the program are stored. The main program is embedded in the main sequence structure, which consists of two frames. In the first frame, the initialization of all instruments is done. The second frame contains of the main loop of the program. The main loop consists of another sequence structure. This will ensure proper program flow.

When program starts, basic settings of instruments are executed automatically, additional settings can be controlled from the front panel. Front panel (figure 2) consists of three main parts. Part 1 (on the left side of the front panel) is for setting up the generator and oscilloscope. The program allows the user to change excitation signal of the generator from classical one-tone methods to non-traditional polyharmonic methods or sin(x)/x signal as well as signal parameters like amplitude, phase and frequency for single tone and frequency span for polyharmonic methods. Time base scale (horizontal resolution) and vertical scale for individual channels, autoset and reset of the scope as well as switching between channels can be changed for the oscilloscope settings.

Part 2 shows the results of processed signal, measured by receiving coil, in complex (hodograph), time and frequency domain. This part consists of different tabs, where users can choose between all types of measurement and signal processing. Data visualization area changes with selected tab (figure 2 - in the middle of the front panel) i.e. selected excitation signal or data processing.

Time domain tab shows time waveform of both signals (transmitted and received). Frequency domain tab shows processed signal using Fast Fourier Transform algorithm (FFT) method [12]. This tab shows amplitude spectrum of the received signal, phase shift between transmitted and received signal and graph with amplitude difference of transmitted and received signal. Hodograph shows complex impedance plane for crack evaluation at individual frequencies.

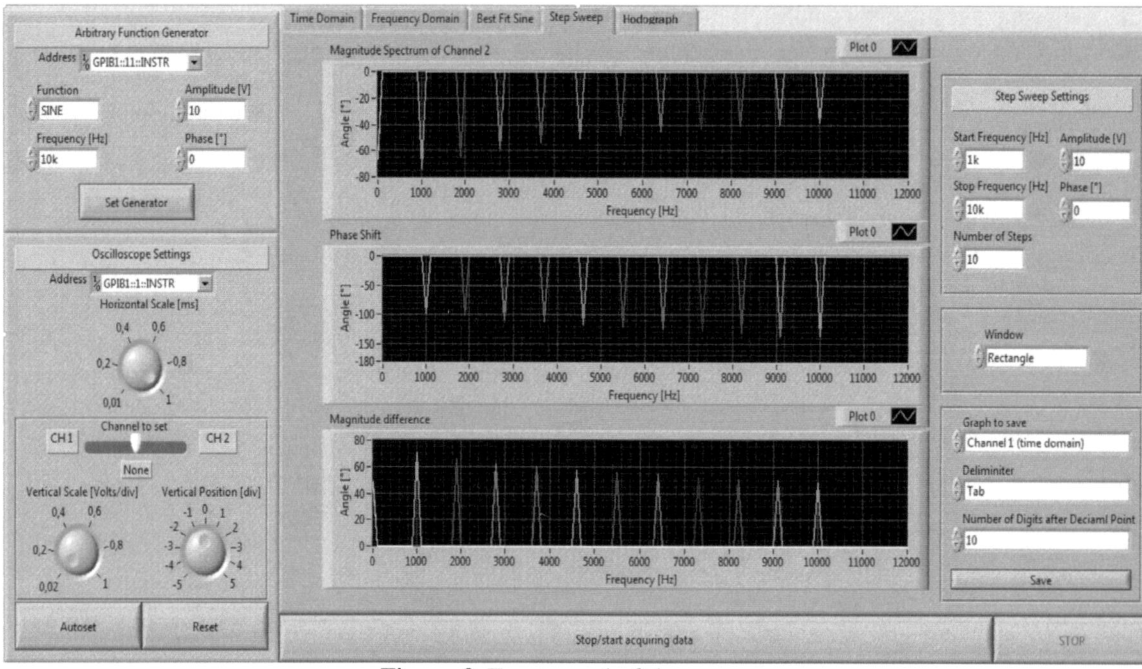

Figure 3. Front panel of the program

As an excitation signal for Frequency domain tab, classical single tone method and non-traditional step sweep sine wave [9] or sin(x)/x signal [13] can be selected. Furthermore, user can find an exact value of magnitude and phase shift at the desired frequency (figure 4, left). There is also a filed for selecting a window [12] witch will be used in FFT.

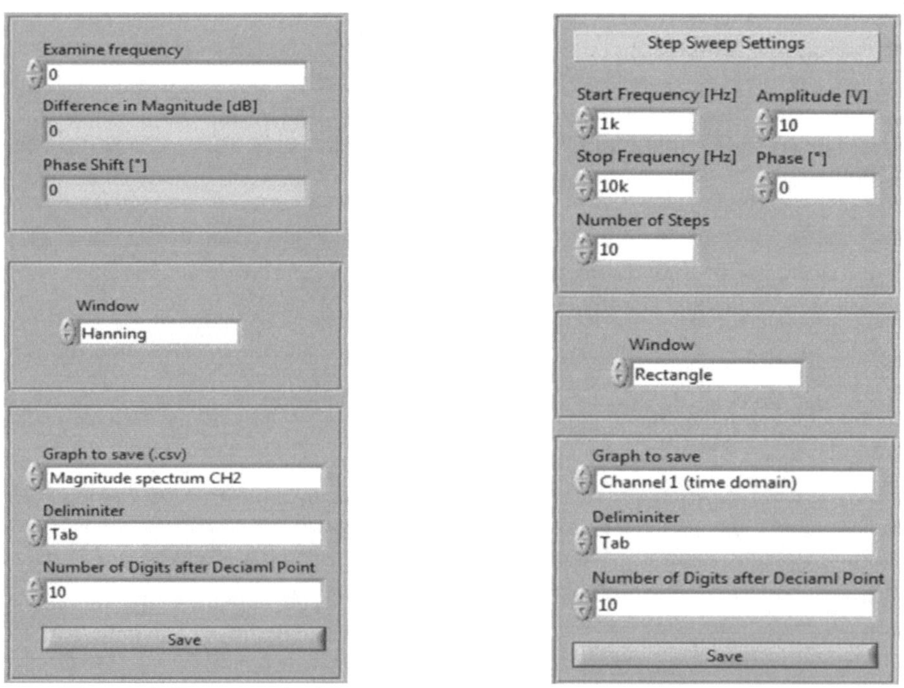

Figure 4. Front panel of Frequency Domain (left) and Sweep (right) tabs

The tab Step sweep shows non-traditional step sweep sine signal. Step sweep sine signal increase uniformly sinusoidal frequency in every step. Thanks to this user obtain frequency response – Response function of the detected object over the wide frequency spectrum. Tab shows graphs with amplitude spectrum of the received step sweep sine signal, phase shift of each frequency and graph with amplitude differences between transmitted and received signal. User can set start and stop frequency, number of steps, amplitude and phase of the signal (figure 4, right).
Best Fit Sine tab uses standardized sine wave fit algorithm [14]. This method could bring the completely new approach to metal detection, but is still at the beginning of the research and is presented here only for compactness.
Hodograph tab shows complex impedance plane of the received signal for individual frequency. Thanks to various frequencies user can inspect different penetration depths, which means cracks in different depth.
All signals can be also saved. Measured data are send form the oscilloscope to the PC as a waveform. Data can be stored to .csv or .mat format for further advanced data processing.
Created programs and methods is designed for education needs of Czech Technical University in Prague, Faculty of Electrical Engineering, Department of Measurement. Program will be introduced to subjects Analog Signal Processing and Digitalization (A3M38ZDS) and Diagnostics and Testing (A3M38DIT) of master degree program.
Crack measurements are made with various calibration standards with different cracks sizes and at various depth to show the dependency of the penetration depth on the frequency.
As a objects for metal detection experiments different size spheres from different ferromagnetic and non-ferromagnetic materials are used. Testing objects covers all sorts of electromagnetic properties (permeability and conductivity) and spheres sizes. Polyharmonic methods shows that induced voltage depends on electromagnetic properties, sizes but also on operating frequency. Low frequency magnetic fields, which penetrate deeper into the ground, are less affected by magnetic or mineralized soils and a skin effect is reduced. On the other hand high frequencies offer better resolution and sensitivity. [15]
Polyharmonic signal also offer possibility to measure the Response function of the detected object and thanks to that better characterise and determined detected material in terms of conductivity and permeability.
In Fig. 7 there is an example measurement using step sweep sine excitation signal without any object present (left) and with testing stainless steel sphere (right).

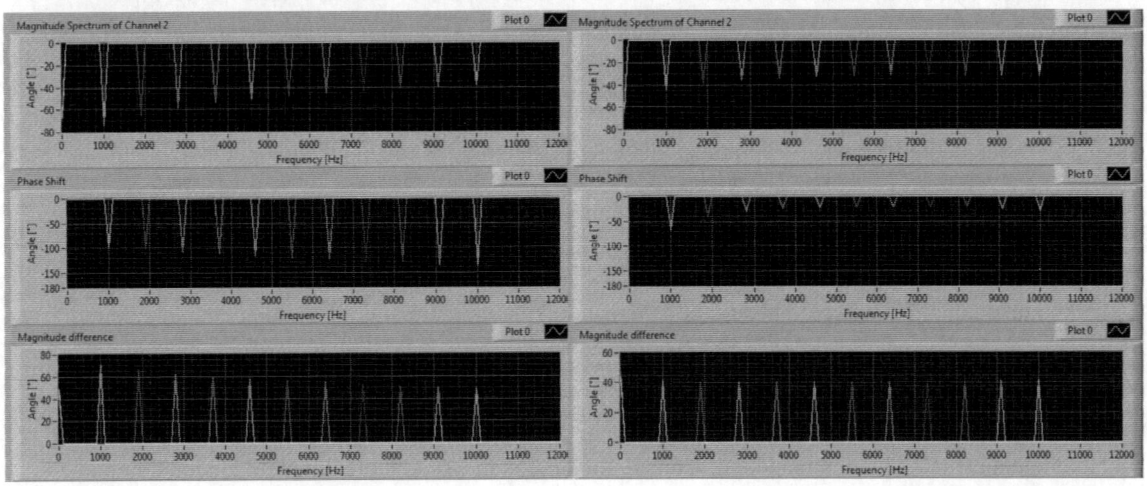

Figure 5. Example of measurement without any object present (left) and with stainless steel sphere (right)

4. Conclusion

A brief introduction to techniques for eddy current non-destructive testing and metal detection as well as designs of measuring chain for this type of eddy current principles has been described. Created program in graphical environment LabVIEW is designed for an education need of Czech Technical University in Prague, Faculty of Electrical Engineering, Department of Measurement. During the implementation of algorithms emphasis on the clear arrangement of the methods, on transparency and readability of the block diagrams has been putted. The graphical user interface has been designed to give a comprehensive view of the program. A program has been created to meet all these criteria. The program is written to minimize the user interaction and allow exporting data for further advanced data processing. Possibility to excite the detector with different signals (single tone, step sweep, sin(x)/x) and signal processing (FFT, best-fit sine method, hodograph) along with visualization, brings new opportunity to better understand the problem of eddy current techniques and principles. Proposed method of single-tone and polyharmonic metal object detection and inspection software in LabVIEW allow the education of the real eddy current principles behavior.

Acknowledgment

This research was supported by the grant SGS 10/207/OHK3/022/16 "Model-based testing methods for automotive electronics systems" by the Grant Agency of the Czech Technical University in Prague.

Reference

[1] Z. Bielecki, J. Janucki, A. Kawalec, J. Mikolajczyk, N. Palka, M. Pasternak, T. Pustelny, T. Stacewicz, and J. Wojtas, "Sensors and systems for the detection of explosive devices - An overview," *Metrol. Meas. Syst.*, vol. 19, no. 1, pp. 3–28, 2012.

[2] J. García-Martín, J. Gómez-Gil, and E. Vázquez-Sánchez, "Non-destructive techniques based on eddy current testing.," *Sensors (Basel).*, vol. 11, no. 3, pp. 2525–65, Jan. 2011.

[3] L. F. Stine and D. L. Shumate, "Metal detecting: An effective tool for archaeological research and community engagement," *North Am. Archaeol.*, vol. 36, no. 4, pp. 289–323, Jul. 2015.

[4] C. Bruschini, "A Multidisciplinary Analysis of Frequency Domain Metal Detectors for Humanitarian Demining," Vrije Universiteit Brussel, 2002.

[5] B. Candy, "Metal Detector Basics And Theory," *Minleab*, 2010. [Online]. Available: http://www.minelab.com/files/KBA_METAL_DETECTOR_BASICS_&_THEORY.pdf.

[6] J. Svatoš, P. Fexa, and J. Vedral, "Metal detector excited by polyharmonic signals," in *2010 International Conference on Applied Electronics, AE 2010*, 2010, pp. 339–342.

[7] J. Svatoš, P. Nováček, and J. Vedral, "Summary of Non-traditional Methods for Metal Detection and Discrimination," *Procedia Eng.*, vol. 47, pp. 298–301, 2012.

[8] J. Svatos and M. Linda, "Polyharmonic metal detector," in *Engineering for Rural Development*, 2015, vol. 14, no. January, pp. 817–822.

[9] J. Svatos and J. Vedral, "The Usage of Frequency Swept Signals for Metal Detection," *IEEE Trans. Magn.*, vol. 48, no. 4, pp. 1501–1504, Apr. 2012.

[10] J. Svatos, J. Vedral, and T. Pospisil, "Advanced Instrumentation for Polyharmonic Metal Detectors," *IEEE Trans. Magn.*, vol. 52, no. 5, pp. 1–4, May 2016.

[11] A. Siegenfeld, "ATMID – Technologie und Schaltungsbeschreibung." Schiebel, 2003.

[12] C. S. S. Burrus and T. W. Parks, *DFT/FFT and Convolution Algorithms: Theory and Implementation*, 1st ed. NY, 1991.

[13] J. Svatoš, P. Nováček, and J. Vedral, "sin(x)/x signal utilization in metal detection and discrimination," *J. Electr. Eng.*, vol. 63, no. 7 SUPPL, pp. 110–113, 2012.

[14] P. Händel, "Evaluation of a standardized sine wave fit algorithm," in *Proceedings Nordic Signal Processing Symposium*, 2000, pp. 453–456.

[15] J. E. McFee, *Electromagnetic Remote Sensing, Low Frequency Electromagnetics*. Ralston: Suffield Special Publication, 1989.

Dual education and industrial cooperation in electrical engineering

A. Váradiné Szarka[1]
Associate professor,
University of Debrecen, Faculty of Science and Technology, Department of Electrical and Electronic Engineering, Bem tér 18/a. Debrecen, 4026 Hungary

E-mail: angela.varadi@science.unideb.hu

Abstract. Dual education in higher education is a new system in Hungary introduced by Mercedes Benz with cooperation of Kecskemét College. In the new system companies support certain number of students and provide them strong practical education in their field. Students applying successfully for dual education study together with non-dual students at the university, so they go through the same university courses as their non-dual colleagues, but while non-dual students' academic year includes 2x14 weeks active semester and 2x6 weeks exam session, all over 40 weeks, dual students have 48 working weeks including study at the university and practicing at the company. The main question of the success which one is the most effective model to be applied. This paper summarises 2 models of dual education with their advantages and disadvantages and also it presents practical realization at the University of Debrecen with special attention to measurement and instrumentation. Dual education in BSc level electrical engineering course cooperates with 6 multinational companies of the region in four specialization. Dual education also has great impact to the modernisation of engineering education. Detailed study of dual education in field of instrumentation and measurement is provided in the paper.

1. Introduction

Dual education is a well-known and successful practical oriented educational form mostly in German speaking countries and around for a long time. This education aims to educate professionals for certain dedicated industries mostly in secondary and technician levels. As this type of education is supported by a defined company, students take part in a shared education, one part of their education runs in the school, this covers mainly the theoretical education and the other part runs in the industrial plant of the cooperating company which makes this education a highly practical oriented one. This type of education seems to be a really good composition of theory and practice, but the lune always has a shine and also a dark side. In this system students will be specialised in a narrow field of one defined industry which is a real advantage in case of being employed by that company, but it can also limit their selection or changing possibilities when they are not employed or don't want to be employed by that company where they were dual students during their studies. The profit oriented companies obviously aims to make a strong selection of employees, so their strategy is supporting education of a relatively high number of students, which is a really good opportunity for those joining this type of education, but at the end of education in most of cases the company employs only few students, the best ones of course, and all the other students are stressed to find job in another industry in which they are not or only partly experienced. Therefore this type of education has strong supporters and oppositions even in secondary level education.

The Mercedes Benz company has founded a large plant in Hungarian town Kecskemét in 2008. They suffered from leak of well specialised employees in all levels including secondary and higher levels. Therefore according to German practice they have introduced a dual education in Kecskemét, but not only in secondary but also on bachelor level.

2. Predecessor of dual education in higher engineering education in Hungary

Before the Mercedes Benz's dual education was introduced in Kecskemét some educational cooperation (like sandwich courses) between industry and universities already existed in Hungary. The most successful type of such education is the so called "cooperative education". This is a classical sandwich course on bachelor level, the main representative of this type of education is University of Óbuda in Budapest which is the biggest engineering educator in Hungary on BSc level. BSc engineering courses in Hungary include 7 semesters. Students wishing to join cooperative education apply for 1 or 2 semesters practical education to industry after they have completed the 6th semester at the university. In the last 1 or 2 semesters they work like interns for the company while they are completing their thesis project which is defined by the company. At the end of their training period they defend their thesis and get their degree. In this system students finish their studies in 7 or 8 semesters depending on number of practical sandwich semesters, and at the end of their education the company has got a ready for work professional. Why this system is good? Universities train and work with students for 6 semesters in theory and in laboratories, students can participate in international mobility programs and/or student scientific or industrial projects, etc. Companies select cooperative students at the very end of their university education, in the 6th semester, when students already have some results and professional knowledge, so practically the usual HR strategy can be applied by the company for selection, and the company still has 1 or 2 semesters to decide if the selection was really for use of company. More than 95% of cooperative students are employed by their company, so we can state that the selection process and practice is excellent and most of the companies prefer this type of educational cooperation. In the official cooperative education university is also interested as the companies provided some additional support to universities. The additional advantage of this system is the close educational cooperation between the university and industry, so the educational plans (theoretical and practical) are harmonized and evaluated, thus the quality assurance, evaluation and feedback are important and successful parts of this system.

The new credit system in higher education starting from the begin of 2000's resulted new trend in intern system and in the practical education in Hungary. Besides of the traditional cooperative education many companies announce shorter or longer intern positions for full-time students. Using the increased freedom of credit system, students have possibility to compose their individual educational plan and they try to share time between the company and university, so number of interns are increased dramatically in engineering education in the last decade. By now more than 70-80% of third year engineering students have intern contract with industry, and practically 100% of MSc students also have such contracts. It means that these students work for industry minimum 20 hours up to 40 hours per week, but the industry and university have no any cooperation or harmonization of education, let' say they the university does not know about the internship of the student at all and there is really weak added value to the students education. The aim of the company in such system is to bring forward the introduction and preparation of new graduates in order to have a ready for work engineer by the time of receiving the diploma. Advantages for the companies are obviously, advantages for students are relative. From one side they are paid by the company, from the other side most of full time students working as intern for companies loose more than 1 semester in studies, so

most of them can receive diploma in 8th, 9th or even 10th and more semesters of their studies, so they work for much lower wages as intern in these extra semesters as they could work as graduated engineers. Also in most of cases educational results of these students are dramatically falling. Universities are suffering from this new trend as they lose the best students to work on projects with, and also they have to spend much more time for education of these students. So we can state that universities are not interested in the individual intern programs of companies and the long-time interest of students is also a question.
On the basis of the above ideas and the Mercedes Benz's practice dual education for higher educational institution of certain field was introduced by law in Hungary.

3. Two models of dual education in Hungary

The general dual education system is introduced for bachelor engineering education lasting 7 semesters. Main points of the system is the timeshare including 26 weeks university/college education plus 22 weeks industrial practice plus 4 weeks holidays in the first three academic years. Students have a contract with the company and they are paid by monthly basis from their very first day of studies in higher education. The monthly payment should be at least 60% of the minimal wage. The university and company also have contract and they can decide how the above described time frames are scheduled within a year.

3.1. Application procedure

Hungarian higher educational application procedure is a fully centralised system with a very complicated calculation of points collected from secondary school results, final exam results and some additional criteria. From the maximum 500 points 200 points come from secondary school results in two final years, 200 points come from the secondary school final exam results and 100 points can be collected by special results like sport results, professional and sport Olympiad results, extra language knowledge, social disadvantage or handicap, etc. Engineering courses require between 280 and 450 points from applicants depending on the institution and specialization. Nowadays the highest level is required at the Technical University Budapest mechatronic engineering course. In general we can state that there is a big gap in required points between Budapest and other parts/towns of the country which is partly a social problem.
All the applicants have to go through the normal application procedure and successful students can apply additionally to become a dual student, but the selection in this stage is made fully by the company.
Why the dual education is important for universities in spite of the fact that there is no financial interest and they lose the best students to work in research projects. The main advantage of the system for universities is the dramatic increase in number of applicants, thus increase of minimum required points which definitely increases the quality.

3.2. Kecskemét model

The first model was introduced by Mercedes and Kecskemét College, so it is called "Kecskemét model". In this model academic year includes 2 semesters with 13 active weeks each, 5 working days per week at the university/college. After the fall semester students spend 4-6 fulltime weeks at the company, and after the spring semester also 16-18 weeks while they receive all together 4 weeks off during winter and/or summer. In this system students go to university lectures and laboratories together with other, non-dual students, they study 13 weeks, but while non-dual students have 6 weeks exam session after each active semester, dual students have to go to work for the company. Dual students have to take exams in parallel with working in 8 hours per day in industry, which means that dual students have no official exam session, which makes their studies really difficult.
Advantage of this system is the clear and continuous time frame for school lessons and for industrial practice with 5 working days of each weeks either at the university or industry. Disadvantage of the

system is the highly separated activity of the two sides and missing of exam session. Figure 1. shows timing plan of Kecskemét model.

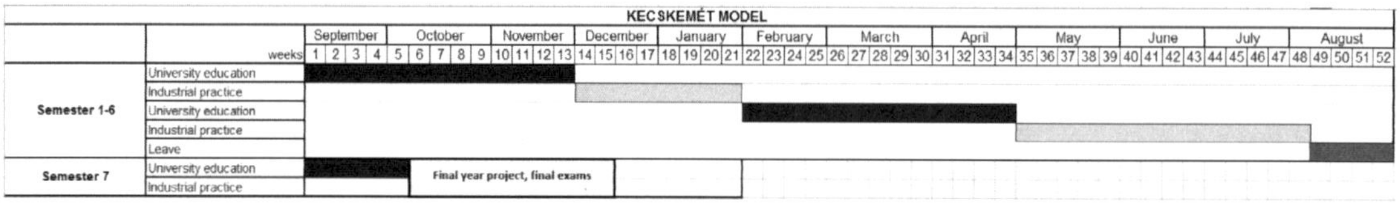

Figure 1. Kecskemét model of dual engineering education

3.3. Debrecen model

If we divide the above described schedule into days, a 52 weeks year including 26 weeks at the university and 22 weeks at the company will result 26x5 days all over 130 teaching days and 22x5 days, all over 110 working days. The 4 weeks leave is the same in both methods.

In the "Debrecen model" one week of the teaching semesters is divided into two parts from which students go to university lessons and laboratory 3 days and they go for the industrial practice to the company 2 days per week. As the fall and spring semesters include 14-14 weeks, students spend 80 days at the university and 60 days at the company. At the end of each active semesters students have 5 weeks (25 days) exam session when they have to spend their time clearly only for being prepared to exams and for taking exams. These 50 days for the exam sessions are included into the university's time frame. Between June and September students have to spend the remained 50 days (10 weeks) at the company.

The main advantages of this system is the continuous contact and work with company from one side and having a regular exam session from the other. The main disadvantage of the system is the shorter working weeks at the university which means that the full class schedule should be included into the first three days of the week. The average contact hours at bachelor courses is 26, so students spend 7-8 hours per day at the university during an active semester. Figure 2. shows Debrecen model. Numbers in checks show number of working days per week to be spent at university/company/leave.

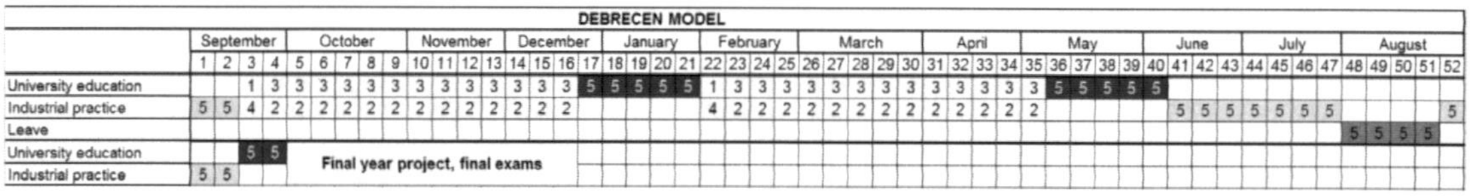

Figure 2. Debrecen model of dual engineering education

As we can see on the diagram the biggest disadvantage of the system is the early start of the practical education. The main question what companies can do with a first year student just leaving the secondary school and not having any engineering or technical knowledge. Most of companies decided to strengthen so called soft-skills of students in this period, like English knowledge, communication and presentation skills, working in team and multicultural environment

4. Dual education in field of instrumentation and measurement at the University of Debrecen

University of Debrecen has accredited dual electrical engineering course in four main specialization: (i) automation and industrial process control; (ii) electrical power networks and supply; (iii) instrumentation and measurement; (iv) electronic industry. Six multinational companies from East Hungary have joint the programme: E.On Hungaria; MSK Ltd., National Instruments; Jabil Circuit; Schaffler Group FAG Hungary Ltd. and RAFI Hungária Ltd. These companies aim to receive 2-5 students per year in dual education system, so all over 15-25 students can get the possibility to be "dual student" at this course. These companies have very high respect in Hungary therefore the number of applicants is very high in 2016, so probably the best students will be in this education.

System of practical education at companies is worked out in close cooperation with the university. In first step companies define departments/professional fields to be included into practical education. In the second step harmonization of company's plans with university's educational plans are worked out. Measurement and instrumentation is included into all the 4 fields but the ratios are obviously different. Focuses in measurement and instrumentation by fields are shown in the table 1.

Table 1. Main practical topics in measurement and instrumentation by dual education fields

Practical fields in dual education (Companies of the field)	Main topics in measurement and instrumentation practical education
Automation and industrial process control (MSK, FAG)	Sensors, instrumentation use and handling, error analysis
Electrical power networks and supply (E.On)	Power and energy meters, smart metering, diagnostics of power quality, electric shock protection, instrumentation for energy efficiency measurements
Instrumentation and measurement (NI)	Development of computer based instrumentation, calibration of instruments, accuracy, gain, offset, linearity, etc. Measuring error estimation, test stands
Electronics industry (Jabil, RAFI)	Testing methods of technology: ICT, flying probe, AOI, AXI, functional testing, JTAG testing, product and production diagnostics, ESD measurements

In the dual education system well defined topics of industrial practice are assigned to certain subjects of the university teaching plan. Thus each industrial partner includes some practice in measurement and instrumentation when student learn it at the university, but differences come of course from the activity orientation of companies. At the electrical engineering courses of the University of Debrecen measurement and instrumentation studies are relatively intensive, as it is shown in table 2. Some courses student have in the second, third, fourth and sixth semesters. In dual education system it means that companies also should focus to this structure and to provide some additional practical knowledge in the certain instrumentation discipline which is also included into the table 2.

5. Summary and Future Plans

Unfortunately in Hungary the higher education and industrial needs were in quite a distance from each other. Very few universities tried to answer companies' needs, and so more and more complains from industry side were directed to certain technical education. The industry is the customer of higher engineering education expressing its demands to have new graduates with up-to-date technical knowledge and abilities to shorten their initial learning period. But how the higher education can be motivated to listen this demand and moreover to do something new. There is definitely not a general good solution but first experiences of dual education introduced on BSc level underline its positive effects to the whole education. Universities are stressed to improve their teaching, to introduce more sophisticated subjects and themes, teaching materials have to be as closer as possible to industrial needs and requirements, some teachers takes part in practical training at their cooperating companies. From the other side 8-10 times more students apply for dual education as to the traditional one, so it seems that only the best students can study in such system which also increases the quality of engineering education.

The next step is introduction of dual education on MSc level.

Table 2. Cooperation between university and industrial studies in instrumentation and measurement

Study semester	Theoretical course	Industrial practice (the most interested companies name in the brackets)
2	***Introduction into instrumentation***	
	Oscilloscope, DMM, function generators	Oscilloscope, DMM, function generators and other field specific basic instrumentation handling. **(All companies)**
3	***Introduction into LabView programming***	
	Basics of programming	Practice in measurement controlling software development. **(NI)**
4	***Measurement science***	
	Measuring errors, error analysis and estimation methods, error distribution and error accumulation	Industrial measurement and analysis methods in specific fields, calibration of instruments, inhouse calibration, test stations of the specific fields, like accelerated life testers, ICTs, power metering, etc. **(All companies)**
6	***Sensors and actuators***	
	Basic active and passive sensing methods. Basic sensors of electrical and non-electrical quantities: temperature, Hall-effect sensors, strain-gauges, etc.	Sensors and transducers of specific fields. In process control field extended practice in mechanical quantities sensing. Electrical motors and drives. **(MSK and FAG)**.
6	***Computer based instrumentation***	
	Digitalization of analogue signals, signal conditioning, signal transfer, data transfer protocols used in instrumentation, multifunctional data acquisition equipment, characteristics of analogue and digital input and output modules	Computer controlled field specific instruments and measuring systems, like ESD, function verification testers, lifetime testers, hazard detection, etc. **(All companies but very different specific field equipment is studied by companies)**

Acknowledgments
Author wishes to acknowledge professional support of the electrical and electronic engineering fields of the University of Debrecen and the fruitful cooperation to industrial partner companies, namely sincere thanks are expressed to NI Hungary, E.On Hungaria, MSK, Jabil Circuit; Schaffler Group - FAG Hungary and RAFI Hungária for providing possibility to publish this paper.

References
[1] Barakonyi, K.: *Rendszerváltás a felsőoktatásban: Bologna-folyamat, modernizáció*. Akadémiai Kiadó, 2004. (Hungarian)

[2] Dinya, László. *"A Bologna Folyamat''a duális képzési rendszer szemszögébol I." The Bologna Process from the point of view of dual education I.).* Magyar Felsooktatás.(Hung. Higher Educ.)(7) (2002): 24-26. (Hungarian)

[3] Tóth, Ákos, István Pap, and István Bársony. *"A duális, és a hagyományos képzésben részt vevőhallgatók tanulással kapcsolatos motivációinak összehasonlító elemzése a Kecskeméti Főiskola GAMF Karán."* Gradus 1.1 (2014): 14-21. (Hungarian)

[4] Ahola, Sakari. *"Different but equal': Student expectations and the finnish dual higher education policy."* European Journal of Education (1997): 291-302.

[5] Thompson, David W. *How valuable is 'short project'placement experience to higher education students?* Journal of Further and Higher Education (2016): 1-12.

Journal of Physics: Conference Series **772** (2016) 012054 doi:10.1088/1742-6596/772/1/012054

Development and characterization of a high speed linear-moving-stage for multispectral measurements

Rosenberger M, Preissler M, Fütterer R, Zhang C, Celestre R, Notni G

Ilmenau University of Technology
Department Quality Assurance and Industrial Image Processing
Gustav Kirchhoff Platz 2, 98693 Ilmenau, Germany

maik.rosenberger@tu-ilmenau.de

Abstract. Multispectral imaging is one of the key topics in the upcoming years, a lot of measurement tasks can be solved using the additional spectral information. For the separation of these additional information bands plenty technologies were developed in the last decades. All of them have characteristic advantages and disadvantages. One working principle is the use of a filter wheel containing different filters between the image sensor and the lens. The additional element inside the light propagation path as well as the lens imperfections lead to chromatic aberrations. To compensate the resulting focus error some investigation were made. One option is the movement of the image sensor stage to a better focus position. Therefore high precision moving stage with high movement speed are characteristic parameters which have to be considered during the construction. In this paper a linear motor stage paired with two high precision linear bearings were connected together to achieve the demanded requirements. For the evaluation of this new approach, several measurements determining the dynamic as well as the static specifications were made to proof the achieved characteristics and will be shown in the paper. With this approach, a miniaturised precision system inside the multispectral imager a precision repeatability of less than four microns can be achieved.

1. Introduction, state of the art and short system description

Image processing is one of the key technologies for optimisation of industrial processes, the option for 100% quality assurance and a lot of other essential industrial and research topics. One thing that some imaging solutions have in common is the movement of parts inside the light propagating path to sample multidimensional images. A major application for these kind of imaging systems is in the field of multi- and hyperspectral imaging. There is no practical way to capture multispectral images in parallel. For most solutions either the sample or some parts of the imaging system have to be moved. There is one solution that will give a good estimation for parallel multispectral imaging namely, the snapshot imager using a multispectral-filter array. To guarantee a high spatial resolution, a high dynamic range over all spectral channels [1], a high spectral accuracy and a high flexibility, a filter wheel system was developed [2]. With this system a rotating filter wheel moves different metal interference filters in front of a CMOS or CCD sensor and captures one image with each different filter. The additional optical element inside the light path as well as the wavelength dependent chromatic aberration of the lens in front of the imager, lead to blurred defocused images. In [3], a solution to overcome these problem was presented. In [3] additional correction glass plates were placed between the filters and the sensor to compensate the chromatic aberrations. An additional way to overcome this problem was given in [4]. The aforementioned approach uses a voice coil actuation for the movement of the imaging sensor. An optical sensor delivers a position feedback to a microcontroller which generates the pulse width modulation for the moving stage. The problem in that solution was the positioning speed and the stability of the position. In the new approach, a linear motor with a Hall position sensing and a high speed movement will be presented. The focusing for the different filters is done along the red arrow is showed in figure 3.

2. Moving system stage for multispectral measurements

The moving stage, which consists out of two high precision linear bearings, a high precision linear motor and a special mechanical mounting (figure 1), were integrated in a self-constructed 12-channel-multispectral imager. For the correction of the chromatic aberration [3], the image sensor is mounted on two high precision linear bearings, which will guarantee a deviation of less than one micron along the specified working range of four millimetres. This enables a wide working range for the chromatic error compensation as well as for the usage of a wide variety of different lenses and filters inside the system, while holding the sensor in a stable lateral position to minimise image shifts. The linear motor is directly connected in the middle of these two bearings (see figure 1, centre and right). With the definition of the image coordinate system in the upper left corner of the images, the rows of the image sensor represent the x-direction and the columns the y-direction (figure 1, left). So a better bearing performance along the x-axis will be expected. During the assembly of the system, the linear motor was actuated along the working range while the mounting was not fixed, so the stresses resulting from the mechanical over-determined construction will be reduced, which translates to a decrease of friction along the moving range. The focusing for the different filters is done along the red arrow shown in figure 3. The electronic implementation into the camera system was done using a serial bus to receive new positions and for sending a feedback of the current position from the positioning stage. The internal controller of the moving stage offers the option to implement codes for some macros for some special movement procedures. To adjust the zero position, the moving stage drives the sensor into the end position and observes the slope of the motor current for this position. After the rise of a defined limit, the counter will be set to zero and the current will be switched down. Additionally, a light barrier or a micro switch can be adapted and read out. With some investigations on the current switch method an uncertainty of less than four micrometres, e.g. less than one increment, was achieved. Besides the automatic detection of the zero position, an automatic dynamic calibration can be activated after the restart of the system. The linear motor offers acceleration of 30000mm/s^2, which leads to very high moving speeds of the image sensor. In the current setup, the moving speed is not the limiting constraint. The moving range in this application can be set with a resolution of 1000 steps with a minimum step size of four micrometres.

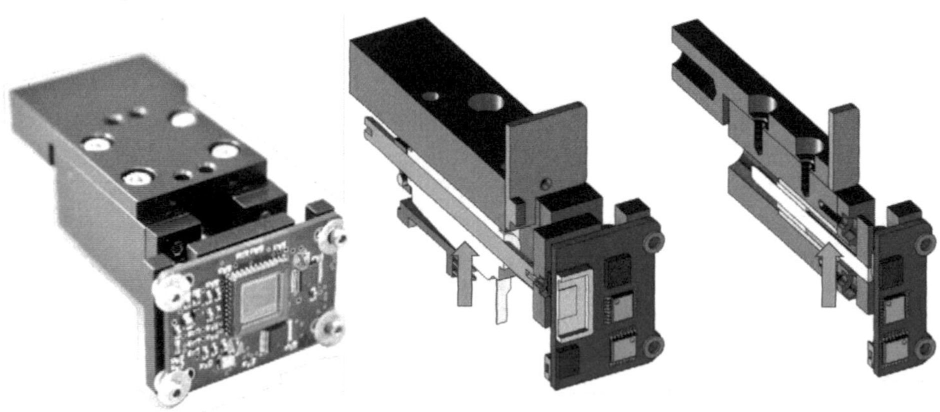

Figure 1: Shows an (left) image of the imaging sensor, its Cartesian coordinates and the moving stage. A sketch shows a cut of the moving stage and its main components with the (centre, purple arrow) linear motor and (right, purple arrow) the linear bearings in evidence.

Journal of Physics: Conference Series **772** (2016) 012054 doi:10.1088/1742-6596/772/1/012054

3. Moving stage investigation static conditions

The first investigations with the new construction were done on an optical coordinate measurement system (OCM) to guarantee high precision results. With the OCM setup, the manufacturer guarantees a maximum permissible error of less than 0.7µm. For the characterization, the camera system was mounted on the OCM (figure 2). Optical probing between the static parts (connected to the camera housing) and the moving part of the actuator stage was done using an image processing toolset, part of the OCM.

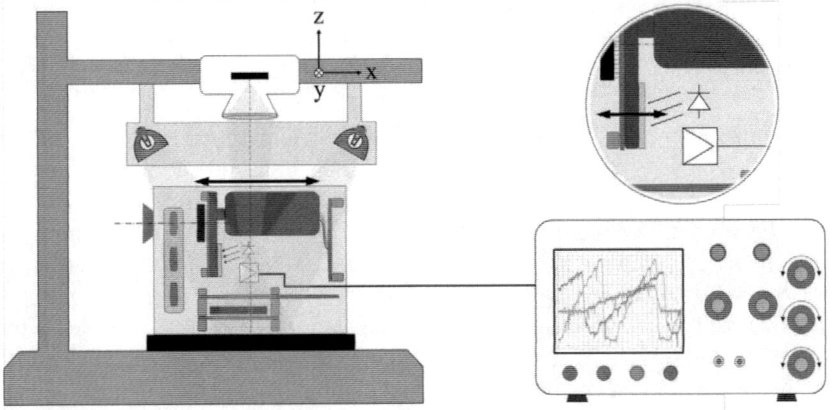

Figure 2: Test setup for dynamic and static investigations on the sensor moving stage.

A typical edge detection algorithm was applied for the displacement measurement. The algorithm is based on 100 search lines, which were placed on both the moving and on the static parts. With this method a high accurate edge detection a can be achieved. For the following description of the experiments, "static condition" means that the measurements were done after the positioning. The results in figure 3 display the moving stage characteristics along the working range of four millimeters. The steps for these experiments were set on 200µm over the working range of 4000µm (1000 counts).

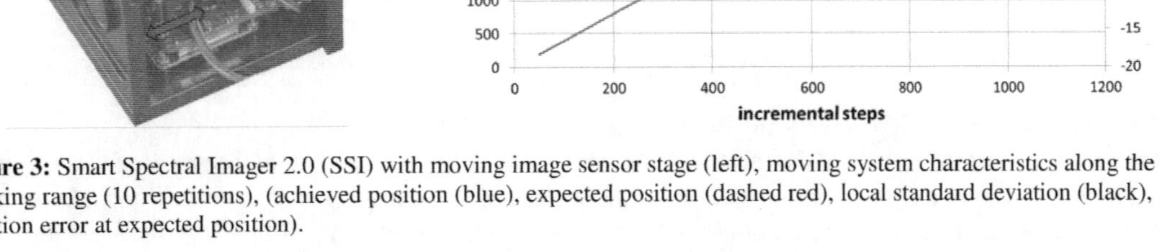

Figure 3: Smart Spectral Imager 2.0 (SSI) with moving image sensor stage (left), moving system characteristics along the working range (10 repetitions), (achieved position (blue), expected position (dashed red), local standard deviation (black), position error at expected position).

The system behavior is nearly linear and therefore can be treated as such. The cumulated error without correction over ten repetitions is 63µm at the upper position limit. The repeatability of the selected positions was calculated with less than one micrometre according to table 1. Additional tests for the characterization of the movement error in x and y direction were made outside the OCM. Therefore the SSI was set into typical working conditions with a lens and a circle calibration target. The center point of the circle target was calculated in a similar way to the edge detection method

Journal of Physics: Conference Series **772** (2016) 012054 doi:10.1088/1742-6596/772/1/012054

mentioned previously. The results of the detected center point coordinates are presented at table 1. The deviation along the full range is less than one micrometre, nearly the same conditions as in a three-chip-camera system.

Table 1: static positioning characteristics integrated in the camera system of the moving stage (standard deviation n=10) with one zero setting and moving range steps (200µm) for z

Range	Deviation x	Deviation y	Deviation z
Full scale raw	+/- 1,605µm	+/- 2,875µm	+/-0,9µm
Full scale corrected	+/- 0,405µm	+/- 0,295µm	+/-0,9µm
C-mount position	+/- 0,28µm	+/- 0,205µm	+/-0,4µm

After the first experiments on full working range, the system characterization on short distances was evaluated. To do so, the moving stage was moved on different positions inside the moving range with step sizes of four, twelve and twenty micrometres to get some information about the stick slip characteristics of the system. Up to the movement of three steps (12µm) nearly nothing will change on the current position. After a movement out of the set position of 50µm and a following reset on the demanded position, leads to some improvement. With that in mind, the investigation of stick slip characteristics the procedure for moving in small ranges were changed. Therefore three different locations along the working range of the moving stage were chosen (start, middle, end positions). On these positions for the movement of one micrometer, the stage was driven 50µm out of set position. From this point a new positioning task was started to set the targeted position. For the movement of two increments, 8µm, this procedure shows no better results in comparison of a direct move. The stability based on the value of uncertainty shows a better bearing condition in the middle of the working range of the moving stage (figure 4).

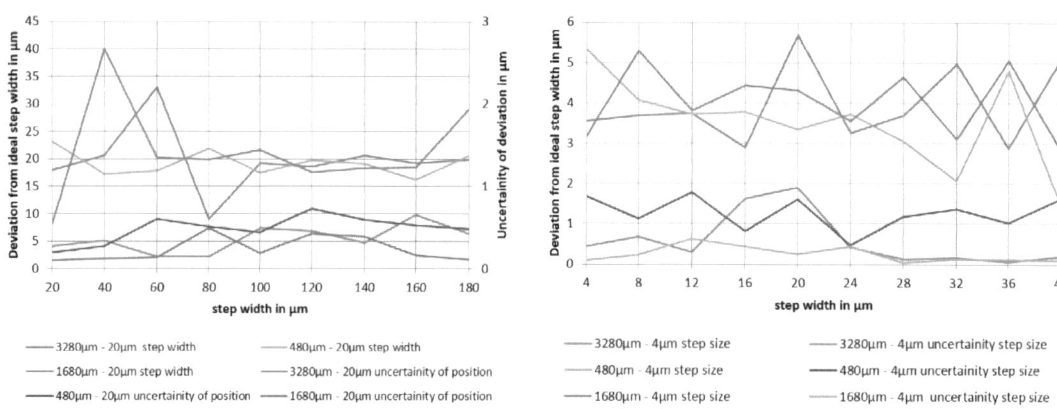

Figure 4: system characteristics in small moving ranges on different positions of the linear bearings.

4. Moving stage investigation dynamic conditions

For the characterisation of the dynamic moving stage parameters, an additional optical distance sensor was adapted in the back of the moving part of the image sensor (showed in details in figure 2, zoomed sketch). The optical sensor was connected to an oscilloscope which samples with a better time resolution than the OCM. The adjustment to the metrological system was realised with a static measurement in the start and end position of the moving program for this test case. So it was possible to evaluate position information with the use of voltage information. The results of these experiments are shown in figure 5. The required timing constraints of 20ms (current sampling speed of the imaging sensor) can be achieved in the incremental positioning mode (figure 5, left). The slightly decrease of the step amplitude versus the position depends on the nonlinear sensor characteristic of the additional optical position sensor and can be ignored.

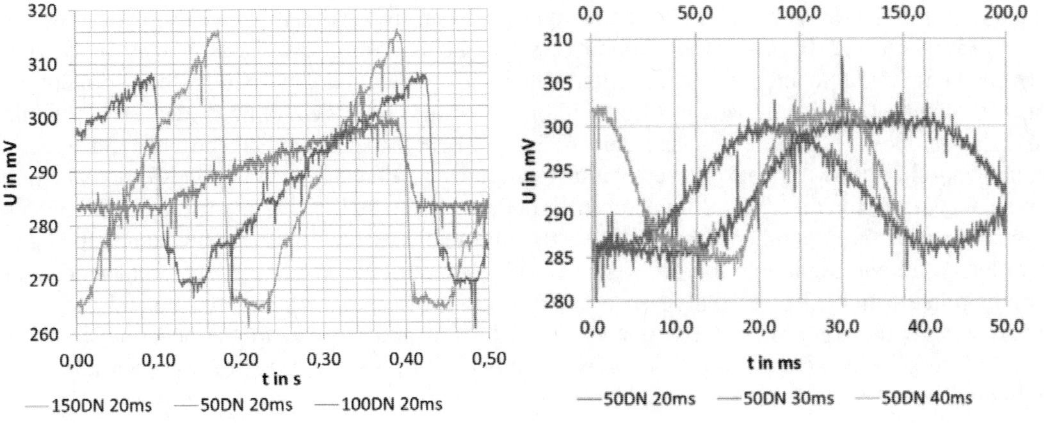

Figure 5: Incremental test mode 20ms settling time (left), ramp mode different settling times (right). DN-digital number equal one increment of four micrometres.

With a distance of 50 DN (200µm), the moving stage reaches the dynamic limits. The transition needs roughly 10ms and the set position will be achieved only on the top for roughly 5ms (figure 5, blue line on the right image). In first experiments, the gap (discontinuity) shown in figure 5 and figure 6 will not happen in typical lenses – typically it is less than 100DN range (see figure 6). Some quasi-dynamic experiments were done additionally. For such, the moving stage was moved along the working range with maximum acceleration. On the position 400µm, 1600µm, 3200µm the stage was stopped and the current position was captured using the OCM - results for ten repetitions the results presented in table 2. Despite of the high requirements measured on this evaluation, the system shows a very good performance with a deviation of less than the specified resolution from manufacturer.

Table 2: quasi static positioning characteristics (n=10)

Position	Position 400µm	Position 1600µm	Position 3200µm
Deviation	+/- 1,314µm	+/- 0,634µm	+/-0,966µm

5. Moving stage investigations on multispectral filter system

In figure 6, the application of the edge-based auto-focus algorithm for three different lenses with three different fields of view (ranging from 3mm per 3mm to 200mm per 200mm) can be seen.

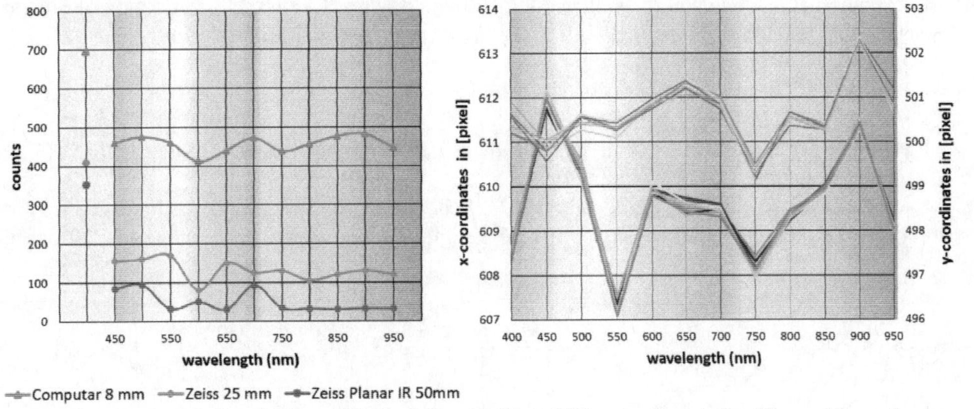

Figure 6: typical lens characteristics of different field of views (left), stability experiment of n=10 repetitions of sensor movement (right) and the influence on image coordinates in x left axis and y right axis.

The yellowish area (left side of the figure 6) shows the focus position for a broad banded light, i.e. illumination with halogen lamps without filtering, where each count corresponds to ca. 4µm. The illumination of the sensor with monochromatic light gives on average a difference on focus of ca. 955μm, 1115μm and 1210μm for the Computar 8mm, Zeiss 25mm and Zeiss Planar IR 50mm, respectively, compared to operation without filter. The biggest deviation was found to be at 600nm for the Zeiss 50mm (1320μm from focus position without filtering). The deviation between values within filtered light (rainbow to grey region in figure 6) is significantly smaller than the values discussed before, showing the following peak to mean value deviations: -10% (-176μm) for the Computar 8mm lens, -40% (-205μm) for the Zeiss 25mm (both at 600nm) and +90% (178μm) for Zeiss Planar IR 50mm at 500nm. This experiment characterizes the chromatic error along the optical path of the objective lens used and highlights the difference between measurements done under broadband light and monochromatic light. Notice that if there were no chromatic error along the optical axis, the curves showed on the right side of the image would be flat and parallel to the abscissa. For this experiment, it was noticed that a resolution of five counts (20μm) is enough to perform measurements with high quality, not being necessary to go beyond it.

6. Conclusion

The result of the system investigation allows a high precision movement of the image sensor. The resolution of four micrometres (according datasheet) delivered out of the linear motor can be achieved in the moving direction. With the definition of a coordinate system in the middle of the imaging sensor with a positive z direction out of the zero point the moving system, the moving stage reaches less than one micrometre deviation in x and y direction and less than two micrometres in z direction. The working range inside the multispectral imager is four millimetres with a resolution of 1000 increments, in contrast to what was observed in [4]: deviation of 16μm was achieved over a working range of 1.5 mm, which shows a great improvement from our previous work/system. With this solution, the focus of the multispectral images can be adjusted with a high precision. In respect to the dynamic investigations, dynamic focussing can be done stable with less than 20ms depending on the way that have to be moved. A dynamic limit for stable sensor positioning of 20μm/ms can be achieved in this current realization.

7. Acknowledgements

We thank the federal ministry of education and research for the strong support of this work. The work is related to the project *Qualimess next generation* (03IPT709X).

References

[1] M. Rosenberger, M. Schellhorn, M. Preissler, R. Futterer, and G. Linss, "Linearization of Silicon CCD-Sensors for Multispectral Imaging," in *Proceedings of the 9th International Conference on Measurement*, 2013, pp. 239–242.

[2] M. Preissler, M. Rosenberger, and G. Linss, "Smart Multispectral Imager," in *IEEE Photonics Conference 2012*, 2012, pp. 591–592.

[3] M. Rosenberger and G. Linß, "Multispectral Image Correction for Geometric Measurements," *J. Phys. Conf. Ser.*, vol. 588, p. 012037, Feb. 2015.

[4] M. Rosenberger, R. Futterer, F. Ziegner, and M. Schellhorn, "Nearfield sensing and actuation for multispectral imaging systems," in *2015 IEEE Sensors Applications Symposium (SAS)*, 2015, pp. 1–6.

Vital Signs Evaluation of Human Behaviour via an Autonomous Body Area Network System

S Hussin, S Takayama

Faculty of Science and Engineering, Ritsumeikan Univ. BKC, 1-1-1 Noji-higashi, Kusatsu, Shiga 525-8577 JAPAN

Email: re0052er@ed.ritsumei.ac.jp

Abstract. Enhancing Quality of Life (QOL) has long been an explicit and implicit goal for individuals, nations, and the world. QOL involves diverse multidimensional factors spanning wealth, physical health, social well-being, and international relationships. This study presents a definition of QOL combining the measurement of health-related QOL with an autonomous Body Area Network System (BANs). A method of evaluating vital signs is performed and linked to physical intensity assistance in exercise. Specifically, BAN acts as a supportive system which can assist a user in monitoring his or her body's parameters, providing real-time feedbacks and dynamically sharing information from any location to one or more users.

1. Introduction

Quality of Life (QOL) is a perceived quality which includes all aspects in individual's daily life. The understanding, measurement and evaluation of human's vital signs and their applications are essential in order to prevent any potentially life-threatening complications, which may reduce one's QOL [1]. An autonomous Body Area Network System (BAN) is comprised of wireless interconnected light-weighted wearable sensors, which provide sensing, monitoring, and processing of human vital signs. Heart Rate (HR), Spontaneous Oxygen Saturation (SpO2), and ambient temperature have been studied under varying daily conditions in order to detect the characteristics of human body behaviors. In physical intensity assistance applications, the relationship between heart rate and exercise intensity has been analyzed in order to achieve an appropriate amount of planned exercise intensity among users. Development of BAN also provides audio-graphical interface feedback as well as real-time information sharing to remote users.

2. System Construction

2.1 Construction of Autonomous Body Area Network System

The BAN system (Figure 1) is comprised of Wireless Sensing Nodes (WSN), Hub Node (HUB), Audio Visual Human Interface (AVHI), and host system. The WSN contains several wearable sensor nodes at the ankle, waist, wrist, and shoulder which collect parametric data. The shoulder module in particular consists of physiological sensors which measures data of human vital signs (HR, SpO2), biokinetic sensor which measures acceleration and ambient sensor which measures environment temperature. Data collection for body parameters and vital signs data processing are performed in the PIC of shoulder module, as well as advice command generation based on heart rate variability. Collected data and commands from WSN are sent wirelessly via ZigBee communication to the hub, consisting of a smartphone and receiver board bridging WSN to AVHI and host system. Command translations are performed in the PIC of hub receiver, then sent to the smartphone's internal storage via Universal Serial Bus (USB). This internal storage acts as a platform for data interchange, providing data fusion based on the needs of further application (refer section 2.2). Using Bluetooth 4.0 communication, data transmission from hub node to AVHI takes place. AVHI, a smart glass from Vuzix, will decode the commands receive within the system, providing users with a real time advice feedback through audio-visual interface. Additionally, the BAN system interacts with host system, enabling real time data access at remote locations via cloud communication. Use of this interactive closed system in physical intensity assistance enables the user to autonomously receive advice for their current body behaviours, thus increasing users' awareness about their physical conditions in daily life.

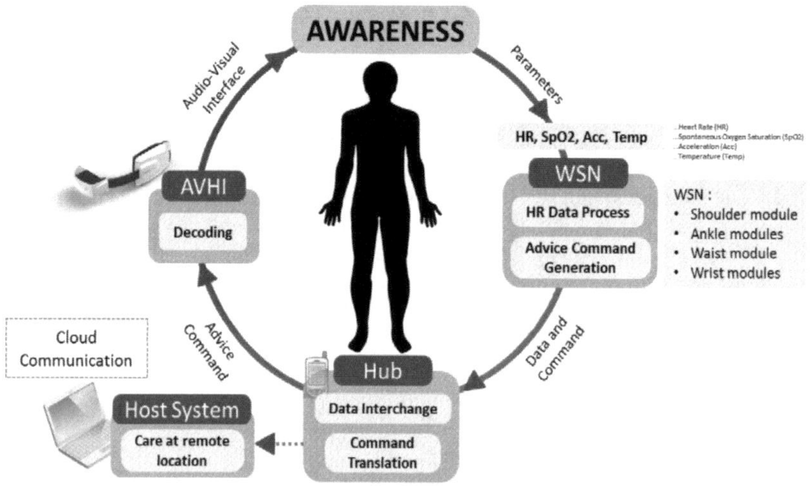

Figure 1. Autonomous Body Area Network System

2.2 Data Interchange in Autonomous Body Area Network System

Data interchange for autonomous BAN (Figure 2) takes place in the hub module, consisting of a transceiver board and smartphone which interconnect the sub systems of WSN, host system, and AVHI feedback systems. Each sub systems are using different types of data format for each parametric data on order to avoid any data congestion or confusion.

The receiver (Application 0), which uses ZigBee wireless communication, receives all data from the WSN and sends it to the internal storage of a smartphone (Application 1) to be stored for further data fusion with other applications. Here, we use Sony Xperia^{TM} Z3 compact, which runs on Android 4.4 "KitKat" operating system. The internal storage of the smartphone is arranged in five folders based on their usages, which can be easily accessed by other applications. Abundance receive data from WSN are converted and saved into CSV files in Data_Collection folder with every minute folder renewal. Besides, the WSN of the shoulder module sends commands in a state of urgency to the Postbox_to_AVHI folders, which enables immediate feedback to user whenever vital signs change or show abnormality (Application 2) .

Additionally, using cloud communication, the HUB is interconnected with host system (Application 3), which will display data and files into computer based interface for remote monitoring. Parametric data from Data_Collection folder are copied to Postbox_to_HS for telemetric usage. Application 3 enables third user of BAN, such as home doctor or family members to request data remotely if they want to know more about WSN user's conditions. Here, the command from host system are saved in the Postbox_to_WSN folder.

Lastly, data of smartphone's condution such as battery conditions, internal storage availability and used internal storage are saved in the Hub_Status folder. This folder is automatically copied to the Postbox_to_HS folder for telemetric data monitoring [2]-[5].

Journal of Physics: Conference Series **772** (2016) 012055 doi:10.1088/1742-6596/772/1/012055

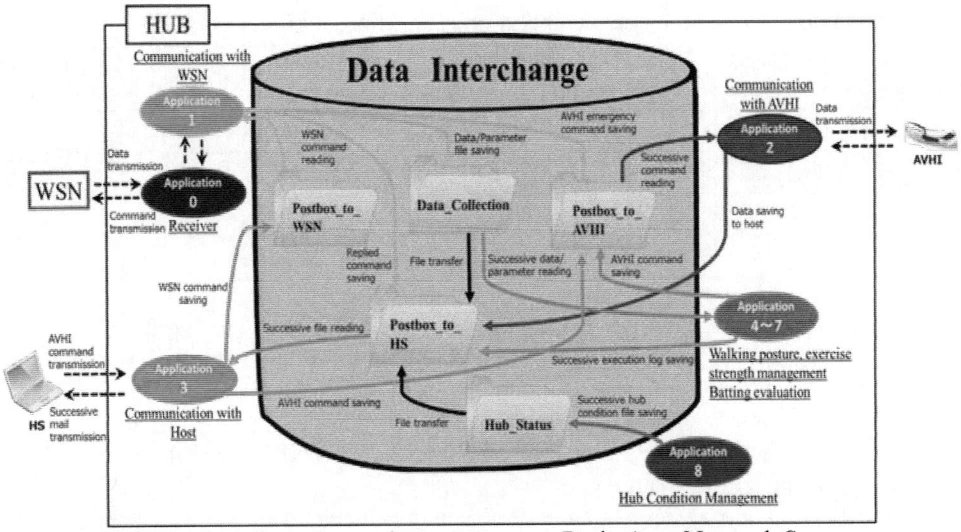

Figure 2. Data Interchange in Autonomous Body Area Network System

3. System Application : Physical Intensity Assistance in Exercise

In order to achieve a maximum health benefits, human is advisable to perform exercises at medium level of intensity [6]. One of the BAN application enables user to manage exercise intensity at their desired level, keeping heart rate at targeted range of threshold, thus ensuring a pleasant and good quality of physical exercise.

3.1 Methodology

Relationship between HR variability and exercise intensity are defined by Karvonen Formula as in equation 1, which is employed in order to determine heart rate threshold ranges and indexes of heart rate trend in generating advices.

$$\mathrm{EI}[\%] = \frac{HR - HR_{min}}{HR_{max} - HR_{min}} \times 100 \qquad (1)$$

Here, HR_{min} represents heart rate in rest, while HR_{max} represents heart rate during performing the heaviest exercise. As the HR_{max} cannot be measured directly, it can be calculated by using $HR_{max} = 220$ – Age formula. Employment of Karvonen Formula enables exercise intensity to be classified into four groups as shown in the table below.

Table 1. Exercise intensity

Exercise Intensity [%]	Exercise Level
20-40	Light
40-60	Medium
70-85	Stressed
85-100	Heavy

Thresholds values are determine by the value of lowest and highest heart rate during certain exercise intensity. For example, when exercise intensity is assigned to medium level, minimum threshold value [Thr_Min] is the heart rate that corresponds to 40%, while maximum threshold value [Thr_Max] is corresponds to 60% of intensity. The targeted heart rate zone while performing physical exercise will be between the minimum and maximum threshold range. Table 2 shows the relation between heart rate existence and indexes of heart rate trend in advice generation.

Table 2. Relation between heart rate existence and indexes of trend

Heart rate existence \ Trend	HR Difference = Trend	HR Difference < Trend	HR Difference >= Trend
HR <= Thr_Min	PACE UP		KEEP PACE
Thr_Min < HR < Thr_Max	KEEP PACE		
HR >= Thr_Max	PACE DOWN		
HR > 180	STOP EXERCISE		

Here, heart rate existence ranges are divided into four types: [HR <= Thr_Min], [Thr_Min < HR < Thr_Max], [HR >= Thr_Max], [HR > 180]. A method of comparing the value of heart rate difference and trend is performed. Here, trend means change of heart rate. The maximum trend calculation of data acquisition for heart rate is three seconds. Heart rate trend are classifieds into three types: falling or flat, mild and steep trend with indexes of [HR Difference = Trend], [HR Difference < Trend] and [HR Difference >= Trend] respectively.

In this application, advice is generated autonomously when heart rate is out of threshold range and is refreshed every 25 seconds, enabling the user to remain aware about their exercise pace. The effectiveness of this system is evaluated by the percentage of heart rate in threshold range as shown in equation 2.

$$Percentage\ of\ heart\ rate\ in\ thr\ [\%] = \frac{Total\ HR\ data\ in\ thr}{Total\ HR\ data\ in\ running\ period} \times 100 \quad (2)$$

Here, a higher percentage of heart rate in threshold range indicates a good quality of running.

3.2 Experiment Method

Experiment is conducted on the treadmill. Two healthy male students with normal heart rate reading were chosen as the subjects with medium level of exercise intensity (40% to 60%).

Table 3. Subjects' information

Subject	Age	Sex	Normal Heart Rate [bpm]	Height [cm]
A	23	Male	73	172.3
B	24	Male	74	173.0

In order to prove the effectiveness of this TBAN system, experiment was conducted under two condition, non-advice running and advice running. Experiment schedules are as follows. Running without advice; A minute of rest, 10 minutes of free running without advice, and 5 minutes of rest. Running with advice; A minute of rest, 10 minutes of running with advice and 5 minutes of rest. Total time for each experiment is 16 minutes.

3.3 Experiment Results

Effectiveness of this system is shown by the comparison of percentage of heart rate in threshold range of advised running and non-advised running. For running with advices, type of advices feedback to the users are present as the coloured background of the graph where grey for pace up, green for keep pace and red for pace down respectively.

Figure 3 shows the comparison results of running for subject A. By using Karvonen Formula, the threshold ranges for the subject A with normal heart rate reading of 73[bpm] was between 122[bpm] and 147[bpm]. When running without advice, percentage of heart rate at threshold was 6.7%, with heart rate value was mostly over the threshold range, compared to running with advice, which increased this value to 45.9%.

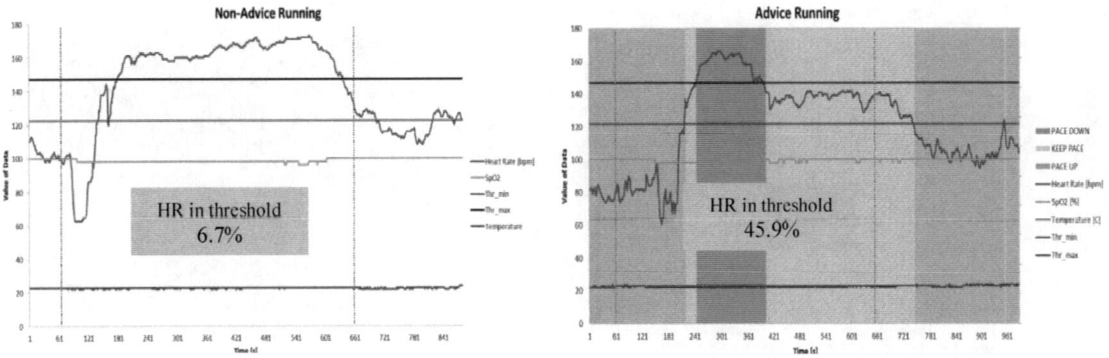

(a) Non-advice running (b) Advice running

Figure 3. Comparison of non-advice and advice running for subject A

Figure 4 shows the comparison results of running for subject B. The threshold ranges for subject B with normal heart rate reading of 74[bpm] was between 122[bpm] and 147[bpm]. When running without advice, percentage of heart rate at threshold was 15.17%, compared to running with advice, which increased this value to 82.15%.

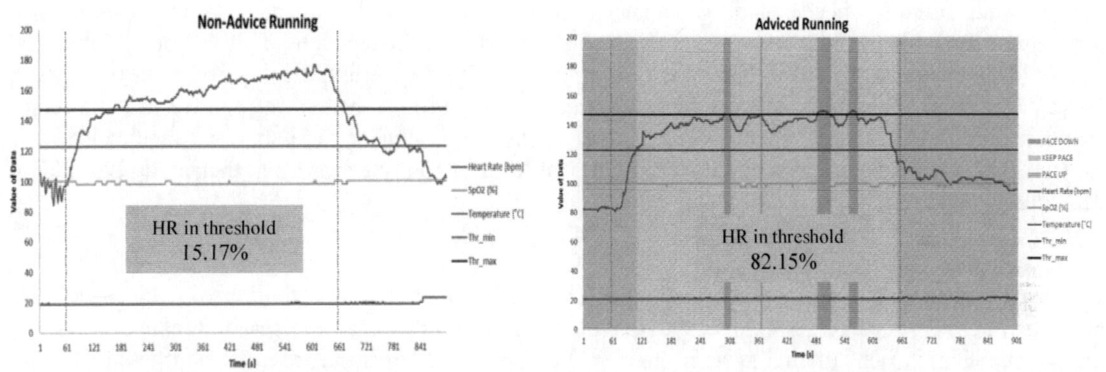

(a) Non-advice running (b) Advice running

Figure 4. Comparison of non-advice and advice running for subject B

These results indicate a positive relationship between percentage of heart rate at the threshold and advice given by the system, ensuring users better-quality exercise. In other words, the application of physical intensity management in exercise is actualized.

3.4 Remote Health Care Monitoring

Host system is capable in accumulating and storing all receive data within the system. The collected data from WSN can be viewed in real time, where parametric data are saved in every seconds and integrated into computer-display interface for easier remote data monitoring. Figure 5 shows the example of interface view for advice running of subject B.

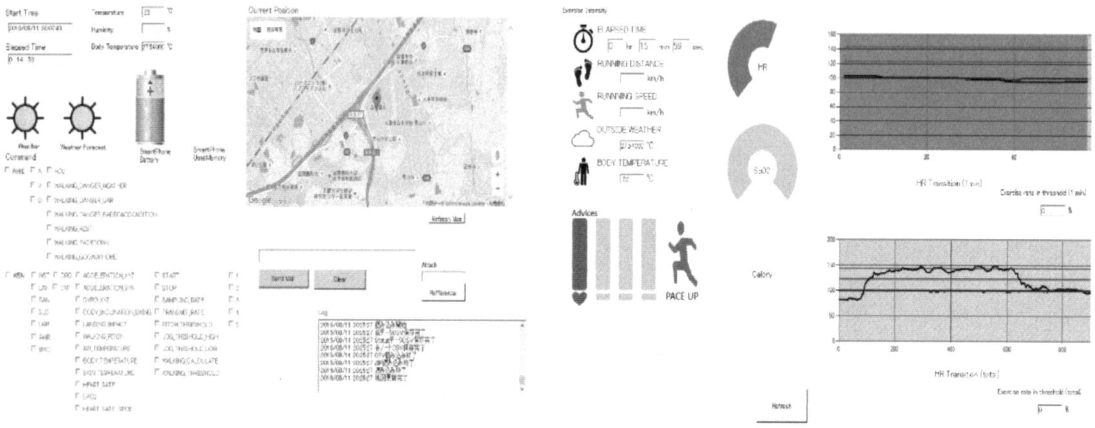

(a) Home view (b) User view

Figure 5. Host system interface

Figure 5(a) shows the home view of host system interface, where current location of user's Global Positioning System (GPS), smart phone status, battery condition, hub node status and weather data are displayed. Figure 5(b) shows the user's interface view during advice running of subject B, where elapsed time, running speed, running distance, heart rate and SpO2 graphs, heart rate in threshold percentage data and advices command are displayed. The evaluated data of heart rate, SpO2 and threshold range are displayed in two type of graphs; every minute graph display and total time graph display. Renewal of graph in every minute enables the user of host system to monitor the vital signs changes in detail. Coloured background of the graphs shows the advices that are feedback to the subject. Besides, data storage in the host system enables user to review or track their body parametric data after exercising.

4. Conclusion

This paper has supported the use of autonomous BAN in daily life as a contributor to a higher QOL. In physical intensity management, as irregular pattern or over limit of exercises may cause unwanted illness, the employment of this system enables the user to perform exercises at a targeted exercise intensity, therefore ensuring both pleasant and safe exercise. BAN acts as a supportive system which provides dynamic Human Interface (HI) and remote data access via host system, which is useful for remote health monitoring. In a conclusion, this system may become the basis of efforts to prevent potentially life-threatening diseases, helping user to achieve an optimal health status, as well as improving users' general QOL.

References

[1] Mark A. Hanson and Harry C. Powell "Body Area Sensor Networks: Challenges and Oppurtunies", *Computer*, Jan.2009, p 58-65.

[2] Ishii T and S Takayama, "Flexible data management by wireless body monitoring system", *SICE International Conference 2014*, Sept.2014, Hokkaido, Japan.

[3] Y.Nakanishi and S.Takayama, "Design of Wearable Sensing Network System for Monitoring Dynamic Physical Condition", *SICE International Conference 2012*, Akita, Japan.

[4] NA Azlan, S.Kobayashi and S.Takayama, "Flexible Data Management in Body Area Network System", *Institute of Electrical Engineers National Convention 2016*, Sendai, Japan.

[5] T. Katsuno, S. Kobayashi, T. Ishii, S. Takayama, "Data Flow Management in Telemetric Body Area Network System", *XXI IMEKO World Congress*, August.2015, Prague, Czech Republic.

[6] Better Health Channel, "www.betterhealth.vic.gov.au/health/healthyliving"

Assessment of the transition strip effect in the transonic flow over the sounding rocket Sonda III

J B P Falcão Filho[1,2], **M L C C Reis**[1,2], **C P F Francisco**[1], **L M Silva**[2]

[1]Instituto de Aeronáutica e Espaço, Pr. Mal. Eduardo Gomes, 50, CEP 12228904, Brasil.
[2]Instituto Tecnológico de Aeronáutica, Pr. Mal. Eduardo Gomes, 50, CEP 12228901, Brasil.

E-mail: marialuisamlccr@iae.cta.br

Abstract. Measurements of normalized pressure distribution are carried out over a 1:8 scale half-model of the Sonda III sounding rocket. The objective is to analyze the effect of the implementation of transition devices on the flow over the vehicle. Measurements show that the presence of the transition devices affect pressure distributions in different Mach numbers around the inter-stage region of Sonda III depending on its location and independently of the turbulent transition method employed. The study of these effects plays a significant role for future developments, since transition phenomena and the modification of the boundary layer behaviour due to the expansion can alter the load distributions and the turbulent structures of the flow. Furthermore, the experimental verification of such phenomena is crucial for the correct implementation of computational fluid dynamics calculations, as they might be able to capture the correct flow behaviour in these regions.

1. Introduction

To perfectly represent steady flight conditions in an experimental facility without significant effect of height variations and heat transfer it is necessary to reproduce the same Mach (Ma) and Reynolds (Re) number conditions. Generally, the flow around small size models have very low Re giving rise to large laminar regions over the model, when compared to real flight conditions. This is even more present in wind tunnels with limited test section dimensions, which is the case of the Pilot Transonic Wind Tunnel (TTP) used in the experiments.

At the full scale Re, it is likely that the boundary layers will be turbulent, so the laminarity of the flow can have a profound impact on the determination of forces and moments over the vehicle. In order to better mimic real flow phenomena in experimental tests, the use of roughness elements adequately located over the model surface is a common practice [1, 2]. These roughness elements, as discussed by Klebanoff *et al.* [3], cause the excitation of traveling wave instabilities or serve as catalysts to stationary crossflow modes and vortex instabilities promoting earlier transition to turbulence in the boundary layer.

Here, we present the results of a transonic test campaign with the Sonda III sounding rocket half-model in 1:8 scale, where two different devices were used to promote transition earlier in the flow after the inter-stage section. Normally, transition to turbulence takes place near the conical nose for

[1] To whom any correspondence should be addressed.

rocket models, but some previous experimental results raised doubts as to the existence of a possible flow relaminarization after the inter-stage region of the Sonda III model. The laminar region is expected to be shorter in length in real flight. The shock wave boundary-layer interactions are governed by different physical processes depending on the condition of the boundary layer, *i.e.*, laminar or turbulent, thereby it is important to use a transition strip during the wind tunnel tests to fix the location of transition from laminar to turbulent. The location on the model and the characteristics of the strip such as thickness and roughness must be carefully chosen to prevent undesirable effects on the pressure distribution over the model [4].

This is the aim of this study: to investigate the impact of changing strip configurations on the pressure distribution over the model surface. Comparison of experimental results is shown in this paper for the subsonic and transonic regimes.

The Sonda III is an 8 m long two stage rocket with a maximum flight altitude of 600 km and liftoff thrust of 102.00 kN. It is part of a family of sounding rocket vehicles developed by the Institute of Aeronautics and Space (IAE) of the Brazilian Air Force.

The objective of this test campaign was to assess the surface pressure distribution over the vehicle, especially in the vicinity of the inter-stage region where the interaction between the compressible flow and the boundary layer is expected to be complex [5].

The Sonda III test half-model was installed on the side wall of the TTP test section of the IAE. The model has 154 pressure taps distributed over the first, second and inter-stage regions in order to provide finely distributed pressure measurements. The experimental tests were conducted with test section nominal Mach number varying from 0.20 to 1.00, stagnation pressure of 94 kPa and 0° angle of attack.

The detailed analysis of the flow behaviour around Sonda III, especially in the inter-stage region, is of great importance for the development of future launchers by IAE, as it experimentally describes the gas dynamics around a critical part of the vehicle. Also, it is expected that the better understanding of the flow behaviour around the vehicle will help improve the IAE computational modeling capability for the development of the next generation of launching vehicles.

2. Sonda III experimental campaign

A 1:8 scale half-model of the Sonda III vehicle was built using aluminum for the experimental campaign. As described by Falcão Filho *et al.* [6], the test model is 0.778 m long and has a first stage diameter of 0.070 m.

The half-model has a total blockage of 2.5 % that combined with the slotted walls of the TTP causes no significant wall effects on the pressure distribution over the model. Furthermore, the model is mounted on the side wall of the test section in a way to leave a 0.004 m of spacing between the model and the wall to impede the boundary layer flow from reaching the model surface.

Figures 1 and 2, adapted from Falcão Filho *et al.* [6], show the drawing views of the test model and its dimensions.

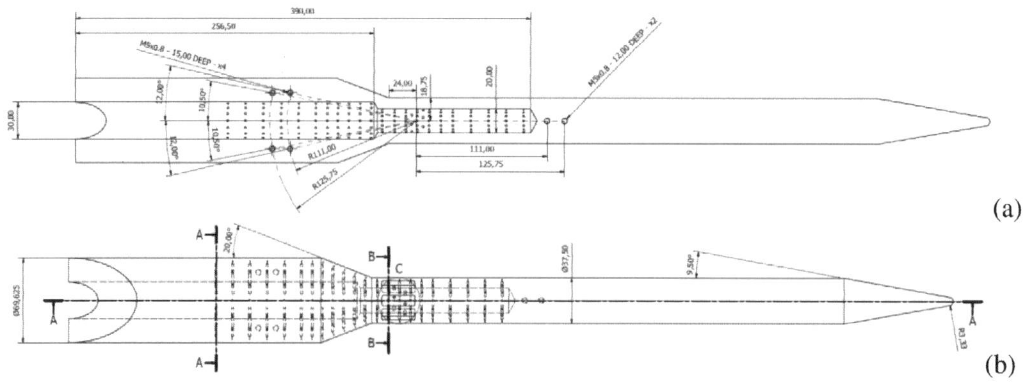

Figure 1. Drawings of the inner (a) and outer part (b) of the Sonda III half-model.

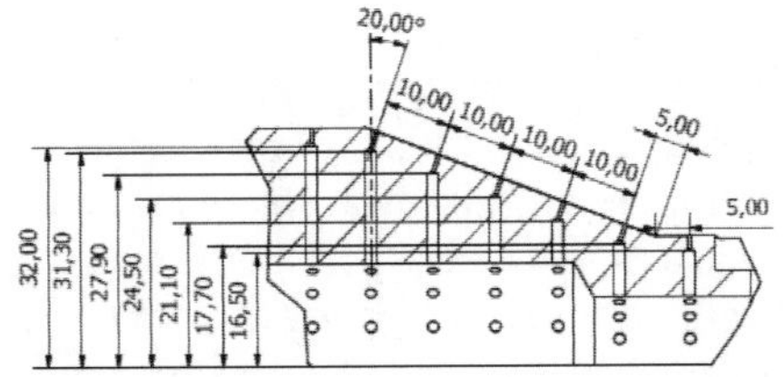

Figure 2. Detailed view of the inter-stage region with 5 stations on the frustum cone. The 0.5 mm pressure taps had to be connected to the 2.0 mm holes with an angle of 20°.

In the inner part of the model, figure 1a, there are 154 holes with 2 mm diameter circumferentially distributed at 22 stations. Figure 1b is a drawing of the outer surface of the model with 0.5 mm of diameter holes connected perpendicularly to the surface, so that the 0.5 mm and 2 mm holes are connected. Figure 2 presents the pressure taps distribution around the inter-stage region of the model.

The stations are distributed in such a way that there are 7 stations in the first stage, 5 in the frustum cone and 10 in the second stage, with a distance of 10 mm between the stations close to the frustum cone and 15 mm between those outside the inter-stage region.

Pressure measurements were performed at 0° angle of attack with and without transition strips placed in two distinct regions upstream of the expansion corner. Different techniques for transition strip placement were employed, *i.e.,* grit (carborundum) strip and nylon wire.

2.1. *Transition strip and nylon wire placement*

As stated above, two different types of transition strips were used, namely, grit strips and nylon wire. The grit is made of carborundum particles spread over the surface and glued with glaze. The carborundum particles used were of the same size for all the strips (120 grit [1]).

Additionally, the size and placement of the strips were varied in order to assess the effect of the different techniques on the flow characteristics around the inter-stage region of the test model.

Two different strip sizes of 2 mm and 4 mm width were used and placed between stations F and G or stations G and H. Figure 3 shows the photographs of the half-model with the different methods used in the experiments. The height of the transition strip is approximately 0.2 mm and the diameter of the nylon wire is 0.5 mm.

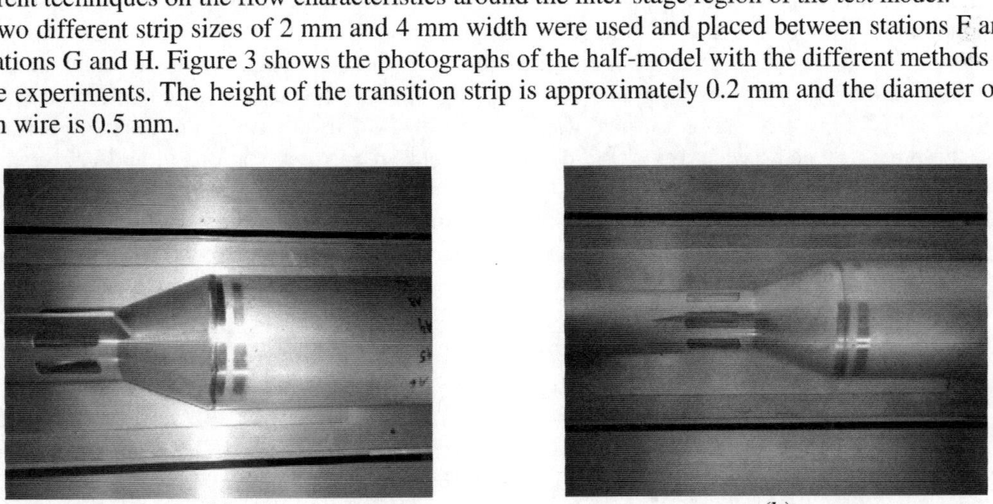

(a) (b)

Figure 3. Photographs of the Sonda III half-model with the implementation of the transition devices, 2 mm transition strip (a) and nylon wire strip (b). The transition strips and the nylon wire in the figures are placed between the stations G and H, but the space between stations F and G are also visible as a bright region after the transition devices.

2.2. Pressure coefficient distribution over Sonda III

The experimental campaign results are shown in figures 4 and 5. Pressure normalized distributions for different configurations but same Mach numbers are shown together for clarity. Pressure scanners were connected to the model pressure taps to measure local static pressures. The scanners are calibrated prior to the tests. The measurements were performed with Mach numbers varying from 0.20 to 1.10, with an error below 1 % in the test section, as shown in Reis *et al.* [7]. The static pressures, p, at the stations over the test model are normalized by the freestream pressure, p_∞, measured at the inlet of the test section. The axial nondimensional distance shown in the graphs corresponds to the position from the beginning of the frustum cone divided by the diameter of the first stage of the test model. The dashed vertical lines highlight the limits of the frustum cone, around where the most complex flow dynamics is akin to occur.

Figure 5 presents subsonic flow results, while in Figure 6 transonic results are presented. For the subsonic cases, the Reynolds numbers are at the limit of transition, while for transonic cases, the boundary layer is expected to be turbulent. The Reynolds numbers are 212786, 531967, 744754, 957541, 1010737, 1063934 for Mach numbers 0.20, 0.50, 0.70, 0.90, 0.95 and 1.00, respectively, based on the linear length of the frustum cone.

For Mach number 0.20, figure 4a, the normalized pressure distributions are the same for all cases, within data error, with no significant difference obtained when using the transition devices. For Mach numbers 0.50 and 0.70, although, the experiments show quite a different behaviour, as can be observed in figures 4b and 4c. The pressure variations are more pronounced in figure 4c.

The results presented in figure 4b show that the pressure distribution is not altered when the transition strip is placed between stations F and G, where the boundary layer is expected to be turbulent, as the pressure values are almost equal to those obtained without the transition device.

Upstream of the end of the frustum cone (station H), in the streamwise direction, the pressure values obtained with the usage of the transition strip and nylon between stations G and H, are slightly above the pressure values observed without the transition strip or with the strips placed between stations F and G. Downstream of the inter-stage corner, at station G, the pressure value obtained for the nylon wire is noticeably below those obtained for the remaining four cases.

In figure 4c, Mach number 0.70, just before the end of the frustum cone (station H), the pressure values measured without the use of transition devices or with the transition strip placed between station F and G are equal, and smaller than the pressure values obtained when the other transition devices are employed, with the pressure values being slightly higher for more intrusive devices (transition strip 4 mm). At station G, just after the end of the frustum cone, the opposite behaviour occurs: the pressure values when transition devices are placed between stations G and H are smaller when compared to the pressure values measured without the use of transition devices or with the transition strip placed between station F and G. Thus, figure 4c shows that the use of transition devices between stations G and H noticeably impacts the pressure values at these sites, as they compress the flow before the end of the frustum cone and expands the flow after the end of the frustum cone.

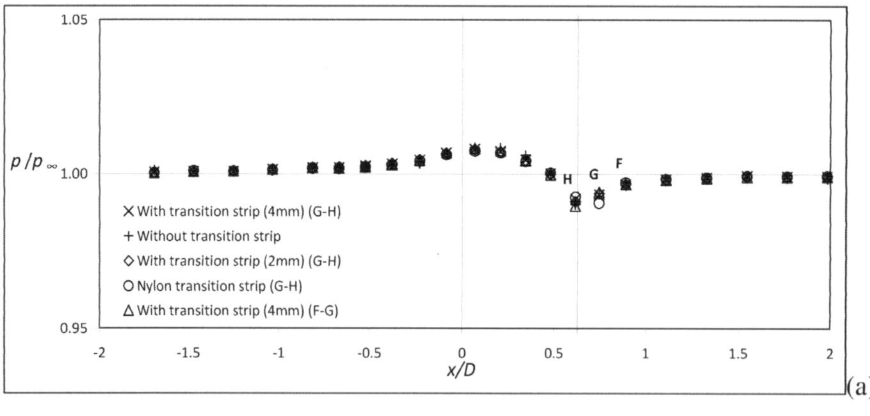

(a)

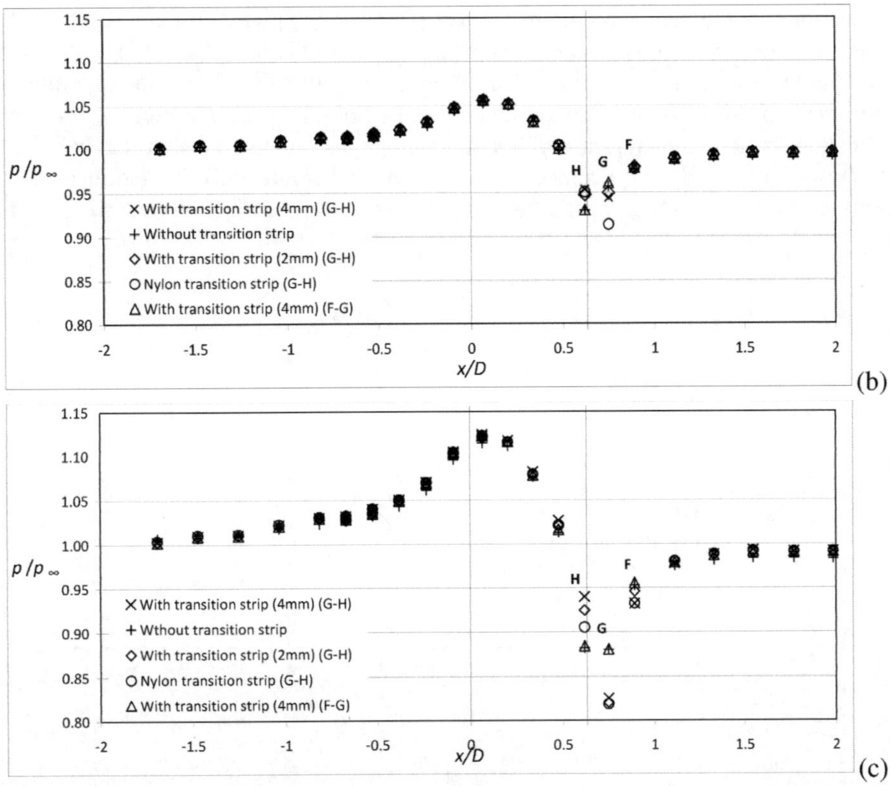

Figure 4. Normalized pressure distribution over the Sonda III for Ma 0.20 (a), 0.50 (b) and 0.70 (c).

Figure 5 shows the normalized pressure distribution over the half-model for transonic Mach numbers, namely, 0.90 (figure 5a), 0.95 (figure 5b) and 1.00 (figure 5c).

For figures 5a and 5b the pressure values measured at station H show the same behaviour observed in figure 4c, with the pressure values measured with use of transition devices between stations G and H slightly higher, when compared to the two other cases.

At station G, the pressure values measured when 4 mm and nylon wire transition devices are placed between stations G and H, are higher when compared to the cases where no transition devices were employed or with the transition strip placed between stations F and G. Otherwise, when the 2 mm transition strip is placed between stations G and H for Mach number 0.95 (figure 5b), a different behaviour is observed at station G, with the normalized pressure presenting the lowest value.

Thus, at the transonic regime depicted in figures 5a and 5b, the presence of 4 mm and nylon wire between the stations G and H compress the flow just after the end of the frustum cone, inversely to what is observed when the 2 mm transition strip is employed, showing that, for Mach number 0.95, this device is less intrusive in the flow. Also, the transition devices alter the values measured at the stations D, E and F which were insensible to the presence of these devices in the subsonic regime.

Comparing figures 5a and 5b, there is a tendency of departure in pressure values obtained in stations D and E when using no transition devices from those obtained in the other four cases, *i.e.*, the pressure is higher without the transition devices. The same tendency of departure is observed in station F, with pressure values being lower for the case without transition device.

For Mach number 1.00, figure 5c, the pressure values do not show significant differences for the five cases studied, the largest difference in pressures values being noted at station E. One possible explanation could be the effect of a thick, full turbulent boundary-layer profile over the model, for which the presence of transition devices is not felt anymore. Another possibility could be the effect of

Journal of Physics: Conference Series **772** (2016) 012056 doi:10.1088/1742-6596/772/1/012056

the flow dilatation that thickens the boundary layer. The same explanations can be applied for the case where the 2 mm transition strip is placed between station G and H, for Mach number 0.95 (figure 5b).

According to Dussauge and Gaviglio [8] and Smith and Smits [9], flow dilatation due to supersonic expansion in high speed flows reduces turbulence in the boundary layer leading to relaminarization of the flow, creating a thick, almost laminar, boundary-layer after the expansion region. Although, Mach number 1.00 is small for the occurrence of full relaminarization, bulk dilatation of the flow could thicken the boundary layer, which would be less sensible to the presence of the transition devices.

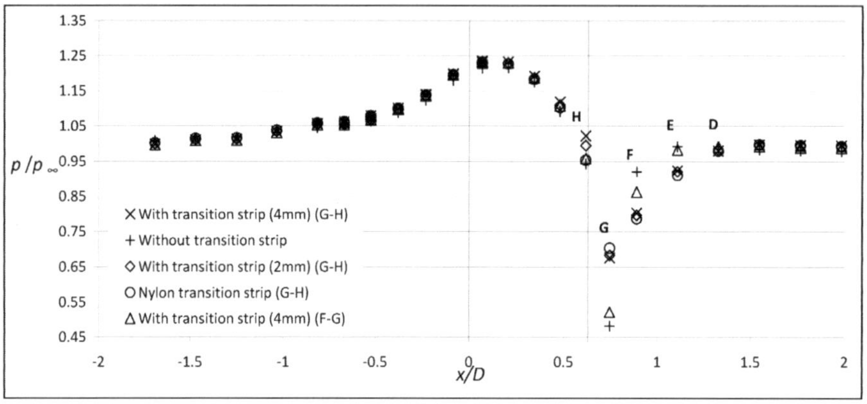

(a)

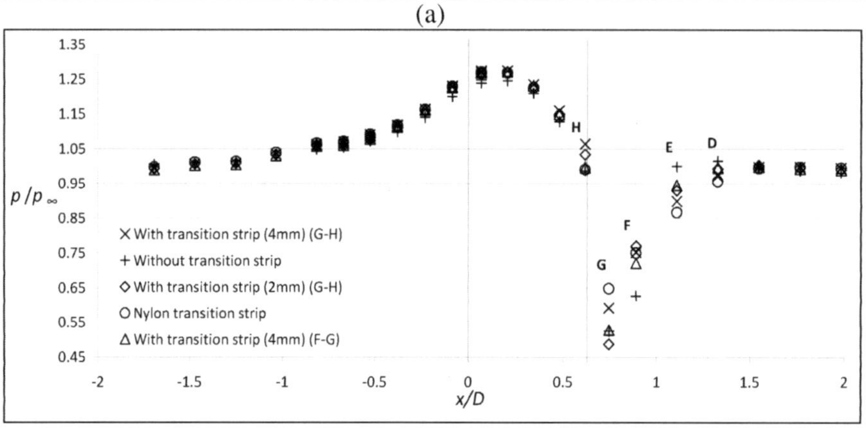

(b)

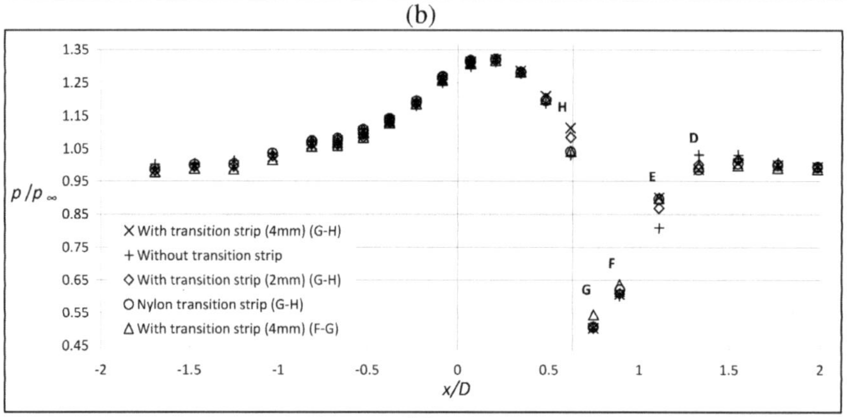

(c)

Figure 5. Normalized pressure distribution over the Sonda III for Mach number 0.90 (a), 0.95 (b) and 1.00 (c).

3. Conclusions

Pressure distribution for subsonic and transonic flow regimes along the SONDA III vehicle half-model were carried at the Pilot Transonic Wind Tunnel Institute of Aeronautics and Space. Particularly, the effects of the usage of transition devices over pressure profiles along the model surface were analyzed.

The transition devices used to promote transition to turbulence of the boundary-layer flow over the model were seen to affect the pressure distribution around the model inter-stage region.

In the experiments performed we are not able to assert if transition to turbulence occurs, or where it takes place without the transition devices, because only pressure measurements were carried out.

We can conclude that the use of transition devices employed after the end of the frustum cone affects the pressure distribution over the model, notably at high subsonic and transonic regimes, altering the flow characteristics when compared to the cases where no transition devices are employed.

This leads to the need to carefully analyze the impact of the usage of transition devices for vehicles such as the Sonda III, for high subsonic and transonic flow experiments to judge if the gain obtained in correctly representing the boundary-layer dynamics is overshadowed by the effect of the devices on the pressure distribution over the model.

As other measurements such as surface shear stress, particle velocimetry or *schlieren* optics were not performed it is difficult to weave accurate conclusions about the flow behaviour downstream of the expansion corner. Surely, such additional information would greatly assist correct interpretation of the pressure behaviour observed in the experiments. Tests for the above mentioned measurements are being planned and are essential for the more accurate aerodynamic analysis to be employed in the future launching vehicle developed by the Institute of Aeronautics and Space.

Acknowlegment

The authors would like to express their gratitude to FAPESP, the São Paulo Research Foundation, through grant FAPESP/CEPID 2013/07375-0 and to CNPq, the Brazilian Counsel of Research and Development, through grant 560200/2010-2 and 384491/2013-8, for their support during the development of this study.

References

[1] Pope A and Goin K L *High Speed Wind Tunnel Testing* 1978 (John Wiley & Sons, New York).

[2] Rhode M N and Chan D T Aerodynamic Testing of the Orion Launch Abort Tower Separation with Hettison Motor Jet Interactions *Proc. of the 29th AIAA Applied Aerodynamics Conference* 2011 Honolulu, Hawaii.

[3] Klebanoff P, Cleveland W G and Tidstron K D On the evolution of a turbulent boundary layer induced by a three dimensional roughness element 1992 *J. Fluid Mech.* **237** 101-187.

[4] Barlow J B, Rae W H, Pope Jr A *Low-Speed Wind Tunnel Testing* 1999 (John Wiley & Sons, New York).

[5] Andrews D C and Carlson D R Shadowgraph study of the upper stage flow fields of some Saturn V study configurations in the transonic Mach number range 1965 NASA Technical Note TN D-2755.

[6] Falcão Filho J B P, Reis M L C C and Ubertini G P A Experimental results of the interstage region of sounding rocket Sonda III in a transonic wind tunnel *Proc. of the 22nd International Congress of Mechanical Engineering - COBEM2013* 2013 Ribeirão Preto, Brazil.

[7] Reis M L C C, Falcão Filho J B P, Paulino G M and Truyts C F Aerodynamics load measurements of a sounding rocket vehicle tested in wind tunnel *Proc. of the 21st XIX IMEKO World Congress, Fundamental and Applied Metrology* 2009 Lisbon, Portugal.

[8] Dussauge J P and Gaviglio J The rapid expansion of a supersonic turbulent flow: role of bulk dilatation 1987 *J. Fluid Mech.* **174** 81-112.

[9] Smith R D and Smits A J The rapid expansion of a turbulent boundary layer in a supersonic flow 1991 *Theor. and Comput. Fluid Dyn.* **2** 319-328.

Journal of Physics: Conference Series **772** (2016) 012057 doi:10.1088/1742-6596/772/1/012057

Selected algorithms for measurement data processing in impulse-radar-based system for monitoring of human movements

Andrzej Miękina, Jakub Wagner, Paweł Mazurek, Roman Z. Morawski[1]

Warsaw University of Technology, Faculty of Electronics and Information Technology, Institute of Radioelectronics and Information Technology, Nowowiejska 15/19, 00-665 Warsaw, Poland

E-mail: r.morawski@ire.pw.edu.pl

Abstract. The importance of research on new technologies that could be employed in care services for elderly and disabled persons is highlighted. Advantages of impulse-radar sensors, when applied for non-intrusive monitoring of such persons in their home environment, are indicated. Selected algorithms for the measurement data preprocessing – *viz.* the algorithms for clutter suppression and echo parameter estimation, as well as for estimation of the two-dimensional position of a monitored person – are proposed. The capability of an impulse-radar-based system to provide some application-specific parameters, *viz.* the parameters characterising the patient's health condition, is also demonstrated.

1. Introduction

The life expectancy has been growing in Europe for many years, while the healthy life expectancy has been slightly diminishing since the last decade of the XXth century (*cf. http://www.healthy-life-years.eu/*). The problem of organised care for elderly and disabled persons is, therefore, of growing importance. Hence the demand for research on new technologies that could be employed in monitoring systems supporting care services for such persons. The capability of those systems to detect dangerous events, such as person's fall, is of key importance [1]. The systems for monitoring of elderly and disabled persons are expected not only to detect dangerous events, but also to predict those events on the basis of acquired data, and therefore contribute to the prevention of such events. The analysis of gait, as well as of the itinerary and timing of activities of the monitored persons, may thus contribute to prevention [2, 3].

Recently, numerous attempts have been made to apply radar technology for monitoring of elderly and disabled persons (*cf.*, for example, the documents [4–6]). They are mainly motivated by the conviction that this technology may be less intrusive, less cumbersome, and less invasive with respect to the home environment than existing solutions. Another attractive feature of the radar-based systems is the possibility of the through-the-wall monitoring of human activity [7, 8].

A consistent set of algorithms, described in this paper, is designed for extraction of key information from measurement data – the information characterising the patient's health condition, and, therefore, enabling the automatic prevention of dangerous events. This set of algorithms constitutes a core of

[1] To whom any correspondence should be addressed.

Journal of Physics: Conference Series **772** (2016) 012057 doi:10.1088/1742-6596/772/1/012057

measurement data processing in impulse-radar-based system for monitoring of human movements; its practical validity is proved here on the basis of a set of real-world experiments.

2. Measurement data acquisition

Detailed technical considerations on the applicability of the impulse-radar sensors for monitoring of human movements can be found in [4]; the principle of operation of such a sensor is explained in figure 1.

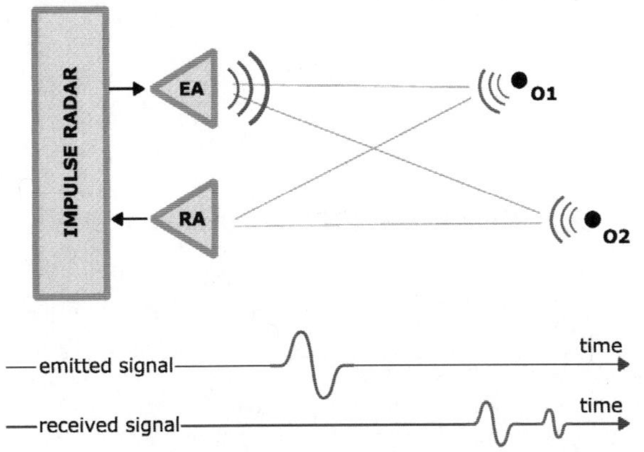

Figure 1. Principle of operation of the impulse-radar sensor.

A short impulse sent by the emitting antenna (EA) is reflected by various objects (O1, O2, ...) present in the space surrounding the sensor. All of the reflected copies of the emitted impulse (called *echoes*) are then captured by the receiving antenna (RA). Depending on the spatial structure of the object, the reflected impulse may have larger or smaller *magnitude*, *i.e.* the received echo can be stronger or weaker; moreover, depending on the distance between the object and the sensor, the reflected impulse may appear sooner or later, *i.e.* may have different *position*, depending on the total time of impulse propagation.

The impulse-radar sensor provides a sequence of raw measurement data representative of the received signal composed of several impulses reflected from all the objects "seen" by the sensor; those data are next subject to preprocessing aimed at extraction of informative parameters characterising the movement of a selected object, *e.g.* position or velocity of a monitored person.

3. Measurement data preprocessing

Measurement data preprocessing consists of the following operations: clutter suppression, estimation of the echo parameters and estimation of the two-dimensional position of the monitored person.

3.1. Clutter suppression

The raw measurement data (shown in figure 2a) are corrupted both with noise and with the so-called clutter, *i.e.* reflections from static objects (such as walls or furniture), and multiple reflections of static objects and moving objects. Therefore, before the extraction of the features characterising the monitored person, noise and clutter have to be suppressed. This can be achieved through the use of a family of algorithms reported by the authors in [9], combined with band-pass filters. The result of this preprocessing operation is shown in figure 2b.

3.2. Echo parameters estimation

After the clutter suppression, the estimation of the parameters of the echo reflected from the observed person becomes a lot easier. Those parameters, *viz.* position and magnitude, can be determined by means of many different algorithms, *e.g.* by means of those described by the authors in [10]. The result of this preprocessing operation is shown in figure 2c.

3.3. Position estimation and tracking

The estimates of the two-dimensional coordinates of the person in the monitored area can be acquired by means of two concurrently working radar sensors. Those estimates can be obtained by completing the following steps:

– calculation of the distances (d_1, d_2) between the monitored person and the radar sensors (figure 2d);
– smoothing of the sequences of the (d_1, d_2) values, *e.g.* by means of the method described in [11] (figure 2e);
– transformation of the pairs (d_1, d_2) in the coordinates (x, y) (figure 2f);
– smoothing of the sequences of the (x, y) coordinates.

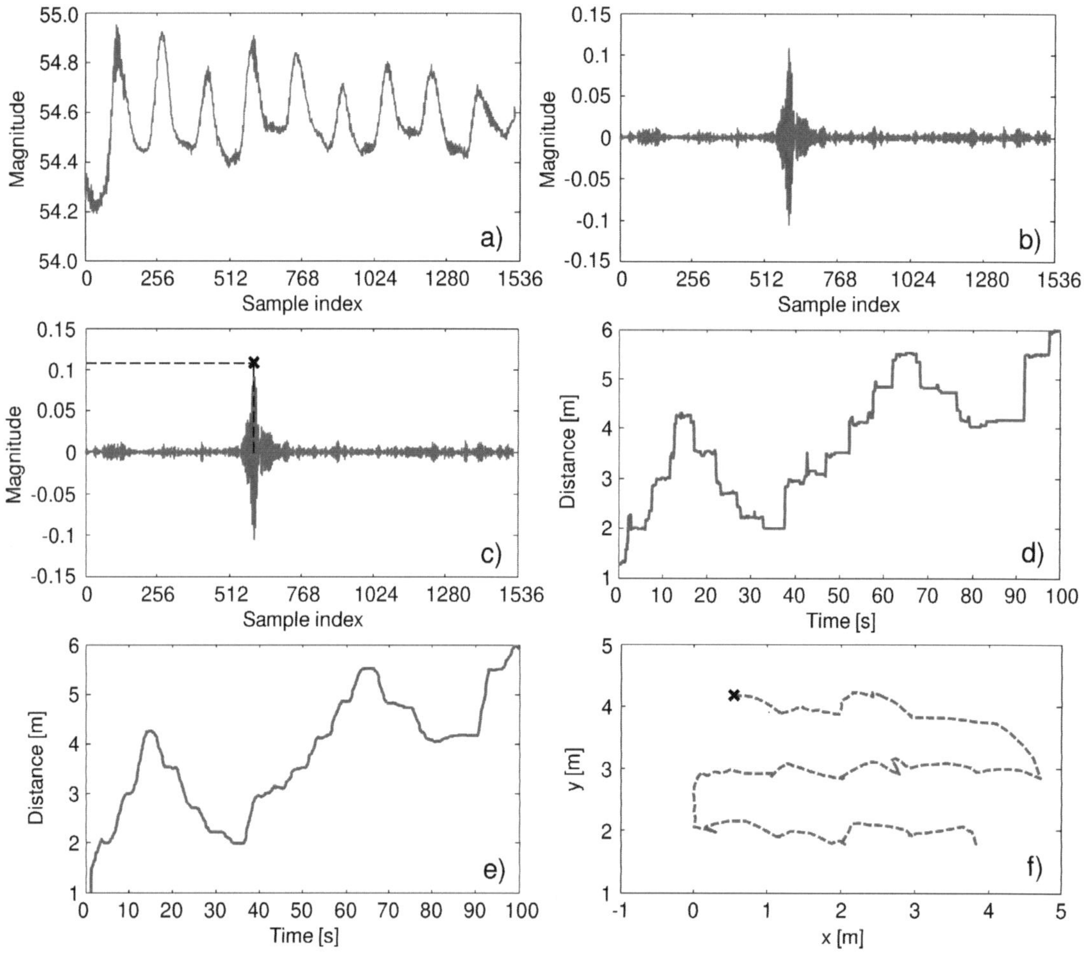

Figure 2. Preprocessing of data from impulse-radar sensors: a) the raw measurement data; b) data after clutter suppression; c) the estimates of useful signal parameters; d) the estimates of the distance between one of the radar sensors and the monitored person; e) the same sequence after smoothing; f) the estimate of the two-dimensional trajectory of the monitored person.

4. Extraction of application-specific parameters

On the basis of the data resulting from preprocessing of the raw measurement data, further parameters characterising the patient's health condition can be derived [2]. Here, due to space limitations, only the estimation of three of them will be introduced, *viz.* detection of motion, estimation of the walking velocity, and estimation of the travelled distance.

4.1. Detection of motion

Motion can be detected by comparing the distance, travelled in a given period of time, T, to a given threshold, ϑ. A Boolean sequence, indicating motion, is computed according to the formula:

$$b_n \equiv 1 \quad \text{if} \quad d_n > \frac{t_n - t_{n_0}}{T}\vartheta \text{, and } b_n \equiv 0 \quad \text{otherwise,} \quad \text{for} \quad n = 1, \ldots, N \tag{1}$$

where:

$$d_n \equiv \sum_{\nu=n_0}^{n} \sqrt{\left(x_\nu - x_{\nu-1}\right)^2 + \left(y_\nu - y_{\nu-1}\right)^2} \text{, with } n_0 \equiv \arg\inf_{\nu} \left\{ \nu \mid t_\nu \geq t_n - T, \nu = 0, \ldots, n-1 \right\} \tag{2}$$

In the above equation $\{x_n\}$ and $\{y_n\}$ are the estimates of the sequences of coordinates, obtained during measurement data preprocessing (*cf.* Section 3.3). The values of T and ϑ should be optimised to prevent small deviations in the position from being considered as motion. To smooth the sequences at the output of the algorithm for motion detection, morphological opening and closing filters are applied [12].

4.2. Estimation of walking velocity

An estimate of the sequence of the instantaneous walking velocity may be obtained by numerical differentiation of the sequence of the position estimates, *e.g.* by means of the central-difference method [13], defined by the following formula:

$$\hat{x}_n^{(1)} \equiv \frac{x_{n+1} - x_{n-1}}{\Delta t_n} \quad \text{for} \quad n = 1, \ldots, N-1 \tag{3}$$

where $\{x_n\}$ is a sequence of data to be differentiated, and $\Delta t_n \equiv t_{n+1} - t_{n-1}$ are the differentiation steps, with t_n denoting the time moments at which the data have been acquired. The central-difference method is, however, very sensitive to errors corrupting the data used for derivative estimation; therefore, it should be regularised through, *e.g.*, optimisation of the differentiation step [14].

Estimation of the instantaneous walking velocity comprises differentiation of the sequences of the x and y coordinates separately, and calculation of the direction-independent velocity magnitude according to the following formula:

$$v_n = \sqrt{\left(\hat{x}_n^{(1)}\right)^2 + \left(\hat{y}_n^{(1)}\right)^2} \quad \text{for} \quad n = 1, \ldots, N-1 \tag{4}$$

An estimate of the average walking velocity can be obtained by averaging the estimates $\{v_n\}$ over the time intervals in which the monitored person is walking; these intervals are determined using the algorithm for motion detection described in the previous subsection.

4.3. Estimation of travelled distance

The estimation of the travelled distance should be preceded by smoothing of the monitored person's trajectory $\{x_n\}$. This operation is aimed at removing outlying values since those values could cause important overestimation of the distance. This can be done by using a median filter: $x'_n = \text{median}\{x_{n-1}, x_n, x_{n+1}\}$. The travelled distance can then be calculated by summing up the distances between consecutive locations of the monitored person:

$$s = \sum_{n=1}^{N} \sqrt{\left(x_n - x_{n-1}\right)^2 + \left(y_n - y_{n-1}\right)^2} \tag{5}$$

5. Results of experiments

In figure 3, exemplary results of motion detection, obtained by means of the procedure proposed in Section 4.1, are presented; while in table 1, some statistics of errors in the motion detection are shown.

Journal of Physics: Conference Series **772** (2016) 012057 doi:10.1088/1742-6596/772/1/012057

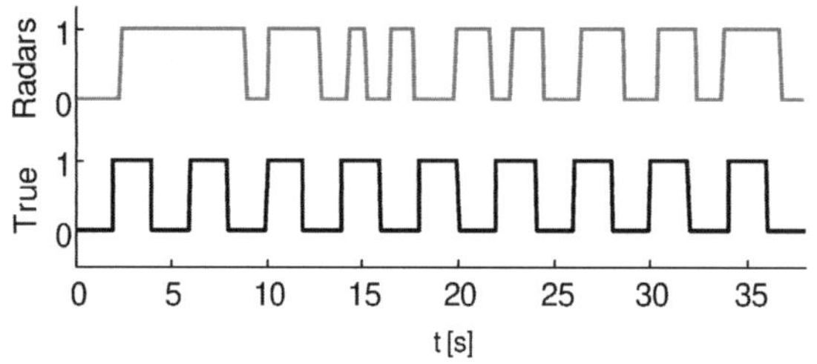

Figure 3. Results of motion detection, based on the data from the radar sensors; '0' means standing still, '1' – moving.

Table 1. Errors in start/stop estimation.

Mean error	0.33 s
Standard deviation	0.64 s
Maximum error	1.92 s

It can be seen that, according to the proposed procedure, the motion of the moving person can be correctly detected, though sometimes two separate moves are not distinguished, or one move is interpreted as two separate, shorter moves. Nevertheless, the mean and maximum error of start/stop estimation provided in table 1 indicate the practical validity of the procedure.

In figure 4a, the results of estimation of the average walking velocity are shown, while in figure 4b – the results of estimation of the travelled distance. The estimation task has been repeated for a number of real-world experiments, in which a person has been walking in the monitored area according to different scenarios and with varying walking velocity. The estimates of both quantities have been obtained by using the methods described in the previous section. In tables 2 and 3, some statistics of the errors of the velocity and travelled distance estimates are provided.

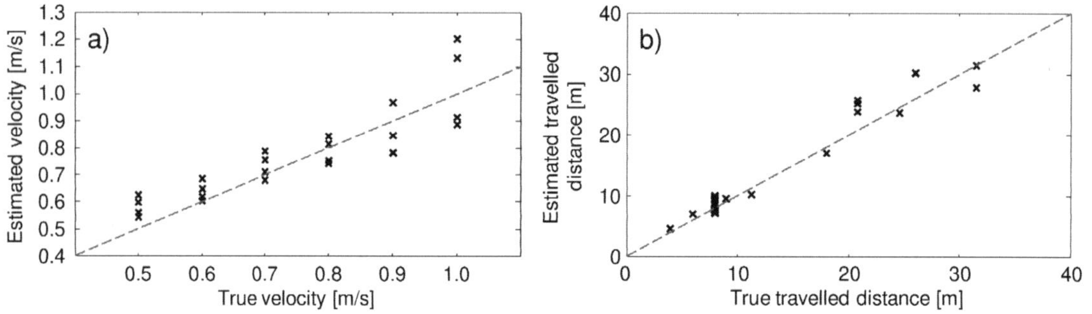

Figure 4. Estimates of the average walking velocity vs. true velocity (a) and estimates of the travelled distance vs. true travelled distance (b).

Table 2. Errors of walking velocity estimates.

Mean error	0.02 m/s
Standard deviation of errors	0.08 m/s
Maximum error	0.21 m/s

Table 3. Errors of travelled distance estimates.

Mean relative error	7.65 %
Standard deviation of relative errors	11.82 %
Maximum relative error	26.32 %

As it can be observed, in case of the velocity estimates, the mean absolute error is 0.02 m/s, with standard deviation of the errors equal to 0.08 m/s; the maximum absolute error reaches 0.21 m/s. In case of the estimation of the travelled distance, the mean relative error is 7.65 % with standard deviation of the errors equal to 11.82 %; here, the maximum relative error reaches 26.32 %.

As suggested in the literature, *e.g.* [15], the walking velocity lower than 0.6 m/s predicts future risk of falls and of hospitalisation of a patient. Taking into account the presented statistics of the estimates errors, it can be seen that mid- and long-term changes in the walking velocity could be reliably detected using the presented estimation procedures.

6. Conclusions and prospects for further work

The applicability of impulse-radar sensors for non-invasive monitoring of human movements has been analysed and demonstrated by means of a consistent set of algorithms for processing of measurement data provided by such sensors. In particular, selected algorithms for measurement data preprocessing – including clutter suppression, echo parameter estimation, and position estimation – have been described.

It has been shown that the presented set of algorithms enables one to quite accurately detect the motion of the monitored person and estimate his/her walking velocity and travelled distance. These are two basic quantities which can be used by the medical and healthcare personnel to assess the overall health condition of the patient; however, some more would be desirable – *e.g.* number of turns made during a walk, or the person's acceleration in three types of situations, *viz.* starting to walk, stopping, and changing the direction.

Further works will focus on the tuning of the algorithms of measurement data processing described in this paper, and on the development of new algorithms for estimation of other healthcare-related quantities to be used in impulse-radar-based system for monitoring of human movements.

Acknowledgments

This work has been accomplished within the project PL12-0001 financially supported by *EEA Grants – Norway Grants* (*http://eeagrants.org/project-portal/project/PL12-0001*).

References

[1] Hamm J, Money A G, Atwal A and Paraskevopoulos I 2016 *Journal of Biomedical Informatics* **59** pp 319–45

[2] Baldewijns G, Debard G, van Den Broeck B, Mertens M, Karsmakers P, Croonenborghs T and Vanrumste B 2016 ed F Florez-Revuelta and A A Chaaraoui (Herts, UK: IET) pp 1–22

[3] Baldewijns G, Luca S, Vanrumste B and Croonenborghs T 2016 *BMC Medical Research Methodology* pp 16–23

[4] Morawski R Z, Yashchyshyn Y, Brzyski R, Jacobsen F and Winiecki W 2014 in *Proc. IMEKO TC-4 International Symposium* (Benevento, Italy), pp 786–91

[5] Su B Y, Ho K C, Rantz M and Skubic M 2015 *IEEE Transactions on Biomedical Engineering* **62** pp 865–75

[6] Jian Q, Yang J, Yu Y, Björkholm P and McKelvey T 2014 in *Proc. 8th European Conference on Antennas and Propagation* (The Hague, The Netherlands), pp 3078–81

[7] Yinan Y, Jian Y, McKelvey T and Stoew B 2012 *International Journal of Antennas and Propagation* **2012** pp 1–11

[8] Zetik R, Crabbe S, Krajnakc J, Peyerl P, Sachs J and Thoma R 2006 *Proceedings of SPIE – 'Sensors, and Command, Control, Communications, and Intelligence (C3I) Technologies for Homeland Security and Homeland Defense'* **6201** pp 1–12

[9] Wagner J and Morawski R Z 2015 in *Proc. XXI IMEKO World Congress "Measurement in Research and Industry"* (Prague, Czech Republic), pp 1–6

[10] Mazurek P, Miękina A and Morawski R Z 2016 *Measurement* **88** pp 45–57

[11] Garcia D 2010 *Computational Statistics and Data Analysis* **54** pp 1167–78

[12] Serra J 1994 *Signal Processing* **38** pp 3–11

[13] Wagner J, Mazurek P and Morawski R Z 2015 in *Proc. XXI IMEKO World Congress "Measurement in Research and Industry"* (Prague, Czech Republic), pp 1–6

[14] Wagner J, Miękina A, Mazurek P, Morawski R Z, Ciamulski T, Jacobsen F F, Øvsthus K, Sudmann T T and Børsheim I T 2016 Technical Report (Warsaw, Poland: Warsaw University of Technology, Faculty of Electronics and Information Technology) pp 1–22

[15] Fritz S and Lusardi M 2009 *Journal of Geriatric Physical Therapy* **32** pp 2–5

Surveillance of muddy river flow by telemetry system

Motoki Takuno, Shigeru Takayama

Faculty of Science and Engineering, Ritsumeikan University BKC, 1-1-1 Nojihigashi, Kusatsu, Shiga 525-8577, Japan

E-mail: re0027hs@ed.ritsumei.ac.jp

Abstract. This paper shows the dynamic characteristic evaluation system to monitor muddy river flow by video image. The dynamic character is evaluated by the flow speed and direction in each small block distinguished evenly on a video frame. By evaluating dynamic character of river surface, water behavior and dangerous area will be distinguished. Aim of this study is to discover the dangerous area, to presume disasters and to inform local or national government and alarm people.

1. Introduction

Heavy rain and typhoon cause muddy river flow. When such flow becomes heavily, various kinds of water disaster are occurred to life and social infrastructures like houses, plants, bridges, roads and railways. And also human lives are lost.

Monitoring system for muddy river flow is necessary and useful to predict the water disasters. As nobody want to be at the river side. Then, that system should be constructed as a telemetry system.

Telemetry system proposed in this paper monitors state of the river surface at onsite system and estimates dangerous area in the river at host system far from the river. Observer can know the dynamic state of dangerousness of the river in parallel of distributions of river flow speed and direction. And also he can inform the dangerousness to people living near the river. In this telemetry system, onsite and host system are communicated through "Cloud" system.

2. Structure of telemetry system of river flow

Figure 1 shows the system construction of telemetry system to monitor the river. The system is constructed by two parts; Onsite and Host system. Onsite system placed near river is the measurement system of river flow. Host system is the monitoring system at remote place from river.

Onsite system has GPS Receiver, Camera, Computer and Mail Interface. GPS Receiver gets parameter of latitude, longitude, absolute time and Circular Error Probability. They are stored in Location Parameter (.csv) and saved to Parameter Folder. Camera takes a picture of river surface firstly. The first shot image is saved to Parameter Folder. Camera also shoots VGA (640pixel*480pixel) size video. Computer calculates vector of river flow by using captures of the video. The parameters of start point and end point of vector is stored in parameter file (.csv). The file is saved to Parameter Folder in each capture. Computer counts the file in Parameter Folder. If the number of parameter files is sufficient, Parameter Folder is compressed (.zip). After the compression, Computer deletes all contents of the folder. Computer attaches the zip file to mail. Mail Interface sends the mail to cloud system.

Host system has Computer, Mail Interface and Monitor Window. Mail Interface receives mail from "Cloud" system. This program gets the zip file and save to Contemporary Parameter Folder and Temporary Parameter Folder. Computer refers Temporary Parameter Folder. Computer extracts the zip file. First shot image and Location Parameters are displayed on Monitor Window. Computer also reads parameter of vector from parameter files. After that, this Computer estimates flow speed and flow direction by using the parameter. From the flow speed and direction, Computer analyzes dangerousness and characteristic on river surface. The analysis is fit together first shot image and displayed on Monitor Window. That is to say, it is possible to observe safely by using this telemetry system.

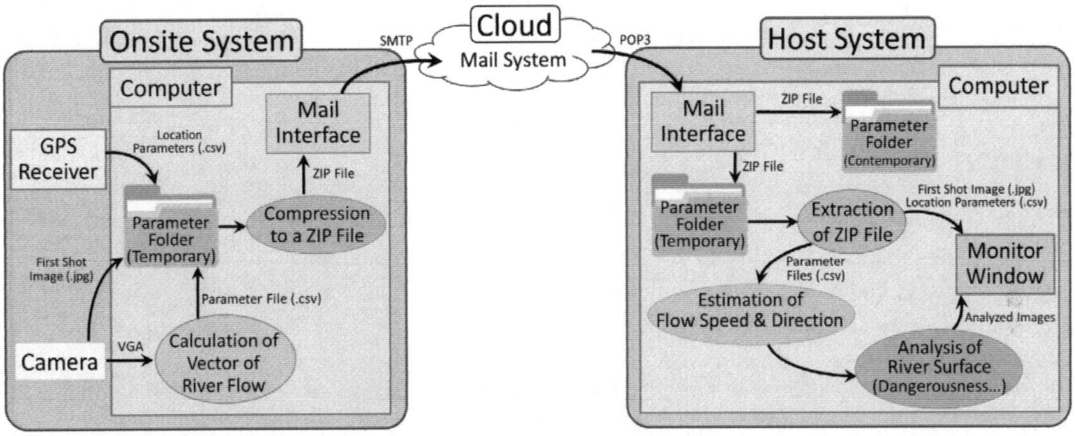

Figure 1. Telemetry System

3. Vector calculation of river flow

To get the vector of river flow, each image are distinguished as small square areas (8pixel*8pixel). Figure 2 shows the original image and segmentation. An image segment block is defined as Region of Interest (ROI). The vector is calculated for each blocks.

Figure 3 shows how to find the movement of ROI blocks. To find the movement of an original ROI block in an image frame, correlation to blocks in next image frame is calculated. The searched range is 24pixel*24pixel centered at same position with the original ROI block in previous image frame. Original ROI block and searched range image are converted from RGB into gray scale. The intensity values the size is 8pixel*8pixel are obtained from upper left corner of searched range image. The correlation is calculated by using the intensity values from searched range image and original image. This calculation is repeated by sliding 1pixel on searched range image. After the calculation, the highest correlation value of the block is discovered. The block is the most resemble block with original ROI block. The vector is decided by connecting the two blocks.

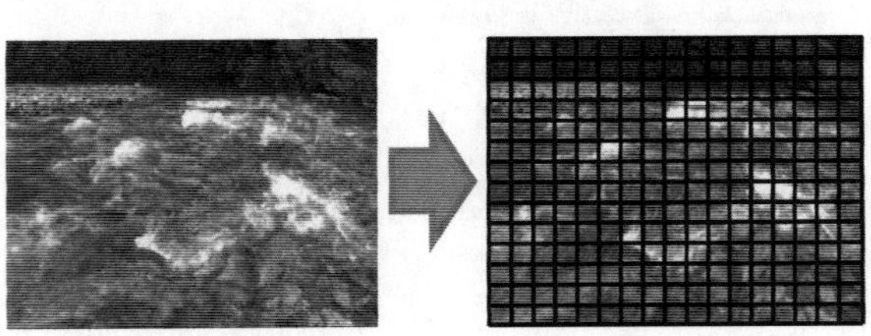

Figure 2. Segmentation of image frame

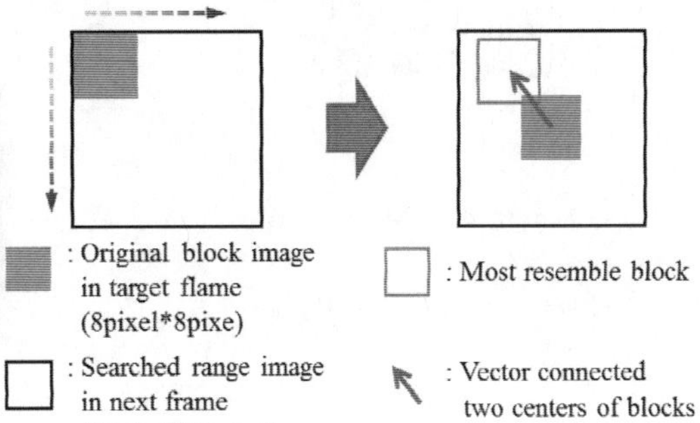

Figure 3. Generation of a movement vector of a ROI block

Journal of Physics: Conference Series **772** (2016) 012058 doi:10.1088/1742-6596/772/1/012058

4. Estimation of dangerousness on river surface

4.1. Estimation of flow speed and direction

The flow speed and direction of original ROI block are estimated by the vector.

To estimate the flow speed, the vertical and horizontal lengths of 1pixel is needed. The length is calculated by actual measured value. Using the vector, lengths and frame rate, the flow speed is calculated (1).

$$V = \sqrt{\{(x' - x) \cdot h\}^2 + \{(y' - y) \cdot v\}^2}/(1/f) \quad (1)$$

V means flow speed [meter/sec], x and y means coordinate of start point on image, while x' and y' means coordinate of end point on image, h means horizontal length [meter/pixel], v means vertical length [meter/pixel], f means frame rate and (1/f) means interval of two images [sec].

Flow direction is classified by the angle of vector. The angle of vector is calculated by the inverse trigonometric function (2).

$$\theta = \tan^{-1}\left(\frac{y' - y}{x' - x}\right) \quad (2)$$

Here, θ means angle of vector, x and y means coordinate of start point on image, while x' and y' means coordinate of end point on image. From the calculation, the angle is classified into 8 directions. (Table1)

Figure 4 and 5 shows the results of flow speed and direction. They are expressed by color in each ROI blocks.

Table1. Direction to color conversion table

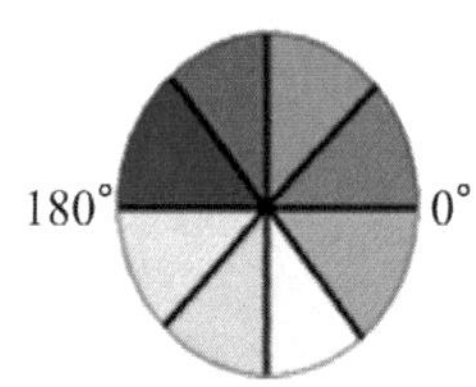

- Direction: 135° $\leq \theta <$ 180°
- Direction: 90° $\leq \theta <$ 135°
- Direction: 45° $\leq \theta <$ 90°
- Direction: 0° $\leq \theta <$ 45°
- Direction: -45° $\leq \theta <$ 0°
- Direction : -90° $\leq \theta <$ -45°
- Direction : -135° $\leq \theta <$ -90°
- Direction : -180° $\leq \theta <$-135°

(a) Flow speed image (b) Color chart of flow speed

Figure 4. Distribution of flow speed

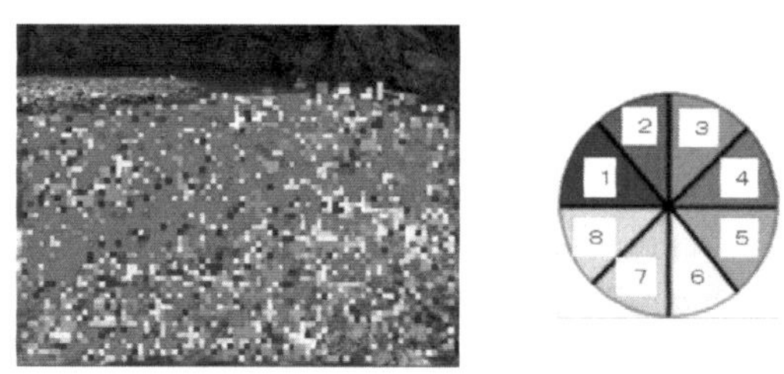

(a) Flow direction image (b) Color chart and number of flow direction

Figure 5. Distribution of flow direction

4.2. Estimation of turbulent flow

To evaluate dangerousness of river surface, it is necessary to estimate the dynamic state of the turbulence of river flow. The turbulence is estimated by the relation of flow directions.

Figure 6 shows how to estimate the turbulent flow. Number in each block shows the direction assigned in Figure 5 (b). 0 means no vector in the block. Figure 6 (a) shows the distribution of flow directions of nine blocks in previous image frame. Figure 6 (b) shows the flow direction of target block which is the center block in same nine blocks in present image flame. Comparing two kinds of distributions, flows of two blocks assigned 1, 8 are same and adjacent direction to the target block. On the other hand, flows of six blocks assigned 6, 7 are different directions to the target block. The number of different direction segments is counted. Turbulence is estimated by the number.

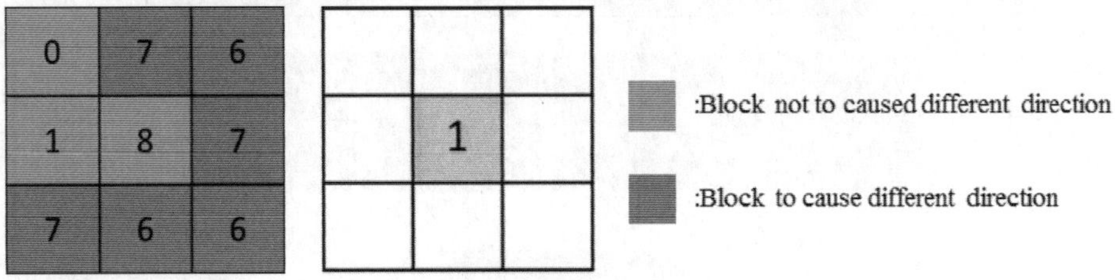

(a) Distribution of flow direction in previous image frame

(b) Flow direction of target block in present image flame

Figure 6. Estimation of turbulent flow

4.3. Estimation of characteristic of river surface

From the flow speed and directions, the characteristic of river surface is estimated.

Table2 shows the relation. The flow speed is distinguished to 3 levels. They are "Extremely high speed", "High speed" and "Low speed". The river flow is also distinguished to 3 levels by the number of different direction blocks compared one block with neighbor blocks. They are "Turbulent flow", "Normal flow" and "Orderly flow". By these level divisions, the characteristic is classified into 5 modes. Mode4 expresses the most dangerous state. Mode3 expresses the dangerous state. Mode2 expresses the careful state. Mode1 expresses the calm state. Mode0 expresses safe state. These levels are estimated each ROI blocks. (above 4 images in Figure 7)

Then, by emphasizing high density area of ROI blocks, the characteristic area is specified into each mode. Figure 7 shows the specification of characteristic area each mode. The arrow shows main flow direction in characteristic area. These characterized images are displayed to observer.

Table2. Classification of characteristic

Speed \ Flow	Turbulent	Normal	Orderly
Extremely High	Mode4	Mode4	Mode3
High	Mode3	Mode2	Mode2
Low	Mode1	Mode1	Mode0

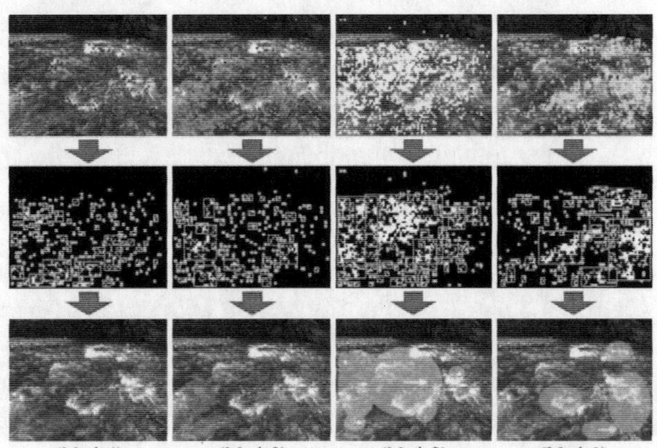

Figure 7. Specification of characteristic area each modes

4.4. Estimation of dangerousness of river surface

Dangerousness area is estimated by the fusion of characteristic modes on river surface. At first, river picture is divided into small segments. A segment is formed by square of 8 pixels. In each segments, weights defined in each modes are added for 1 minute. The weight of Mode4, Mode3, Mode2 and Mode1 is 4, 3, 2 and 1. Figure 8 shows the distribution of dangerous area colored by sum of weights. High scores are colored by red. These are estimated as dangerous areas. And Low scores are colored by green. These are estimated as safe areas relatively.

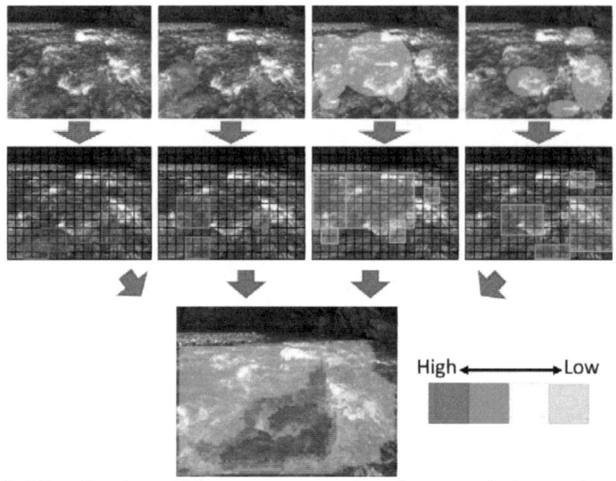

Figure 8. Distribution of dangerousness by characteristic modes fusion

5. Evaluation of river flow

5.1. Results of muddy river flow

Figure 9-12 show original image and results of evaluation in case of muddy river flow. In Figure 9, human feels this river is clearly dangerous. In Figure 10, flow speed is fast and flow direction is turbulent. In Figure 11, Mode4 area is the largest. Mode2 is so few and Mode3 and Mode1 do not spread. In Figure 12, dangerous level is so high. From these results, this river is estimated so dangerous by image processing. This evaluation is identical with human sense. Then, it is confirmed that this system has found invisible dangerous area.

Figure 9. Original image of muddy river flow

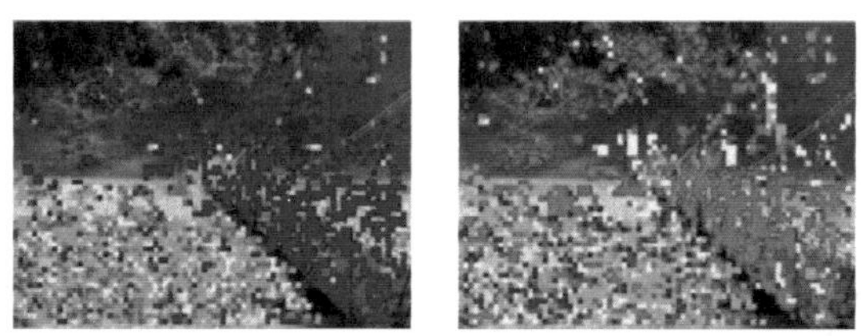

(a) Flow speed distribution (b) Flow direction distribution

Figure 10. Flow speed and direction of muddy river

Figure 11. Characteristic area of muddy river flow

Figure 12. Distribution of dangerousness on muddy river

5.2. Results of swollen river flow

Figure 13-16 show original image and results of evaluation in case of swollen river flow. In Figure 13, human feels less danger. In Figure 14, flow direction is turbulent around rock. In Figure 15, Mode2 area is the largest. Mode1 area is less than Mode2. Mode3 area is so few, and Mode4 area do not spread. In Figure 16, dangerous level around rock is higher than other places. From these results, the area around rock is dangerous by the estimation of image processing. This evaluation is different with human sense. It is also confirmed that this system has found invisible dangerous area.

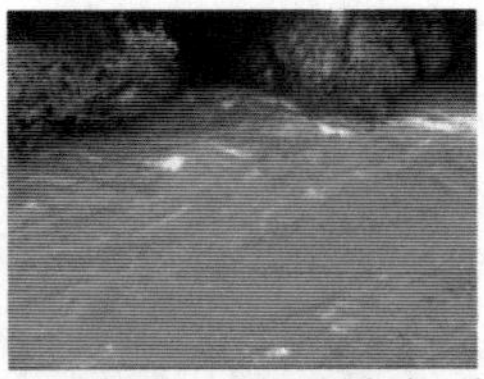

Figure 13. Original image of swollen river flow

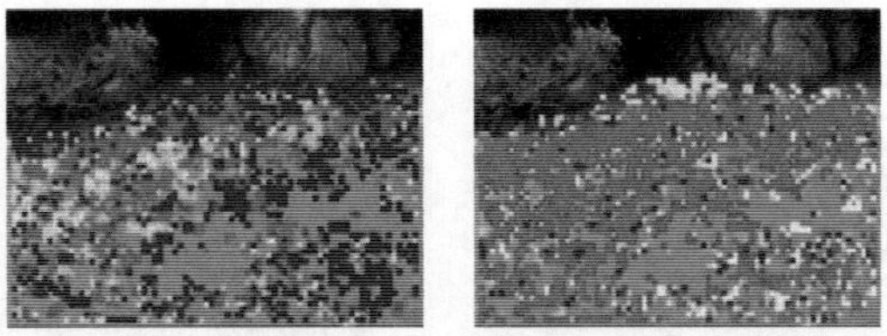

(a) Flow speed distribution (b) Flow direction distribution

Figure 14. Flow speed and direction of swollen river

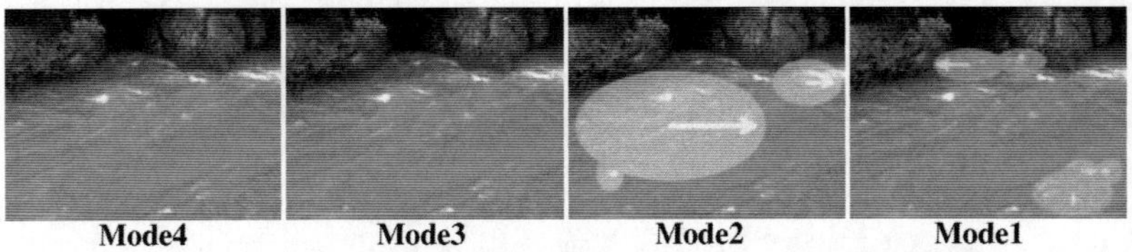

Figure 15. Characteristic area of swollen river flow

Figure 16. Distribution of dangerousness on swollen river

5.3. Results of curve river flow

Figure 17-20 show original image and results of evaluation in case of winding river flow. In Figure 17, human feels this river is calm. In Figure 18, flow speed is so slow. In Figure 19, only Mode1 area spread on river surface. In Figure 20, dangerousness is expressed but the level is low. From these results, this river is not dangerous by the estimation of image processing. This evaluation is identical with human sense.

Figure 17. Original image of calm river flow

(a) Flow speed distribution (b) Flow direction distribution

Figure 18. Flow speed and direction of winding river

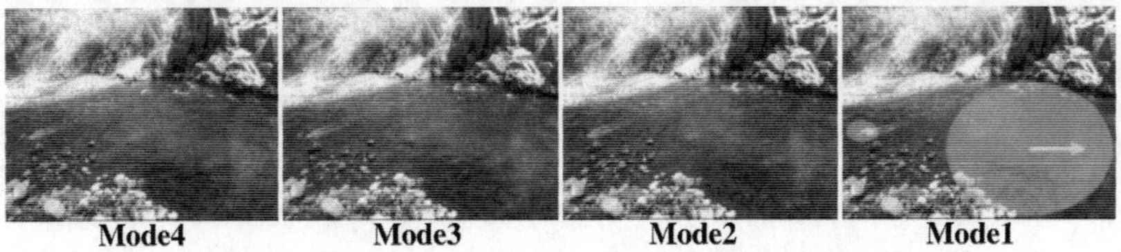

Figure 19. Characteristic area of calm river flow

Figure 20. Distribution of dangerousness on winding river

6. Conclusion

This paper has shown the telemetry system to surveil river flow. This system estimates dangerous area of the river by data fusion of flow speeds and flow directions of small segments divided river surface. This information becomes so helpful for people living near the river.

References

[1] M.Takuno, S.Takayama, "Onsite Mesurement System for Dyanamic Characteristic Evaluation of River Flow", XXI IMEKO World Congress2015, Prague, Czech Republic, Aug. 30-Sep. 4.2015

[2] T.Tomoo, S.Takayama, "Dynamic Measurement of River Surface for Monitoring Muddy Flow", Proc. of SICE International Conference 2011, Tokyo, JAPAN, pp.64-69, 2011.9

Journal of Physics: Conference Series **772** (2016) 012059 doi:10.1088/1742-6596/772/1/012059

Fault Injection Software Tools and Robust Design Principles for Reliability and Safety in Measurement Science Education

Lisa-Marie Faller, Hubert Zangl and Juliana P Leitzke

Institute of Smart System Technologies, Sensors and Actuators, Alpen-Adria-Universitaet Klagenfurt, Universitaetsstrasse 65-67, 9020 Klagenfurt, Austria

E-mail: lisa-marie.faller@aau.at

Abstract. In the design of measurement systems we face the fact that parameters are subject to (measurement-) uncertainties. Additionally, components may behave entirely different from what is specified, which is then considered a fault. Consequently, both uncertainty as well as probability of failure should be considered in education on robust design and reliability. In this paper we present a teaching concept based on hardware fault injection using a simple level sensor system as an example. Learning objectives are faults, errors, failures, false alarms versus misses as well as advantages and disadvantages of redundancy.

1. Introduction

In our modern society, measurement devices, i.e. sensors, the respective electronic evaluation and signal processing chains, are indispensable. Measurement science is employed to fulfil manifold tasks in consumer goods such as mobile phones and personal computers or tablets as well as in cars, (automated) production lines or safety and security systems. All electronic devices, society relies on in everyday private as well as working life, are equipped with sensors and measurements electronics. When educating engineers to become the future designers of such measurement chains, critical analysis of sensor systems with respect to robust design and reliability thus seems mandatory. It is necessary to enhance the awareness with respect to uncertainty, safety and fault handling for such systems and teach proper usage and adoption of the available methodology to prevent undesired system states. This is even more important when all of us rely on electronic measurement systems where, in, e.g., automated production chains with a cooperation and collaboration with human as well as robotic workers or autonomous vehicles, even lives depend on them.

In order to provide an introduction to the methods that are used to cope with this challenges, we provide a course on robust design and reliability for our Master students. Most of them are enrolled in the Master study program Information and Communication Engineering with a specialization in autonomous systems and robotics. It is assumed that the students are well familiar with the concept of measurement uncertainty as explained in [1], probability theory, estimation and detection theory [2] as well as electrical measurements and sensors. We consider both, qualitative and quantitative robust design methodologies [3], in order to provide a conceptual view of the topic as well as to acquaint students with the principal approach to design

1. Introduction and Defintions	Requirement Specification
	Fault, Error, Failure
	Probability of Failure, Failure Density, Failure Rate
	Failure Modes and Distributions
2. Reliability Modeling and Analysis	Hazard Assessment and Risk Analysis
	Failure Mode and Effects Analysis
	Fault Tree Analysis, Reliability Block Diagram
	Fault Injection
3. Design Methods	Design Critera
	Model Based Optimal Design
	Design of Experiments
	Model Based Optimal Design

Figure 1. Coarse outline of the course on Robust Design and Reliability.

a robust measurement chain (including principles from, e.g. [4] and [5]). Fault tree analysis [6] and simulation based fault injection [7] are shown on the example of a safety critical fill level detection system.

2. General Teaching Concept

The respective course where this concept is applied is a combination of lecture and exercise under the title of robust design and reliability. A course outline is shown in Figure 1. Since this course is suggested as successor to the measurement and signal processing course where advanced stochastic concepts are covered to yield, in a certain sense, optimal estimators for measurements subject to various sources of noise, students are assumed familiar with probability theory and statistics. Hence, mathematical background to deal with measurement uncertainty is assumed known, only necessary probability distributions used to derive component failure rates are introduced at the beginning [8].

3. Principle of Fault Injection

Fault injection is a method that is widely used in evaluating the dependability of software and computer systems [7], [9] and [10]. By means of introducing faults into the system under test, error handling can be tested. However, the approach can also be used for hardware and indeed the first experiments with fault injection were proposed for hardware. As with software, the approach is used to test the dependability of the hardware system. Practically, hardware fault injection is nothing more than replacing good components in the system with faulty components, breaking connections, introducing shorts between nets and to ground. This is done in network simulations.

In order to demonstrate the idea to students, we have developed a simple Matlab toolbox for fault injection that works with the currently freely available simulator LTspice [11]. It works with the original netlist, introduces faults, calls the simulator and aggregates the results. The results are then displayed and compared to the reference. This process flow is illustrated in Figure 2.

4. Fault Injection - Educational Example

In order to emphasize the difference between faults, errors and failures as illustrated in Figure 3, as well as the concept of redundancy (e.g. [12], [13] and [14]), we use a simple measurement

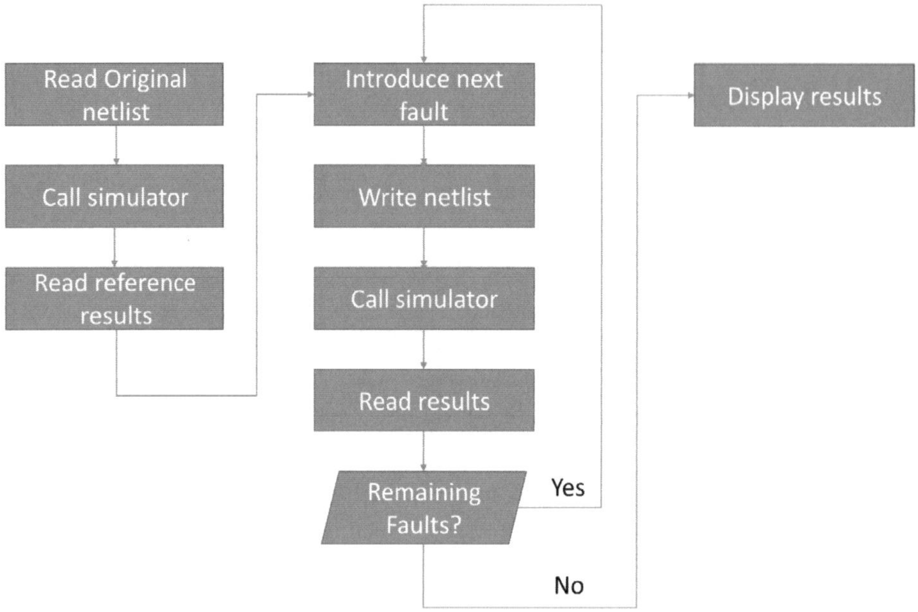

Figure 2. Flow diagram of fault injection as implemented in our educational toolbox.

system as an example and apply fault injection to it. The example application is illustrated in Figure 4 and the respective circuitry is illustrated in Figure 5. The circuitry is introduced to the students and, employing the fault injection toolbox for Matlab, the circuit-netlist is modified in order to inject faults belonging to the category of either 'open','short' or 'short to ground'. Based on this first evaluation, the distinction between faults, errors and failures can be illustratively given. Also, the basis of detection for the considered system, i.e. hypothesis testing, the quality of which is characterized by its false alarm and miss rates, is explained. In a next step, redundancy concepts to aid this design in order to avoid dangerous situations, i.e. undetected excess of level (miss), are developed. For one of these extended systems, a parallel configuration, again fault injection is applied and results are illustrated using Matlab (compare Figures 6 and 7). The outcomes of the redundant and original configurations are then compared with respect to fault tolerance. While the redundant design is capable of equalizing some system faults, it is important to point out that others, such as the short to ground of the voltage source for the respective system, can not be covered.

4.1. Parallel Redundancy Concept

Since there is not just a single solution for system redundancy, it is important to point out differences and advantages as well as disadvantages of the different concepts. A simple way to system redundancy is adoption of a second system in parallel which is a duplicate of the first. This was done for the illustrating example with results shown in Figure 7. The supply voltage source is chosen to be the same for both and the connection is done via diodes, to avoid feedback. Such a parallel configuration reduces the overall miss rate and its probability of failure can be generally given as

$$F_{par_n}(t) = \prod_{i=1}^{n} F_{simple_i}(t) \tag{1}$$

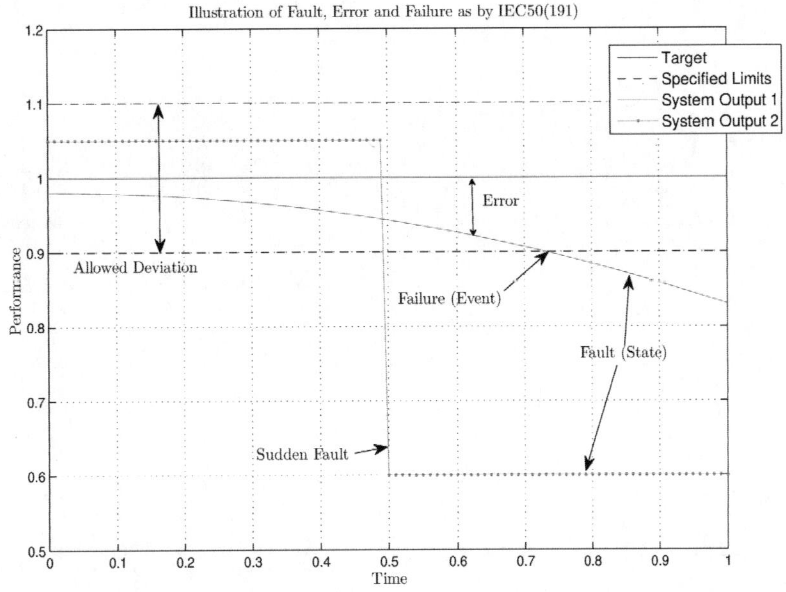

Figure 3. Illustration of fault, error and failure as defined by the IEC 60050-191:1990 [15], compare also [3]: The error is defined as the deviation between the target and the actual system output, a failure occurs if the system output violates the pre-defined limits, while a failure, according to this definition is transient, the term fault is used to describe a static system behaviour, i.e. when the system output is out of specification and will not return to within specifications without external input (e.g. repair). While in the norm the term 'event' is used to describe the transient nature of failure, this is not to be confused with an event in the probabilistic sense.

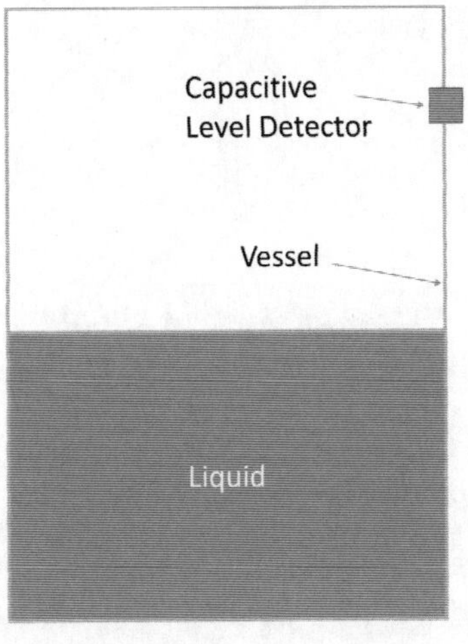

Figure 4. Schematic illustration of the example application.

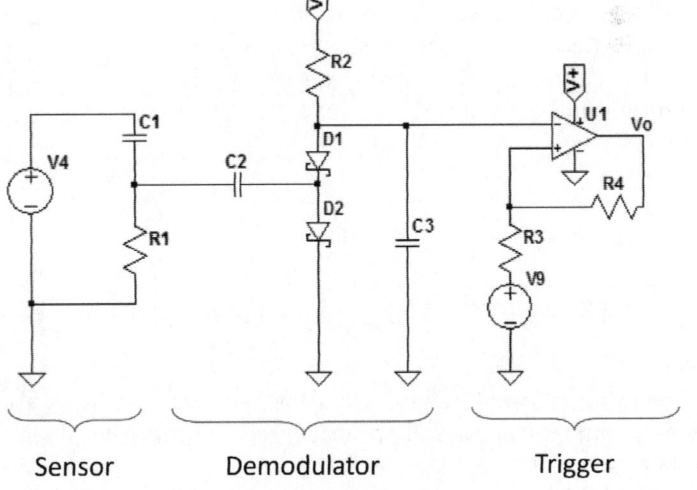

Figure 5. Simple (non-redundant) measurement system for illustration of fault injection. The system comprises a capacitive sensor front end, a demodulation circuitry and a trigger. If the capacitance is high due to the proximity of a liquid, the output indicates that the fill level is critical.

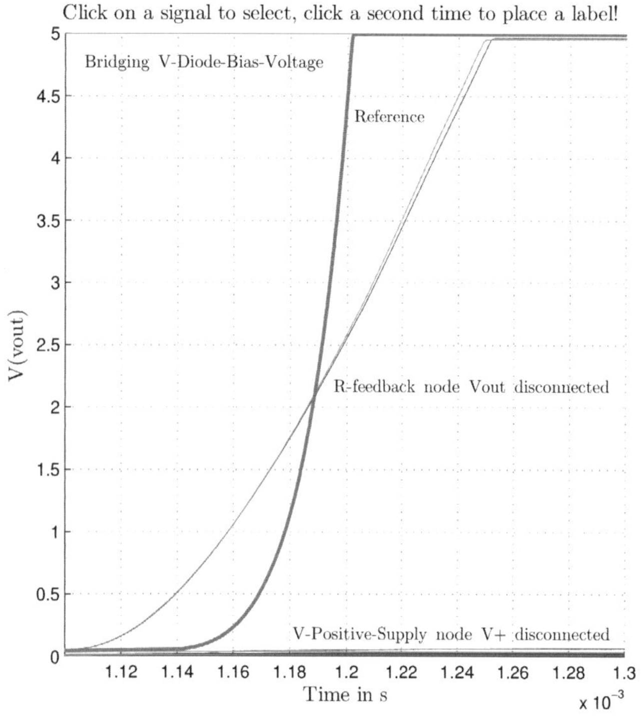

Figure 6. Results for the output signal as obtained using the fault injection toolbox on the non-redundant system shown in Figure 5. In the tool, the results are easily annotated by clicking on the traces. A succeeded detection is indicated by the rise of the voltage level to 5 V. System failures are caused by the faults resulting in non-rising curves (here: 16 curves).

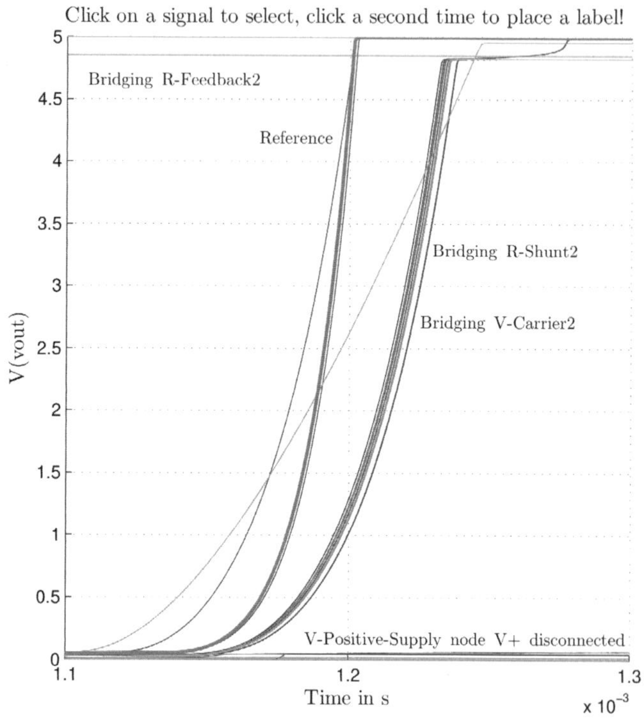

Figure 7. Results for the output signal as obtained using the fault injection toolbox on the redundant system. In the redundant design, the number of non-rising curves could be reduced to seven when the detection threshold is lowered to 4.5 V.

and the reliability can subsequently be given by

$$R_{par_n}(t) = 1 - F_{par_n}(t) = 1 - \prod_{i=1}^{n} F_{simple_i}(t) \tag{2}$$

Assuming constant failure rates for the two parallel systems, i.e. the probability of failure follows an exponential distribution, we can find the Mean Time To Failure (MTTF), which is the expected failure time, as [16]

$$MTTF_{par_n}(N, n) = \int_0^\infty R_{par_n}(t)dt \tag{3}$$

With N the number of series components (of the respective subsystems) and n the number of parallel systems. As is outlined in [16] a sufficiently accurate approximation of this integral for systems with components exhibiting identical failure rates λ is

$$MTTF_{par_n}(N,n) = \frac{1}{(\lambda N)}\left[1 + \sum_{i=1}^{n}\sum_{j=1}^{i}\left(\frac{n+1-j}{N+j}\right)\right] \quad 0 < n \tag{4}$$

While these calculations are valid for systems without common components, in the example system, the performance is poorer due to, e.g. the common voltage supply. In our example, the improvement in performance can be quantified by the reduction in the misses (16 versus seven) for the redundant system. This is achieved although the number of components and consequently also of possible faults is increased.

4.2. Further Take-Home-Messages

While dual (parallel) systems already provide for increased system reliability, various other configurations are possible which will add safety in critical applications. A possible alternative is to introduce parallel units with subsequent voting, so-called 'k-out-of-n' systems. Also partitioning can be adopted to build systems with redundant paths to provide increased system reliability. For all applications, the way to the design of such hierarchies should follow the 'Keep It Small and Simple' (KISS) concept. Since, although we achieve better performance in case of misses and provide for an increased MTTF, it is important to be aware of a major drawback of such extended topologies: increase in power consumption, weight and number of parts. Even for designs with dynamic redundancy, where a switch is employed to change between either of two parallel systems in case of one of them failing, additional maintenance as well as detection mechanisms are necessary.

5. Conclusion

Besides the consideration of uncertainties, robust design also has to consider possible faults. In this paper we introduce a simple toolbox for Matlab that allows to perform fault injection in spice simulations of measurement systems. Using this in a course on robust design and reliability, it is aimed to increase the awareness with respect to faults, errors and failures as well as advantages and possible shortcomings of redundancy.

References

[1] Zangl H, Zine-zine M and Hoermaier K 2015 Utilisation of software tools for uncertainty calculation in measurement science education *Journal of Physics: Conference Series* vol 588
[2] Kay S M 1993 *Fundamentals of Statistical Signal Processng, Estimation Theory* (Prentice-Hall, Inc.)
[3] Rausand M and Hoyland A 2004 *System Reliability Theory* (John Wiley & Sons, Inc., Hoboken, New Jersey)
[4] Brasseur G 2003 *IEEE Transactions on Instrumentation and Measurement*
[5] Bergman B, de Mar J, Lorn S and Svensson T (eds) 2009 *Robust Design Methodology for Reliability* (John Wiley & Sons, Ltd.)
[6] Dhillon B S 2007 *Applied Reliability and Quality* (Springer-Verlag London Limited)
[7] Benso A and Prinetto P (eds) 2003 *Fault Injection Techniques and Tools for Embedded Systems Reliability Evaluation* (Springer US)
[8] Kececioglu D B 2002 *Reliability Engineering Handbook, Volume I* (DEStech Publications, Inc.)
[9] Piscitelli R, Bhasin S and Regazzoni F 2015 Analog circuit fault simulation approach based on pspice engine *10th International Conference on Design & Technology of Integrated Systems in Nanoscale Era (DTIS)*
[10] Barenghi A, Hocquet C, Bol D, Standaert F, Regazzoni F and Koren I 2014 *IEEE Transaction on Emerging Topics Comp.* **2**
[11] 2016 URL `http://www.linear.com/designtools/software/`
[12] Hecht H 2004 *System Reliability and Failure Prevention* (Artech House, Boston)
[13] Ushakov I A (ed) 1994 *Handbook of Reliability Engineering* (John Wiley & Sons, Inc.)
[14] Pham H (ed) 2003 *Handbook of Reliability Engineering* (Springer-Verlag London Limited)
[15] Committee I T 1990 Iec 60050-191:1990 Tech. rep. International Electrotechnical Commission
[16] Sweet A L, Lam Y and Bott J 1986 *IEEE Trans. on Reliability* **R-35** 539–540

Journal of Physics: Conference Series **772** (2016) 012060 doi:10.1088/1742-6596/772/1/012060

Methods for estimating comparable prevalence rates of food insecurity experienced by adults in 147 countries and areas

Mark Nord, Carlo Cafiero and Sara Viviani
Statistics Division, Food and Agriculture Organization of the United Nations, Viale delle Terme di Caracalla 00153, Rome, Italy.

E-mail: mark.nord@fao.org

Abstract. Statistical methods based on item response theory are applied to experiential food insecurity survey data from 147 countries, areas, and territories to assess data quality and develop methods to estimate national prevalence rates of moderate and severe food insecurity at equal levels of severity across countries. Data were collected from nationally representative samples of 1,000 adults in each country. A Rasch-model-based scale was estimated for each country, and data were assessed for consistency with model assumptions. A global reference scale was calculated based on item parameters from all countries. Each country's scale was adjusted to the global standard, allowing for up to 3 of the 8 scale items to be considered unique in that country if their deviance from the global standard exceeded a set tolerance. With very few exceptions, data from all countries were sufficiently consistent with model assumptions to constitute reasonably reliable measures of food insecurity and were adjustable to the global standard with fair confidence. National prevalence rates of moderate-or-severe food insecurity assessed over a 12-month recall period ranged from 3 percent to 92 percent. The correlations of national prevalence rates with national income, health, and well-being indicators provide external validation of the food security measure.

1. Introduction

Food security is said to exist when all people, at all times, have physical, social and economic access to sufficient safe and nutritious food that meets their dietary needs and food preferences for an active and healthy life [1]. Food security is inherently multi-dimensional, but one critical dimension is consistent economic access to adequate food. The United Nations Food and Agriculture Organization (FAO) has undertaken the Voices of the Hungry (VoH) project to develop and support a survey-based experiential measure of this dimension of food security. The methodology is similar to that used in the United States, Brazil, Canada, Mexico, and several other countries [2]. The project has taken on increased importance as the experiential-based measure has been endorsed by the United Nations Statistical Commission to monitor Target 2.1 of the recently agreed 2030 Agenda for Sustainable Development, which states: "By 2030, end hunger and ensure access by all people, in particular the poor and people in vulnerable situations, including infants, to safe, nutritious and sufficient food all year round" [3].

The VoH methodology measures the severity of food insecurity (lack of economic access to adequate food), based on conditions and behaviours reported in response to an 8-question survey module, the Food Insecurity Experience Scale Survey Module (FIES-SM; table 1). Response data are assessed and combined to create a measure using statistical methods based on Item Response Theory, in which the severity of food insecurity experienced by an individual or household is modelled as a latent trait—a characteristic not directly observable. Observable conditions assumed to be caused by the latent trait and elicited in response to the FIES-SM questions provide the information for constructing the measure. Based on the measured severity of food insecurity of a nationally representative sample of individual adults or households, a national prevalence rate of food insecurity beyond a specified threshold of severity is calculated. Thresholds and, thus, prevalence rates of food insecurity are made comparable across countries using information on the severity each item from a statistical model of each country's scale.

Table 1. English version of the Food Insecurity Experience Scale

Num.	Short reference	Question wording
1	WORRIED	During the last 12 MONTHS, was there a time when you were worried you would not have enough food to eat because of a lack of money or other resources?
2	HEALTHY	Still thinking about the last 12 MONTHS, was there a time when you were unable to eat healthy and nutritious food because of a lack of money or other resources?
3	FEWFOODS	Was there a time when you ate only a few kinds of foods because of a lack of money or other resources?
4	SKIPPED	Was there a time when you had to skip a meal because there was not enough money or other resources to get food?
5	ATELESS	Still thinking about the last 12 MONTHS, was there a time when you ate less than you thought you should because of a lack of money or other resources?
6	RANOUT	Was there a time when your household ran out of food because of a lack of money or other resources?
7	HUNGRY	Was there a time when you were hungry but did not eat because there was not enough money or other resources for food?
8	WHOLEDAY	During the last 12 MONTHS, was there a time when you went without eating for a whole day because of a lack of money or other resources?

2. Data

FAO expects that national prevalence rates of food insecurity for monitoring progress toward the Sustainable Development Goal Target 2.1 will eventually be based on data from national surveys conducted by national statistical agencies in each country. However, to obtain initial baselines on a large number of countries and to provide data for developing methods to make prevalence rates across countries comparable, FAO contracted with the Gallup Organization, as a data collection service provider. The 8-question FIES-SM was thereby added as a client module to the Gallup World Poll beginning in 2014. The Gallup World Poll has been conducted annually since 2006 in about 150 countries, interviewing nationally representative samples of 1,000 adults (ages 15 and older) in each country on a range of topics including family economics, employment, human development, and well-being [4]. In 2014, FIES data were collected in the Gallup World Poll in 147 countries, areas, and territories.

This paper reports on the psychometric assessment of the FIES data collected in the 2014 Gallup World Poll, the development of a methodology to make prevalence estimates comparable across countries, and initial estimated prevalence rates of food insecurity at two levels of severity.

3. Psychometric Assessment of FIES Data

The Item Response Theory measurement model used for FIES data assessment and scale construction is the single-parameter logistic measurement model commonly known as the Rasch Model [5]. The Rasch model assumes that the log-odds of respondent r saying "yes" to item i is a linear function of the difference between the severity of food insecurity experienced by r and the severity of item i. The severity of an item is defined as the severity experienced by respondents who are equally likely to affirm or deny the item. Two further important assumptions of the model are: 1) that all items discriminate equally—that is, that the slopes of the curves of log-odds of affirmation against respondent severity are equal for all items; and 2) that items are conditionally independent—that is, that correlations between items are entirely due to their common association with the latent trait. Given these assumptions, and a few additional assumptions less likely to be problematic for these data, the raw score of a respondent (the number of affirmative responses to the set of items) is a sufficient statistic for, and an ordinal indicator of, the severity of food insecurity he or she has experienced.

Maximum likelihood methods are used to estimate the severity of each item and the severity experienced by each respondent. Model estimation also includes calculation of item and respondent fit statistics, conditional correlations across items, and measurement reliability of the scale. Item severity parameter estimation, item-fit statistics, respondent-fit statistics, and inter-item conditional correlations are based on conditional maximum likelihood (CML) estimation. That is, the likelihood of item affirmation is calculated conditional on raw score. Respondent severity parameters and measurement error (uncertainty) are calculated as the maximum likelihood values given the item severity parameters.

The Rasch Model assumption of equal discrimination of items was assessed primarily by the item infit statistic, an information-weighted chi-square-type statistic that compares observed with expected misfit of each item. The expected value of all infits is 1.0; higher values indicate weaker discrimination; and values between 0.7 and 1.3 were considered to adequately meet the assumption of equal discrimination. In spite of the wide range of cultures and languages in which the FIES-SM was administered and the attendant challenges of translation, the fit of all the items to the measurement model was remarkably good in almost all countries. In 93 percent of countries, infit statistics for every item were in the acceptable range. Only 7 countries had any item with an infit higher than 1.4, and with one exception, those were countries with small numbers of non-extreme cases and sizeable resulting estimation errors for item-infits. (Only non-extreme cases—those with raw scores of 1 to 7—are used to estimate the measurement model. In highly food-secure populations, a large majority of respondents say "no" to all 8 questions, and the estimation analysis sample is small.)

Conditional correlations were not found to be excessive for any pairs of items in countries with sufficient sample size of non-extreme cases to produce reliable assessments.

Rasch reliability is the proportion of total variance in the population that is accounted for by the measurement model. A modified Rasch reliability, in which calculations of error variance and variance accounted for by the model were weighted equally across raw scores, was used in this study. The standard

Rasch reliability statistic is affected both by model fit and by the distribution of severity of food insecurity in the sample. The equal-weighted reliability statistic avoids the confounding effect of differences in the distribution of severity of food insecurity across countries. Mean Rasch reliability across countries (using the modified statistic) was .740, with a range from .676 to .847. Reliability was between .70 and .80 for 79 percent of countries. These levels of reliability for a scale comprising just eight items reflect reasonably good model fit, and result in measurement errors in national prevalence estimates that are small compared with sampling errors.

4. Making the Measure Comparable across Countries

Measures were made comparable across countries by first calculating a global standard scale based on item severity parameters of almost all countries; second, specifying thresholds for moderate and severe food insecurity on the global scale, and third, adjusting each country's scale to the global standard based on item severity parameters in order to transform the global thresholds to that country's scale. These processes are based on the assumption that, with identifiable exceptions, the objective conditions elicited by responses to FIES questions are the same in all countries and languages. However, due to differences in linguistic nuances and differences in the ways in which food insecurity is experienced and managed in different cultures, one or more items in a country's data collection may, in fact, not represent the same objective conditions as in the majority of countries. Therefore, in both the calculation of the global standard scale and the adjustment of each country's scale to the global standard, an item or items in a country's scale were considered unique to that country if they differed in severity from the global standard in excess of a specified tolerance (about 0.5 logits). Unique items were retained in the country's scale, but were not considered equivalent to the corresponding items in other countries' scales or in the global standard. Making these adjustments is complicated by the fact that Rasch-model based scales are undefined with respect to location (i.e., they have no natural zero point) and that dispersion of items is affected by the extent of memory and reporting error, which can differ from country to country.

The global standard scale was calculated based on item severity parameters of almost all countries. A few countries were omitted due to small estimation sample sizes or poorly fitting data. An iterative process of estimating the global standard item parameters allowed up to 3 items to be considered unique in each country. In the first iteration, each country's scale was divided by the standard deviation of its item parameters to equate the dispersion of items in all countries; no items were considered to be unique, and the global standard parameter for each item was calculated as the median parameter for that item across countries. In the second iteration, each country's scale was adjusted by a linear mapping to equate the mean and standard deviation of all items to those of the global standard. Items were specified as unique if their differential item function (DIF) from the global standard exceeded a set tolerance. In each succeeding iteration, each country's scale was adjusted by a linear mapping to equate the mean and standard deviation of the non-unique items to the mean and standard deviation of the corresponding items in the global standard. Unique items were also omitted from the calculation of the median for that parameter—the global standard parameter to be used in the following iteration. Iteration continued until the DIF of all non-unique items in all countries was within the set tolerance. Countries in which more than three items were unique were removed from the set of countries used to calculate the global standard.

The threshold for moderate food insecurity—the level of severity beyond which respondents would be classified as having moderate or severe food insecurity—was specified at the severity of the item ATELESS on the global standard scale. The threshold for severe food insecurity was specified at the severity of the item WHOLEDAY on the global standard scale (see Table 1 for the full wording of these questions). These specifications give substantive meaning to the main statistics reported for each country. Individuals with "moderate or severe food insecurity" have typically eaten less than they thought they should at times during the year because they lacked sufficient resources for food, and most will have experienced more severe conditions. Those with severe food insecurity have typically gone whole days without eating at times due to lack of resources.

For country scales with three or fewer unique items (93 percent of countries), the linear mapping required to adjust the country scale to the global standard was already specified in the final step of the development of the global standard. By reversing that mapping, the thresholds for moderate and severe food insecurity on the global standard scale were back-transformed to the metric of each country. Adjustment of the remaining countries to the global standard—those with DIF of one or more items still out of tolerance after specifying the most discrepant three items as unique—was conducted on a case by case basis, and greater uncertainty about prevalence rates is noted in published results. For four countries, no defensible adjustment was found, and raw-score parameters as well as thresholds were based on the global standard scale.

Finally, two prevalence rates were calculated for each country: the percentage of the population with moderate-or-severe food insecurity and the proportion with severe food insecurity. These calculations are based on the assumption that within each raw score, the true severity of food insecurity is normally distributed with mean at the value of the respondent parameter for that raw score and standard deviation equal to the estimated measurement standard error (uncertainty) of severity associated with that raw score. The proportion of the population beyond a threshold was then calculated by summing across raw scores the proportion of the sample with that raw score multiplied by the proportion of the normal distribution of severity for that raw score falling beyond the threshold. Survey sampling weights were used for these calculations so that the prevalence rates are representative of the national population.

Special considerations had to be given to respondents with raw scores 0 and 8. The conditional maximum likelihood methods used to estimate the measurement model do not provide statistically rigorous severity parameters or measurement errors for these extreme scores. Respondents with raw-score 0 were classified as highly food secure. Although no severity parameter or measurement error was specified for those respondents, their classification as highly food secure implicitly assumes that the proportion with severity beyond the moderate threshold is negligible. For respondents with raw-score 8, a severity parameter was calculated based on a pseudo raw score that ranged from 7.5 to 7.7, depending on the ratio of respondents with raw-score 8 to respondents with raw scores 1 to 7. Measurement error (uncertainty) for these respondents was calculated based on a pseudo raw score of 7.5. These ad-hoc specifications were developed and assessed by analysis of additional, more severe items available for a few countries in the 2014 data and a larger number of countries in the 2015 data. Any imprecision due to imperfect specification of the distribution of severity in raw-score 8 affects only a relatively few countries that have a substantial proportion of cases with raw score 8.

5. National Prevalence Rates

National prevalence rates of moderate-or-severe food insecurity assessed over a 12-month recall period ranged across countries, areas, and territories from 3 percent to 92 percent. [6]. The prevalence of severe food insecurity (characterized by individuals not eating for a whole day at times during the year) ranged from below 0.5 percent to 76 percent.

Regional prevalence rates for FAO statistical regions were calculated as population-weighted means across countries, areas, and territories for which data are available (more than 90 percent of population in most regions). Food insecurity was most prevalent in Sub-Saharan Africa and least prevalent in East Asia and Developed Regions (figure 1).

Preliminary external validation was provided by associations of national prevalence rates of food insecurity with expected correlates such as the World Bank Poverty Rate (Pearson $r = .76$), Human Development Index ($r = -.82$), FAO Prevalence of Undernourishment ($r = .76$), and the World Health Organization Under-5 Mortality Rate ($r = .83$). To date, few of these associations have been explored in detail, but the association with Under-5 Mortality Rate was found to remain substantial and statistically significant with controls for Poverty Rate and Prevalence of Undernutrition.

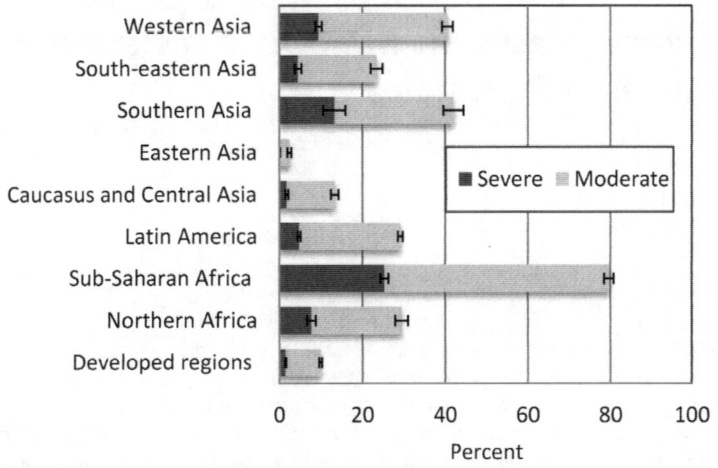

Figure 1. Prevalence rates of food insecurity by FAO Statistical Region, 2014

Acknowledgements
Financial support for the Voices of the Hungry project has been provided by the United Kingdom Department for International Development (DfID) and by the Kingdom of Belgium through the FAO Multipartner Programme Support Mechanism (FMM). The comments, opinions and judgments expressed in this paper are those of the authors and do not imply any official position by FAO, the FAO Statistics Division, or the funding partners.

References

[1] Declaration of the World Summit on Food Security, World Summit on Food Security, Rome, 16-18 November 2009

[2] Coleman-Jensen A, Rabbitt M, Gregory C and Singh A 2015 *Household Food Security in the United States in 2014* (Washington, DC: U.S. Department of Agriculture, Economic Research Service, Economic Research Report No. ERR-194)

[3] United Nations 2015 *Transforming Our World: the 2030 Agenda for Sustainable Development* (United Nations, Resolution adopted by the General Assembly on 25 September 2015)

[4] Gallup World Poll. http://www.gallup.com/services/170945/world-poll.aspx

[5] Fischer G H and Molenaar I W 1995 *Rasch Models: Foundations, Recent Developments, and Applications* (New York: Verlag Springer)

[6] United Nations Food and Agriculture Organization 2016 *Methods for Estimating Comparable Rates of Food Insecurity Experienced by Adults throughout the World.* (Rome: FAO)

Realizing a Rasch measurement through instructionally-sequenced domains of test items.

E Matthew Schulz[1]
561 Junipero Ave, Pacific Grove, CA, 93950, USA

E-mail: mattschulz54@gmail.com

Abstract. This paper presents results from a project in which instructionally-sequenced domains were defined for purposes of constructing measures that that conform to an ideal in Guttman scaling and Rasch measurement. A fundamental idea in these measurement systems is that every person higher on the measurement scale can do everything that lower-level persons can do, plus at least one more thing. This idea has had limited application in educational measurement due to the stochastic nature of item response data and the sheer number of items needed to obtain reliable measures. However, it has been shown by Schulz, Lee, and Mullen [1} that this ideal can be can be realized at a higher level of abstraction -- when items within a content strand are aggregated into a small number of domains that are ordered in instructional timing and difficulty. The present paper shows how this was done, and the results, in an achievement level setting project for the 2007 Grade 12 NAEP Economics Assessment.

1. Conceptual Background

Cliff remarked that a Guttman scale is one of the best examples of a good idea in all of psychometric measurement [1]. A Guttman scale typically consists of a relatively small number of ordered tasks or observational trials where each task represents a level of proficiency [2]. Persons in the population to which the scale applies can generally be expected to have mastery of tasks up to and including their assigned level, and non-mastery of tasks representing higher levels. The universal ordering of task difficulty for all persons allows one to predict a person's mastery of each of the levels defining the scale, solely from the person's assigned level. Moreover, from observing a person's performance on just one level, one can predict performance on any lower level (if the performance was successful) or on any higher level (if the performance was not successful).

Using an example of a four-item test, Andrich [3] showed that the Rasch model [4] is a probabilistic version of a Guttman scale. In the Rasch model, Guttman levels correspond to points on the latent proficiency scale, and there are as many levels as there are binary items or rating scale categories in the assessment (as in traditional Guttman scaling). The Rasch model provides a probability for success as a function of the difference between the measure of examinee ability and the measure of task, or level, difficulty. Mastery of a level can be viewed as a matter of degree, quantified by the probability of success, rather than as an all-or-none phenomenon. Importantly, however, Andrich showed that in order for items to be Guttman-scalable their item characteristic curves on the latent proficiency scale must not cross. That is, the ordering of levels must be the same at all levels of ability and for all probabilities of success.

[1] To whom any correspondence should be addressed.

Schulz, Kolen & Nicewander [5] noted two characteristic features of educational tests, besides the stochastic nature of test items pointed out by Andrich [3], that has made it difficult to put Guttman scaling ideals into practice in education. One is the sheer number of items on educational tests. Guttman scales typically consist of 4 to 7 levels, each represented by a single task that can be observed and scored with near-perfect reliability. The distance between Guttman levels on a Rasch scale would be very large and there would be little chance that their characteristic curves would overlap. Large numbers of items are required for educational testing because their reliability is so low. Items are so closely spaced on the measurement scale that their characteristic curves do tend to cross. The sheer number of crossings, however, discourages educators from attaching a great deal of meaning to this particular violation of a measurement ideal.

The other way that educational test items differ from Guttman tasks is in that they are universally regarded as exchangeable, random sampling units of larger domains. A general area of skill, such as mathematics, or even a specific skill such as working with fractions, may be represented by hundreds, if not thousands, of test items. Virtually any skill that is a target of assessment and general inference is assessable with multiple, exchangeable test items, at least in theory. The tasks or observations comprising a Guttman tasks are typically treated as essential components of the scale. It is not common practice to exchange one task for another or for there to be two separate versions, or forms, of a Guttman assessment, each consisting of a different set of tasks measuring the same thing.

Recognizing test items as exchangeable sampling units of a broader domain of skill, Schulz, et al., [5] argued that item parameters such as the difficulty and discrimination parameters in item-response theory (IRT) models should be treated as random variables underlying similar parameters of domains. In subsequent studies, Schulz, et al., [6, 7, 8, 9] developed and applied a technique for defining a relatively small number of instructionally-relevant, difficulty-ordered domains within the broader domains of educational achievement tests including the National Assessments of Educational Progress (NAEP) in Grades 8 and 12 mathematics. Expected percent correct scores on the domains supported the notion that the domains had the same order of difficulty for all persons and that higher-level persons could do what lower-level persons could do, plus at least one more thing were realized at the level of percent correct scores on domains. Item parameters were considered only to the extent that they were expected to be normally distributed around the domain parameters with which they were associated and outlier status in this regard could cause items to lose their association with a domain.

In this paper, the domain definition process developed by Schulz et al., [7] is illustrated in a subject area, economics, that is not generally regarded as containing naturally-ordered progressions of skill. In mathematics, skills such as addition, subtraction, multiplication, and division seem to have a natural ordered in difficulty. In a subject such as economics, the prospect of defining difficulty-ordered domains based on instructional sequence is less clear. The domain development work in economics conducted by ACT for the NAEP economics standard setting project [10] has not been reported in detail previously except in technical reports and presentations delivered by the contractor to its technical advisory committee and to the committee on standards, design and methodology of the National Assessment Governing Board (NAGB).

2. Application

The achievement level setting project for the 2007 NAEP in grade 12 economics included a process of defining difficulty-ordered domains within the grade 12 economics assessment. The domains were intended to support the same Guttman-scale relationships as domains previously developed using items in the Grade 8 and Grade 12 NAEP mathematics assessments [6,7,8,9]. A key step in the domain-development process used in those studies was is to quantify the instructional timing of the test items. In a mathematics assessment covering primary grades, or even high school mathematics, instructional timing can be associated with grade levels or courses in a standard high school sequence. In economics, the rating scale shown in Figure 1 was used. Five curriculum and content experts used the rating scale to rate the NAEP Grade 12 economics items. An average rating was computed for each item.

Journal of Physics: Conference Series **772** (2016) 012061 doi:10.1088/1742-6596/772/1/012061

The content experts then used the average ratings of items to organize the items into a series of "teacher" domains that they expected to be coherent, instructionally useful, and ordered ininstructional timing. Item-response-theory (IRT) estimates of item difficulty were available, but were not used at this stage of domain definition. The researchers leading the project felt it was felt important to define domains primarily on the basis of instructional timing and content and to consider less substantive characteristics of the items, such as difficulty only later.

Item Rating Scale	
Rating	Meaning
5	The knowledge, skills, and abilities (KSAs) required to get full credit on this item are usually mastered after the vast majority of other KSAs in an Economics curriculum have been mastered, the goal of which is mastery of all the benchmarks in the NAEP framework.
4	Mastery of the KSAs required to get full credit on this item typically follow mastery of the majority of other KSAs in Economics.
3	The KSAs required to get full credit on this item are mastered about midway through mastery of all the KSAs in Economics.
2	The KSAs required to get full credit on this item typically require mastery of some earlier KSAs in Economics.
1	Most of the KSAs needed to get full credit on this item are mastered early in a learning sequence in Economics.
DOES NOT APPLY	The KSAs required by this item are mastered in an Economics instruction in no particular sequence in relation to other KSAs. They may occur early in some Economics curricula and late in others.

Figure 1. Rating scale for quantifying the instructional timing of the knowledge, skill, and/or ability (KSAs) needed to correctly answer NAEP grade 12 economics assessment items.

Domains were expected to exhibit a positive, but not necessarily perfect, correlation between instructional timing and difficulty. Item difficulty was expected to be less variable within than across domains, but substantial overlap in item difficulty across domains was also expected. Figure 2 shows the dispersion of IRT difficulty parameters within and across domains that were ultimately defined within one of the three main strands of the NAEP Grade 12 economics assessment, National Economy. The other two strands are Market Economy and International Economy. The domains are ordered from left to right by the average instructional timing of their items. The correlation between mean instructional timing and mean item difficulty among the domains met expectations. The variability of item difficulty within and across domains also met expectations.

Figure 3 shows percent correct curves of the ten teacher domains defined within the National Economy strand. It can be concluded from this plot that the number of domains is too large to exhibit the desired Guttman and Rasch patterns of widely-spaced, non-crossing percent correct curves. The emphasis at this stage, however, was on defining domains that were meaningful and useful to teachers (teacher domains) as evidence by the ability of teachers to classify the items reliably into the domains using only a brief narrative description and approximately three exemplar items for each domain. The study did in fact find high agreement among teachers independently classifying items into the domains. To some extent, the degree of crossing and variability in the slopes of the percent correct curves may be due to the small number of items and differences in item types comprising the domains at this stage. A teacher domain might consist of as few as three items due to the limited total number of items available and the breadth of the assessment.

The four circles superimposed on the plot in Figure 3 shows how the teacher domains were combined to form a smaller set of domains for supplying relatively stable, Guttman-style descriptions of growth and differences in achievement. Teacher domains were combined through primarily through consideration of similarities among the domains in instructional timing and content.

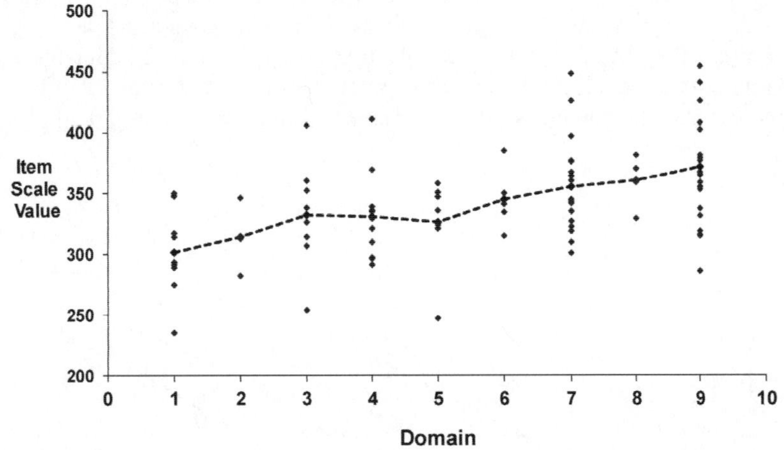

Figure 2. Item scale values within teacher domains defined for the National Economy strand of the 2007 Grade 12 NAEP Economics Assessment.

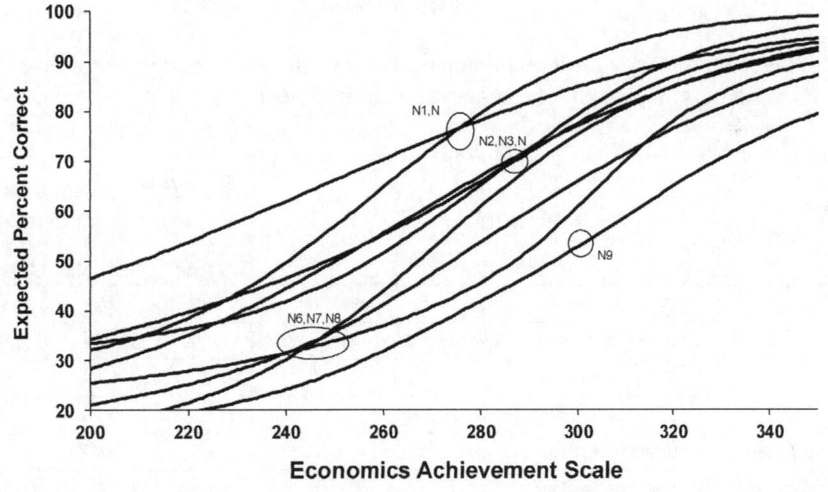

Figure 3. Percent correct on teacher domains within the National Economy substrand of the 2007 NAEP Grade 12 Economics assessment as a function of achievement.

Figure 4 shows that the resulting 'reporting' domains were relatively widely-spaced on the achievement scale and that their percent correct curves do not cross. Importantly, the domains will be mastered in the same order not matter what percent correct criterion for mastery is used. This pattern was felt to be stable enough for reporting and descriptive purposes. The expected stability of the mastery pattern is due to the spacing between the curves and to the larger (relative to the teacher domain) number of items comprising each domain.

Finally, Figure 5 shows how the teacher domains were used in an achievement level setting pilot study. Percent correct scores conditional on the lower boundaries of the cut scores being considered

Journal of Physics: Conference Series **772** (2016) 012061 doi:10.1088/1742-6596/772/1/012061

for the achievement levels helped the panellists understand what students at the cut scores could and could not do and what students at a higher achievement level could do that students at a lower achievement level could not do. The brief titles representing the teacher domains in Figure 5 may convey to the reader a sense of growing complexity and skill with as students move up the scale. In addition to these titles, the panellists read brief narrative descriptions of the domains and studied the content of items within the domains. Process evaluations during the course and at the conclusion of the workshop showed that the panellists felt the domain-information was very helpful in the standard setting process.

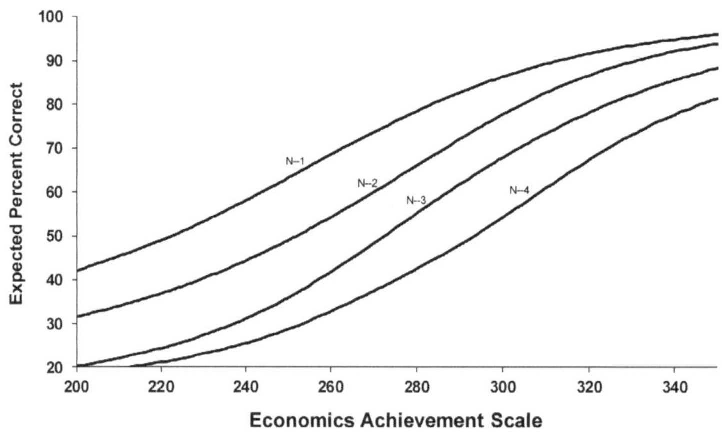

Figure 4. Percent correct on reporting domains in the National Economy substrand of the 2007 NAEP Grade 12 Economics assessment as a function of achievement.

Content Area	Domain	Expected Percent Correct at Lower Borderline of...		
		Basic	Proficient	Advanced
National	N1. Money, Loans, and Interest Rates	56%	74%	91%
	N2. Spending, Income, and Related National Measures	41%	71%	97%
	N3. Resource Allocation	42%	59%	89%
	N4. Economics Growth and Productivity	38%	60%	87%
	N5. Government Programs and Taxes	38%	57%	92%
	N6. Real Interest Rates	24%	51%	87%
	N7. Inflation and Unemployment	27%	46%	79%
	N8. Money Supply	29%	40%	82%
	N9. Fiscal and Monetary Policy	21%	36%	68%

Figure 5. Domain percent correct scores at lower boundaries of achievement levels in a pilot study for 2007 NAEP Economics achievement-level setting.

3. Educational Significance

The results of this, and previous work by Schulz, et al. adds something to the currently hot topic of learning progressions, where growth in achievement is seen as a sequential mastery of skills. Mastery in educational achievement testing is typically based on an arbitrary criterion percentage correct score, sometimes by policy and sometimes established through standard setting workshops. In Figure 5, the percent correct scores highlighted in yellow are above the 67% criterion for mastery used in the NAEP grade 12 economics standard setting workshop. If the order in which skills are mastered (and by implication the order in which skills should be taught) is to be independent of an arbitrarily-set mastery criterion, percent correct curves representing performance on the skills in the sequence must not cross. If descriptions of growth in student achievement can become independent of arbitrary mastery criteria and at the same time capture the simplicity and inferential power of Guttman scales, teachers, parents and policy makers will place more confidence in the results and implications of educational achievement tests for curriculum and instruction.

4. References

[1] Cliff, N. (1983). Evaluating Guttman Scales: Some old and new thoughts. In H. Wainer & S. Messick (Eds.), *Principles of modern psychological measurement: A festschrift for Frederick M. Lord.* Hillside, NJ.

[2] Guttman, L. (1950). The basis for scalogram analysis. In S. A. Stouffer, L. Guttman, E. A. Suchman, P. A. Lazarsfeld, S. A. Star, & J. A. Clausen (Eds.), *Measurement and Prediction* (pp 60-90). Princeton: Princeton University Press.

[3] Andrich, D. (1985). An elaboration of Guttman scaling with Rasch models for measurement. In N. Tuma (Ed.), Sociological Methodology (pp. 33-80). Jossey-Bass.

[4] Rasch, G. (1980). Probabilistic models for some intelligence and attainment tests. Danish Institute for Educational Research. Chicago: The University of Chicago Press. (Original work published 1960).

[5] Schulz, E. M., Kolen, M., & Nicewander, W. A. (1999). A rationale for defining achievement levels using IRT-estimated domain scores. *Applied Psychological Measurement, 23* (4), 347-362.

[6] ACT, Inc. (September, 2003). *Defining Multiple Content Domains for Describing Achievement Levels on the NAEP Grade 8 Mathematics Assessment. A preliminary report to the National Center for Education Statistics.* Iowa City, iA: Author.

[7] Schulz, E. M., Lee, W., & Mullen, K. (2005). A domain-level approach to describing growth in achievement. *Journal of Educational Measurement. 42*, 1-26.

[8] ACT, Inc. (April, 2005). Developing achievement levels on the 2005 national assessment of educational progress in grade twelve mathematics: Domain Development Report. Iowa City, IA: Author.

[9] ACT, Inc. (April, 2005*). Developing achievement levels on the 2005 national assessment of educational progress in grade twelve mathematics: Process Report.* Iowa City, IA: Author.

[10] ACT, Inc. (April, 2007). *Developing achievement levels on the 2007 national assessment of educational progress in grade twelve economics: Special Studies report.* Iowa City, IA: Author.

An alternative to Rasch analysis using triadic comparisons and multi-dimensional scaling

C Bradley[1], R W Massof[2]
[1]Research Associate, Wilmer Eye Institute, Johns Hopkins University, Baltimore, MD, USA
[2]Professor of Ophthalmology, Johns Hopkins University, Baltimore, MD, USA

E-mail: cbradley05@gmail.com

Abstract. Rasch analysis is a principled approach for estimating the magnitude of some shared property of a set of items when a group of people assign ordinal ratings to them. In the general case, Rasch analysis not only estimates person and item measures on the same invariant scale, but also estimates the average thresholds used by the population to define rating categories. However, Rasch analysis fails when there is insufficient variance in the observed responses because it assumes a probabilistic relationship between person measures, item measures and the rating assigned by a person to an item. When only a single person is rating all items, there may be cases where the person assigns the same rating to many items no matter how many times he rates them. We introduce an alternative to Rasch analysis for precisely these situations. Our approach leverages multi-dimensional scaling (MDS) and requires only rank orderings of items and rank orderings of pairs of distances between items to work. Simulations show one variant of this approach - triadic comparisons with non-metric MDS - provides highly accurate estimates of item measures in realistic situations.

1. Introduction

Rasch analysis is often used to estimate the magnitude of some common, underlying property shared by a set of items[1,2,3]. The inputs to Rasch analysis are ordinal ratings assigned to the items by a group of people. The outputs are estimated measures of both the items and the people. These estimated item and person measures are placed on the same invariant scale, allowing differences between item and person measures at different points on the axis to be directly comparable because they share the same units. A common application is to take ordinal ratings assigned to items on a questionnaire and estimate the relative difficulties of the items (e.g. if the items are tasks) and the relative abilities of the people that rated the items.

In order for Rasch analysis to provide good estimates of person and item measures, it is necessary for there to be sufficient variation in the observed responses. The reason for this is because Rasch analysis assumes that there is always a probability associated with any person assigning a given rating to any item, even if the true person and item measures are known. This variability in responses is quantified by probability distributions centered on both the person measures and item measures, as well as on any thresholds the person uses to define the boundaries of the rating categories. These assumptions are realistic because neither a person's ability, nor his perception of item difficulty, nor the precise thresholds used to categorize items remain constant from trial to trial. These probability distributions - also interpretable as error distributions around some mean value - allow Rasch analysis to map observed frequencies in assigned ratings to distances between item measures on an axis.

For most applications the required variance for Rasch analysis to be applicable is supplied by the differing responses people have to the same item. That is, the variance comes from a population. We run into trouble if the goal is to estimate item measures based on responses from just a single person. When only one person is providing the responses, it is possible the person will assign the same rating to an item no matter how many times he rates that item. For example, if there are only two possible rating categories 0 and 1 and the person is asked to rate items as either "difficult" or "easy", we might obtain the same answer each time for items such as "create a website" or "swim 1 km". Even when there are multiple rating categories we may run into this lack of sufficient variance problem (e.g. rating a movie as "great", "fairly good", "not that good", or "terrible").

2. Multi-dimensional scaling

The purpose of this paper is to outline an alternative to Rasch analysis in cases where there is insufficient variance in the observed responses, specifically in the case where only a single person is rating all the items and for whatever reason the person's responses show little to no variance over time. Our approach leverages the well-known mathematical technique of multidimensional scaling (MDS)[4,5,6,7]. MDS estimates coordinates of points in a low dimensional space (in our case a 1D space) from a dissimilarity matrix whose entries represent some dissimilarity measure between pairs of points (items). Depending on the types of responses we choose to obtain from the person in question, we will either use classical MDS or non-metric MDS. Classical MDS treats the entries in the dissimilarity matrix as actual distances and attempts to minimize a loss function of those distances. Non-metric MDS assumes much less about the nature of the input data, essentially interpreting the entries in the dissimilarity matrix as rank orderings of the distances between pairs of items. Which method we use will depend on the nature of the input data, which in turn depends on how difficult it may be for a person with limited cognitive abilities to provide internally consistent responses. Importantly, MDS is in general not a substitute for Rasch analysis because it assumes all inputs are in the same units whereas Rasch analysis allows each person to assign ratings in his own units.

Suppose our goal is to estimate the relative utilities of different events or outcomes for a single person. Since we are interested in estimating utilities for just one person, information about how a population assigns utilities to the items is in principle irrelevant. Thus, our only source of relevant information is the person himself. Our task is certainly simplified if the person solves the problem for us. That is, if a person can estimate his subjective utilities for different items in an internally consistent way, we are done. We just need a method for testing whether his responses are internally consistent. One way to do this is to ask the person to not only directly estimate item utilities, but to also estimate the distances in utility units between pairs of items. Since these distances are measures of dissimilarities between item pairs, they serve as inputs to classical MDS, which can be used to estimate item measures on a 1D axis. If the estimated item measures are highly similar to the person-estimated item utilities, we are done. The problem is that these two sets of estimated item measures may very well differ from each other, necessitating a different approach.

3. Paired and triadic comparisons

If person-estimated item measures and classical MDS estimated item measures differ too much from each other, it is almost certainly because the task of providing internally consistent estimates of item measures or distances between them is too challenging from a cognitive point of view. Humans are limited in their cognitive abilities and it is important that we require the person to perform a task that is easy enough so that his responses have good enough internal consistency. In general, it is easier for humans to simply rank order a set of items than to assign real numbers to them that accurately represent their relative magnitudes. This is true even if we are talking about rank-ordering distances between items. If $d(A, B)$ and $d(C, D)$ represent the distances between items A and B, and items C and D, respectively, then it is much easier for a person to say whether $d(A, B) > d(C, D)$ or $d(A, B) < d(C, D)$ compared to numerically estimating $d(A, B)$ and $d(C, D)$. Thus, our goal should be to estimate item measures using only rank orderings, either of items or of distances between items.

In the simplest case, we could ask the person to rank order pairs of distances between items. On each trial, the larger distance could be assigned a rank of 2 while the smaller distance could be assigned a 1. Such a "paired comparison" approach, if done on sufficiently many pairs, will over time provide us with accurate estimates of the average rank value for the distance between any two item measures. The average ranks for all item pairs is precisely the type of input non-metric MDS can be applied to, giving us a way of estimating item measures using only rank information.

If it turns out that rank ordering $d(A, B)$ and $d(B, C)$, where there is one common item between the pairs, is easier than rank ordering $d(A, B)$ and $d(C, D)$, a third approach can be taken. First, we ask the person to rank order a triplet of items, say items A, B, and C. If the person responds with $B > A > C$, then we automatically know that $d(B, C)$ is the largest distance between all possible item pairs. The second question is which of $d(B, A)$ and $d(A, C)$ is larger. Assign rank values of 3, 2 and 1 for the largest to smallest distances in this "triadic comparisons" approach and we will over many trials obtain estimates of the average rank value for the distance between any two item measures. Non-metric MDS can once again be applied to these average ranks to estimate item measures.

One key difference between the paired and triadic comparisons approaches is that paired comparisons can in principle compare every item pair to every other, ensuring that accurate estimates of average rank values are obtained, while triadic comparisons cannot because every comparison involves a common item. Thus, with triadic comparisons, we need evidence through simulations that estimated item measures are linearly related to true item measures. Because the number of unique triplets for N items is $N(N-1)(N-2)/6$, it makes little sense to apply this approach if N is large. One potential real world application where N is small involves estimating utilities of different possible treatment outcomes for individual patients. For example, 3200 patients were administered the Activity Inventory (AI), a questionnaire that asks patients to rate the importance and difficulties of 50 goals in daily life, from driving to recognizing people's faces to preparing meals without assistance[8]. They rated the importance of each goal on an integer scale from 0 to 3 and the difficulty of each goal from 0 to 4. While it is not clear precisely how importance and difficulty ratings map to utilities, one hypothesis is that utility is the product of the two. Because there is no need for an intervention if either the importance or difficulty of a goal or task is zero, we can try to estimate the utilites of $N = 12$ items, the possible non-zero products of importance and difficulty ratings. With 12 items, there are 220 unique item triplets, leading to 220 triadic comparisons, which is within manageable size for a patient questionnaire.

We simulated the triadic comparisons approach by first bootstrapping utility values from the distribution of importance times difficulty ratings obtained from the 3200 patients. Noise was added on each of 1000 simulations. For each simulation, all 220 possible triadic comparisons were performed, with random noise added to item distances to prevent rank orderings from being too deterministic. The average ranks for each item pair were computed and non-metric MDS applied to the resulting dissimilarity matrix to estimate item measures. As long as the standard deviation of the noise added to the distances was less than 1 (utility values ranged from 1 to 12 before noise was added), the estimated item measures had a highly linear relationship with the true item measures, with r^2 averaging around 0.97 over 1000 simulations. Separate simulations with 12 randomly chosen points from a uniform or normal distribution also yielded r^2 values averaging around 0.97 or higher. Skewing a uniform distribution by squaring the values led to an r^2 of around 0.95, while greater amounts of skewing or larger amounts of added noise progressively reduced r^2 (e.g. increasing the standard deviation on the distances to half the range at 6 or skewing the values by a power of 4 both lowered r^2 to the 0.85 range). These simulations show that the method of triadic comparisons with non-metric MDS provides good estimates of item measures for a single person when insufficient variance exists for Rasch analysis to be applicable.

References

[1] Rasch G 1960 *Probabilistic models for some intelligence and attainment tests* (Copenhagen: The Danish Institute of Educational Research)
[2] Andrich D 1978 A rating formulation for ordered response categories *Psychometrika* **43** 561-573
[3] Masters G N 1982 A Rasch model for partial credit scoring *Psychometrika* **47** 149-174
[4] Torgerson W S 1952 Multidimensional scaling, 1: theory and method *Psychometrics* **17** 401-419
[5] Kruskal J B 1964 Multidimensional scaling by optimizing goodness of fit to a nonmetric hypothesis *Psychometrics* **29** 1-27
[6] Kruskal J B 1964 Nonmetric multidimensional scaling: a numerical method *Psychometrics* **29** 115-129
[7] Borg I and Groenen P Kruskal J B 2005 *Modern multidimensional scaling: Theory and applications* 2nd ed (New York: Springer Verlag)
[8] Massof R W, Ahmadian L, Grover L L, Deremik J T, Goldstein J E, Rainey C, Epstein C, Barnett G D 2007 The Activity Inventory (AI): An adaptive visual function questionnaire *Optom Vision Sci* **84** 763-774

Journal of Physics: Conference Series **772** (2016) 012063 doi:10.1088/1742-6596/772/1/012063

Emotion models for textual emotion classification

O. Bruna[1], **H. Avetisyan, J. Holub**
Department of Measurement, Faculty of Electrical Engineering, Czech Technical University in Prague, Technicka 2, Prague 6, 166 27, Czech Republic

E-mail: brunaond@fel.cvut.cz

Abstract. This paper deals with textual emotion classification which gained attention in recent years. Emotion classification is used in user experience, product evaluation, national security, and tutoring applications. It attempts to detect the emotional content in the input text and based on different approaches establish what kind of emotional content is present, if any. Textual emotion classification is the most difficult to handle, since it relies mainly on linguistic resources and it introduces many challenges to assignment of text to emotion represented by a proper model. A crucial part of each emotion detector is emotion model. Focus of this paper is to introduce emotion models used for classification. Categorical and dimensional models of emotion are explained and some more advanced approaches are mentioned.

1. Introduction

Emotion detector is software used for detection and recognition of emotion from the input. Its goal is to help with improving customer experience, removing inappropriate posts from social networks, discover opinions about products and many other applications. One of the most recent applications being human–agent interaction [1].

Textual emotion recognition is the youngest branch of affective computing which should be able to recognize emotions mainly from text. The present textual emotion detectors developed from facial [2]–[4] and voice analysis of emotions as a branch of natural language processing.

This tool has grown in importance in the last years, where computer mediated communication plays greater role than ever before. Textual recognition can be applied to statuses, posts from social networks, microblogs, blogs, online newspapers, forums, reviews, and many others. It can help with removal of inappropriate or abusive content of social network, recognition of blog post author stands in certain matter, gathering negative opinions of users about specific product. Information gained from these sources can serve as a starting point to improving services, providing feedback, fixing errors and avoiding mistakes.

Emotion recognition is very challenging since the information in text can be very limited, extremely ambiguous, and largely depends on context and common sense. Therefore there are resources with information about the word's purpose and meaning in different contexts. Along them there are many different emotion models. There does not seem to be a consensus on how to distinguish these models and some ambiguity in terms of labeling emotions. This papers aims to provide an overview of existing approaches. Emotion models are divided into three categories and explained one by one with

[1] To whom any correspondence should be addressed.

Journal of Physics: Conference Series **772** (2016) 012063 doi:10.1088/1742-6596/772/1/012063

examples of use. This paper distinguished three major areas of emotion models – categorical, dimensional and extended, which are described in the chapter two.

2. Classifying emotions

Ekman found evidence for a group of six basic emotions [3], [5] and since then there were other researchers taking this approach and creating new emotion categories [6]. Nevertheless, Ekman's basic six emotions remain the most popular ones in terms of use in natural language processing [7], [8]. Based on this thought the emotions can be classified according to two main branches:

- Categorical
- Dimensional
- Extended

Categorical divides emotions into several groups and assigns the emotion into a proper discrete category. These methods are based on works of Ekman, who introduced this concept [3].
Dimensional methods see emotions as a place in two or even multi-dimensional space, where the emotion is expressed as a point in the plane or space defined by variables taken into account.
To determine the emotion from a textual input, it is crucial to have some framework to use. In order to assess one's emotion, lexicons are used where each of the records is marked with proper emotional content or a classifier is trained with a database of well annotated data stating the nature of emotions. What emotions are used, and how they are treated is not standardized. There are many different approaches to define and classify emotions. Next two sections will focus on describing available models for emotion classification used in related works.

2.1. Categorical Emotion Models

Categories which are used in the research of textual emotion recognition are based on the idea of discrete emotion theory. This theory argues that there is only a limited set of emotion which a person can experience.
The simplest categorical classification is to decide whether the emotion is positive or negative one. Some of the papers are working with similar idea, but use "happy" and "sad" state instead, as can be seen for example in the work of J.T. Hancock [9]. They have based their emotional recognition on leveraging the Social Information Processing model (SIP).
These models are also used in linguistic resources. To annotate the records with the emotional information, very simple categorical labels were used: positive, negative, objective for SentiWordNet. Labels positive, negative, ambiguous, and neutral were used for WordNet-Affect.

Table 1. Categories of emotions.

Author	**Count**	**Emotions**
Ekman	6	anger, disgust, fear, joy, sadness, surprise
Parrot	6	anger, fear, joy, love, sadness, surprise
Frijda	6	desire, happiness, interest, surprise, wonder, sorrow
Plutchik	8	acceptance, anger, anticipation, disgust, joy, fear, sadness, surprise
Tomkins	9	desire, happiness, interest, surprise, wonder, sorrow
Izard	12	interest, joy, surprise, sadness, anger, disgust, contempt, self-hostility, fear, shame, shyness, guilt
Extended Ekman	18	anger, disgust, fear, joy, sadness, surprise, amusement, contempt, contentment, embarrassment, excitement, guilt, pride in achievement, relief, satisfaction, sensory pleasure, and shame

Many works are building up on a research done by prof. Ekman. His work is based on recognizing emotions from facial expressions. Cross-cultural study was conducted concluding that in case of facial expressions there are six basic emotions expressed the same way regardless of the subject place of origin [3], [2]. Emotions recognized by Ekman are happiness, sadness, anger, fear, disgust, and surprise.

For the purposes of textual emotion recognition, there is also a neutral state considered as one of the categories to reflect neutral statements, such as newspaper headlines. Some of the models based on Ekman's basic six also distinguish between positive and negative surprise [10].

Similar approach is taken by prof. Izard from the University of Delaware in [11] where he considers 12 basic emotions: interest, joy, surprise, sadness, anger, disgust, contempt, self-hostility, fear, shame, shyness, and guilt.Summary of possible emotional labels, which were used in other works is in Table1.

Using two emotional categories is very often used to examine novel designs and test that the algorithm operates as expected. Decision on a specific emotion can be then perceived as a selection from the category of positive or negative emotions.

However, there are some drawbacks. Simple discrete categories are unable to reflect the valence or arousal of a particular emotion, unless somehow modified. The labels given to emotions may have a slightly different meaning when translated into other language, hence reducing its information value. Therefore dimensional emotion models are used. Table 2 shows how often were specific models used in studied literature. Table suggests that there is so far no unified way of assigning labels to emotions.

Table 2. Overview of used models

Paper	Data	Input granularity	Emotional model
[12]	SentiFul, WordNet-Affect	fine-grained	Anger, Disgust, Fear, Guilt, Interest, Joy, Sadness, Shame, and Surprise
[13]	OMCS, WordNet-Affect	N/A	Ekman
[14]	N/A	N/A	Ekman
[15]	Blog posts annotated with Ekman's model, sentences	Sentences	Ekman, 2 states
[16]	WordNet-based Lexicon, Lexicon of Emoticons, Abbreviations	Sentence-Level	Ekman
[17]	N/A	N/A	Ekman - only 5 emotions
[18]	Aman 2007, Twitter - hashtag annotated	N/A	Ekman
[19]	WordNet-Affect, Twitter annotated by hashtags	N/A	Ekman + neutral
[20]	WordNet-Affect	Word, Sentence	Ekman
[21]	BAWE Corpus	N/A	Anger, Fear, Joy, Sadness
[22]	Ren-CECps	Sentence, fine-grained	Positive emotions:expect, joy, love, surprise; Negative emotions: anxiety, sorrow, anger, hate. <expect, joy, love, surprise, anxiety, sorrow, anger, hate> range 0.0 to 1.0
[23]	N/A	Sentence-Level	Ekman + neutral extended of three intensity states.
[24]	Political articles	Sentence-Level, Document-Level	Binary: Palestinian, Israeli
[25]	AutoTutor Log Files	Sentence-Level, Chat exchange level	boredom, confusion, eureka, frustration
[26]	AutoTutor Log Files, Video annotated by trained judges, Experiment participants cross annotation	Sentecne-Level, Chat exchange level	boredom, flow/engagement, confusion, frustration
[27]	LiveJournal Blog posts, YouTube Videos (tagged like/dislike)	Blog Level, Video level	Hourglass model (10 emotions effectively)

2.2. Dimensional Emotion Models

In dimensional classification methods emotions are presented as a point or a region within a two-dimensional or multi-dimensional space. Emotions are therefore not a subject of assignment to a single category, but to many variables. Consider Ekman's basic six emotions. The categorical principle would assign to a particular word one single emotion label. Extending the categorical classification by adding a scale to each of the emotions and the result is a more flexible emotion vector. With emotion scales for each category, the final classification forms a vector, where each item corresponds to a certain emotion and each value corresponds to emotion intensity.

Another transformation of categorical to dimensional is that instead of using positive or negative emotion categories, a single emotion classification with the range from -1 to 1 is used. Less than zero refers to negative emotions and above zero refers to positive emotions. The value expresses the arousal.

Some special approaches were also developed to suit the needs of affective computing. The examples of such approaches are positive activation and negative activation model (PANA), Circumplex model (Valence, Arousal, Power), Plutchik emotion model, Hourglass model.

Each model has its own advantages and disadvantages in terms of use or implementation. PANA and Circumplex mode are very powerful and simple tools to evaluate emotional state. Emotion is modelled as a point in a plane where axis are assigned to a variable of interest (e.g. arousal, valence). Such approach is easy to navigate and understand. Plutchik and Hourglass model use similar method, with the difference being the number of variables. There are more axis along which the emotion is modelled. On top of it, authors explicitly assign specific labels to certain regions, providing some sort of compatibility with categorical models. These models are more advanced than categorical in terms of complexity and ability to express complex emotions. Despite their complexity, mentioned models do not take into account the context and personal typology of the reader or author. Therefore there is a third group, which is within this paper recognized as extended models.

2.3. Extended Models

On top of above mentioned models, there are additional models considered in the field of textual emotion recognition. Two examples are Ortony, Clore and Collins model and Five Factors Model. Both take a little bit different approach than previous ones. Firstly, these models are used in papers, where the personality, targets and desires of the presenting or receiving the text are taken into account. For example in the work of Kao et al. is used a FFM and in the paper from Li et al. is used OCC together with FFM [28], [29], [30]. OCC's main features are a decision tree to determine emotion, considering the event triggering the emotion, and evaluating the relation between the feeling and the situation. In work of Li et al. the FFM is used to establish a personal typology of the reader/writer of the text. This model is employed to improve the efficiency of the emotion recognition for a particular individual.

3. Conclusions

This paper presented the emotion models, which are used for emotion recognition. Emotion detectors will play a significant role in the future and therefore it is important to provide an overview of existing approaches and to understand their tradeoffs.

The models presented here were categorical, dimensional, and extended. Categorical models, summarizes in table 1, are extensively used for their simplicity and unique output. Dimensional models are used less (for comparison see table 2), but provide more flexibility in assigning emotions, where there are no labels. The process is harder to calculate. Extended models build on top of two previous categories and aim to provide a better assignment of emotions based on i.e. personal typology.

Especially the ability to use personal traits seem to be very important in upcoming years, where the human computer interaction may get extremely personal and the computer may need to extend its field of situation awareness to be emotion aware. Thanks to ability to recognize, model, and understand human emotions, the use, communication, and overall efficiency when dealing with electronic device might improve.

4. Acknowledgements
This work was supported by the Grant Agency of the Czech Technical University in Prague, grant No. SGS16/095/OHK3/1T/13

5. References

[1] C. Clavel and Z. Callejas, "Sentiment analysis: from opinion mining to human-agent interaction," *IEEE Trans. Affect. Comput.*, vol. 3045, no. c, pp. 1–1, 2015.
[2] P. Ekman, "Universals and cultural differences in facial expressions of emotion," *Nebraska symposium on motivation*. pp. 207–282, 1972.
[3] P. Ekman, W. V Friesen, M. O'Sullivan, A. Chan, I. Diacoyanni-Tarlatzis, K. Heider, R. Krause, W. A. LeCompte, T. Pitcairn, P. E. Ricci-Bitti, K. Scherer, and M. Tomita, "Universals and cultural differences in the judgments of facial expressions of emotion.," *Journal of personality and social psychology*, vol. 53, no. 4. pp. 712–7, 1987.
[4] R. W. Picard, "Affective Computing," *Pattern Recognit.*, vol. 20, no. 321, p. 304, 1995.
[5] P. Ekman, "An argument for basic emotions," *Cogn. Emot.*, vol. 6, no. 3, pp. 169–200, Jan. 1992.
[6] J. L. Tracy and D. Randles, "Four Models of Basic Emotions: A Review of Ekman and Cordaro, Izard, Levenson, and Panksepp and Watt," *Emot. Rev.*, vol. 3, no. 4, pp. 397–405, 2011.
[7] L. Teijeiro-Mosquera, J.-I. Biel, J. L. Alba-Castro, and D. Gatica-Perez, "What Your Face Vlogs About: Expressions of Emotion and Big-Five Traits Impressions in YouTube," *IEEE Trans. Affect. Comput.*, vol. 6, no. 2, pp. 193–205, Apr. 2015.
[8] R. Amini, C. Lisetti, and G. Ruiz, "HapFACS 3.0: FACS-Based Facial Expression Generator for 3D Speaking Virtual Characters," *IEEE Trans. Affect. Comput.*, vol. 6, no. 4, pp. 348–360, Oct. 2015.
[9] J. T. Hancock, C. Landrigan, and C. Silver, "Expressing emotion in text-based communication," in *Proceedings of the SIGCHI conference on Human factors in computing systems - CHI '07*, 2007, pp. 929–932.
[10] C. O. Alm, D. Roth, and R. Sproat, "Emotions from text: machine learning for text-based emotion prediction," *Proc. Hum. Lang. Technol. Conf. Conf. Empir. Methods Nat. Lang. Process.*, no. October, pp. 579–586, 2005.
[11] C. E. Izard, D. Z. Libero, P. Putnam, and O. M. Haynes, "Stability of emotion experiences and their relations to traits of personality.," *J. Pers. Soc. Psychol.*, vol. 64, no. 5, pp. 847–860, 1993.
[12] A. Neviarouskaya, H. Prendinger, and M. Ishizuka, "Affect Analysis Model: novel rule-based approach to affect sensing from text," *Nat. Lang. Eng.*, vol. 17, no. 01, pp. 95–135, Sep. 2011.
[13] C. Ma, H. Prendinger, and M. Ishizuka, "Emotion Estimation and Reasoning Based on Affective Textual Interaction," *Lect. Notes Comput. Sci. (including Subser. Lect. Notes Artif. Intell. Lect. Notes Bioinformatics)*, vol. 3784 LNCS, pp. 622–628, 2005.
[14] H. Binali, C. Wu, and V. Potdar, "Computational Approaches for Emotion Detection in Text," *4th IEEE Int. Conf. Digit. Ecosyst. Technol. - Conf. Proc. IEEE-DEST 2010, DEST 2010*, vol. 37, no. 5, pp. 498–527, 2010.
[15] S. Aman and S. Szpakowicz, "Identifying Expressions of Emotion in Text," in *Text, Speech and Dialogue*, vol. 4629, Berlin, Heidelberg: Springer Berlin Heidelberg, 2007, pp. 196–205.
[16] U. Krcadinac, P. Pasquier, J. Jovanovic, and V. Devedzic, "Synesketch: An Open Source Library for Sentence-Based Emotion Recognition," *IEEE Trans. Affect. Comput.*, vol. 4, no. 3, pp. 312–325, 2013.

[17] S. Gupta, A. Mehra, and Vinay, "Speech emotion recognition using SVM with thresholding fusion," in *2015 2nd International Conference on Signal Processing and Integrated Networks (SPIN)*, 2015, pp. 570–574.

[18] S. Shaheen, W. El-Hajj, H. Hajj, and S. Elbassuoni, "Emotion Recognition from Text Based on Automatically Generated Rules," in *2014 IEEE International Conference on Data Mining Workshop*, 2014, pp. 383–392.

[19] G. Valkanas and D. Gunopulos, "A UI Prototype for Emotion-Based Event Detection in the Live Web," *Human-Computer Interact. Knowl. Discov. Complex, Unstructured, Big Data*, vol. 7947 LNCS, pp. 89–100, 2013.

[20] C. Ma, H. Prendinger, A. Osherenko, M. Ishizuka, Chunling Ma, A. Osherenko, H. Prendinger, and M. Ishizuka, "A chat system based on emotion estimation from text and embodied conversational messengers (preliminary report)," in *Proceedings of the 2005 International Conference on Active Media Technology, 2005. (AMT 2005).*, 2005, vol. 2005, no. i, pp. 546–548.

[21] D. T. Ho and T. H. Cao, "A High-Order Hidden Markov Model for Emotion Detection from Textual Data," 2012, pp. 94–105.

[22] F. Ren and C. Quan, "Linguistic-based emotion analysis and recognition for measuring consumer satisfaction: an application of affective computing," *Inf. Technol. Manag.*, vol. 13, no. 4, pp. 321–332, 2012.

[23] Y. Baimbetov, K. Ismail, M. Steinbauer, and A.-K. Gabriele, "Using Big Data for Emotionally Intelligent Mobile Services Through Multi-modal Emotion Recognition," 2015, vol. 9102, pp. 127–138.

[24] W. H. Lin, T. Wilson, J. Wiebe, and a Hauptmann, "Which side are you on?: identifying perspectives at the document and sentence levels," *Proc. Tenth Conf. Comput. Nat. Lang. Learn.*, pp. 109–116, 2006.

[25] S. K. D'Mello, S. D. Craig, J. Sullins, and A. C. Graesser, "Predicting Affective States expressed through an Emote-Aloud Procedure from AutoTutor's Mixed-Initiative Dialogue," *International Journal of Artificial Intelligence in Education*, vol. 16, no. 1. pp. 3–28, 2006.

[26] S. D'Mello, N. Dowell, and A. Graesser, "Cohesion relationships in tutorial dialogue as predictors of affective states," *Front. Artif. Intell. Appl.*, vol. 200, no. 1, pp. 9–16, 2009.

[27] E. Cambria, M. Grassi, A. Hussain, and C. Havasi, "Sentic Computing for social media marketing," *Multimed. Tools Appl.*, vol. 59, no. 2, pp. 557–577, 2012.

[28] H. Li, N. Pang, S. Guo, and H. Wang, "Research on textual emotion recognition incorporating personality factor," in *Robotics and Biomimetics, 2007. ROBIO 2007. IEEE International Conference on*, 2007, pp. 2222–2227.

[29] E. C.-C. Kao, C.-C. Liu, T.-H. Yang, C.-T. Hsieh, and V.-W. Soo, "Towards Text-based Emotion Detection A Survey and Possible Improvements," in *2009 International Conference on Information Management and Engineering*, 2009, pp. 70–74.

[30] A. Ortony and T. J. Turner, "What's basic about basic emotions?," *Psychol. Rev.*, vol. 97, no. 3, pp. 315–331, 1990.

Representation of interval data by weak orders yields robustness of the data fusion outcomes

S V Muravyov, L I Khudonogova and I A Marinushkina
Department of Computer-aided Measurement Systems and Metrology
National Research Tomsk Polytechnic University
Tomsk, 634050, Russia

E-mail: muravyov@tpu.ru

Abstract. It is shown in the paper that an interval data fusion procedure can be carried out by means of representation of intervals on the real line by weak order relations (or rankings) over set of discrete values belonging to these intervals. It is possible to determine a consensus ranking for collection of discrete values rankings, corresponding to initial intervals. The highest ranked value, accepted as a result of the fusion, guarantees improved accuracy and robustness of the interval data fusion procedure outputs.

1. Introduction

Description of measurement results in form of *intervals* with bounds defined by experimentally obtained or given beforehand uncertainty values is rather common both in theory and practice of measurement. Moreover, interval data are typical kind of information in such areas as distributed computations, data bases, systems and networks of data acquisition, etc. [1,2].

Data fusion is a process of joint processing of data on some object obtained from multiple sources aiming to acquire fuller, more objective and accurate knowledge of a characteristic under investigation than knowledge derived from a single source. List of data fusion methods usually includes mathematical statistics and probability theory, fuzzy sets theory, possibility theory, Dempster-Shafer evidence theory [3], Bayesian approach, different artificial intelligence methods [4], methods of voting and preference aggregation [5].

A simple example of data fusion utility can be taken from a classical analysis of repeated observations that is considerably older than comparatively new concept of data fusion. Indeed, from mathematical statistics point of view, *single* measurement result x is characterised by standard deviation S, while estimation $\bar{x} = \sum_{i=1}^{n} x_i$ of expectation in a form of arithmetic mean of n normally distributed observations has the standard deviation $S(\bar{x}) = S/\sqrt{n}$. Therein lies the only useful effect of arranging repeated observations instead of a single one.

Procedure of interval data fusion consists in shaping an interval to be consistent with maximal number of given initial intervals (not necessary consistent among each other) and to be with maximal likelihood including a value x^* that can serve as representative of all the given intervals.

Interval data fusion procedures are applied in interlaboratory comparisons where main task is to establish a reference value that characterizes a largest subset of consistent (reliable) measurement data [6]. Other applications are prediction of fundamental constant values on the base of different measured values [7], conformity testing [8] and enhancement of multisensor's readings accuracy in sensor networks [9].

In this paper, it is proposed an interval data fusion procedure that represents intervals on the real line by means of weak order relations (or rankings) over set of belonging to these intervals discrete values. One may determine a consensus ranking for collection of discrete values rankings, corresponding to initial intervals. The highest ranked value, accepted as a result of the fusion, guarantees improved accuracy and robustness of the interval data fusion procedure outputs.

2. Terms and notations

2.1. Intervals

We will consider a collection of m closed intervals $\{I_k\}$, $k = 1, \ldots, m$, in the real line. Each interval is characterized by *lower bound* l_k, *upper bound* u_k and *middle point* x_k, so that $I_k = [l_k, u_k]$; $l_k < x_k$; $x_k = 0{,}5 \cdot (u_k + l_k)$; $l_k, u_k, x_k \in \mathbb{R}$ (see figure 1).

Frequently, the middle point x_k is a result of some measurement, *uncertainty interval* of which is $[x_k - 0{,}5 \cdot (u_k - l_k), x_k + 0{,}5 \cdot (u_k - l_k)]$ = $[x_k - \varepsilon_k, x_k + \varepsilon_k]$. In this case, k-th interval can be represented by pair $<x_k, \varepsilon_k>$.

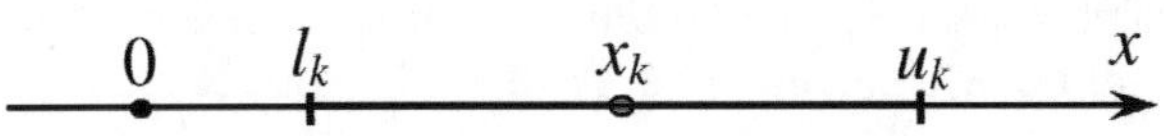

Figure 1. An interval on the real line.

Any two intervals do not necessary have non-empty intersection, i.e. there can be such two intervals I_j and I_k, $j \neq k$; $j, k = 1, \ldots, m$, that $I_j \cap I_k \neq \varnothing$. Such the intervals we will call *inconsistent*.

2.2. Range of actual values and its partition

Notice that aim of interval fusion is a selection of a point x^* on the real line that belongs to maximal number of intervals from $\{I_k\}$. We are going to choose the point from finite number of elements connected to the intervals, though, generally speaking, each of closed intervals in the real axis is infinite. For this purpose, let us introduce a *range of actual values* (RAV) $A = \{a_1, a_2, \ldots, a_n\}$, over that there exists (inherited from the real line) *binary relation of linear (full) order* $a_1 < a_2 < \ldots < a_n$, i.e. transitive, antisymmetric and linear binary relation.

The RAV could be shaped by means of union of the intervals $\{I_k\}$ and following partition of the union result into equal subintervals. However, existence of inconsistent intervals can produce a disjoint output of the union. In order to exclude possible breaks, in the capacity of the RAV's lower bound a_1 a least lower bound for all the intervals is chosen, i.e.

$$a_1 = \min\{l_k \mid k = 1, \ldots, m\}; \qquad (1)$$

and as the upper bound a_n a largest upper bound for these intervals is taken, i.e.

$$a_n = \max\{u_k \mid k = 1, \ldots, m\}. \qquad (2)$$

To generate elements $a_2, a_3, \ldots, a_{n-1}$ let us partition the obtained interval $[a_1, a_n]$ into $n - 1$ equal subintervals of length

$$c = (a_n - a_1)/(n-1). \qquad (3)$$

The length c will be called *norm* (or *mesh*) of the partition. Clear that, after partitioning, the norm is defined by the formula

$$c = |a_i - a_{i-1}|,\ i = 2, \ldots, n, \qquad (4)$$

and i-th RAV's element is

$$a_i = a_{i-1} + c,\ i = 2, \ldots, n. \qquad (5)$$

Notice that choice of an appropriate number $n-1$ of partition subintervals must guarantee necessary and sufficient accuracy of values a_i, $i = 1, \ldots, n$, as one of them will be taken as interval fusion result x^*.

2.3. Representation of intervals by rankings

Ranking of n elements of a set $A = \{a_1, a_2, \ldots, a_n\}$ in the form $\lambda_k = (a_1 \succ a_2 \ldots \sim a_s \sim a_t \succ \ldots \sim a_n)$ specifies a relation of weak order (preference relation) on the set A. The preference relation λ is a union of two following relations: strict preference relation ρ, i.e. $a_i \succ a_j$, and equivalence relation τ, i.e. $a_i \sim a_j$, that is $\lambda = \rho \cup \tau$. Set of m rankings $\Lambda = \{\lambda_1, \lambda_2, \ldots, \lambda_m\}$ of n elements is called *preference profile* for m and n given.

Some k-th interval of $\{I_k\}$, $k = 1, \ldots, m$, can be represented by ranking λ_k using the following rules at $i, j = 1, \ldots, n$:

$$a_i \succ a_j, \text{ if } a_i \in I_k \text{ and } a_j \notin I_k; \quad (6)$$

$$a_i \sim a_j, \text{ if } a_i, a_j \in I_k \text{ or } a_i, a_j \notin I_k; \quad (7)$$

$$a_i \prec a_j, \text{ if } a_i \notin I_k \text{ and } a_j \in I_k. \quad (8)$$

Notice that k-th ranking consists of two equivalence classes, one of which includes elements of A belonging to the interval I_k, and another includes all other elements of A. At that, the former class is strictly preferred over the latter one. Hence, each ranking contains a single symbol of strict order $\succ$ and $n-1$ symbols of equivalence $\sim$. Thus, the interval collection $\{I_k\}$, $k = 1, \ldots, m$, can be represented in accordance with equations (6), (7) and (8) by the preference profile $\Lambda = \{\lambda_1, \lambda_2, \ldots, \lambda_m\}$.

3. Fusion of intervals

By *interval data fusion* we will understand a procedure of determination of a point x^* position in the real line that belongs to maximal number of intervals from $\{I_k\}$, $k = 1, \ldots, m$, and that can be considered to represent all middle points x_k at minimal uncertainty $\pm\varepsilon_k$.

Notice that fusion methods based on classic mathematical statistics work on continuous intervals. The same is, for example, procedure A [6] applied to fuse measurement results uncertainty intervals provided by laboratories when processing comparison data. In this procedure, to estimate a reference value x^* one uses a weighted mean value y, and consistency of intervals is examined by criterion χ^2 calculated for y. Procedure A can be reasonably applied if the measurement results are characterized with normal probability distribution. In the cases where measurement results distribution differs from normal or unknown it is desirable to use robust methods for interval data fusion.

The property of robustness is inherent to the procedure proposed below that will be called *interval fusion with preference aggregation* (IF&PA). This procedure works on the set A of discrete values obtained in the way described in section 2.2 that allows to represent intervals by rankings.

Main stages of the proposed procedure IF&PA are as follows.

1. Generation of range of actual values $A = \{a_1, a_2, \ldots, a_n\}$ of initial intervals $\{I_k\}$, $k = 1, \ldots, m$.

2. Representation of the intervals by rankings and composing preference profile $\Lambda = \{\lambda_1, \lambda_2, \ldots, \lambda_m\}$.

3. Determination of value x^* as the best alternative in consensus ranking for the profile Λ. For this aim, any algorithm can be used that implements some of voting (preference aggregation) rules, for example, simple majority, Condorcet rule, Borda count, Kemeny rule, approval voting [10], etc. Authors prefer to use the Kemeny rule at this stage [11]. The rule produces multiple consensus rankings [12] which should be transformed into a single final consensus ranking. All the consensus rankings are strict preference relations and the final consensus ranking β is a weak order [13].

4. Deletion of inconsistent intervals in set $\{I_k\}$, i.e. those not including the value x^*. Then a cardinality of the output set becomes to be equal to m'.

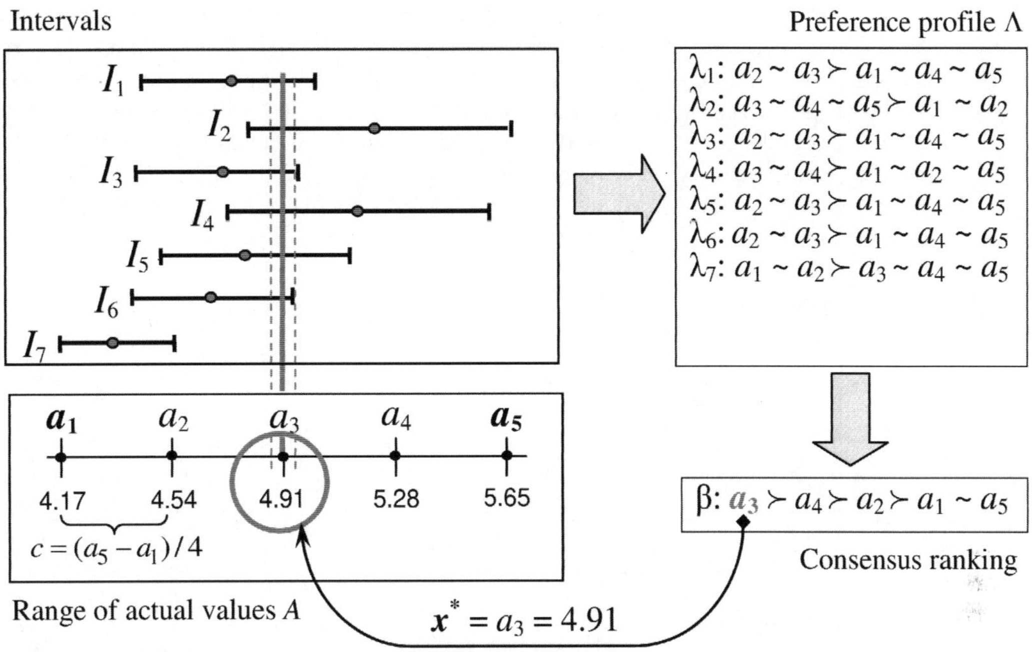

Figure 2. Example of initial collection of intervals $\{I_k\}$, $k = 1, \ldots, 7$; range of actual values $A = \{a_1, a_2, \ldots, a_5\}$; its partition into four meshes; corresponding preference profile $\Lambda = \{\lambda_1, \lambda_2, \ldots, \lambda_7\}$; and the interval fusion result $x^* = 4.91$.

5. Determination of uncertainty ε^* of the value x^* as by the formula

$$\varepsilon^* = \min(\max_{k=1,\ldots,m'} \{l_k \le x^*\}, \min_{k=1,\ldots,m'} \{u_k \ge x^*\}) . \tag{9}$$

The stages outcomes are illustrated in figure 2 for an example collection of seven initial intervals.

4. Robustness of fusion of intervals represented by rankings

Robustness, that is independence on particular law of interval data distribution, has been examined for the procedure IF&PA (with Kemeny rule at stage 3, see Section 3) by means of numerical experimentation. There were generated $N = 100$ uniformly and normally distributed interval collections (individual problems) in the vicinity of a fixed nominal value $x_{\mathrm{nom}} = 3$; number of intervals $m = 15$. An example of generated data for 15 intervals is shown in table 1.

Table 1. Generated data for individual problem #33.

k	1	2	3	4	5	6	7	8	9	10	11	12	13	14	15
x_k	2.918	3.085	3.225	2.491	2.989	3.222	2.738	3.069	2.559	3.100	2.918	2.894	2.626	2.899	3.077
ε_k	0.190	0.244	0.289	0.053	0.213	0.288	0.133	0.239	0.075	0.249	0.191	0.183	0.097	0.184	0.242

We will judge a procedure's performance by deviation

$$\xi = |x^* - x_{\mathrm{nom}}| \tag{10}$$

of value x^* from x_{nom}. The deviation ξ can serve as a measure of a data fusion process robustness [14] since if, for some individual problem, values ξ at normal and uniform data distributions are close enough to each other then the corresponding procedure should be deemed as a robust one. Some fragment of the obtained unsorted numerical experimental results is shown in table 2.

In figure 3 graphs of deviations ξ are presented that were obtained by means of procedure A and procedure IF&PA for normal and uniform distributions. Values ξ are shown for each of 100 individual problems and sorted in ascending order. One can see from Figure 3 that discrepancy by ξ between normal and uniform distributions (curves 3 and 4), in case of using IF&PA, does not mainly exceed 0.08. Thus, the procedure IF&PA is explicitly more robust than the procedure A that produced considerably larger discrepancy about 0.46 (curves 1 and 2). It follows from the presented in Figure 3 experimental data that values x^* obtained by the procedure IF&PA are essentially closer to x_{nom} than those obtained by the procedure A.

Table 2. A fragment of numerical experimental results (unsorted).

Problem number	Procedure IF&PA						Procedure A					
	Normal distribution			Uniform distribution			Normal distribution			Uniform distribution		
	x^*	ε^*	ξ	x^*	ε^*	ξ	x^*	ε^*	ξ	x^*	ε^*	ξ
1	2.68	0.07	0.32	2.97	0.04	0.03	2.77	0.13	0.23	2.96	0.05	0.04
2	2.99	0.02	0.01	2.95	0.03	0.05	2.97	0.04	0.03	2.32	0.01	0.68
3	2.96	0.01	0.04	3.01	0.07	0.01	2.99	0.11	0.01	2.88	0.01	0.12
4	2.93	0.29	0.07	2.85	0.01	0.15	2.99	0.12	0.01	2.46	0.01	0.54
5	3.03	0.03	0.03	2.98	0.20	0.02	2.90	0.05	0.11	2.84	0.02	0.16
6	2.99	0.05	0.01	3.07	0.01	0.07	2.97	0.07	0.03	2.89	0.08	0.11
7	2.97	0.03	0.03	2.95	0.20	0.05	2.80	0.11	0.20	2.83	0.03	0.17
8	2.94	0.01	0.06	3.11	0.05	0.11	2.97	0.02	0.03	3.07	0.04	0.07
9	2.98	0.02	0.02	3.23	0.10	0.23	2.94	0.07	0.06	2.88	0.11	0.12
						…						
96	2.96	0.02	0.04	2.98	0.20	0.02	2.92	0.09	0.08	2.30	0.05	0.70
97	2.94	0.07	0.06	2.97	0.01	0.03	2.93	0.08	0.07	2.72	0.05	0.28
98	2.92	0.24	0.08	2.88	0.03	0.12	2.89	0.14	0.11	3.10	0.20	0.10
99	2.99	0.01	0.01	2.94	0.09	0.06	2.99	0.04	0.01	2.73	0.08	0.27
100	2.97	0.03	0.03	2.96	0.20	0.04	2.91	0.09	0.09	2.89	0.12	0.11

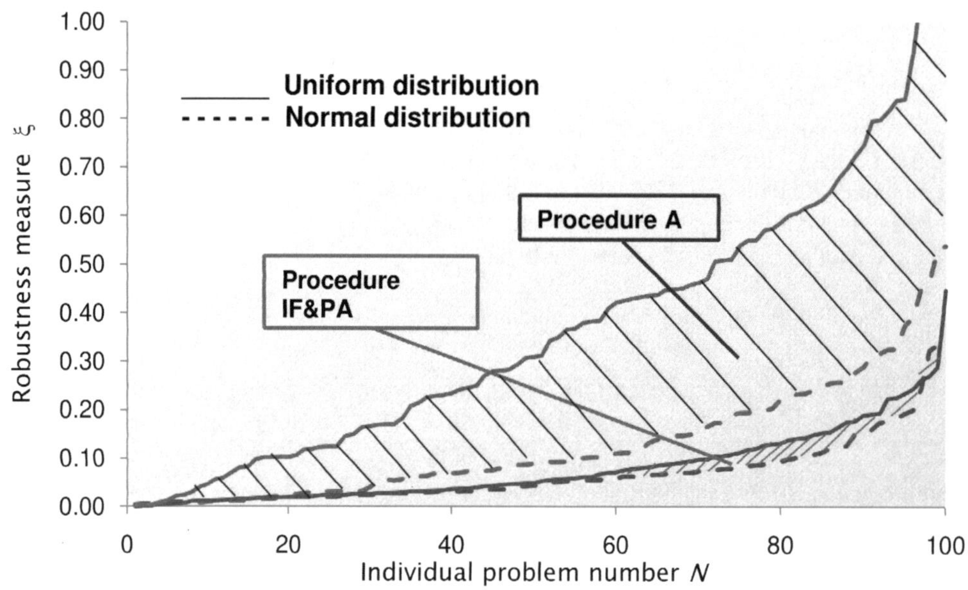

Figure 3. Deviations ξ obtained with procedure A and procedure IF&PA for uniform and normal distributions of interval data.

5. Conclusion

It was proposed an interval data fusion procedure IF&PA on the base of representation of intervals on the real line by weak order relations (or rankings) over set of belonging to these intervals discrete values. Numerical experimental investigations have shown that using as a result of the fusion the best value in a consensus ranking found for rankings collection, corresponding to initial intervals, guarantees improved accuracy and robustness of the interval data fusion procedure outputs.

The proposed procedure starts to work on interval data obtained in ratio scale, transforms the data into multiple ordinal scale entities, determines a best entity with aggregating multidimensional preferences, and the best entity is again a point in ratio scale. We should conclude that the transition from ratio to ordinal scale facilitates revealing essential information in input interval data what, in turn, yields property of resilience (robustness and accuracy) of the procedure IF&PA.

Acknowledgment

This work was supported by the Russian Science Foundation, Project # 14-19-00926. Part of this work was performed while the senior author was at Vignan University, India. Vignan University's, particularly Prof. B. Ramamoorthy's, support is gratefully acknowledged.

References

[1] Parhami B 2007 Distributed Interval Voting with Node Failures of Various Types *Proc 12th IEEE Workshop on Dependable Parallel, Distributed and Network-Centric Systems (DPDNS '07 Long Beach, California, USA)* pp 1-7

[2] Dubois D and Prade H 2001 Possibility Theory in Information Fusion *Data fusion and perception (Eds. G. Della Riccia, H.-J. Lenz, R. Kruse)* (Berlin: Springer) 53-76

[3] Hall D L and McMullen S A H 2004 *Mathematical Techniques in Multisensor Data Fusion* (Norwood: Artech House Publishers)

[4] Halpern J 2003 *Reasoning about Uncertainty* (Cambridge: MIT Press)

[5] Parhami B 1994 Voting algorithms *IEEE Trans. Reliab.* **43** 617–29

[6] Cox M G 2007 The evaluation of key comparison data: determining the largest consistent subset *Metrologia* **44** 187–200

[7] Mana G, Massa E and Predescu M 2012 Model selection in the average of inconsistent data: an analysis of the measured Planck-constant values *Metrologia* **49** 492–500

[8] Pendrill L R 2014 Using measurement uncertainty in decision-making and conformity assessment *Metrologia* **51** 206–18

[9] Muravyov S V and Khudonogova L I 2015 Multisensor accuracy enhancement on the base of interval voting in form of preference aggregation in WSN for ecological monitoring *Proc. 7th Int. Cong. on Ultra Modern Telecommunications and Control Systems (ICUMT 2015, Brno, Czech Republic, 6–8 October 2015)* pp 293-297

[10] Berg D E, Norine S, Su F E, Thomas R and Wollan P 2010 Voting in agreeable societies *Amer. Math. Monthly* **117** 27–39

[11] Muravyov S V 2013 Ordinal measurement, preference aggregation and interlaboratory comparisons *Measurement* **46** 2927–35

[12] Muravyov S V and Marinushkina I A 2013 Intransitivity in multiple solutions of Kemeny Ranking Problem *J. Phys.: Conf. Ser.* **459** 012006

[13] Muravyov S V 2014 Dealing with chaotic results of Kemeny ranking determination *Measurement* **51** 328–34.

[14] Muravyov S V and Marinushkina I A 2015 A software tool for simulation of interlaboratory comparison data evaluation method using preference aggregation *Proc. 21st IMEKO World Cong. (Prague, Czech Republic, 30 August–4 September 2015)*

Development of non-conventional instrument transformers (NCIT) using smart materials

Bojan Nikolić, Sanowar Khan and Nikita Gabdullin

City University London, Northampton Square, London EC1V 0HB
United Kingdom

E-mail: bnikolic@ieee.org

Abstract. In this paper is presented a novel approach for current measurement using smart materials, magnetic shape memory (MSM) alloys. Their shape change can be controlled by the application of magnetic field or mechanical stress. This gives the possibility to measure currents by correlating the magnetic field produced by the current, shape change in an MSM-based sensor and the voltage output of a Linear Variable Differential Transducer (LVDT) actuated by this shape change. In the first part of the paper is presented a review of existing current measurement sensors by comparing their properties and highlighting their advantages and disadvantages.

1. Introduction

Under the term conventional instrument transformers is usually considered measurement of current and voltage using transformers with an iron core. Their design has reached its optimum and has not changed significantly over the last few decades. With the development of electronics and application of electronic circuits in all spheres of today's life, one of their main disadvantages came to the fore - incompatibility with modern measurement equipment.

This is one of the main reasons why the so-called non-conventional instrument transformers (NCIT) have gained considerable attention in recent years. NCITs are usually transformers without ferromagnetic core or those, which, instead of the standard iron core use other materials that have a better response to rapidly changing signals. As a part of measurement systems, different electrical circuits are used with NCITs. Moreover, they are often integrated in the NCIT. This design flexibility provides full compatibility with modern digital measuring devices and opens up possibilities for many applications.

Today, there are many different devices and principles for current measurement. The existence of a significant number of papers dealing with this subject shows popularity and importance of this topic, but also shows that there still exist many unsolved problems and many challenges that need to be overcome. Some of the proposed measurement principles have already found industrial application, while others are still explored for such applications.

This paper presents a review of existing principles for sensors for current measurement and proposes a novel approach using magnetic shape memory (MSM) smart alloys, which change shape when subjected to magnetic fields.

2. Sensors for current measurement

2.1. Rogowski coils

One of the most popular sensors for current measurement is Rogowski coil. These coils do not have any ferromagnetic core, which allows measurement of current over a wide range. Using the same Rogowski coil, it is possible to measure currents larger than 1 MA and as small as few mA [1]. Furthermore, current measurement over this wide range is linear since there is no core that can get saturated.

Their measurement principle is based on “sensing” the magnetic field produced by current, meaning that they cannot be used for measurement of DC currents. The measurement frequency range is approximately 0.1 Hz-1 GHz [1], so they can be successfully used for the measurement of AC currents as well as transients.

Rogowski coils are small, compact, lightweight and their power consumption is very low, which makes current measurements in high-voltage circuits easy and safe owing to the absence of any electrical connections with the measured circuit. There are also the so-called flexible designs of Rogowski coils which allow measurements without interrupting the current circuit. Rogowski sensors cause no disturbances to the measurement system since they draw no power from the main circuit and the inductance added into the circuit due to the presence of the coil is only a few pH [1].

Moreover, these coils require almost no maintenance, except for periodical examination for physical damage if the coils are exposed to some harsh environment. They can be calibrated at any convenient current level and later used for any current level, including very large currents. They cause no damage by large overload.

Furthermore, considering that the same device can be used for current measurement, control and protection and that it has low price, it is more than clear why Rogowski coils found wide applications in industry. However, there are several drawbacks which points to other instruments for current measurement.

One of their main disadvantages is sensitivity to surrounding electromagnetic fields, meaning that some kind of shielding is needed. Their output is very small which necessitates proper grounding and shielding. The voltage on the coil’s output is in order of μV for primary currents bellow 100A. However, at same time, such low sensitivity can actually be their advantage in some applications. Measurement errors are typically in the range 5-20% of rated current [2], but there are some solutions which increased their accuracy [3] where the error is in the range of 0.5-5%. This problem is partly solved by the design of printed circuit board Rogowski coils, which have better accuracy (0.2-0.5%) and higher immunity to external fields compared to classical Rogowski coils, but at same time their output voltage is lower [4].

Another drawback of Rogowski-coil based current measurement is the need for an integrator for vast majority of measurements. In addition to the fact that the integrator affects the accuracy and stability of Rogowski sensor, the measurements conducted usually require that sampling, A/D conversions and data transmissions are done at measurement locations, which can be impractical in some situations. This also means that if all these electrical components are placed at a high voltage site, a separate power supply is needed.

2.2. Optical current sensors

Optical current sensors usually use one of the two principles for measurement of current: change of fibre-grating wavelength or the Faraday’s effect. There are two main types of fibre gratings: short-period gratings or fibre Bragg gratings (FBG) and long-period gratings (LPG). It should be noted at the outset that current sensors based on Faraday’s effect (effect of plane rotation of polarised light when it travels through a magnetic field) have too low sensitivity for practical applications, are extremely difficult to implement in an accurate and stable way and may be only affordable for high voltage lines [5], [6]; this means more attention will be paid to sensors based on the first of the above optical principles.

One of the main advantages of optical sensors is the fact that there is an optical signal at their output which makes them very convenient for use in modern systems where a lot of fibre optics is used. Unlike Rogowski sensors optical sensors are not sensitive to surrounding electromagnetic fields and they are smaller and more compatible to the requirements of modern smart grids. Their nonconductive nature makes them electrically passive which, along with their chemical inertness make them an excellent choice for measurement in corrosive and explosive environments. Generally, optical sensors don't saturate and have linear outputs comparable to Rogowski sensors.

Optical sensors are lightweight which helps to produce portable equipment, but also a collector, which contains all accompanying electrical components and, hence does not need to be placed near the measurement location when fibre optics is used. This is very useful for current measurement in the presence of high voltages. This is not only because it is easier to place equipment and manipulate later with it, but also because sensor electronics at the ground level is separated from sensing fibre optic at high voltage. They can be easily used for remote operations and inspections over long distances.
Even though it is possible to add to the list of very positive features of optical sensors features as large bandwidth and possibility to achieve highly accurate current measurements, there are also several drawbacks of this technology. They are very sensitive to vibrations, temperature and other environmental effects making outdoor installation complex. The temperature sensitivity of FBG is typically around 6 pm/°C and for LPG can be as high as 3.4 nm/°C [7]. Their output data are more unstable than those of the Rogowski coils [2].

The time delay of fibre optic current sensor's output signal is generally around 100μs [8], while some magneto-optical sensors (sensors which use the Faraday's effect) with GHz response have been produced [9]. One of the main disadvantages of current measurement using FBG or LPG is often very complex demodulation system needed for the wavelength shift measurement. High production costs due to high demand for materials and complex production process make it difficult for them to achieve mass production.

2.3. Magnetic sensors

Magnetic sensors attracted a lot of attention in recent years. This primarily attributable to the fact that these sensors are seen to be better in a good number of theoretical papers dealing with these materials and proposing different operating principles. The proposed measurement systems using magnetic sensors usually involve some other transducers and other technologies (very often optical fibre grating), what makes these systems very complex and susceptible to errors. So far many such magnetic sensors have been proposed. Among them are magnetoresistive sensors, which work on the principle that magnetic field affects the resistance of the sensor. This could be based on anisotropic magnetoresistance (AMR), giant magnetoresistance (GMR), magnetic tunnel junction (MTJ) and giant magnetoimpedance (GMI) sensors. Sensors which use magnetic force produced by measured current have been also proposed, as well as sensors made of magnetostrictive materials. Of particular interests are magnetostrictive materials, which belong to the group of giant magnetostrictive materials (GMM). All magnetic sensors are generally very small and light which makes them interesting for many applications.

Magnetoresistive magnetometers can be energised by applying a constant current and the output voltage is proportional to the strength of magnetic field. They are especially attractive for low cost applications. Magnetoresistive sensors are generally energy efficient devices (for example, AMR sensors' power consumption is less than 1W) [5]. They can sense very low magnetic fields (for AMR sensors this can be as low as 0.1nT) as well as a quite strong fields (GMR sensors can be used in fields as large as 0.1T) [9].

Increased interest in magnetostrictive sensor technology results from improvements in the performance of magnetostrictive materials and led to an increased use of these sensors for a wide range of applications.

Despite what has been said earlier, these magnetic sensors have several serious drawbacks that prevent their even wider application. They tend to be highly nonlinear and suffer from hysteresis which usually requires some compensation methods to be used to linearise the sensor. Besides, they are also sensitive to temperature changes and, often to vibrations. For example, magnetostrictive materials have a temperature dependence in the order of 5ppm/°C [6].

Magnetostriction is a unipolar phenomenon which means that a positive strain is produced for opposite directions of the magnetic field. Furthermore, relatively large magnetic field is needed to make use of the magnetostrictive property. Also since magnetostrictive sensors do not transduce current directly into an electrical form, at least one more step is needed for current measurement.

For magnetoresitive materials the change of resistance caused by an external magnetic field is very small. For example, for permalloy, which is the most common used material for AMR sensors, this is less than 4% [9].

The impedance of GMI sensors depends strongly on both the magnetic field strength and its frequency. To benefit from the GMI effect, high frequency current of the order of GHz is needed which limits their application as sensors [9].

2.4. Hall effect sensors

Another very popular method of measuring current is by the use of Hall effect sensors. These sensors have easier fabrication process and lower cost compared to Rogowski coils and optical sensors. They can measure currents over a relatively wide frequency range (from DC-1MHz [9]) provided that the current values are between 10mA and 35kA [10]. They are usually very small and linear (linearity errors less than 1% over significant operating range) [11] and have modest power consumption between 0.1-0.2W [9]. It is possible to install them without interrupting current circuit.

In general, Hall effect sensors are sensitive to temperature, stress and ageing. Consequently these sensors have non-negligible offset and sensitivity drifts. Without calibration, the typical temperature drift is 500ppm/°C and ageing drift of up to 2% [12]. They are also affected by external magnetic fields and can be very sensitive to the relative positioning of the current-carrying conductor and the sensor. Hall effect sensors are usually made of semiconductor materials which are highly piezoresistive.

3. Novel sensing principle based on 'smart alloys'

Magnetic shape memory (MSM) alloys are relatively new "smart" materials which change their shape when subjected to external magnetic fields. The Ni-Mn-Ga alloys are the most studied MSM alloys which have been shown to have enormous potential to be used in actuators and sensors. MSM alloys have a remarkable magnetic field induced (MFI) strain, maximum theoretical value of which varies, depending on material's microstructure from 6% for the most studied five-layered modulated (10M) to 12% for non-modulated (NM) [13], [14]. These are more than one order of magnitude higher than maximum strains produced by conventional magnetostrictive and piezoelectric materials reaching strains of well-known thermal shape memory alloys. However, MSM alloys respond to magnetic field excitation significantly faster and are predicted to have very long fatigue life if certain operational conditions are met. As it is obvious from their reversible properties MSM alloys can be used both as position and stress sensors [15].

The basic principle of a novel current sensor based on MSM alloys, proposed in this paper relies upon the proportionality of the strain produced by an MSM element which is subjected to a magnetic field produced by the current whose magnitude is being measured. This is shown schematically in figure 1 in which the output voltage, V generated by the LVDT is proportional to the strain, ε produced by the MSM element. This MSM strain is in turn proportional to the magnetic field B which is ultimately proportional to the current in the conductor, I (measurand).

Figure 1. Schematic showing the proposed current-measurement system based on magnetic shape memory (MSM) alloys.

The sensor system consists of an MSM element and a suitably coupled LVDT. The magnetic field (B) produced by a current (I) in the conductor is converted into displacement of the MSM element (ε) which changes the output voltage (V) of the LDVT. This approach allows sensing of very high currents remotely essentially by sensing the magnetic field produced by these currents.

The main disadvantages of the proposed MSM-based current sensors are attributable to temperature sensitivity and magneto-mechanical hysteresis of currently available MSM alloys. The former is associated with the upper temperature limit determined by the austenite-martensite transformation temperature of 60-80˚C [16] for conventional alloys and by the temperature-dependent twinning stress. The hysteresis is also associated with this twinning stress. The twinning stress can be seen as an energy barrier to be overcome for propagating twin boundaries in the MSM material and hence, changing the shape of the MSM element. Consequently the sensitivity of the proposed sensor is inversely proportional to the magnitude of the twinning stress. Twinning stress is still considerably high for conventional type I twin MSM elements. However, less studied type II twin crystals can be very promising for sensor applications due to low twinning stress which does not depend on temperature [17]. This results in very low hysteresis together with good temperature stability.

Due to larger strain range MSM alloys can be used in applications where GMM materials are not adequate. For instance, a typical 20mm MSM element with 5M microstructure elongates by up to 1.2mm under no larger than 0.6T magnetic field if the mechanical load is small [18]. Furthermore, relatively high magnetic field is needed to “activate” an MSM element which essentially suggests their possible application for measuring only large currents.

4. Conclusions

It seems the incompatibility of conventional measurement transformers with modern equipment is one of the main reasons for searching novel current-sensing and measurement solutions based on different alternative principles. The first part of this paper gives a brief overview of already existing sensing principles of current NCITs by comparing their properties and focusing on their advantages and disadvantages.

Although Rogowski coils, Hall sensors, conventional optical sensors as well as measurement principles based on FBG are dominant at the moment, there are still many unsolved problems.

In this paper, though at an early stage we have proposed the concept of a novel alternative approach for current measurement using magnetic shape memory (MSM) smart materials. The main idea behind this approach is to correlate the magnetic field produced by the current to be measured with the output voltage of an LVDT actuated by the shape change of an MSM element under the magnetic field.

There are, however many challenges to be overcome before a viable prototype can be designed, built and tested under laboratory and ‘field’ conditions. These include magneto-mechanical and thermo-mechanical stability of MSM material under variable ambient conditions, achievement of high degree of linearity at every stages of transition between the physical sub-systems shown in figure 1. This essentially means ensuring linearity of interdependent variables I, B, ε and V. This work is being undertaken at present by numerical finite element modelling and design.

References

[1] Metwally I A 2010 Self-integrating Rogowski coil for high-impulse current measurement *IEEE Trans. on Instrumentation and Measurement* **59** No. 2 353-360

[2] Cao L 2010 Application Comparison of Rogowski-type and Optical-type Electronic Current Transformer *CICED Int. Conf. (Nanjing)* pp. 1-5

[3] Frolov G, Grudin O and Warland T 2008 Compensating Rogowski coils for current measurement *Electronic Engineering Times* pp. 31-32

[4] Kojovic Lj A et al. 2010 Practical Aspects of Rogowski Coil Applications to Relaying *IEEE PSRC Special Report*

[5] Kirkham H 2009 Current Measurement Methods for the Smart Grid *IEEE PES General Meeting* pp. 1-7

[6] Mora J, Díez A, Cruz J L and Andrés M V 2000 A magnetostrictive sensor interrogated by fiber gratings for DC-current and temperature discrimination *IEEE Photonics Technology Letters* **12**, No. 12 1680-82

[7] Thyagarajan K and Ghatak A 2007 *Fiber Optic Essentials* (John Wiley & Sons Inc.).

[8] Viawan F A, Wang J, Wang Z and Ying Yang W 2009 Effect of Current Sensor Technology on Distance Protection *IEEE/PES Power Systems Conf. and Exposition* pp 1-7

[9] Lenz J and Edelstein A S 2006 Magnetic sensors and their applications *IEEE Sensors J* **6** No. 3 631-649

[10] Tsai Y P, Chen K L, Chen Y R and Chen N 2014 Multifunctional coreless Hall-effect current transformer for the protection and measurement of power systems *IEEE Trans. on Instrumentation and Measurement* **63** No. 3 557-565.

[11] Ramsden E 2006 *Hall-Effect Sensors: Theory and Application* (Elsevier Inc.)

[12] Pastre M and Kayal M 2005 A Hall Sensor-Based Current Measurement Microsystem With Continuous Gain Calibration *IEEE PhD Research in Microelectronics and Electronics* **2** 95-98

[13] Murray S J, Marioni M, Allen S M, O'Handley R C and Lograsso T A 2000 6% magnetic-field-induced strain by twin-boundary motion in ferromagnetic Ni–Mn–Ga *Appl. Phys. Lett.* **77** No. 6 886

[14] Sozinov A, Lanska N, Soroka A and Zou W 2013 12% magnetic field-induced strain in Ni-Mn-Ga-based non-modulated martensite *Appl. Phys. Lett.* **102** No. 2 021902

[15] Stephan J M, Pagounis E, Laufenberg M, Paul O, Ruther P and Reorientation A M 2011 A novel concept for strain sensing based on the ferromagnetic shape memory alloy NiMnGa *IEEE Sens J* **11** No. 11 2683–89

[16] Pagounis E, Chulist R, Szczerba M J and Laufenberg M 2014 High-temperature magnetic shape memory actuation in a Ni–Mn–Ga single crystal *Scr. Mater.* **83** 29–32

[17] Straka L, Soroka A, Seiner H, Hanninen H and Sozinov A 2012 Temperature dependence of twinning stress of Type I and Type II twins in 10M modulated Ni-Mn-Ga martensite *Scr. Mater.* **67** No. 1 25–28

[18] Pagounis E and Müllner P 2014 Recent Developments in Magnetic Shape Memory Actuation *Proc. Actuator* 86–91

AUTHOR INDEX

AUTHOR INDEX